2024 IEEE International Symposium on Defect and Fault Tolerance in VLSI and Nanotechnology Systems (DFT 2024)

Didcot, United Kingdom
8-10 October 2024

IEEE Catalog Number: CFP24078-POD
ISBN: 979-8-3503-6689-1

**Copyright © 2024 by the Institute of Electrical and Electronics Engineers, Inc.
All Rights Reserved**

Copyright and Reprint Permissions: Abstracting is permitted with credit to the source. Libraries are permitted to photocopy beyond the limit of U.S. copyright law for private use of patrons those articles in this volume that carry a code at the bottom of the first page, provided the per-copy fee indicated in the code is paid through Copyright Clearance Center, 222 Rosewood Drive, Danvers, MA 01923.

For other copying, reprint or republication permission, write to IEEE Copyrights Manager, IEEE Service Center, 445 Hoes Lane, Piscataway, NJ 08854. All rights reserved.

*** This is a print representation of what appears in the IEEE Digital Library. Some format issues inherent in the e-media version may also appear in this print version.

IEEE Catalog Number:	CFP24078-POD
ISBN (Print-On-Demand):	979-8-3503-6689-1
ISBN (Online):	979-8-3503-6688-4
ISSN:	2576-1501

Additional Copies of This Publication Are Available From:

Curran Associates, Inc
57 Morehouse Lane
Red Hook, NY 12571 USA
Phone: (845) 758-0400
Fax: (845) 758-2633
E-mail: curran@proceedings.com
Web: www.proceedings.com

TABLE OF CONTENTS

Range Restriction to Harden CNNs Against Hardware Faults: A Broad Empirical Analysis.............................. 1
Cristiana Bolchini, Luca Cassano, Antonio Miele, Dario Passarello

Impact of Image Sensor Input Faults on Pruned Neural Networks for Object Detection.................................... 7
Yizhi Chen, Wenyao Zhu, Dejiu Chen, Omar Mohammed, Parthib Khound, Zhonghai Lu

Towards Biology-Inspired Fault Tolerance of Neuromorphic Hardware for Space Applications 13
Shadi Matinizadeh, Sarah Johari, Arghavan Mohammadhassani, Anup Das

An AI-Assisted Connection Weight Prediction for Regression Testing of Integrated Circuits.......................... 20
*Abishaan Ravikumar, Xiaohan Yang, Rajendra Prasad, Rajanataraj Sivaraj, Alexander Rast,
Abusaleh Jabir*

BayWatch: Leveraging Bayesian Neural Networks for Hardware Fault Tolerance and Monitoring................. 26
*Julian Hoefer, Matthias Stammler, Fabian Kreß, Tim Hotfilter, Tanja Harbaum, Juergen
Becker*

Zero-Memory-Overhead Clipping-Based Fault Tolerance for LSTM Deep Neural Networks 32
*Bahram Parchekani, Samira Nazari, Mohammad Hasan Ahmadilivani, Ali Azarpeyvand, Jaan
Raik, Tara Ghasempouri, Masoud Daneshtalab*

Inferred Fault Models for RISC-V and Arm: A Comparative Study 36
Ihab Alshaer, Ahmed Al-Kaf, Valentin Egloff, Vincent Beroulle

Image Degradation Due to Interacting Hot Pixels and SEUs 42
Glenn H. Chapman, Alireza Farahmandpour, Amit Chakma, Israel Koren, Zahava Koren

A Novel Digitisation Method for Pulse Switchable Memristive Chemical Sensors.. 48
Meenakshi Devi, Saurabh Khandelwal, Marek Vidiš, Tomas Plecenik, Abusaleh Jabir

Implementation and Reliability Evaluation of a ChaCha20 Stream Cipher Hardware Accelerator 54
Wesley Grignani, Khalil G. Q. Santana, Douglas A. Santos, Luigi Dilillo, Douglas R. Melo

Digital Generation of RF Phase-Modulated Test Stimuli: Application to BPSK Modulation Scheme 58
K. Tahraoui, R. Burelle, T. Vayssade, F. Lefevre, L. Latorre, F. Azaïs

A Flexible FPGA-Based Test Equipment for Enabling Out-Of-Production Manufacturing Test Flow
of Digital Systems 64
*Nicola Di Gruttola Giardino, Francesco Angione, Paolo Bernardi, Tommaso Foscale,
Claudia Bertani, Vincenzo Tancorre*

Large Language Model-Based Optimization for System-Level Test Program Generation 70
*Denis Schwachhofer, Peter Domanski, Steffen Becker, Stefan Wagner, Matthias Sauer, Dirk
Pflüger, Ilia Polian*

An Automated and Effective Approach for SBST Generation Targeting RISC-V CPUs 76
*Endri Kaja, Nicolas Gerlin, Jad Al Halabi, Ares Tahiraga, Sebastian Prebeck, Dominik
Stoffel, Wolfgang Kunz, Wolfgang Ecker*

A Low Power Oriented Multiple Target Test Generation Method for 2-Cycle Gate-Exhaustive
Faults 80
Toshinori Hosokawa, Momona Mizota, Masayoshi Yoshimura, Masayuki Arai

A Structural Testing Approach for SRAM Address Decoders using Cell-Aware Methodology 86
X. Xhafa, E. Faehn, P. Girard, A. Virazel

Special Session: Security and RAS in the Computing Continuum 90
*Martí Alonso, David Andreu, Ramon Canal, Stefano Di Carlo, Odysseas Chatzopoulos,
Cristiano Chenet, Juanjo Costa, Andreu Girones, Dimitris Gizopoulos, George
Papadimitriou, Enric Morancho, Beatriz Otero, Alessandro Savino*

RAPPER: **R**obust and **APP**roximate **ER**ror Tolerating Communication 96
Somayeh Sadeghi-Kohan, Sybille Hellebrand, Hans-Joachim Wunderlich

Optimizing Waveform Accurate Fault Attacks using Formal Methods 102
Devanshi Upadhyaya, Ilia Polian

Malware Detection on Linux using Runtime Opcode Tracing 108
Martí Alonso, Juan José Costa, Enric Morancho

Special Session: Impact of Compiler Optimizations on the Reliability of a RISC-V-Based Core 114
Pegdwende Romaric Nikiema, Marcello Traiola, Angeliki Kritikakou

An Experimental Comparison of RISC-V Processors: Performance, Power, Area and Security
(Special Session Paper) 120
Elia Lazzeri, Bruno Endres Forlin, Gianluca Furano, Marco Ottavi, Luca Cassano

Special Session: Reliability and Performance Evaluation of a RISC-V Vector Extension Unit for
Vector Multiplication 126
Carolina Imianosky, Douglas A. Santos, Douglas R. Melo, Felipe Viel, Luigi Dilillo

Special Session: Software-Based Self-Test Generation for RISC-V – Stuck-At Generation,
Functional Cell-Aware Untestability, and FPGA Demonstration – 132
*Tobias Faller, Nikolaos I. Deligiannis, Riccardo Cantoro, Matteo Sonza Reorda, Bernd
Becker*

Special Session: A Mixed Simulation-, Emulation-, and Formal-Based Fault Analysis Methodology
for RISC-V 138
*Endri Kaja, Nicolas Gerlin, Ares Tahiraga, Jad Al Halabi, Sebastian Prebeck, Dominik
Stoffel, Wolfgang Kunz, Wolfgang Ecker*

Special Session: In-Field ML-Assisted Intermittent Fault Localization and Management in RISC-V
SoCs 144
Hardi Selg, Konstantin Shibin, Anton Tsertov, Maksim Jenihhin, Peeter Ellervee, Jaan Raik

Exploring Total Ionizing Dose Radiation Effects Across Generations of NVIDIA Jetson Devices: A
Comparative Analysis 150
*Ivan Rodriguez-Ferrandez, Maris Tali, Leonidas Kosmidis, Alessandra Costantino, David
Steenari*

An Enhanced Fault Injection Framework for FPGA-Based Soft-Cores 156
*Tijmen T. Smit, Bruno Endres Forlin, Kuan-Hsun Chen, Ioanna Souvatzoglou, Mihalis
Psarakis, Marco Ottavi*

Using High-Level Profiling Data to Early Assess the Robustness of Digital Systems 162
Luc Noizette, Florent Miller, Youri Helen, Régis Leveugle

Reliability Analysis of a Low-Cost CCSDS 123 Hyperspectral Image Compressor 168
Wesley Grignani, Douglas A. Santos, Luigi Dilillo, Douglas R. Melo

Dependable Systems and AI in Critical Infrastructures: A Case Study in European Earth Observation Missions 172
Valentina Zancan, Filomena Decuzzi, Gianluca Furano, Lorenzo Canese

Safe Satellite Electronics Design Utilizing COTS Components, FDIR Techniques, LCL Protections, and Thorough Qualifications 178
Bojan Kotnik, David Selcan, Matic Erker, Tomaž Rotovnik, Dejan Gacnik, Gianluca Furano, Iztok Kramberger

Special Session: Exploring the Potential of Versal ACAP: Advancing Onboard Edge AI for Spacecraft 182
Carlo Ciancarelli, Davide Di Ienno, Renato Trois, Luca Scandelli, Catriel De Biase, Paolo Serri, Antonio Leboffe, Dario Pascucci, David Steenari, Gianluca Furano

Special Session: SE-UVM, an Integrated Simulation Environment for Single Event Induced Failures Characterization and Its Application to the CV32E40P Processor 187
Marcello Barbirotta, Marco Angioli, Antonio Mastrandrea, Francesco Menichelli, Abdallah Cheikh, Mauro Olivieri

Special Session: Testing of Digital Computing-In Memories with MAC Function 193
Jin-Fu Li

Special Session: Overcoming Transient Faults and Aging Effects in Digital Computing-In-Memory Architectures: Detection, Tolerance, and Mitigation Strategies 198
Yu-Guang Chen, Ting-Yi Wu

Special Session: Enhancing Reliability in Digital Computing-In-Memory Architectures Through Approximation and Fault Tolerance Methods 204
Shih-Hsu Huang, Chih-Li Hsiao, Wei-Che Cheng

Special Session: Architecture-Level DCIM Technologies for Edge AI Computing Applications.................... 210
Chun-Lung Hsu, Hsuan-Yu Chen, Yi-Lin Chen

Dual-Modular-Redundancy Voting Circuits for Single-Event-Transient Mitigation....................................... 216
Marcello Barbirotta, Marco Angioli, Antonio Mastrandrea, Francesco Menichelli, Abdallah Cheikh, Mauro Olivieri

A Novel Self-Repair Mechanism for Tiled Matrix Multiplication Unit 222
Chandra Sekhar Mummidi, Sandeep Bal, Sandip Kundu

An Effective TMR Approach for Low-Latency Configurable-Accuracy Adders... 228
Ioannis Tsounis, Dimitris Agiakatsikas, Mihalis Psarakis

Author Index

37th IEEE International Symposium on Defect and Fault Tolerance in VLSI and Nanotechnology Systems

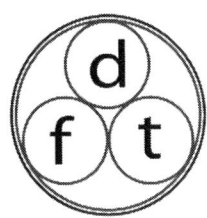

Harwell campus in Oxfordshire, Didcot, UK
October 8th - 10th, 2024

Partners

Best Paper Awards Sponsor

Technical Sponsors

37th IEEE International Symposium on Defect and Fault Tolerance in VLSI and Nanotechnology Systems (DFT 2024) Harwell Campus, Oxfordshire, UK, October 8th - 10th, 2024

Foreword

On behalf of the organizing committee and the program committee, we welcome you to the thirty seventh edition of the IEEE International Symposium on Defect and Fault Tolerance in VLSI and Nanotechnology Systems (DFT 2024 held at the Harwell Campus in Oxfordshire, UK, from October 8th-10th , 2024). The Symposium is sponsored by the IEEE Computer Society, IEEE Fault-Tolerant Computing Technical Committee, and IEEE Test Technology Technical Council. Over the last 36 years, DFT has served as an international forum for research in the field of defect and fault tolerance in VLSI systems inclusive of emerging technologies. The Symposium brings together researchers from academia, industry, and federal agencies. Topics of interest span manufacturing or environmental sources of defects and their impact on design, manufacturing, test, system reliability, safety, and availability. Design and manufacturing methods to mitigate the impact of faults and defects, and hardware security approaches are also topics of interest for DFT.

This year the symposium features 24 long presentations (6 pages each) and 8 short presentations (4 pages each) with authors hailing from 21 different countries. Moreover, we propose four keynote speeches from industrial experts, which cover hot research and engineering topics in the dependability and fault tolerance of computing systems. The first keynote speaker is Dr. Vilas Sridharan from the AND. His talk is titled *Addressing Emerging Fault Modes with Testing and Reliability*. The second keynote speaker is. Nicolas Ganry of Microchip. His talk is titled *High Performance 64-bit HPSC Microprocessor (MPU) for a New Era of Autonomous Space Computing*. The third keynote speaker is Dr. Michael Seidl of Texas Instruments. His talk is titled *Benefits of Using Functional Safety in Commercial Space Applications*. And the final keynote talk is from Dr. David Bacon from Google whose talk is titled, *Preventing Silent Data Corruption in a Hyper-scale Database*. Taken together these keynote talks will provide new insight into the reliability challenges faced by industry, both in large scale terrestrial compute installations, as well as emerging space applications which require ever more compute capacity.

This year we have continued our tradition of hosting special sessions on promising new areas. Five special sessions have been accepted entitled: 1) *Adopting RISC-V, How Far are We?*, 2) *Architecture, Reliability, and Testing of Digital Computing-in Memories with MAC Function*, 3) *Software-Based Self-Test Generation for RISC-V*, 4) *In-Field ML-Assisted Intermittent Fault Localization and Management in RISC-V SoCs*, 5) *A mixed simulation-, emulation-, and formal-based fault analysis methodology for RISC-V*. These special sessions aim at providing a complementary experience with respect to the regular sessions by focusing on hot and emerging topics of interest to the DFT community, as well as on cross-disciplinary topics. This year's special sessions highlight, the importance of RISC-V in research activities around safety and reliability as well as the emergence of near-memory compute solutions.

An event of this nature and dimension is only possible due to contributions of many individuals and institutions. These include technical contributions from the authors, the constructive feedback from the technical program committee members and the highly experienced team of reviewers, and the session chairs who moderate the discussions and keep the schedule on track. We thank them all for their efforts and time. We are also grateful to Dr. Mario Barbareschi for organizing the special sessions, Prof. Pedro Reviriego and Dr. Shanshan Liu for publicizing the event, Dr. Marcello Traiola for the preparation of the proceedings, and Bruno Endres Forlin for updating the DFT website.

We hope that you will find DFT 2024 rewarding and exciting. We wish you all a productive and enjoyable stay in Oxfordshire, UK and hope that you will continue to make DFT a success through technical participation, assisting in its organization, and providing feedback to make it even better.

37th IEEE International Symposium on Defect and Fault Tolerance in VLSI and Nanotechnology Systems (DFT 2024) Harwell Campus, Oxfordshire, UK, October 8th - 10th, 2024

Welcome to DFT 2024!

General Co-Chairs : Filomena Decuzzi, European Space Agency, Netherlands; Carlo Cazzaniga, UKRI-STFC, Chiplr, United Kingdom

Program Co-Chairs : Jaume Abella, Barcelona Supercomputing Center, Spain; Adrian Evans CEA/LIST, France.

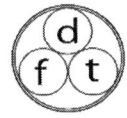

37th IEEE International Symposium on Defect and Fault Tolerance in VLSI and Nanotechnology Systems
(DFT 2024)
October 8th - 10th, 2024

Committees

Organizing Committee

General co-Chairs	Filomena Decuzzi	European Space Agency, The Netherlands
	Carlo Cazzaniga	UKRI-STFC, ChipIr, United Kingdom
Program co-Chairs	Adrian Evans	CEA, France
	Jaume Abella	Barcelona Supercomputing Center, Spain
Special Session	Mario Barbareschi	University of Naples, Italy
Publicity Chair	Pedro Reviriego	Universidad Politécnica de Madrid, Spain
	Shanshan Liu	University of Electronic Science and Technology of China, China
Publication Chair	Marcello Traiola	INRIA, France
Web Chair	Bruno Endres Forlin	University of Twente, The Netherlands

Steering Committee

Cristiana Bolchini	Politecnico di Milano, Italy
Glenn Chapman	Simon Fraser University, Canada
Luigi Dilillo	IES/UM/CNRS, France

Israel Koren	University of Massachusetts Amherst, United States
Prashant Joshi	Intel, United States
Fabrizio Lombardi	Northeastern University, United States
Marco Ottavi	University of Rome Tor Vergata, Italy and University of Twente, The Netherlands
Mihalis Psarakis	University of Piraeus, Greece

Technical Program Committee

Adrian Evans	CEA
Alberto Bosio	Ecole Centrale de Lyon
Alessandro Savino	Politecnico di Torino
Almudena Lindoso	University Carlos III Madrid
Amlan Ghosh	Intel Corporation
Anees Ullah	University of Engineering and Technology, Peshawar, Pakistan
Antonio Miele	Politecnico di Milano
Carles Hernández	Universitat Politècnica de València
Cheng Liu	Institute of Computing Technology, Chinese Academy of Sciences
Chih-Tsun Huang	National Tsing Hua University
Corrado De Sio	Politecnico di Torino
Costas Argyrides	AMD
Cristiana Bolchini	Politecnico di Milano
Daniele Rossi	University of Pisa
David Hely	Grenoble INP
Dimitris Agiakatsikas	University of Piraeus
Domenic Forte	University of Florida
Douglas Melo	University of Vale do Itajaí
Eduardo Bezerra	Universidade Federal de Santa Catarina
Fabrizio Lombardi	Northeastern University

Fernando Fernandes dos Santos	INRIA
Gherman Valentin	CEA
Gianluca Furano	European Space Agency
Gulay Yalcin	Abdullah Gul University
Hideyuki Ichihara	Hiroshima City University
Huifang Jiao	Huawei Technologies
Ilia Polian	University of Stuttgart
Ioannis Sourdis	Chalmers University of Technology
Jaan Raik	Tallinn University of Technology
Jaume Abella	Barcelona Supercomputing Center (BSC)
Jiafeng Xie	Villanova University
Jie Han	University of Alberta
Josep Balasch Masoliver	KU Leuven
Kazuteru Namba	Chiba University
Kuan-Hsun Chen	University of Twente
Luca Cassano	Politecnico di Milano
Luis Entrena	Universidad Carlos III de Madrid
Maksim Jenihhin	Tallinn University of Technology
Marcello Traiola	INRIA
Marco Ottavi	University of Rome Tor Vergata, University of Twente
Mehran Mozaffari Kermani	University of South Florida
Mihalis Psarakis	University of Piraeus
Mohammadhashem Haghbayan	University of Turku
Naghmeh Karimi	University of Maryland, Baltimore County
Nicola Nicolici	McMaster University
Pedro Revriiego	Universidad Politécnica de Madrid
Petr Fišer	Czech Technical University in Prague
Ramon Canal	Universitat Politècnica de Catalunya
Sebastian Huhn	Siemens Electronic Design Automation GmbH, Germany

Shanshan Liu	New Mexico State University
Siting Liu	Shanghai Tech University
Stephan Eggersglüß	Mentor Graphics
Zhen Gao	Tianjin University

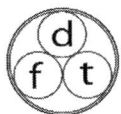 37th IEEE International Symposium on Defect and Fault Tolerance in VLSI and Nanotechnology Systems

Best Paper Award

Range restriction to harden CNNs against hardware faults: A broad empirical analysis

Antonio Miele, Cristiana Bolchini, Luca Cassano and Dario Passarello

Outstanding Student Paper Award

to: Somayeh Sadeghi-Kohan

RAPPER: Robust and APProximate ERror Tolerating Communication

Somayeh Sadeghi-Kohan, Sybille Hellbrand and Hans-Joachim Wunderlich

General Co-Chairs : Filomena Decuzzi, European Space Agency, Netherlands; Carlo Cazzaniga, UKRI- STFC, Chiplr, United Kingdom

Program Co-Chairs : Jaume Abella, Barcelona Supercomputing Center, Spain; Adrian Evans CEA/LIST, France.

DFT 2024 Keynotes

Addressing Emerging Fault Modes with Testing and Reliability
Speaker: Vilas Sridharan, Ph.D.
AMD

This talk will cover the challenges, current state, and future directions of addressing emerging fault modes with testing and reliability features.

Vilas Sridharan is currently an AMD Senior Fellow where he leads the RAS (Reliability, Availability and Serviceability) Architecture team. His research focuses on the modeling of hardware faults and architectural and micro-architectural approaches to reliability and fault tolerance in high-performance microprocessors. Vilas received his Ph.D. and M.S.E. from the Department of Electrical and Computer Engineering at Northeastern University, and his B.S.E. in Computer Engineering from Princeton University in 2000. From 2000 - 2004, he worked in the SPARC server division at Sun Microsystems. Since 2010, he has been on AMD's RAS Architecture team.

High Performance 64-bit HPSC Microprocessor (MPU) for a New Era of Autonomous Space Computing
Speaker: Nicolas Ganry
Microchip

PIC64-HPSC MPU integrates 8 high-performance RISC-V® CPUs augmented with vectorprocessing instruction extensions to support Artificial Intelligence/Machine Learning (AI/ML) applications. The MPUs also feature a suite of features and industry-standard interfaces and protocols not previously available for space application like Time Sensitive Networking (TSN) Ethernet and PCIe® Gen 3 with Compute Express Link® (CXL®) 2.0.

To address mission critical applications, PIC64-HPSC includes comprehensive features to enable endto-end partitioning, isolation, fault-tolerance, and mitigation. To counter existing and emerging security threats, PIC64-HPSC implements as well defense-in-depth approach to security including support for post-quantum cryptography and anti-tamper features. The PIC64-HPSC is complemented by a broad portfolio of space solutions with Radiation-Tolerant (RT) and Radiation-Hardened (RH) MCUs, FPGAs and Ethernet PHYs, power devices, RF products, timing, as well as discrete components

Nicolas Ganry is Senior Product Marketing Manager for Aerospace & Defense applications at Microchip Technology. He joined the company in 2012, bringing with him 15 years of experience in various electronic systems (hardware and software development).
Nicolas is leading Aerospace & Defense product line in France delivering space qualified processors, microcontrollers, memories, and communication interfaces. He is also driving business development activities in Europe for all Microchip products. Located in Nantes (ex-Atmel site), Nicolas is representing Microchip in Europe at ESA CTB working closely with European Space agencies (ESA, CNES, DLR, ...) and space industry primes.

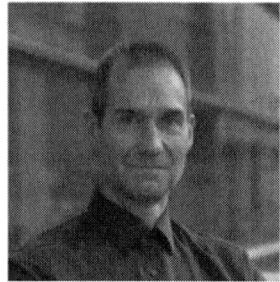

Benefits of Using Functional Safety in Commercial Space Applications
Speaker: Michael Seidl
Texas Instruments

This presentation delves into the evolving landscape of space exploration with a focus on the growing influence of NewSpace companies. It explores the critical concept of system-level resiliency, offering a fresh and comprehensive perspective on how to build robust and dependable space systems. Specifically, the presentation will shed light on the connections between two key frameworks: Reliability, Availability, Maintainability, and Safety (RAMS) and IEC 61508 functional safety. By comparing and contrasting these frameworks, the presentation will highlight the unique advantages that functional safety principles bring to space systems based on the example of TMS570LC4357-SEP: Space-enhanced product lock-step MCU.

Michael received his Dipl. Ing. (FH) degree in communication technologies from Fachhochschule Munich in Germany. Michael has 29 years of experience in semiconductors and held positions in DSP software design, applications, product marketing, business development and system engineering.
Michael is a Systems Engineer for Aerospace Applications at Texas Instruments. He

supports customers in their decision making with in-depth system knowledge, combined with expertise on TI's product offerings.

Preventing Silent Data Corruption in a Hyper-scale Database
Speaker: David F. Bacon
Google

Spanner is Google's database of record, storing over 15 exabytes of data for the majority of Google's infrastructure and Cloud systems, including multiple applications with over a billion users (Gmail, Search, YouTube, Photos, etc). As Google's most mission-critical and by some metrics largest system, we have been at the forefront of detecting and preventing silent data corruption (SDC) from its earliest manifestations. In this talk I will describe the evolution of our approach to "silent" data corruptions, covering a broad range of software techniques, debugging, tooling, and operational principles. I'll present case studies of how we found and fixed failures that occurred less than once per million CPU-years, and how our approach has moved from preventing individual corruptions to implementing fail-fast properties that protect Google's entire fleet. Finally, I'll talk about what the future holds, covering hardware, library, and language techniques.

David F. Bacon is a Principal Engineer at Google, where he leads the design and evolution of the Spanner storage engine (Ressi) along with the exploitation of new hardware technologies in databases. David received his A.B. from Columbia University in 1985 and his Ph.D. from U.C. Berkeley in 1997. His prior work includes compilation and run-time systems for object-oriented programming (used in most JVMs and modern compilers), hardware compilation, and real-time garbage collection (for which he was made a Fellow of the ACM). He has served on the governing boards of ACM SIGPLAN and SIGBED.

Range restriction to harden CNNs against hardware faults: a broad empirical analysis

Cristiana Bolchini, Luca Cassano, Antonio Miele, Dario Passarello

Dipartimento di Elettronica, Informazione e Bioingegneria
Politecnico di Milano, Italy
first_name.last_name@polimi.it

Abstract—Due to the increasing use of Deep Learning in mission/safety-critical application contexts, in the recent past several techniques have been designed to harden the system to guarantee its correct behavior even in presence of faults affecting the hardware. Often, such new techniques are evaluated on a reduced set of Convolutional Neural Network (CNN) models and/or data sets, such that their generality and robustness could actually be limited. This paper presents a broad and systematic experimental evaluation of a state-of-the-art range restriction technique presented in literature, i) by applying it to a large set of CNNs, implementing different functional tasks, and ii) by using multiple datasets. The obtained results demonstrate that the effectiveness of the technique highly depends on the complexity of the considered task; in particular, classification CNNs benefit the most, while regression, image segmentation, and object detection are subject to different levels of benefits.

Index Terms—Deep Learning, Error Simulation, Reliability Analysis, Hardening Techniques.

I. INTRODUCTION AND RELATED WORK

Nowadays, Deep Learning (DL) applications, such as CNNs, are deployed in a large number of system for perception functionalities. A very popular example is the Advanced Driver Assistance System (ADAS), that is increasingly employed in the automotive scenario [1], used for lanes and tracks detection, pedestrians and obstacles identification and road signs and traffic light recognition [2].

Beyond the automotive scenario, DL is used for autonomous driving and control also for unmanned aerial vehicles and space exploration systems, other examples of safety-/mission-critical systems. Indeed, a failure of the system may lead to the failure of the overall mission or to catastrophic consequences, therefore it is paramount to guarantee the correct behavior of the systems also in presence of hardware faults by introducing specific fault detection and management mechanisms.

DL applications are extremely compute-intensive; in fact they are generally composed by a pipeline of steps performing elaborations, such as convolutions, on large multidimensional data matrices called tensors. At the same time, most autonomous systems require very short processing latency to guarantee real-time responses to external events; this exacerbates even more the quest for computing power. As a consequence, classical fault detection and mitigation techniques, such as Duplication with Comparison (DWC) or Triple Modular Redundancy (TMR), may be unfeasible because of an excessive overhead on the performance, power consumption, area and other design drivers [3].

In the recent past, the research community has devoted a considerable effort in defining new techniques specifically tailored for DL applications capable at guaranteeing robustness against faults with a limited overhead [4]. Since DL applications demonstrated to be intrinsically resilient to faults because of the relevant information redundancy in the DL models, a first class of works proposed a selective or approximate DWC or TMR to the most vulnerable portions of the DL application. For instance, in [5] error simulation is used to identify the most vulnerable layers to replicate. With the same purpose, the work in [6] exploits Explainable AI to identify the most vulnerable layer, while in [7] replication is performed at the finer grain of neurons. Other approaches [8], [9] define Algorithm-Based Fault Tolerance techniques specifically tailored to matrix elaborations performed within convolutions to obtain a high robustness at a reduced cost.

A second class of works aims at exploiting the mathematical properties and the intrinsic information redundancy of the DL models to mitigate and correct the effects of possible faults in the elaborations. In [10] the training process is enhanced by introducing faults so that the DL model can learn to correctly carry out its nominal task even when a fault occurs. The approach in [11] uses *range restriction* operators inserted in the model; they force the values of the tensor to fit into a specific range of values, so that high-magnitude errors are mitigated. Finally, the work in [12] redefines the ReLU operator to define a similar clipping strategy.

Most solutions introducing novel hardening strategies for DL applications are analyzed and evaluated considering only a limited number of CNN models in the experimental campaigns, to investigate the effectiveness of the approach. Yet, a CNN can be used for several tasks, such as image classification or object detection, while many works [6], [7], [10], [12] only consider CNNs for image classifications. On the other hand, only a limited number of papers consider also the CNN implemented for the other tasks, such as regression (as in [5], [11]) or object detection (as in [8], [9]). Finally, image segmentation is never explored. In our opinion, such common practice in the experimental evaluations may represent a limitation in the robustness of the proposed solution, since it may not allow to identify weaknesses. Indeed, CNNs implementing different tasks have variable characteristics and structures, that, in turn, may result in different resilience. Finally, as already claimed in [13], the adoption of error models not accurately representing the effects of physical faults may also lead to over-optimistic outcomes.

Given these considerations, this works aims at performing a systematic experimental evaluation of a popular state-of-the-art hardening technique for CNNs called *Ranger* [11]. To this purpose, we have defined a comprehensive benchmark suite with CNNs performing four different tasks, and trained on different datasets. We used an accurate error simulation framework [13] to carry out the reliability evaluation. The results of our analysis may be summarized in the following points: i) Ranger demonstrated to achieve interesting results with some CNNs for image classification and regression, however it is frequent to find cases where its performance is quite low when increasing the number of analyzed models; ii) when a fault corrupts a large number of values in a tensor, e.g., when the fault affects the control flow, Ranger's effectiveness in recovering the error is limited; and iii) Ranger's efficacy decreases when decreasing the size of the corrupted tensor.

The rest of this paper is organized as follows. Section II sets the background on the Ranger, the analysed hardening technique. Section III presents the experimental framework used for the extensive evaluation of Ranger's effectiveness, whose results are discussed in Section IV, and Section V draws concluding remarks.

II. BACKGROUND: RANGE RESTRICTION

Ranger is based on the intuition that i) nominal values in the processed tensors stay in a limited numerical range, and ii) faults causing high-magnitude errors are the ones that cause the CNN to fail. The technique consists in introducing in the CNN models a new operator performing a range restriction on the input tensor, replacing each x value with the new $r(x)$ based on the following formula[1]:

$$r(x) = \begin{cases} 0 & \text{if } x > T_{up} \text{ or } x < T_{low}, \\ x & \text{otherwise} \end{cases} \quad (1)$$

where the two bounds T_{low} and T_{up} are empirically characterized. This new operator is introduced in the activation layers of the CNN, and on Pooling layers immediately following the activation layers.

The technique has been experimentally evaluated by considering 6 CNNs for image classification and 2 ones for regression, demonstrating a resilience to Silent Data Corruption (SDC) close to 100%, with the only exception of a regression model due to specific model characteristics. However, a larger testing scenario with different models may produce less optimistic results. Moreover, the employed error simulation framework considers only an error model causing a single value corruption, that does not accurately represent the effects of physical faults, as discussed in [13]. Therefore, this paper presents a systematic analysis of Ranger as it has proposed in [14] without any aim of enhancing or fine-tuning it.

III. EXPERIMENTAL FRAMEWORK

This section presents the characteristics of the extensive experimental campaigns carried out to evaluate Ranger's effectiveness, starting from the selected models and data sets, the adopted error simulation framework, and the specifics of the injection campaigns.

A. Tasks, CNNs, Training Sets and Metrics

We selected four typical tasks, for each one of them two or more CNN models, implemented in Pytorch, and two data sets. For each model/dataset we report the accuracy of the *golden* implementation, the execution in a fault-free situation, computed according to typical metrics, that compare the obtained output against the *ground truth* output as labelled in the dataset. Robustness is used to measure the effectiveness of the Ranger technique. We put into relation the output of the *corrupted* execution with the one of the *golden* one and set context-specific thresholds to discriminate between corrupted but *usable* outputs and *unusable* ones (according to the method in [15]). The final robustness is computed as the ratio between the usable results and the overall number experiments. A detailed discussion of the so defined *benchmark suite* follows, and is summarized in Table I.

Classification is the most popular and considered task also in the literature on DL reliability. A CNN implementing this task considers a set of classes, and, for a given input image, it provides a vector of values expressing the probability of the image to belong to each one of the considered classes; the final output is the class having the highest probability. The metric generally used to evaluate the accuracy of the classification is called *top-1* which reports average percentage of correct classifications, in terms of the class with the highest probability, w.r.t. the ground-truth. Robustness is evaluated with the same metric; the output is classified as usable if the $top\text{-}1_{corrupted}$ class is the same as its golden counterpart $top\text{-}1_{golden}$, unusable otherwise. We selected three CNNs typically employed for classification, namely VGG16 [16], ResNet50 [17] and MobileNet_v2 [18], each one trained with two datasets, namely MNIST[2] and GTSRB[3].

Regression predicts, given an image, a specific quantity. Examples here considered are the prediction of the age of a person in a picture and the steering angle computed from images taken with the on-board car camera. The accuracy of the prediction is measured with the Mean Absolute Error (MAE) w.r.t. the actual measure taken as ground-truth. For the robustness we define a simpler and more stringent metric based on the absolute difference between the corrupted execution output $y_{corrupted}$ and the golden counterpart y_{golden}; when such a value is below thr_r value set by an application-context expert, the output is classified as usable. We considered two CNNs for regression, namely VGG11 [16] and ResNet18 [17],

---

[1] In the original proposal [11], $r(x)$ is assigned to one of the two bounds. We here consider the more refined proposal of the technique [14], showing that it is more effective to set $r(x)$ to 0.

[2] http://yann.lecun.com/exdb/mnist/
[3] https://benchmark.ini.rub.de/

Table I: The considered set of benchmarks

Task	Model	Dataset	Accuracy Metric	Value	Robustness Metric
Classification	VGG16	MNIST	top-1	98.8%	$\text{top-1}_{corrupted} = \text{top-1}_{golden}$
	VGG16	GTSRB		98.8%	
	ResNet50	MNIST		97.1%	
	ResNet50	GTSRB		97.1%	
	MobileNet_v2	MNIST		96.8%	
	MobileNet_v2	GTSRB		97.6%	
Regression	VGG11	Utk Face Age	MAE	5.43	$|y_{corrupted} - y_{golden}| \leq thr_r$
	VGG11	Car Steering Angle		1.32	
	ResNet18	Utk Face Age		5.09	
	ResNet18	Car Steering Angle		1.09	
Image segmentation	Unet	Oxford-IIIT Pet	mIoU	0.675	$\forall_{classes} \text{IoU}_{class}(golden, corrupted) \geq thr_s$
	Unet	Cityscapes		0.710	
	Deeplab_v3	Oxford-IIIT Pet		0.757	
	Deeplab_v3	Cityscapes		0.739	
Object detection	YOLO_v3	Aerial Maritime	mAP	0.619	$\text{all BBs}_{corrupted} \equiv \text{all BBs}_{golden}$
	YOLO_v3	KITTI		0.707	
	SSD (VGG16)	Aerial Maritime		0.635	
	SSD (VGG16)	KITTI		0.625	

each one trained with two datasets datasets, namely Utk Face Age[4] and Car Steering Angle[5]. Threshold thr_r values are set to 2 years and 2 degrees, respectively.

Image segmentation assigns each pixel in the input image to one class from a set of predefined classes. An example application is street images analysis to identify pedestrians, vehicles, or roads. The standard metric for evaluating the accuracy of segmentation tasks refers to the Intersection over Union (IoU) between the *golden* and *ground truth* outputs, and is computed as the mean across all classes, mean Intersection over Union (mIoU). Similarly, for robustness we take the IoU between the golden output and the corrupted one; the output is considered usable if, for each class, the IoU is above a context-specific threshold thr_s. In this class, we considered two CNNs, namely Unet [19] and Deeplab_v3 [20], each one trained with two datasets, namely Oxford-IIIT Pet[6] and Cityscapes[7]. The context-specific threshold is set to 70%.

Object detection identifies objects belonging to specific classes in an input image. The CNN annotates the input image by drawing rectangular Bounding Boxes (BBs) around each identified object. The BBs are overlapped on the golden ones to compute the IoU and the BBss are classified as true positives, false positives or false negatives by applying a typical threshold on the obtained IoU. Based on this classification, the mean Average Precision (mAP) metric evaluates the performance by averaging the areas under the precision-recall curves for all classes, providing a comprehensive measure of accuracy. For this kind of task, we consider the output to be usable the resulting set of BBs in the corrupted and golden executions coincide. We consider two CNNs, namely Yolo_v3 [21], and SSD (based on the VGG16 architecture) [22], each trained

with two datasets, namely Aerial Maritime[8], and KITTI[9].

B. Error Simulation Framework

To perform the resilience analysis of the nominal and Ranger-hardened models, we exploited the CLASSES[10] environment [23]. It adopts a cross-layer approach divided into the following two steps:

- **Architecture-level fault injection** is applied on each DL operator of the considered CNN to build a set of representative errors models. An error model is defined as a corruption pattern of the values of a tensor and it is applied on the output of the operator under analysis; moreover, each error model for a given operator is also annotated with its occurrence probability.

- **Application-level error simulation** is then executed on the entire CNN by means of the previously identified error models to analyze the overall robustness of the entire CNN model by observing the propagation of the simulated errors to the final outputs at the functional level.

We use the error model repository targeting a Graphics Processing Unit (GPU) device built by using the NVBitFI environment [24] (extended as discussed in [25]), as discussed in [13]. For each DL operator we define two different sets of error models[11]:

- **Data errors**: representing the functional effects of microarchitectural faults corrupting the result of a single GPU thread while running an instruction. Indeed, when corrupting a single thread, NVBitFI mimics the occurrence of an physical fault in the datapath of a single lane of a GPU multiprocessor[12].

---

[4]https://susanqq.github.io/UTKFace/

[5]https://www.kaggle.com/datasets/asrsaiteja/car-steering-angle-prediction

[6]https://www.robots.ox.ac.uk/vgg/data/pets/

[7]https://www.cityscapes-dataset.com/

[8]https://public.roboflow.com/object-detection/aerial-maritime

[9]https://www.cvlibs.net/datasets/kitti/

[10]CLASSES is freely downloadable from https://github.com/D4De

[11]The probability of a fault to generate a data error or a control error is not available. Therefore, we keep results of these two classes separate.

[12]For further information on the GPU architecture, please refer to [26] or subsequent white papers of more recent NVIDIA GPU architectures.

Table II: Error simulation campaigns' characteristics

Model	Layers	Control errors			Data errors		
		Errors / layer	Input images	Total	Errors / layer	Input images	Total
VGG16	18	256	1024	4.72M	256	1024	4.72M
ResNet50	107	256	512	14.0M	256	512	14.0M
MobileNet_v2	104	256	512	13.6M	256	512	13.6M
VGG11	13	512	256	1.70M	256	128	426k
ResNet18	41	512	256	5.37M	256	128	1.34M
UNet	37	512	256	4.85M	64	128	303k
DeepLab_v3	122	256	128	4.00M	64	64	500k
Yolo_v3	157	256	64	2.57M	128	32	643k
SSD	40	512	64	1.31M	128	32	164k

- **Control errors**: representing the functional effects of microarchitectural faults corrupting the outputs of the entire warp, i.e., a group of threads working in lock-step with a Single Instruction Multiple Data scheme, during the execution of an instruction. In this way, NVBitFI mimics the occurrence of a physical fault in the control flow of a GPU multiprocessor.

The two sets of error models include the same value corruption patterns, which occur with different probabilities. The corruption patterns are:

(a) **Single Point**: one value is corrupted in the output tensor.

(b) **Same Row**: a subset of elements belonging the same row of a single channel are corrupted.

(c) **Bullet Wake**: a set of elements having same coordinates and belonging to different channels are corrupted.

(d) **Skip X**: an error is introduced every *X* positions in the linearized tensor.

(e) **Single Block**: a single channel is corrupted by modifying a set of adjacent values.

(f) **Single Channel Alternated Blocks**: blocks of at least 16 consecutive erroneous values regularly interleaved with non-corrupted blocks.

(g) **Full Channel**: almost an entire channel is corrupted.

(h) **Multichannel Block**: the *Single Block* model is applied on multiple channels.

(i) **Shattered Channel**: all corrupted channels have at least have a corrupted value in the same coordinates along with some other random corrupted values.

(j) **Random**: either a single channel or multiple ones are corrupted with an irregular pattern.

C. Error simulation campaigns

To conclude the experimental framework setup, we here describe the injection campaigns. In particular, Table II reports for each CNN:

- **Layers** the number of layers we considered for error simulation, which only include convolution, batch normalization, max pooling and average pooling layers. Other layers, e.g., ReLu and fully connected ones, have not been considered due to their extremely limited execution time and consequent negligible contribution to the unreliability of the overall CNN.

- **Errors/layer** the number of error simulations for each one of the considered layers.

- **Input images** for every pair layer and simulated errors, we indicate the used number of images from the entire data set.

- **Total** reports the overall total number of executed experiment per CNN.

The results of the error simulations are reported for both **Control** and **Data errors**.

It this worth mentioning that we ran the very same experiments (in terms of corrupted layer, simulated error and input image) twice, one for the *Plain*, nominal CNN, and the second one for the *Hardened* CNN, protected with Ranger.

We conducted from 164k up to 14M error injection experiments depending on the considered CNN and errors (control or data errors). For some CNNs we ran a significantly smaller number of data error simulations w.r.t. the control error ones because we observed that in most cases the effect of data errors is almost completely absorbed after the application of Ranger (refer to the subsequent discussion for more details).

IV. RESULTS DISCUSSION

For every CNN and dataset pair, Table III reports the achieved robustness, i.e., the percentage of error simulations that led to a correct result, for the plain CNN and the Ranger-hardened one when considering control and data errors.

As expected, errors have always a disruptive effect on the plain CNNs. Data errors (which generally involve a reduced number of erroneous tensor values) cause a robustness drop from about 8% (for MobileNet_v2 with GTSRB) to about 40% (for Yolo_v3 with Aerial Maritime). On the other hand, the effect of control errors is dramatic due to the generally very large number of erroneous values in the corrupted tensor. We observe a robustness drop from about 37% (again for MobileNet_v2 with GTSRB) to about 87% (for SSD with Aerial Maritim).

When analyzìng the robustness of the CNNs protected with Ranger, data errors are completely mitigated in almost all cases. Indeed, in this fault/error scenario the protected CNN always achieves more than 99% robustness except for MobileNet_v2 with MNIST and for SSD with both datasets. This outcome actually aligns with what reported in [11], as expected, since the for the data errors correspond to the error models adopted in that study. However, data errors only partially cover the entire set of possible faults, and the outcomes when injecting control errors offer a different perspective. Ranger still largely mitigates the effects of the errors, however the overall robustness of the CNN never gets higher than 96%, with several cases where the final value is between 80% and 90%. There are two critical cases (e.g., SSD and MobileNet_v2) where the final robustness is below 80%.

Dissecting the obtained results, we draw further more detailed considerations for each one of the considered CNNs. For the sake of space, we report only the graphs related to VGG11 applied to face age regression. First, we analyzed the impact of errors occurring in every layer on the final robustness, in the plain CNN and in the hardened one. Figure 1 reports the *layer vulnerability* [5] of every layer of the plain and hardened

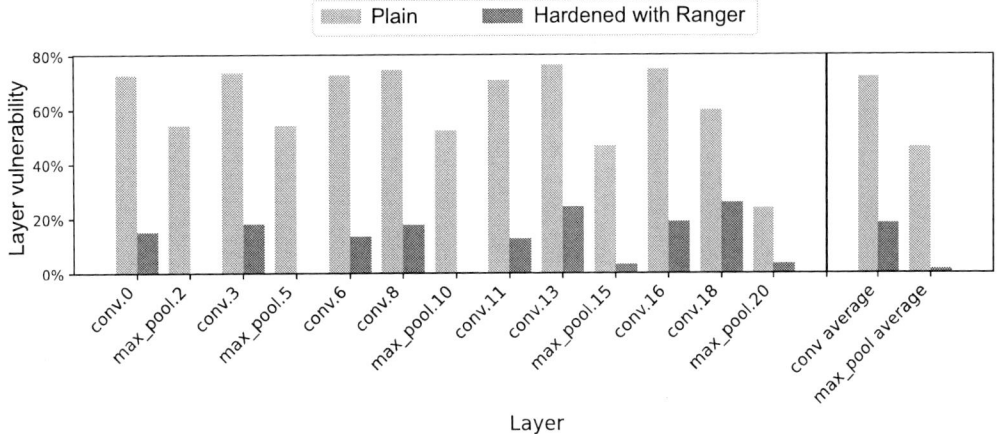

Figure 1: VGG11: layer vulnerability for all layers of the plain and hardened versions.

Table III: Robustness of the plain and hardened CNNs

Model	Dataset	Control errors		Data errors	
		Plain	Hard.	Plain	Hard.
VGG16	MNIST	42.5%	90.6%	75.7%	99.9%
VGG16	GTSRB	31.1%	88.6%	73.4%	99.9%
ResNet50	MNIST	45.1%	95.0%	75.4%	99.4%
ResNet50	GTSRB	35.4%	93.5%	74.5%	99.9%
MobileNet_v2	MNIST	53.8%	69.4%	80.5%	90.2%
MobileNet_v2	GTSRB	67.2%	83.8%	92.1%	99.4%
VGG11	Utk Face Age	31.3%	83.7%	78.2%	99.9%
VGG11	Car Steer. Angle	33.1%	87.9%	78.2%	100%
ResNet18	Utk Face Age	34.3%	91.6%	75.5%	99.9%
ResNet18	Car Steer. Angle	36.8%	92.3%	75.9%	99.9%
UNet	Oxford-IIIT Pet	36.7%	81.0%	77.1%	100%
UNet	Cityscapes	39.3%	89.4%	82.4%	100%
DeepLab_v3	Oxford-IIIT Pet	37.1%	95.7%	77.2%	100%
DeepLab_v3	Cityscapes	32.6%	88.4%	78.0%	99.9%
Yolo_v3	Aerial Maritime	25.5%	96.2%	59.8%	99.6%
Yolo_v3	KITTI	39.7%	96.0%	76.3%	99.2%
SSD	Aerial Maritime	13.4%	75.6%	72.2%	95.6%
SSD	KITTI	25.5%	73.6%	71.7%	95.6%

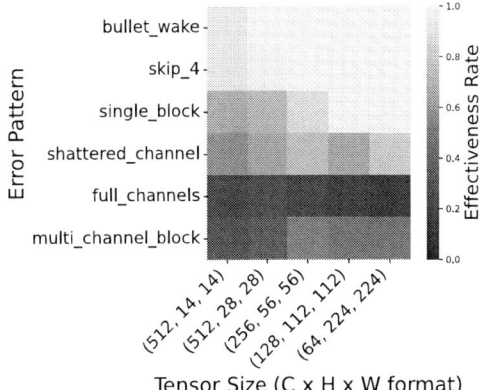

Figure 2: Ranger's effectiveness rates correlated with the error patterns and the size of the corrupted tensors in VGG_11 for face age regression.

versions of VGG11 computed as the percentage of CNN failures due to errors injected in the layer. The chart clearly shows that Ranger is actually able to mitigate the effects of almost all the errors affecting the pooling layers, while several errors occurring in the convolutional layers are not handled. Indeed, when looking at the average failure probabilities across all layers (right-hand side of the figure), for convolutions Ranger is not effective in 20% of the cases.

Similar trends emerge from the analysis of the other model/dataset combinations, where the specific robustness values vary from pair to pair. As such, it is advisable to carry out extensive experiments that encompass different application scenarios.

Then, to further detail the analysis we defined Ranger's *effectiveness rate* as the percentage of error simulations that led to an unusable result in the plain and hardened versions. Figure 2 correlates the effectiveness rate with the size of the corrupted tensor and with the pattern of the simulated error, where the lighter the color the higher the effectiveness rate.

Three considerations can be drawn based on the observation of this graph: i) *small* errors, e.g., bullet wake and skip 4, (that are more likely to appear when the fault affects the dataflow) are almost always corrected by Ranger while *large* errors (generally caused by faults in the control flow) are more difficult to be mitigated; ii) for all error patterns except for full channel, the larger is the corrupted tensor the higher is Ranger's effectiveness: this is because the effect of the fault in large tensors is diluted thus allowing Ranger to be effective; iii) in an opposite trend w.r.t. the previous one, the larger the corrupted tensor is, the less effective Ranger is when dealing with full channel errors. This is because full channel errors do not involve a fixed number of corrupted values but the entire channel, therefore there is no error mitigation when the size of the corrupted pattern increases.

Similar considerations can be drawn when looking at Figure 3 where Ranger's effectiveness is correlated with the size of the corrupted tensor and with the number of corrupted values, irrespective of the simulated error pattern. It can be

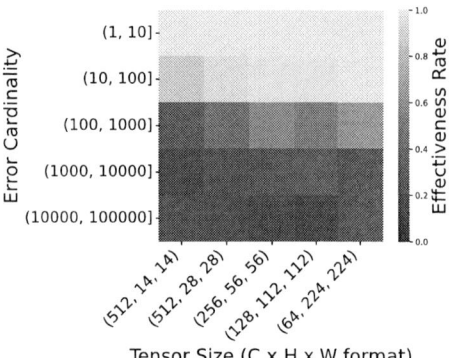

Figure 3: Ranger's effectiveness rates correlated with the error cardinalities and the size of the corrupted tensors in VGG_11 for face age regression.

observed that when the number of erroneous values is below 100, Ranger is able to correct almost all the errors irrespective of the size of the corrupted tensor. On the other hand, as soon as the number of erroneous values exceeds 100, Ranger's effectiveness drops dramatically, below 40% or even below 20% for large tensors.

The entire experimental campaign consists of a total of 176M experiments, accounting for those injecting errors in the plain and hardened versions, and excluding those run with the plain CNNs in a fault free situation to set the golden models used as baselines. Besides the planning, the execution of the campaign took 408hrs running on an x86 machine hosting a RTX 4060Ti GPU.

V. CONCLUSIONS

This paper presented an extensive experimental campaign for the analysis of the effectiveness of a recently proposed hardening technique, i.e., Ranger, to explore its generality and validity across different application context, CNN models and datasets. Indeed, a diverse experimental framework allowed us to better evaluate the real benefits and limitations of this technique, highlighting situations where it might not be as effective as expected, or not as necessary (when fault impacts are really limited and hardening costs can be avoided). The experimental framework definition and setup is indeed time/effort consuming but can be reused in subsequent campaigns, allowing for a broader understanding of benefits/costs trade-offs in the design of resilient DL application.

REFERENCES

[1] M. Campbell, M. Egerstedt, J. How, and R. Murray, "Autonomous driving in urban environments: approaches, lessons and challenges," *Philosophical Trans. of the Royal Society A: Mathematical, Physical and Engineering Sciences*, vol. 368, no. 1928, pp. 4649–4672, 2010.

[2] Z. Ouyang, J. Niu, Y. Liu, and M. Guizani, "Deep CNN-Based Real-Time Traffic Light Detector for Self-Driving Vehicles," *IEEE Trans. Mobile Computing*, vol. 19, no. 2, pp. 300–313, 2020.

[3] F. F. dos Santos, L. Carro, and P. Rech, "Kernel and layer vulnerability factor to evaluate object detection reliability in GPUs," *IET Computers & Digital Techniques*, vol. 13, no. 3, pp. 178–186, 2019.

[4] C. Bolchini, L. Cassano, and A. Miele, "Resilience of deep learning applications: a systematic survey of analysis and hardening techniques," 2023. [Online]. Available: https://doi.org/10.48550/arXiv.2309.16733

[5] C. Bolchini, L. Cassano, A. Miele, and A. Nazzari, "Selective Hardening of CNNs based on Layer Vulnerability Estimation," in *Proc. Int. Symp. Defect and Fault Tolerance in VLSI and Nanotechnology Systems*, 2022, pp. 1–6.

[6] M. Sabih, F. Hannig, and J. Teich, "Fault-Tolerant Low-Precision DNNs using Explainable AI," in *Proc. Int. Conf. Dependable Systems and Networks Workshops*, 2021, pp. 166–174.

[7] A. Ruospo, G. Gavarini, I. Bragaglia, M. Traiola, A. Bosio, and E. Sanchez, "Selective Hardening of Critical Neurons in Deep Neural Networks," in *Proc. Int. Symp. Design and Diagnostics of Electronic Circuits and Systems*, 2022, pp. 136–141.

[8] F. F. dos Santos, P. F. Pimenta, C. B. Lunardi, L. Draghetti, L. Carro, D. R. Kaeli, and P. Rech, "Analyzing and Increasing the Reliability of Convolutional Neural Networks on GPUs," *IEEE Trans. Reliability*, vol. 68, no. 2, pp. 663–677, 2019.

[9] K. Zhao, S. Di, S. Li, X. Liang, Y. Zhai, J. Chen, K. Ouyang, F. Cappello, and Z. Chen, "FT-CNN: Algorithm-Based Fault Tolerance for Convolutional Neural Networks," *IEEE Trans. Parallel and Distributed Systems*, vol. 32, no. 7, pp. 1677–1689, 2021.

[10] N. Cavagnero, F. F. dos Santos, M. Ciccone, G. Averta, T. Tommasi, and P. Rech, "Transient-Fault-Aware Design and Training to Enhance DNNs Reliability with Zero-Overhead," in *Proc. Symp. On-Line Testing and Robust System Design*, 2022, pp. 1–7.

[11] Z. Chen, G. Li, and K. Pattabiraman, "A low-cost fault corrector for Deep Neural Networks through range restriction," in *Proc. Int. Conf. Dependable Systems and Networks*, 2021, pp. 1–13.

[12] L.-H. Hoang, M. A. Hanif, and M. Shafique, "FT-ClipAct: Resilience Analysis of Deep Neural Networks and Improving Their Fault Tolerance Using Clipped Activation," in *Proc. Design, Automation and Test in Europe Conference & Exhibition*, 2020, p. 1241–1246.

[13] C. Bolchini, L. Cassano, A. Miele, A. Nazzari, and D. Passarello, "Analyzing the Reliability of Alternative Convolution Implementations for Deep Learning Applications," in *Proc. Int. Symp. Defect and Fault Tolerance in VLSI and Nanotechnology Systems)*, 2023, pp. 1–6.

[14] F. Geissler, S. Qutub, S. Roychowdhury, A. Asgari, A. Peng, A. Dhamasia, R. Graefe, K. Pattabiraman, and M. Paulitsch, "Towards a Safety Case for Hardware Fault Tolerance in Convolutional Neural Networks Using Activation Range Supervision," in *Proc. AI Safety Workshop*, 2021, pp. 1–9.

[15] M. Biasielli, C. Bolchini, L. Cassano, E. Koyuncu, and A. Miele, "A Neural Network Based Fault Management Scheme for Reliable Image Processing," *IEEE Trans. Computers*, vol. 69, no. 5, pp. 764–776, 2020.

[16] K. Simonyan and A. Zisserman, "Very deep convolutional networks for large-scale image recognition," 2014.

[17] K. He, X. Zhang, S. Ren, and J. Sun, "Deep residual learning for image recognition," 2015. [Online]. Available: https://arxiv.org/abs/1512.03385

[18] M. Sandler, A. Howard, M. Zhu, A. Zhmoginov, and L.-C. Chen, "Mobilenetv2: Inverted residuals and linear bottlenecks," 2018.

[19] O. Ronneberger, P. Fischer, and T. Brox, "U-net: Convolutional networks for biomedical image segmentation," 2015.

[20] L.-C. Chen, G. Papandreou, F. Schroff, and H. Adam, "Rethinking atrous convolution for semantic image segmentation," 2017.

[21] J. Redmon and A. Farhadi, "Yolov3: An incremental improvement," *arXiv*, 2018.

[22] W. Liu, D. Anguelov, D. Erhan, C. Szegedy, S. Reed, C.-Y. Fu, and A. C. Berg, "Ssd: Single shot multibox detector," 2015.

[23] C. Bolchini, L. Cassano, A. Miele, and A. Toschi, "Fast and Accurate Error Simulation for CNNs Against Soft Errors," *IEEE Trans. on Computers*, vol. 72, no. 4, pp. 1–14, 2022.

[24] O. Villa, M. Stephenson, D. Nellans, and S. W. Keckler, "NVBit: A Dynamic Binary Instrumentation Framework for NVIDIA GPUs," in *Proc. Int. Symp. Microarchitecture*, 2019, p. 372–383.

[25] F. F. dos Santos, J. E. Rodriguez Condia, L. Carro, M. Sonza Reorda, and P. Rech, "Revealing GPUs Vulnerabilities by Combining Register-Transfer and Software-Level Fault Injection," in *Proc. Int. Conf. Dependable Systems and Networks*, 2021, pp. 292–304.

[26] E. Lindholm, J. Nickolls, S. Oberman, and J. Montrym, "NVIDIA Tesla: A Unified Graphics and Computing Architecture," *IEEE Micro*, vol. 28, no. 2, pp. 39–55, 2008.

Impact of Image Sensor Input Faults on Pruned Neural Networks for Object Detection

Yizhi Chen[*], Wenyao Zhu[*], Dejiu Chen[†], Omar Mohammed[§], Parthib Khound[§], and Zhonghai Lu[*]

[*]School of Electrical Engineering and Computer Science, KTH Royal Institute of Technology, Stockholm, Sweden
[†]School of Industrial Engineering and Management, KTH Royal Institute of Technology, Stockholm, Sweden
[§]Chair of Reliability of Technical Systems and Electrical Measurement, University of Siegen, Siegen, Germany

Abstract—Object detection is one of the most fundamental problems in computer vision, and image sensors are commonly used for this. In this paper, we present the impact of image sensor faults on pruned neural networks for object detection. We compare the error sensitivities of networks after network slimming, networks after magnitude-based pruning, and native compact models. We also explore different spatial fault types with three intensities. Furthermore, we have developed a temporal error model based on realistic aging image sensor faults. The results illuminate that the performance on clean images is important as the mean Average Precision (mAP) experiences a decrease with an increase in injected faults. Additionally, we demonstrate that the size of the model does not invariably yield a decisive impact on error tolerance when comparing small models such as pruned models and native compact models.

Index Terms—Image Sensor Fault, Network Pruning, Error sensitivity, Network Slimming, Object Detection

I. INTRODUCTION

As one of the most challenging and fundamental problems in computer vision, object detection has absorbed researchers' great interest in recent years [1]. The goal of object detection is to determine whether objects exist in an image and the corresponding spatial location [2]. Object detection is the foundation for solving complex computer vision tasks such as semantic segmentation, scene understanding, and object tracking. Object detection also supports a wide range of applications, including robot vision, autonomous driving, and augmented reality [1], [2].

Many popular neural networks (NN) have been developed to solve object detection tasks, such as R-CNN [3], YOLO [4], and YOLOv3 [5]. These complex networks have high performance and high demand for computation resources. To implement NN in edge devices with limited resources, techniques like pruning [6], [7] and quantization [8] are proposed to reduce the model size while maintaining high accuracy.

The error sensitivity of pruned neural network models has been a hot topic. Mittal [9] indicates that the majority of research assessing the reliability of DNN is directed towards assessing the error in the model's layers [10], [11] or hardware [12]. However, for pruned networks, they are not focusing on faults occurring in input to the pruned NNs.

The occurrence of faults in image sensors is a prevalent phenomenon and the image sensor outputs are the most widely used inputs to object detection NN. Numerous studies investigate input error sensitivities in autonomous driving like blur [13], darkness [14], and rain [15], [16]. These image sensor faults or environmental impacts are examined on a complete model rather than a pruned neural network.

In this study, we investigate the impact of various types and intensities of image sensor faults on pruned neural networks. Our contributions can be summarized as follows:

- We investigate the impacts of faults on pruned neural networks and native compact models, utilizing both state-of-the-art (SOTA) pruning method, network slimming, and classic magnitude-based pruning.
- We focus on faults on inputs rather than faults inside model layers or hardware in object detection applications. Our research examines image sensor faults across a wide and representative range of fault types injected into a real street-view data set, KITTI [17].
- We model the distribution and linear growth of temporal faults to reflect the gradual evolution of image sensor faults over time with usage.

The rest of the paper is organized as follows. **Sec. II** covers the related work; **Sec. III** introduces network pruning and object detection neural network; **Sec. IV** describes the image sensor faults and environmental impacts; **Sec. V** presents the experimental results; **Sec. VI** concludes the paper.

II. RELATED WORK

After 2014, deep learning has been widely used in the object detection field. The two-stage method, like R-CNN [3] and Faster R-CNN [18], is known to be powerful but computationally expensive. One stage involves extracting objects' regions and the other stage is used for classification and location refinement. YOLO [4] is the first one-stage detector in the deep learning object detection area and it applies a single network for the whole image to detect objects. YOLO and YOLOv3 [5] have a fast detection speed but relatively lower accuracy than the two-stage method.

Pruning is essential for implementing NN on edge devices and the error sensitivity of the pruned DNN is attractive. Gao *et al.* [10] measure the reliability of the networks with different pruned rates. The faults are injected into three types of parameters in the model, including the weights, bias, and BN parameters. Ibrahim *et al.* [19] conduct a thorough analysis

979-8-3503-6689-1/24 $31.00 © 2024 IEEE

and evaluation of the error resilience of ResNet models. Zhang *et al.* [12] propose two strategies, fault-aware pruning and fault-aware pruning+retraining to deal with permanent faults that occur in MACs. These studies on pruned NNs are limited to faults in the layers or the hardware.

It is noteworthy that not only are faults within the model important, as highlighted by Su and Chen [20], but faults in the input images are also prevalent and widespread. Khound *et al.* [21] analyze the performances of a lane keeping system with faults in the image sensors. Rashed *et al.* [14] study darkness and aims to compensate for the loss of RGB image quality using LiDAR signals. Chapman *et al.* [22] show that two hot pixels located near each other can interact, leading to significant image degradation. None of these image sensor faults are investigated on pruned networks.

III. NEURAL NETWORK PRUNING

A. Network slimming

A SOTA neural network pruning method called *network slimming* was introduced in [6] to prune the network with little accuracy loss and simultaneously reduce the model size. Network slimming re-defines a loss function, which forces the γ scaling factors of the batch normalization (BN) layer to approach value '0' as much as possible while keeping high accuracy. A channel with a small γ factor has a small output and we identify it as an insignificant channel. These channels are removed as they have a slight influence on the whole network.

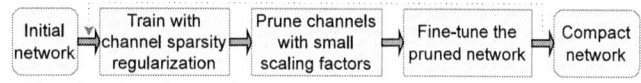

Fig. 1: Flowchart of the network slimming procedure [6].

The process of network slimming [6] is shown in **Fig. 1**. We perform sparse training, following the approach outlined in [6] to force the scaling factors to approach the value '0'. It trains the network with sparsity regularization to decrease some γ factors. As pruned network has a lower accuracy than the original network, fine-tuning is applied to get an improved accuracy than the pruned network.

B. Magnitude-based network pruning

For comparative analysis, we implement unstructured magnitude-based pruning as it is one of the most fundamental and classic pruning techniques. We use the L1 norm in PyTorch as a pruning criterion. It prunes weights according to their magnitudes as weights containing small absolute values have minimal impacts on the output values. Magnitude-based network pruning method replaces weights with zeros instead of cutting off whole channels.

C. Object detection NN: YOLOv3 and YOLOv3-tiny

1) YOLOv3: YOLO series is one of the most popular object detection algorithms that achieve high speed and attractive accuracy. YOLOv3 [5] uses a network containing 53 convolutional layers called Darknet-53 shown in **Fig. 2**(a). There are some latest versions of YOLO but YOLOv3 is still one of the versions that have attracted most of the researchers' eyes.

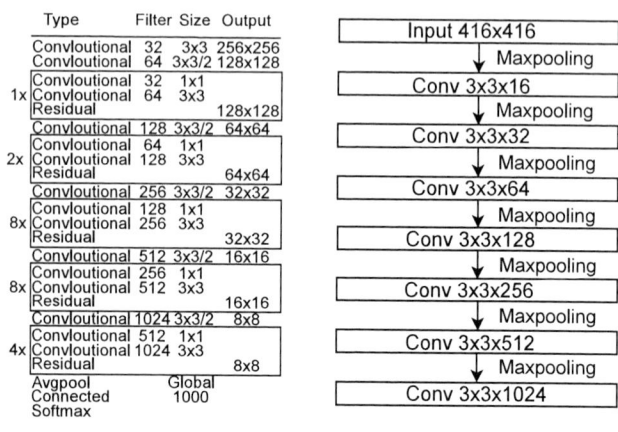

(a) Structure of YOLOv3 [5]. (b) Structure of YOLOv3-tiny [23]

Fig. 2: Structures of YOLOv3 and YOLOv3-tiny.

2) YOLOv3-tiny: YOLOv3-tiny is a native compact version of YOLOv3 and contains only seven convolutional layers along with six pooling layers shown in **Fig.** 2(b) [23], which are much less than the 53 convolutional layers in YOLOv3. This tiny network decreases the demand for a large amount of memory and reduces the hardware cost.

IV. FAULTS AND IMPACTS ON INPUT IMAGES

A. Spatial fault model

Based on impact granularity, we categorize image sensor faults adopted in [21] into global faults and pixel faults. Additionally, we investigate the common environmental impact, raindrops. To examine the influences of strength, we use three intensity levels: light (l), medium (m), and extreme (e).

1) Global faults: Blur is a common image quality issue which means that the image has low sharpness or clarity. [24] classifies blur into two different types: near-isotropic blur (including out-of-focus blur) and directional motion blur. We exclude directional motion blur from our study as it is typically a result of high-speed object movement, not an image sensor fault. Our emphasis is on the out-of-focus blur, which may occur due to defects in the focusing system. We apply blur faults across the whole picture as shown in **Fig.** 3.

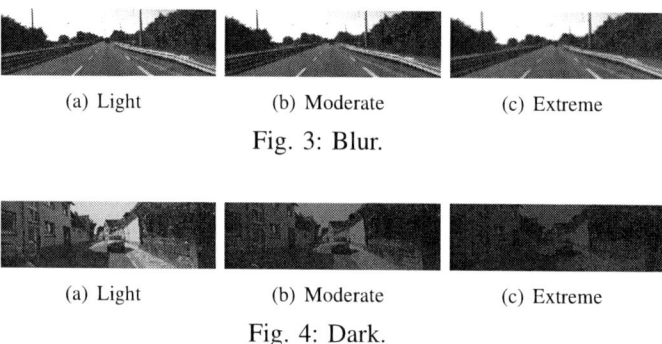

(a) Light (b) Moderate (c) Extreme

Fig. 3: Blur.

(a) Light (b) Moderate (c) Extreme

Fig. 4: Dark.

If faults occur in the camera exposure meter, for example, the camera exposure meter is not operating correctly at low temperatures [25], the camera cannot obtain the correct exposure. Insufficient or excessive exposure time during image capture leads to dark or bright images in **Fig.** 4 and **Fig.** 5.

(a) Light (b) Moderate (c) Extreme

Fig. 5: Brightness.

2) Pixel faults: Speckle noise, one type of multiplicative noise, is shown in **Fig.** 6. Contrasting with additive noise, where the random term is independent of the original state, speckle noise involves a random term that is multiplied by the original image [26].

(a) Light (b) Moderate (c) Extreme

Fig. 6: Speckle noise.

3) Impacts on partial area : As raindrops are uniformly distributed in space [15], we inject uniform random distributed raindrops into each spatial area of the images as shown in **Fig.** 7. Brophy *et al.* [15] concludes that the drop size distribution is correlated with rainfall intensity. To simulate an increase in drop size, we increase raindrop lengths at a higher intensity.

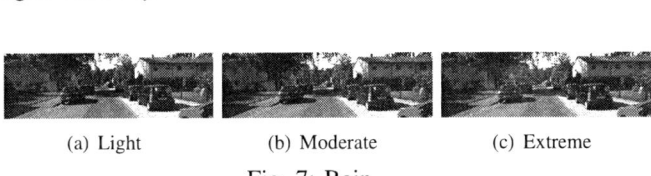

(a) Light (b) Moderate (c) Extreme

Fig. 7: Rain.

4) Combined (blur_dark_rain): Rain fault not only distorts the image but also blurs it [27]. Inspired by this, our approach integrates blur and rain effects and additionally incorporates darkness. Example figures are shown in **Fig.** 8.

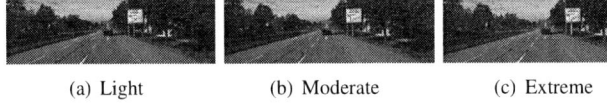

(a) Light (b) Moderate (c) Extreme

Fig. 8: Combined fault: blur_dark_rain.

B. Temporal fault model

To explore the consequences of sensor aging for pruned networks, we adopt a temporal fault model to simulate that sensor performance deteriorates progressively over time.

1) Partially stuck fault type and positions: Fully and partially stuck pixels are the most common type of defective pixels occurring over time [28]. The neutrons of the cosmic rays are the main sources leading to in-field image sensor flaws and the spatial distribution of defective pixels is random [29].

2) Partially stuck fault number and value : With a longer usage lifetime, the number of fault pixels increases linearly [29]. To simulate the progression of aging over time, we inject different numbers of faults from small numbers to big numbers and we show several example rates in **Fig.** 9.

For the defect pixel value, [28] adopts the fault model in which the output of defective pixels is caused by C (a light-independent offset), I (intensity of incoming light), I*K (photo-response non-uniformity), D (dark current), and some modeling noise as shown in **Eq.** (1).

$$Y = C + I + I * K + D + noise \qquad (1)$$

Following the characteristics of fully stuck pixel faults, we build a simplified fault model. We set the stuck offset of pixel output to be a uniform random value in the range of image sensor output, for example, 0 to 255 for an 8-bit image sensor.

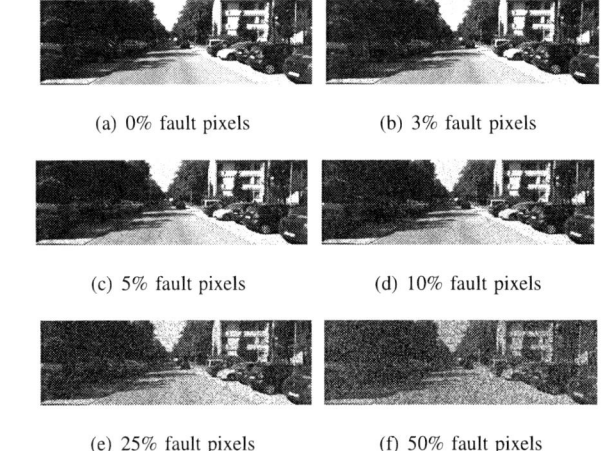

(a) 0% fault pixels (b) 3% fault pixels

(c) 5% fault pixels (d) 10% fault pixels

(e) 25% fault pixels (f) 50% fault pixels

Fig. 9: Images with different temporal fault levels.

The injected fault value and its arbitrary location remain constant because the defective pixels in lower fault rate images are unable to recover and are still flaws in the pictures with higher fault rates. In **Fig.** 10, the normal pixels are labeled as 'N' while those defective pixels are labeled as 'R', which stands for 'random'. Three pixels are defective at a fault rate of 10%. At a fault rate of 20% or 30%, previous defective pixels remain damaged while new faults occur.

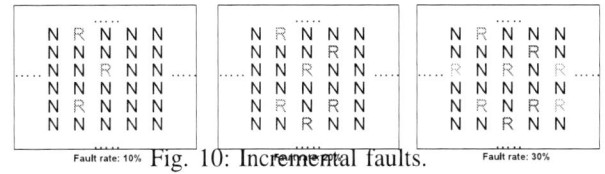

Fig. 10: Incremental faults.

V. EXPERIMENTAL RESULTS

A. Overview

We present our experiments in **Fig.** 11. The input images are injected with different fault types and intensities. We investigate the error sensitivity of NNs before and after network pruning.

Fig. 11: Overview of error sensitivity experiments.

B. Experimental setup

1) Data set: This paper utilizes KITTI's 2D-object detection subsets comprising a total of 7481 labeled images. We split the data set into a training subset and a test subset which contains 80% and 20% images, respectively.

2) Baseline object detection models: We train 1000 epochs for both YOLOv3 and YOLOv3-tiny and we stop the training as there is no further significant improvement. Baseline NNs are trained on images without faults and are not pruned.

3) Pruning with network slimming: We deploy sparse training [6] on baseline models for an additional 1,000 epochs. To collect data and make a more detailed comparison, we record the models of 500 extra epochs as an intermediate step. These models are called YOLOv3-500, YOLOv3-tiny-500, YOLOv3-1000, and YOLOv3-tiny-1000.

After getting YOLOv3-1000 and YOLOv3-tiny-1000, we pruned 80% scaling factors of YOLOv3 and didn't further prune YOLOv3-tiny, as the results in **Fig.** 16 indicate minimal benefit for YOLOv3-tiny. The pruned model is YOLOv3-1000-stemp. We fine-tune YOLOv3-1000-stemp for 100 epochs to recover accuracy and the refined model is YOLOv3-1000-s.

4) Pruning with magnitude-based pruning: Similarly, we deploy magnitude-based pruning with a ratio of 80% and fine-tune it for 100 epochs to get YOLOv3-1000-m.

5) Spatial input sensor error injection: We inject faults into 18 testing datasets representing six different fault conditions: blur, darkness, brightness, rain, speckle noise, and combined faults at three intensities.

6) Temporal image sensor fault injection: We generate 15 testing image sets consisting of six sets of fault levels from 0% to 5% at an interval of 1% and the left sets with fault levels ranging from 10% to 50% at intervals of 5%.

7) Metircs: We use mean $AP^{IoU=0.5}$ (Average Precision) as a metric because it is simple, efficient, and widely used.

C. Experiment results without faults

1) Baseline network results: Full version YOLOv3 achieves a higher overall mAP, 0.789, than YOLOv3-tiny, 0.603, as it has a more complex network architecture. This is consistent with the extensively studied complexity-accuracy trade-off in DNN-based models mentioned in [30].

2) Pruning results of network slimming: The weight distributions are shown in **Fig.** 12 and **Fig.** 13. For YOLOv3, the average mean value of whole γ factors is decreasing from

0.9127 to 0.0162 in **Fig.** 12 while it is slightly decreasing from 2.48 to 2.19 in **Fig.** 13 for YOLOv3-tiny. This indicates that the YOLOv3 network contains many sparsity after sparse training while YOLOv3-tiny is not influenced heavily.

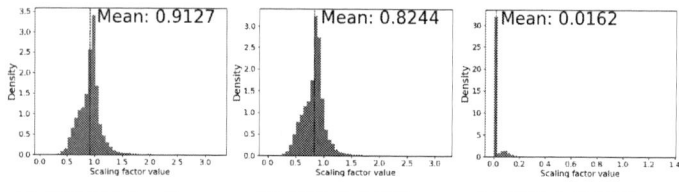

(a) YOLOv3 baseline   (b) YOLOv3-500   (c) YOLOv3-1000

Fig. 12: Scaling factor distribution of YOLOv3.

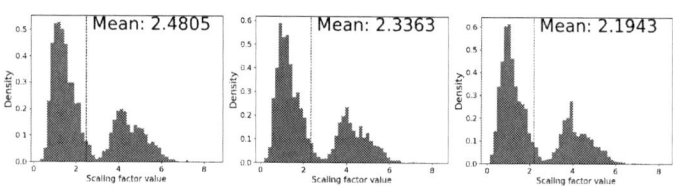

(a) YOLOv3-tiny baseline   (b) YOLOv3-tiny-500   (c) YOLOv3-tiny-1000

Fig. 13: Scaling factor distribution of YOLOv3-tiny.

3) Pruning results of magnitude-based pruning: The whole model size is still 61.62 M as unstructured pruning didn't change the network structure, but the zero parameters (weights equal to zero) are significantly increased from 0 to 49.21 M as shown in **Tab.** I.

TABLE I: Model parameters.

Model	Whole model		Convolution layers	
	All parameters	Zero parameters	All parameters	Zero parameters
YOLOv3 baseline	61.62M	0	61.51M	0
YOLOv3-tiny baseline	8.69M	0	8.68M	0
YOLOv3-m	61.62M	49.21M	61.51M	49.21M
YOLOv3-1000-s	11.12M	0	11.06M	0

D. Experiment results on spatial faults

We present results of all spatial faults in **Fig.** 14. For the same fault type, the accuracy of the same network consistently decreases as the number of faults increases. However, the extent and severity of the performance drop depend on the fault type. For example, for blur faults, all models under extreme blur conditions have a significant mAP drop compared to models under blur_l conditions. For raindrops, different intensities have slight impacts.

The sparse trained models and pruned models suffer mAP drop. In YOLOv3-500 and YOLOv3-1000, we observe a notable trend that the longer the sparse training duration, the greater the drop in mAP. But fine-tuning effectively compensates for the loss, while it still is slightly worse than the non-pruned models.

When looking into the native compact models and pruned models, YOLO-v3 tiny baseline performs worse than the fine-tuned pruned network YOLOv3-1000-s in most scenarios. In the case of speckle fault under extreme intensity (speckle_e), we observe that the result is different. This observation shows that in a comparative analysis between pruned networks and

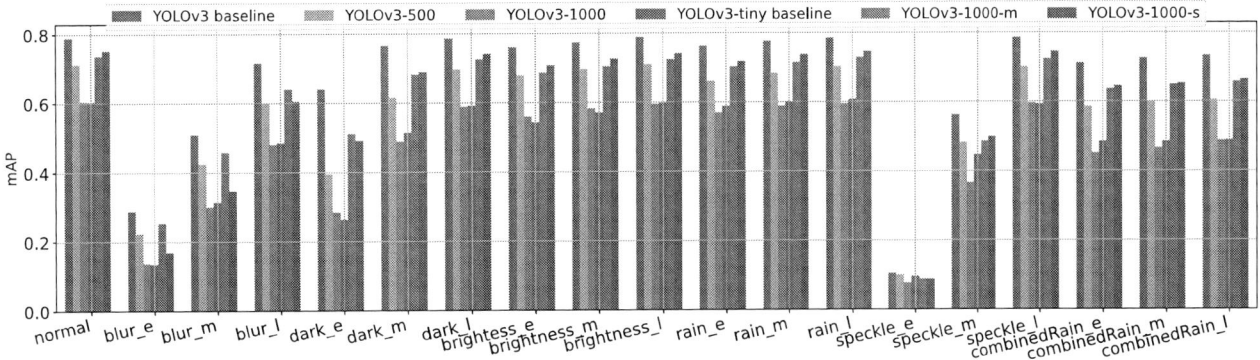

Fig. 14: Spatial fault sensitivity of all evaluated networks.

a native compact model, the number of parameters does not always exert a significant impact on its error sensitivity.

We also evaluate different pruned methods. YOLOv3-1000-s with 11.2 M parameters attains higher mAP in 15 out of 19 fault conditions than YOLOv3-1000-m, which has 12.41 M non-zero parameters. However, in specific scenarios involving three levels of blur and an extreme level of dark fault, YOLOv3-m demonstrates better results. This indicates that even though they have very near non-zero parameters, different internal connections within the model still impact the neural network performance in handling image faults.

E. Experiment results on temporal pixel faults

1) Temporal fault sensitivity of baseline model without pruning: We present the temporal error sensitivity of YOLOv3 and YOLOv3-tiny in **Fig.** 15 as the baseline result.

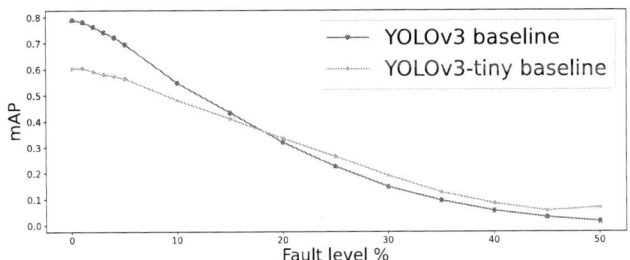

Fig. 15: Error sensitivity of two baseline models.

When the fault level increases from 0% to 50% in **Fig.** 15, the mAP of YOLOv3 decreases from 0.789 to 0.015, and the mAP of YOLOv3-tiny drops from 0.603 to 0.067. A complex model such as the YOLOv3 baseline exhibits superior mAP at low fault levels compared to a significantly simpler network like the YOLOv3-tiny baseline. This aligns with the trend that larger networks are likely to converge to less sensitive functions [31]. This trend is reversed at high fault levels, but the resulting mAP becomes too low to be meaningful.

2) Temporal error sensitivity with pruning: YOLOv3 models with more sparsity have worse performance in the left figure of **Fig.** 16. The performance gap among YOLOv3-tiny baseline, YOLOv3-tiny-500, and YOLOv3-tiny-1000 models is minimal, showing little room for improvement for fine-tuning due to the absence of significant mAP losses.

The temporal fault sensitivity results of native and pruned models are shown in **Fig.** 17. Our results not only indicate that network performance decreases as the fault rate increases but also suggest that the starting point significantly impacts the error sensitivity at different fault levels. The mAP drops from 0% faults to 10% faults are primarily determined by the start point at 0% faults.

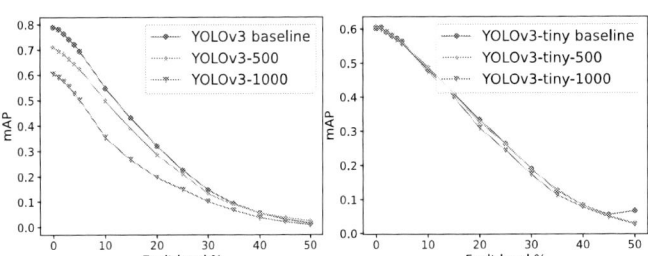

Fig. 16: Results of sparse training (left: YOLOv3; right: YOLOv3-tiny).

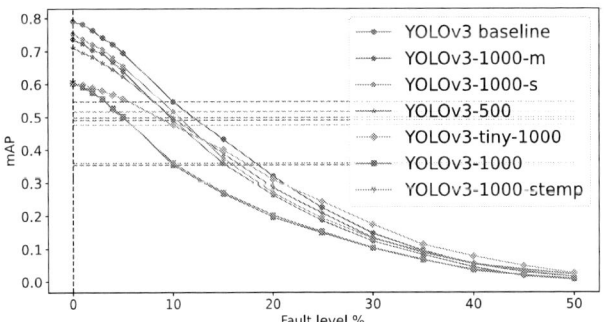

Fig. 17: Temporal fault sensitivity of YOLOv3, pruned YOLOV3, YOLOv3-tiny, and pruned YOLOV3-tiny.

YOLOv3-1000-s, which has 11.12 M parameters, has a slightly higher mAP than YOLOv3-1000-m which contains 12.41 M non-zero parameters. As a native compact model, the YOLOv3-tiny baseline contains 8.69 M parameters which are less than YOLOv3-1000-s. But it outperforms larger networks when the fault level is larger than 10%. These results indicate that the overall size of a small pruned or native compact model is not always a dominating factor in determining its tolerance to errors. These results are consistent with those in spatial analyses, showing that not only does the number of parameters,

979-8-3503-6689-1/24 $31.00 © 2024 IEEE

indicative of model size, matter, but the internal connections are also important when comparing small networks.

VI. CONCLUSIONS

With a focus on the pruned network, we investigate the impacts of image sensor faults across various types and intensities. In addition to spatial faults, we simulate temporal faults due to sensor aging by incrementally injecting faults. We observe that the pruned models are less robust than the original models.

We highlight the significance of the starting point in influencing error sensitivity while observing a consistent decrease in mAP as the fault rate increases, emphasizing the importance of network performance on clean inputs. In comparing small pruned and native compact models, we find that model size does not have a deterministic impact on error sensitivity. With similar parameter numbers, different pruned or compact networks demonstrate varying degrees of performance declination when faults are injected, highlighting the importance of considering architectural choices rather than solely basing decisions on model size for resource-constrained devices. In the future, we will further explore fault-tolerant and compensation methods across a broader range of DNN networks, including transformers. Various pruning techniques including depth pruning will be conducted.

ACKNOWLEDGMENTS

The research was supported in part by VINNOVA (Sweden's Innovation Agency) through the Trust-E project of the Eureka PENTA and EURIPIDES programs.

REFERENCES

[1] Z. Zou, K. Chen, Z. Shi, Y. Guo, and J. Ye, "Object detection in 20 years: A survey," *Proceedings of the IEEE*, vol. 111, no. 3, pp. 257–276, 2023.

[2] L. Liu, W. Ouyang, X. Wang, P. Fieguth, J. Chen, X. Liu, and M. Pietikäinen, "Deep learning for generic object detection: A survey," *International journal of computer vision*, vol. 128, no. 2, pp. 261–318, 2020.

[3] R. Girshick, J. Donahue, T. Darrell, and J. Malik, "Rich feature hierarchies for accurate object detection and semantic segmentation," in *Proceedings of the IEEE conference on computer vision and pattern recognition*, 2014, pp. 580–587.

[4] J. Redmon, S. Divvala, R. Girshick, and A. Farhadi, "You only look once: Unified, real-time object detection," in *2016 IEEE Conference on Computer Vision and Pattern Recognition (CVPR)*, 2016, pp. 779–788.

[5] J. Redmon and A. Farhadi, "Yolov3: An incremental improvement," *arXiv preprint arXiv:1804.02767*, 2018.

[6] Z. Liu, J. Li, Z. Shen, G. Huang, S. Yan, and C. Zhang, "Learning efficient convolutional networks through network slimming," in *2017 IEEE International Conference on Computer Vision (ICCV)*, 2017, pp. 2755–2763.

[7] S. Han, J. Pool, J. Tran, and W. Dally, "Learning both weights and connections for efficient neural network," *Advances in neural information processing systems*, vol. 28, 2015.

[8] E. Michaud, Z. Liu, U. Girit, and M. Tegmark, "The quantization model of neural scaling," *Advances in Neural Information Processing Systems*, vol. 36, 2024.

[9] S. Mittal, "A survey on modeling and improving reliability of dnn algorithms and accelerators," *Journal of Systems Architecture*, vol. 104, p. 101689, 2020.

[10] Z. Gao, X. Wei, H. Zhang, W. Li, G. Ge, Y. Wang, and P. Reviriego, "Reliability evaluation of pruned neural networks against errors on parameters," in *2020 IEEE International Symposium on Defect and Fault Tolerance in VLSI and Nanotechnology Systems (DFT)*, 2020, pp. 1–6.

[11] M. Sabbagh, C. Gongye, Y. Fei, and Y. Wang, "Evaluating fault resiliency of compressed deep neural networks," in *2019 IEEE International Conference on Embedded Software and Systems (ICESS)*. IEEE, 2019, pp. 1–7.

[12] J. J. Zhang, T. Gu, K. Basu, and S. Garg, "Analyzing and mitigating the impact of permanent faults on a systolic array based neural network accelerator," in *2018 IEEE 36th VLSI Test Symposium (VTS)*. IEEE, 2018, pp. 1–6.

[13] L. Yahiaoui, J. Horgan, B. Deegan, S. Yogamani, C. Hughes, and P. Denny, "Overview and empirical analysis of isp parameter tuning for visual perception in autonomous driving," *Journal of Imaging*, vol. 5, no. 10, p. 78, 2019.

[14] H. Rashed, M. Ramzy, V. Vaquero, A. El Sallab, G. Sistu, and S. Yogamani, "Fusemodnet: Real-time camera and lidar based moving object detection for robust low-light autonomous driving," in *2019 IEEE/CVF International Conference on Computer Vision Workshop (ICCVW)*, 2019, pp. 2393–2402.

[15] T. Brophy, D. Mullins, A. Parsi, J. Horgan, E. Ward, P. Denny, C. Eising, B. Deegan, M. Glavin, and E. Jones, "A review of the impact of rain on camera-based perception in automated driving systems," *IEEE Access*, vol. 11, pp. 67 040–67 057, 2023.

[16] P. Su, F. Warg, and D. Chen, "A simulation-aided approach to safety analysis of learning-enabled components in automated driving systems," in *26th IEEE International Conference on Intelligent Transportation Systems (ITSC 2023)*, 2023.

[17] A. Geiger, P. Lenz, and R. Urtasun, "Are we ready for autonomous driving? the kitti vision benchmark suite," in *Conference on Computer Vision and Pattern Recognition (CVPR)*, 2012.

[18] S. Ren, K. He, R. Girshick, and J. Sun, "Faster r-cnn: Towards real-time object detection with region proposal networks," *Advances in neural information processing systems*, vol. 28, 2015.

[19] Y. Ibrahim, H. Wang, M. Bai, Z. Liu, J. Wang, Z. Yang, and Z. Chen, "Soft error resilience of deep residual networks for object recognition," *IEEE Access*, vol. 8, pp. 19 490–19 503, 2019.

[20] P. Su and D. Chen, "Using fault injection for the training of functions to detect soft errors of dnns in automotive vehicles," in *International Conference on Dependability and Complex Systems*. Springer, 2022, pp. 308–318.

[21] P. Khound, O. Mohammed, P. Su, D. Chen, and F. Gronwald, "Performance index modeling from fault injection analysis for an autonomous lane-keeping system," in *the 33rd European Safety and Reliability Conference*. Research Publishing Services, 2023.

[22] G. H. Chapman, K. J. C. Silva Meneses, I. Koren, and Z. Koren, "Image degradation due to interacting adjacent hot pixels," in *2022 IEEE International Symposium on Defect and Fault Tolerance in VLSI and Nanotechnology Systems (DFT)*, 2022, pp. 1–6.

[23] Z. Yi, S. Yongliang, and Z. Jun, "An improved tiny-yolov3 pedestrian detection algorithm," *Optik*, vol. 183, pp. 17–23, 2019. [Online]. Available: https://www.sciencedirect.com/science/article/pii/S003040261930155X

[24] R. Liu, Z. Li, and J. Jia, "Image partial blur detection and classification," in *2008 IEEE Conference on Computer Vision and Pattern Recognition*, 2008, pp. 1–8.

[25] K. B. Newbery and C. Southwell, "An automated camera system for remote monitoring in polar environments," *Cold Regions Science and Technology*, vol. 55, no. 1, pp. 47–51, 2009.

[26] J. M. Bioucas-Dias and M. A. T. Figueiredo, "Multiplicative noise removal using variable splitting and constrained optimization," *IEEE Transactions on Image Processing*, vol. 19, no. 7, pp. 1720–1730, 2010.

[27] A. Cord and N. Gimonet, "Detecting unfocused raindrops: In-vehicle multipurpose cameras," *IEEE Robotics & Automation Magazine*, vol. 21, no. 1, pp. 49–56, 2014.

[28] T. Bergmüller, L. Debiasi, A. Uhl, and Z. Sun, "Impact of sensor ageing on iris recognition," in *IEEE International Joint Conference on Biometrics*. IEEE, 2014, pp. 1–8.

[29] C. Kauba and A. Uhl, "Sensor ageing impact on finger-vein recognition," in *2015 International Conference on Biometrics (ICB)*. IEEE, 2015, pp. 113–120.

[30] Y. Ding, J. Liu, J. Xiong, and Y. Shi, "Revisiting the evaluation of uncertainty estimation and its application to explore model complexity-uncertainty trade-off," in *Proceedings of the IEEE/CVF Conference on Computer Vision and Pattern Recognition Workshops*, 2020, pp. 4–5.

[31] R. Novak, Y. Bahri, D. A. Abolafia, J. Pennington, and J. Sohl-Dickstein, "Sensitivity and generalization in neural networks: an empirical study," *arXiv preprint arXiv:1802.08760*, 2018.

Towards Biology-Inspired Fault Tolerance of Neuromorphic Hardware for Space Applications

Shadi Matinizadeh, Sarah Johari, Arghavan Mohammadhassani, and Anup Das
Electrical and Computer Engineering, Drexel University, Philadelphia, USA
Email: {sm4884,sj984,am4774,anup.das}@drexel.edu

Abstract—High-energy particles in space can induce single- and multibit upsets in random electronic components of FPGA-based neuromorphic systems. We propose NeuFT, a low overhead biology-inspired architecture for fault tolerance of these systems. NeuFT draws inspiration from astrocytes, which are star-shaped glial cells in a mammalian brain that facilitate the self-repair of failed neurons by sending a closed-loop retrograde feedback signal. Our fault-tolerant design methodology is as follows. First, we partition a spiking neural network (SNN) model into synaptic islands and place an astrocyte in each of them. Each astrocyte is trained using spike trains of neurons on its synaptic island to optimize fault tolerance capabilities. Next, we propose a practical astrocyte-inspired controller that integrates astrocyte feedback as bias of a leaky integrate and fire neuron with minimal design changes. Finally, we introduce software to partition and train an astrocyte-enabled SNN and map it to hardware. We evaluate NeuFT for five datasets with varying degrees of bit error rates considering single- and multibit upsets. Our results show a significant reduction in fault tolerance overhead while delivering a high classification accuracy with respect to state-of-the-art.

Index Terms—fault tolerance, astrocytes, spiking neural networks, neuromorphic hardware

I. INTRODUCTION

Neuromorphic computing describes the hardware implementation of biological neurons that use discrete spikes to compute and transmit information [1]. These architectures improve the energy efficiency of machine learning applications designed using spiking neural networks (SNNs) [2]. Due to their reconfigurability, FPGA platforms are recently used to prototype neuromorphic designs for space-oriented applications [3]. In fact, some are even making it to space [4].

FPGAs consist of densely packed static random access memory (SRAM) cells, which implement flip-flops, lookup tables (LUTs), and block RAMs (BRAMs). These electronic components are used to design neuron and synapse circuits for neuromorphic computing. Unfortunately, SRAM-based electronic components are vulnerable to single-event upsets (SEUs) from radiation generated by high-energy particles (e.g., thermal neutrons, alpha particles, and heavy ions) emitted from a variety of sources both within and beyond our solar system. Moreover, given their high integration density, SEUs can lead to both single-bit upsets (SBUs), e.g., a single bit-flip in a SRAM cell, and multi-cell upsets (MCUs), affecting multiple adjacent SRAM cells at once. NASA recently reported an SEU rate of 10^{-7}-10^{-8} errors/bit-hour, even under low-to-moderate radiation conditions [5]. Conventionally, SEU-induced faults are mitigated using the following techniques.

1) **Error Correction Coding (ECC)** involves adding a few parity bits to detect and correct data corruption in FPGA memory [6]. ECC has a high area and power overhead.
2) **Radiation-Hardened Design (Rad-Hard)** involves using technologies such as fully depleted silicon on insulator SRAM instead of the bulk SRAM [7] or using specially designed latches inside an SRAM cell to break cross-coupling between its internal inverters [8]. These technology-oriented solutions tend to be expensive.
3) **Triple Modular Redundancy (TMR)** involves triplicating each logic and using a majority voter to obtain the output [9]. TMR has a high area and power overhead.
4) **Scrubbing** involves storing the FPGA configuration in a rad-hard memory and using it to periodically reprogram the FPGA or a portion of the FPGA affected by SEUs [10]. Scrubbing introduces high latency overhead.

Our **objective** is to improve the fault tolerance of neuromorphic designs implemented on FPGA so that they can mitigate SEUs in space. We achieve this by drawing inspiration from biological astrocytes [11]. These are star-shaped glial cells in the mammalian brain that facilitate self-repair of failed neurons by sending a closed-loop retrograde feedback signal. Our key idea is to partition a machine learning model into synaptic islands and integrate an astrocyte into each of them to facilitate its self-repair. Following are our key **contributions**.

- Instead of implementing the exact dynamics of an astrocyte which introduces a high area and power overhead, we propose a simple control design. Here, each astrocyte maintains (1) an internal potential, which is a function of spike activities in neurons on its synaptic island, and (2) a firing threshold to generate a feedback signal in the event of a decrease in its internal potential due to faults.
- We propose a methodology to train each astrocyte to optimize its specific restoration capabilities for neurons on the corresponding synaptic island.
- We extend the design of a leaky integrate and fire neuron (LIF) to incorporate the astrocyte feedback signal as its bias. In this way, we introduce a marginal overhead.
- We propose a design modification involving ECC to protect both the synaptic weights stored in memory and the control logic of each astrocyte, thereby making the entire design tolerant to SEU induced faults.
- We propose a system software to partition and map an astrocyte-infused SNN to hardware.

Our biology-inspired methodology, called NeuFT, provides

979-8-3503-6689-1/24 $31.00 © 2024 IEEE

a low overhead fault tolerance of neuromorphic computing compared to rad-hard, TMR, and scrubbing mechanisms alone. NeuFT can also be used in combination with these mechanisms to reduce their overhead. We evaluate NeuFT using five datasets and a varying degree of bit-error rates. Our results show significant improvements over the state-of-the-art.

II. INTRODUCTION TO ASTROCYTES

Astrocytes derive their name due to their star-like appearance. These are a subtype of glial cells in the human central nervous system (CNS) [12]. Fig. 1 shows an astrocyte with its retrograde feedback signaling mechanism.

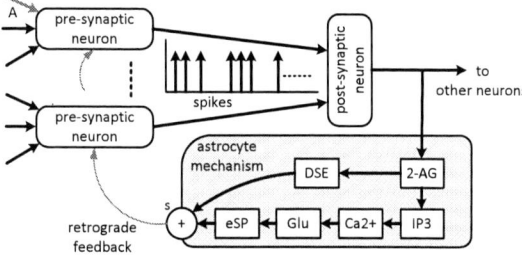

Fig. 1: Astrocyte retrograde signaling mechanism.

If a post-synaptic neuron is depolarized due to the arrival of a sufficient number of spikes from its pre-synaptic neurons, it releases 2-arachidonyl glycerol (2-AG), a retrograde messenger (endocannabinoid), whose dynamics is governed by

$$\frac{dAG}{dt} = \frac{-AG}{\tau_{AG}} + r_{AG} \cdot \delta(t - \tau), \tag{1}$$

where AG is the quantity of 2-AG, τ_{AG} is the rate of decay and r_{AG} is the rate of production of 2-AG.

The retrograde messenger travels to pre-synaptic neurons along two pathways. In the *direct pathway*, the 2-AG binds to the pre-synaptic cannabinoid receptors, suppressing the transmission probability, called Depolarization-induced Suppression Excitation (DSE), whose dynamics is governed by

$$DSE = -K_{AG} \cdot AG(t), \tag{2}$$

where K_{AG} is a constant. In the *indirect pathway*, the retrograde messenger triggers a series of chemical reactions in the astrocyte, leading to the production of eSP, which increases the synaptic transmission at the pre-synaptic sites.

Retrograde messenger 2-AG increases the release of 1,4,5-triphosphate (IP_3), which stimulates Ca^{+2} induced Ca^{+2} release (CICR). The IP_3 dynamics is given by

$$\frac{dIP_3}{dt} = \frac{IP_3(0) - IP_3(t)}{\tau_{IP_3}} + r_{IP_3} \cdot AG(t) \tag{3}$$

Simultaneously, there is an increase in Ca^{+2} due to the Sarco Endoplasmic Reticulum Ca^{+2}-ATPase (SERCA) pumping of $Ca^{+2}(cyt)$ from the cytoplasm (cyt) to the Endoplasmic Reticulum (ER) of an astrocyte. The intracellular astrocytic calcium dynamics control the glutamate (Glu) release from the astrocyte, which is governed by

$$\frac{dGlu}{dt} = \frac{-Glu}{\tau_{Glu}} + r_{Glu}(t - t_{Ca}), \tag{4}$$

where τ_{Glu} is the rate of decay and r_{Glu} is the rate of production of glutamate, and t_{Ca} is time at which Ca^{2+} crosses

the release threshold. The glutamate generates the eSP, the indirect signal to the post-synaptic site. eSP is related to Glu using the following ordinary differential equation (ODE).

$$\frac{deSP}{dt} = \frac{-eSP}{\tau_{eSP}} + \frac{m_{eSP}}{\tau_{eSP}} Glu(t), \tag{5}$$

where τ_{eSP} is the decay rate of eSP and m_{eSP} is a scaling factor. The overall probability of transmission $PR(t)$ at the synaptic site is the sum of the two pathways.

$$PR(t) = PR(t_0) + PR(t_0)\left(\frac{DSE(t) + eSP(t)}{100}\right) \tag{6}$$

A. Self-Repair Mechanism

Consider that the number of spikes from a pre-synaptic neuron is reduced due to a fault in its synaptic connections (annotation A in Fig. 1). This reduces 2-AG production (Eq. 1). On one hand, eSP production reduces in the indirect pathway (Eqs. 3-5). On the other hand, the probability of transmission on the direct pathway increases (Eq. 2). Combining the two pathways, the overall probability of transmission at the synaptic site increases, which leads to an increase in the spikes output from the pre-synaptic neurons.

B. Differences from Existing Works on Astrocytes

Existing works can be organized into two key research areas. For each of them, we highlight important differences.

1) **Accurate astrocyte modeling.** Recent works have modeled the calcium dynamics of astrocytes and their retrograde signaling in biological details [13]. There are also attempts to replicate this accurate dynamics using digital and analog electronic components [14]–[16]. We use approximation to simplify hardware implementation, yet retain the relevant self-repair property.

2) **Self-repair in unsupervised learning.** Recent works have demonstrated restoring spike firing frequency of neurons involved in unsupervised spike timing dependent plasticity (STDP) learning [17]–[19]. We use astrocytes for fault tolerance of inference hardware in the context of supervised learning.

III. HIGH-LEVEL OVERVIEW

Fig. 2 shows the integration of astrocytes in a layer-based neuromorphic hardware [20]–[22]. To simplify the discussion and without loss of generality, we consider a NeuFT design in which a single astrocyte is used to tolerate faults in a layer. Therefore, the number of astrocytes used in NeuFT is the same as the number of layers in baseline neuromorphic hardware.

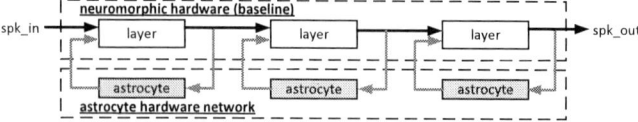

Fig. 2: Astrocyte integration in a neuromorphic hardware.

An astrocyte (see Fig. 7) is implemented as a closed-loop controller with a firing threshold V_{th_a} specific to a given dataset. The concept of thresholding is illustrated in Fig. 4.

Fig. 3 shows the internal architecture of a neuromorphic layer and our fault-tolerant design methodology. NeuFT uses astrocytes to mitigate faults in neuron circuits implemented on the programmable logic of an FPGA and ECC to mitigate faults in synaptic weights stored in the memory (e.g., BRAM).

Mitigating faults in neurons using astrocytes. We insert astrocytes into an SNN model and train it to optimize both synaptic weights and astrocyte firing thresholds. Alternatively, our methodology can also import a trained model, insert astrocytes into it, and perform a fast retraining of the model to optimize astrocyte firing thresholds. The firing threshold is programmed in the astrocyte hardware to generate the closed-loop feedback signal which is used as a bias for neurons while providing fault tolerance (see Sec. V).

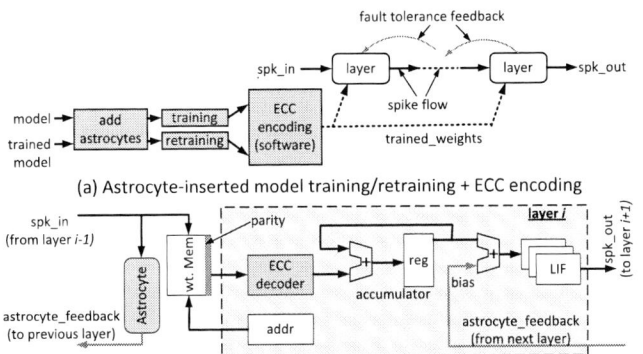

(a) Astrocyte-inserted model training/retraining + ECC encoding

(b) Fault tolerant design of neuromorphic hardware layer

Fig. 3: Proposed fault tolerant design methodology.

Mitigating faults in synapses using ECC. NeuFT uses single error correction and double error detection (SECDED) based Hamming scheme to encode each weight. This is performed in software after training/retraining of an SNN model (Fig. 3a). Upon receiving each parity-encoded weight, NeuFT first decodes the weight and then uses this decoded weight to compute the activation as shown in Fig. 3b.

In the following, we describe the details of NeuFT.

IV. DATASET-SPECIFIC ASTROCYTE TRAINING

This is the first work to highlight the importance of training astrocytes for fault tolerance of supervised SNNs. To understand the need for such training, Fig. 4a shows the spike output of a neuron. Consider a fault in its pre-synaptic connections (neurons or synapses) at time t, which causes the neuron to be silent. Without any fault tolerance mechanism, the neuron does not generate spikes after time t (Fig. 4b).

Fig. 4c shows the spike output from the neuron using the astrocyte design of [14]. We see that three sets of spikes are generated by astrocyte feedback. While spike set 3 is intended to mitigate the impact of fault at time t, sets 1 & 2 are generated because the astrocyte cannot distinguish the dataset-specific inter-spike interval from the one created due to faults. We show that extra spike sets lead to accuracy loss.

Fig. 4d shows our proposed idea. An astrocyte maintains an internal potential, mimicking the 2-AG concentration in Fig. 1. This potential is modulated using the spikes from the neuron. A firing threshold is defined as shown in the figure. Retrograde feedback from an astrocyte is generated when the astrocyte potential falls below this firing threshold. Therefore, learning in an astrocyte involves (1) finding the firing threshold

(V_{th_a}) such that the astrocyte can distinguish between dataset-specific inter-spike interval and the interval created due to faults, and (2) finding the value of astrocyte feedback (f_a) needed to restore the spike firing frequency of the neuron.

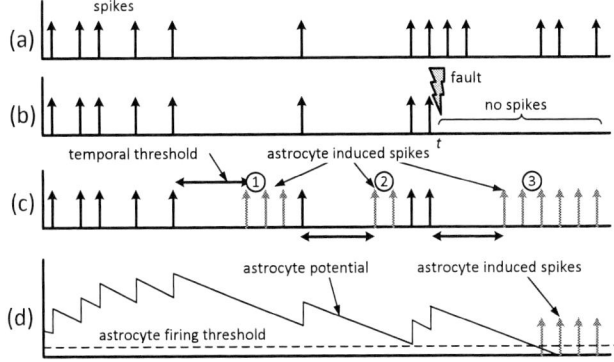

Fig. 4: Motivating threshold-based astrocyte design. (a) Spike profile of a neuron. (b) Impact of fault at time t. (c) Integrating astrocytes without training leading to accuracy loss. (d) Proposed threshold-based astrocyte training.

To address this, we introduce the following definition.

Definition 1: (SNN) *An SNN is represented as a dataflow graph $G = (\mathcal{L}, \mathcal{E})$ consisting of a finite set $\mathcal{L}$ of vertices and a finite set $\mathcal{E}$ of edges. Each vertex $L \in \mathcal{L}$ is a layer in the SNN and it represents some transformation of its input (*`convolution`*, *`pooling`*, *`linear`*, etc). Edge $E_{i,j} \in \mathcal{E}$ connects the layer L_i to L_j. An SNN vertex (layer) $L_i \in \mathcal{L}$ is a collection of neurons, i.e., $L_i = \{n_{i,0}, n_{i,1}, \cdots, n_{i,|L_i|}\}$.*

A. Learning Astrocyte Firing Threshold

We consider neurons in layer L_i to be partitioned into a finite number of clusters. Each cluster is called a **synaptic island**. We introduce an astrocyte for each synaptic island. Without loss of generality, we describe the training mechanism for the first cluster of L_i consisting of k neurons $n_{i,0}, n_{i,1}, \cdots, n_{i,k-1}$. Our discussions can be generalized to other clusters.

Let $s_{i,j}(t) \in \{0, 1\}$ with $j \in \{0, 1, \cdots, k-1\}$ be the digital representation of the output of neuron $n_{i,j} \in L_i$ in a trained SNN model at time t. The astrocyte potential $a(t)$ is formulated considering an exponential behavior (Eq. 1) as

$$a(t) = a(t-1) + \omega_a \cdot \sum_{j=0}^{k-1} s_{i,j}(t) - \beta_a \cdot a(t-1) \qquad (7)$$

The astrocyte potential increases linearly (proportional to ω_a) for each spike event and decays at a rate β_a when there is no spike. To enable efficient hardware implementation of an astrocyte, we set $\omega_a \in \mathbb{N}$ and $\beta_a = 0.5$. The former constraint enables the use of a simple adder to increase the astrocyte potential, while the latter constraint enables division by 2, which can be implemented in hardware using bit shifts (see Fig. 7). The astrocyte firing threshold is defined as the minimum astrocyte potential across all time steps, i.e.,

$$V_{th_a} = \min_t a(t) \qquad (8)$$

Fig. 5 shows the astrocyte potential in a synaptic island in the first layer of a $(256 \times 128 \times 10)$ SNN trained on the MNIST dataset using $\omega_a = 1$ and $\beta_a = 0.5$. The synaptic island consists

979-8-3503-6689-1/24 $31.00 © 2024 IEEE

of 16 neurons. Results are for a single batch with a batch size of 128 and a time step of 100 ms.

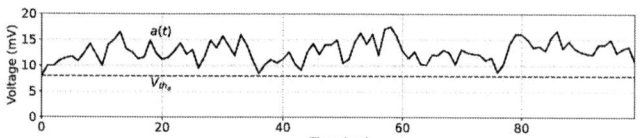

Fig. 5: Astrocyte potential and firing threshold.

We observe that the minimum astrocyte threshold is 8 mV, which is used as the firing threshold V_{th_a}.

B. Learning Astrocyte Feedback for Fault Tolerance

The fundamental idea behind retrograde signaling of astrocytes is a closed-loop feedback control system. To understand the implementation, Fig. 6 illustrates the behavior of a leaky integrate and fire (LIF) neuron for three scenarios – no spike events ❶, spike events leading to exceeding the threshold ❷, and spike events in the presence of faults ❸. [1]

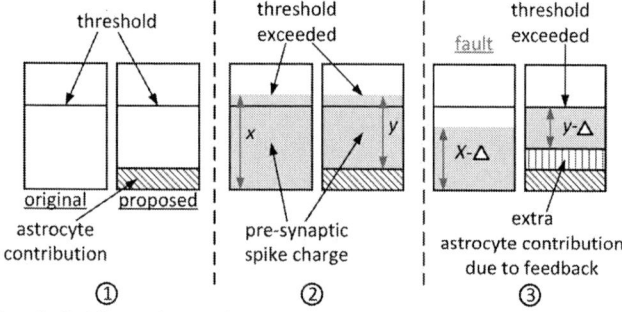

Fig. 6: Self-repair mechanism in a LIF neuron with feedback.

For each scenario, we illustrate the baseline design (left) and our new design (right) introduced due to the astrocyte, which involves adding a residual bias (b_a) to a neuron. To achieve this, we consider a neuron's input current defined as

$$I_{in}(t) = b + \sum_j w_j \cdot x_j(t) \tag{9}$$

where b is the bias of the neuron, w_j's are its pre-synaptic weights, and $x_j(t)$ is the spike at time t from the jth pre-synaptic connection. Equation 9 can be re-written as

$$I_{in}(t) = b_a + \widehat{I_{in}(t)}, \text{ where } \widehat{I_{in}(t)} = (b - b_a) + \sum_j w_j \cdot x_j(t) \tag{10}$$

Equation 10 is illustrated in the scenario ❷ with the action potential due to spike events $x = I_{in}(t)$ (left) and $y = \widehat{I_{in}(t)} < x$ (right). In both the baseline and the new design, the threshold is exceeded. Observe that $x > V_{th} \implies (y + b_a) > V_{th}$.

Consider a fault in the pre-synaptic connection leading to a reduction of the action potential by an amount Δ (scenario ❸). In the original design, the threshold is not exceeded. So, no spike is generated. In the new design, the reduction in the action potential leads to an increase in the astrocyte contribution by an amount b_e. The total action potential $(b_a + b_e - \Delta)$ is greater than the threshold. So, the neuron generates a spike.

---

[1] A LIF model represents a neuron as a parallel combination of a leaky resistor and a capacitor. Input current charge up the capacitor to produce an action potential. A spike is generated when this potential exceeds a threshold.

To implement this exact feedback mechanism, we need the following. First, we consider astrocyte operation where it generates a trigger signal when $a(t)$ is below the firing threshold V_{th_a}. We represent the astrocyte output as

$$a_{out}(t) = \max\left(0, \min\left(a(t), V_{th_a}\right)\right) = \begin{cases} V_{th_a} & \text{if } a(t) > V_{th_a} \\ a(t) & \text{otherwise} \end{cases} \tag{11}$$

Second, we represent the astrocyte feedback as

$$f_a(t) = b_a + g_a \cdot e(t) = b_a + g_a\left(V_{th_a} - a(t)\right) \tag{12}$$

where the error signal $e(t)$ is obtained from the astrocyte output and g_a is the gain parameter of the astrocyte feedback.

V. BIOLOGY-INSPIRED ASTROCYTE DESIGN

Fig. 7 shows the proposed astrocyte design that implements a closed-loop feedback to provide fault tolerance.

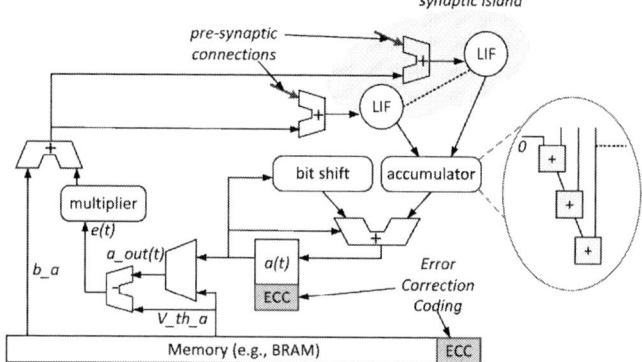

Fig. 7: Astrocyte with closed-loop feedback.

The memory elements required are the following. First, the astrocyte potential is implemented using the n-bit register $a(t)$, where n is the number of bits needed to represent the maximum value of the astrocyte potential, i.e., $n = \log_2 \max_t a(t)$. Second, the astrocyte threshold (V_{th_a}) and the baseline bias b_a are both implemented in memory (e.g., BRAM). The number of bits needed for these constants is a user-defined value q. The total memory overhead of each astrocyte is $(n + 2q)$ bits.

Finally, since astrocytes are critical elements in our design, ECC is enabled for its memory elements and TMR is enabled for all other logic blocks. This is done to prevent faults from affecting the operation of each astrocyte.

A. NeuFT: A Hybrid End-to-End Fault Tolerance Solution

NeuFT is the first work to facilitate end-to-end fault tolerance of neuromorphic inference hardware implemented on FPGA for space applications. Table I summarizes our fault tolerance mechanisms included in NeuFT.

A key aspect of our solution is that it can be combined with other fault tolerance mechanisms. For instance, scrubbing can be optionally enabled in NeuFT when the accuracy loss exceeds a pre-defined threshold. Furthermore, such on-demand scrubbing essentially reduces the latency associated with scrubbing-only solutions [10].

Overall, NeuFT is a hybrid end-to-end fault tolerance solution. We evaluate NeuFT in Sec. VII.

979-8-3503-6689-1/24 $31.00 © 2024 IEEE

TABLE I. Summary of fault tolerance of neuromorphic hardware.

	Fault Tolerance Mechanism			
Synapses	Error Correction Coding (ECC)			
neurons	astrocytes	logic	TMR	scrubbing (optional)
		memory	ECC	

VI. SNN PARTITIONING AND ASTROCYTE TRAINING

Algorithm 1 provides the pseudo-code for clustering the neurons of each SNN layer into synaptic islands for astrocyte training. This is roughly based on the approaches proposed to partition SNNs and map to many-core hardware [23]–[26].

First, it initializes an empty island array and a starting island (line 1). For each layer $L_i \in \mathcal{L}$, it executes lines 3-14. For each neuron in the layer (line 3), the algorithm checks if the current number of neurons on the island is less than the constraint K (line 4). If successful, the neuron is placed on the island (line 5). Otherwise, the island is closed. An astrocyte is instantiated and trained with the spike trains of neurons placed on the island, using Eqs. 7-12 (line 8). The astrocyte is inserted into the island and the island is inserted into the island array $\mathcal{I}$ (line 9). Subsequently, a new island is instantiated (line 10). Once all neurons in the layer are placed into islands, any open island is closed by first training an astrocyte and then inserting it into the island. The island is inserted into the island array (lines 13). The algorithm creates a new island for the next layer (line 14). Finally, the island array is returned (line 16).

Algorithm 1: Clustering SNN graph into synaptic islands.

Input: SNN graph $G = (\mathcal{L}, \mathcal{E})$, neurons per island K
Output: Synaptic islands $\mathcal{I}$

```
1  I = I = ∅;          /* Initialize the island array and a
      starting island. */
2  for Lᵢ ∈ L do              /* For each layer of SNN */
3    for nᵢ,ⱼ ∈ Lᵢ do    /* For each neuron of the layer */
4      if |I| < K then        /* Island has space */
5        I.append(nᵢ,ⱼ);  /* Append the neuron to the
            island. */
6      end
7      else                       /* Island is full */
8        a.train(I); /* Train the astrocyte using the
            spike trains of the neurons in I. */
9        I.append(a) and I.append(I);     /* Append the
            astrocyte to the island and append island
            to the island array. */
10       I = ∅;    /* Create another empty island. */
11     end
12   end
13   Perform lines 8 & 9;   /* Train astrocyte for the island
          and append to island array. */
14   I = ∅;           /* Create another empty island. */
15 end
16 return I
```

VII. RESULTS

We evaluate NeuFT using five classification datasets [27]. These are summarized in Table II. We use snnTorch [28] to train the model and obtain baseline accuracy. Our astrocyte configurations include $\omega_a = 1, \beta_a = 0.5, b_a = 0.85, g_a = 0.75$. Of these parameters, $\omega_a, b_a, \& g_a$ are specific to the proposed astrocyte model and the specific setting of $\beta_a = 0.5$. For other settings of β_a, the following changes will be required. First, the model needs to be re-trained to obtain the new actrocyte parameters (Sec. IV). Second, a floating-point (or fixed-point) multiplier will be needed to implement Eq. 7.

We implement NeuFT on AMD's Virtex UltraScale FPGA development board, whose resource availability is
- **Lookup Table (LUT):** 537,600
- **Flip Flops (FF):** 1,075,200
- **Digital Signal Processing Blocks (DSP):** 768
- **Block RAMs (BRAM):** 1728

TABLE II. Evaluated datasets and corresponding model and accuracy.

	input/output	training/test	model	accuracy
MNIST	input: (28×28)	training: 60,000	MLP	99.8%
	output: 10	test: 10,000		
Fashion MNIST	input: (28×28)	training: 60,000	CNN & MLP	73.4%
	output: 10	test: 10,000		
CIFAR10	input: (32×32)	training: 50,000	CNN	60.7%
	output: 10	test: 10,000		
AG_NEWS	dimension: 128	training: 114,000	MLP	71.8%
	output: 4	test: 7600		
Speech-to-Spike	input: (28×28)	training: 60,000	MLP	80.7%
	output: 10	test: 10,000		

A. Fault Injection Setup

We use a linear feedback shift register (LFSR)-based setup to inject faults into random locations in these models, similar to [14]–[16]. This is illustrated in Fig. 8.

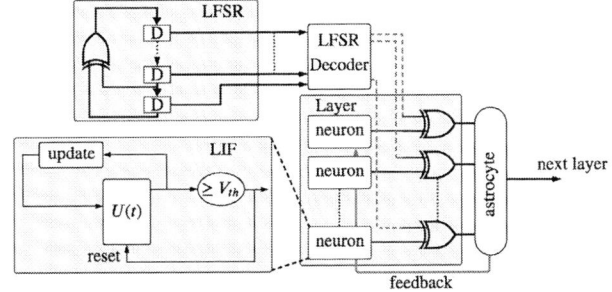

Fig. 8: Fault injection setup (for verification).

The setup consists of LFSR logic, one for each layer of the design. Each LFSR logic consists of shift registers. Two of the outputs are tapped and XORed to generate the input as shown in the figure. The shift register outputs are fed to an LFSR decoder which generates control signals to induce a fault using the XOR gate placed at the output of a neuron. The fault simulation works as follows. When the control signal of an XOR gate is '0', it passes a neuron output to the next layer. Otherwise, the output of a neuron is inverted and sent to the next layer. For completeness, we show the neuron logic to the left. The proposed setup simulates a logic inversion error. By replacing the XOR gate with an AND gate or an OR gate, stuck-at faults can also be simulated using the setup. To this end, we note that NeuFT is independent of the type of error induced in a model. Its astrocyte-based fault tolerance approach can mitigate both logic inversion and stuck-at errors.

To simulate single bit upsets, we use m shift registers, where $m = \log_2 N$ and N is the number of neurons in the layer. In this way, an LFSR signature results in a single control signal being active at any given time, inverting the corresponding output.

Finally, we note that NeuFT is independent of other fault injection setups such as [29]. Furthermore, the area of the proposed LFSR setup is not included in the implementation results reported in this section.

B. Evaluated Metrics

We evaluate the following metrics.

- **Accuracy:** The is computed as the ratio of the number of correctly predicted inputs to the total number of inputs presented to NeuFT, expressed as a percentage.
- **Area:** This is reported in terms of FPGA resource utilization, expressed in terms of LUT, FF, DSP, & BRAM.
- **Power:** Power numbers are reported using AMD's Vivado tool after implementing the design on the FPGA platform and by performing power simulation using the switching activities recorded during inference.

C. Fault-Tolerance

1) Single Bit Upsets: Fig. 9 plots the impact of single bit upsets (SBUs) on the accuracy of 5 evaluated datasets as the bit error rate (BER) increases from 10^{-10} errors per bit hour (under low radiation) to 10^{-3} errors per bit hour (under high radiation). We make the following observations.

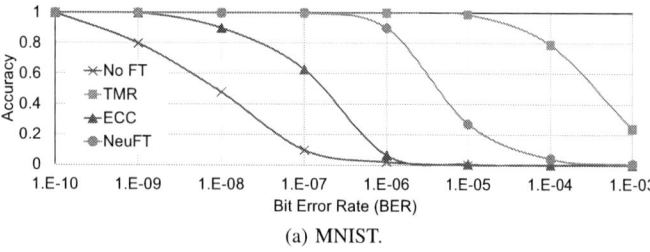

(a) MNIST.

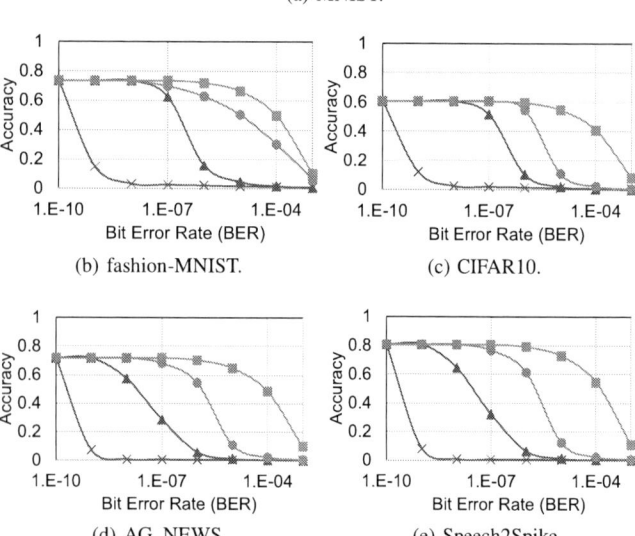

(b) fashion-MNIST.   (c) CIFAR10.

(d) AG_NEWS.   (e) Speech2Spike.

Fig. 9: Impact of SBUs with varying bit error rate (BER).

First, the accuracy loss is significant for all datasets if no fault tolerance mechanism is implemented (labeled as *No FT* in the figures). Second, accuracy is better than before when ECC is used to mitigate SBUs in synaptic memory. However, SBUs that affect neuron logic are not mitigated, causing accuracy to be considerably lower than desired. Third, TMR (with ECC) achieves the best result amongst all fault tolerance mechanisms as it can mitigate the impact of SBUs in both logic and memory. Finally, the accuracy using NeuFT is similar to TMR for moderate to low radiation environments, e.g., in LEO. It can still deliver acceptable accuracy for high-radiation

environments. This is due to the NeuFT's biologically inspired mechanism, which can provide fault tolerance of multiple neurons at once (within a synaptic island) without having to triplicate the logic associated with these neurons.

2) Multi-Cell Upsets: Fig. 10 reports the impact of multicell upsets (MCUs) for different BERs for the MNIST dataset. Essentially, we inject two adjacent faults for each error event.

We observe that for a given BER, the accuracy with MCUs is lower than with SBUs (Fig. 9a) for all techniques evaluated. Although TMR can mitigate an MCU in neuron logic as long as there are two correct copies, it cannot correct an MCU in synaptic memory when the MCU induces more than one bit flips. In the case of NeuFT, such synaptic errors cause the astrocyte potential to fall below the firing threshold, triggering the negative feedback action to restore the firing frequency of a neuron. So, NeuFT can mitigate MCU in both neuron logic and synaptic memory, leading to higher accuracy.

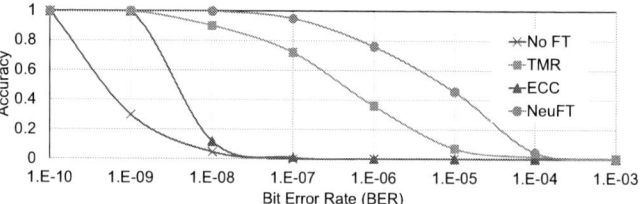

Fig. 10: Impact of MCUs with varying BER in MNIST.

D. Area and Power Overhead

We utilize a combination of lookup tables (LUTs), flip flops (FFs), digital signal processing (DSP) blocks, and Block RAMs (BRAMs) to implement NeuFT. These are the primary resources available on the target FPGA platform. Table III compares the utilization of these resources in the implementation of different fault tolerance solutions. For reference, we have also included utilization for the baseline design [21], which does not incorporate fault tolerance. Table III reports the utilization of a cluster (synaptic island) of 16 and 64 neurons. These numbers were obtained using AMD's Vivado tool after implementation on the FPGA board. We observe the following.

TABLE III. Implementation Results.

		Baseline	Cluster Size = 16				Cluster Size = 64				Min. Gain
			TMR	[15]	[16]	NeuFT	TMR	[15]	[16]	NeuFT	
Neuron	LUT	80	3840	1280	1280	1280	15360	5120	5120	5120	–
	FF	23	1104	368	368	368	4416	1472	1472	1472	–
	DSP	0	0	0	0	0	0	0	0	0	–
	BRAM	0	0	0	0	0	0	0	0	0	–
Astrocyte	LUT	–	–	1345	97	10	–	1345	97	10	–
	FF	–	–	2368	32	16	–	2368	32	16	–
	DSP	–	–	4	4	0	–	4	4	0	–
	BRAM	–	–	4	0	1	–	4	0	1	–
Total Area	LUT	–	3840	3970	2832	1290	15360	10500	11328	5130	2.0×
	FF	–	1104	5104	880	384	4416	10944	3520	1488	2.3×
	DSP	–	0	8	64	0	0	16	256	0	–
	BRAM	–	0	8	0	1	0	16	0	1	–
Power (mW) 100 MHz		1	43	38	27.5	15	210	187	142	72	1.8×

First, a detailed astrocyte model [15] requires more resources, leading to overheads similar to a TMR for a cluster size of 16 and only marginally better for a cluster size of 64. Second, although simple designs such as [16] improve resource utilization for each astrocyte, overall utilization remains

979-8-3503-6689-1/24 $31.00 © 2024 IEEE

significantly higher because they implement one astrocyte per neuron. Finally, NeuFT achieves the best result, providing $\approx$ a 2x improvement in resource utilization. This is due to the intelligent control design of each astrocyte that provides fault tolerance to all neurons on a synaptic island.

For power consumption, we observe a similar trend. NeuFT provides a minimum $1.8\times$ improvement in dynamic power compared to existing approaches.

E. Size of Synaptic Islands

Fig. 11 plots the accuracy and resource overhead trade-off for the MNIST dataset as the number of neurons per synaptic island (K in Alg. 1) increases from 4 to 256.

We observe that using a smaller synaptic island, the accuracy is higher because each astrocyte is able to provide the feedback required to fully restore the firing frequency when mitigating faults. However, more astrocytes are needed to tolerate faults in all neurons in the model. So, the fault tolerance resource overhead is also higher. With more neurons per synaptic island, each astrocyte can only partially restore the firing frequency to mitigate faults. So, the accuracy is lower. However, fewer astrocytes are needed in the entire design. So, the resource overhead is the least. We perform a similar search for all datasets. Our results show that setting $K = 16$, i.e., 16 neurons per synaptic island, gives the best resource and accuracy trade-off.

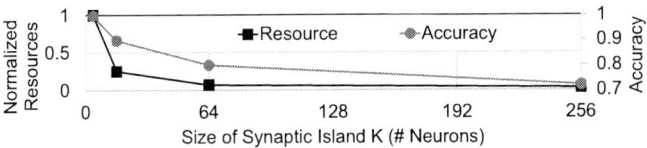

Fig. 11: Resource and accuracy tradeoff.

F. Complexity

The complexity of Algorithm 1 is as follows. For each layer, the algorithm executes lines 3-14. Within each loop iteration, the algorithm iterates for the number of neurons in a layer, performing training (spike accumulation) for a time duration of time steps (line 8) and other unit operations. Therefore, the complexity of Algorithm 1 is $\mathcal{O}(N \times T)$, where N is the total number of neurons and T is the number of time steps.

VIII. CONCLUSIONS

We propose NeuFT, a bio-inspired fault tolerance mechanism for neuromorphic inference hardware implemented on FPGA for space. Our key idea is to use astrocytes, which generate a retrograde feedback signal to restore the spike firing frequency of a failed neuron. NeuFT partitions a spiking neural network into synaptic islands and integrates an astrocyte in each of them. Each astrocyte is trained on the spike trains of neurons in its synaptic island. Through NeuFT we provide end-to-end fault tolerance in which synaptic memory is encoded with ECC, neurons are protected using astrocytes, and logic inside each astrocyte is protected by combining TMR with ECC. Using five datasets and varying bit error rates with single bit upsets and multi-cell upsets, we show significant improvements in both accuracy and resource utilization.

ACKNOWLEDGMENT

This work is supported by United States NSF CCF-1942697, Accenture LLP, and the DOE DE-SC0022014.

REFERENCES

[1] C. Mead, "Neuromorphic electronic systems," *Proc. of the IEEE*, 1990.
[2] W. Maass, "Networks of spiking neurons: The third generation of neural network models," *Neural Networks*, vol. 10, no. 9, pp. 1659–1671, 1997.
[3] AFRL, "Neuromorphic Computing For Space," 2024. [Online]. Available: https://afresearchlab.com/technology/nics
[4] BrainChip, "BrainChip Boosts Space Heritage with Launch of Akida into Low Earth Orbit," 2024.
[5] NASA, "Space Radiation Effects on Electronic Components in Low-Earth Orbit," 1999. [Online]. Available: https://llis.nasa.gov/lesson/824
[6] A. Alacchi, E. Giacomin, S. Temple, R. Gauchi, M. Wirthlin, and P.-E. Gaillardon, "Low latency SEU detection in FPGA CRAM with in-memory ECC checking," *IEEE TCAS I: Regular Papers*, 2023.
[7] Z. Li, C. Elash, C. Jin *et al.*, "Efficacy of transistor stacking on flip-flop SEU performance at 22-nm FDSOI node," *IEEE TNS*, 2023.
[8] C. Spindeldreier, B. Ustaoglu, U. Kulau, and J. Rust, "Performance evaluation of space-grade FPGA architectures," in *EDHPC*, 2023.
[9] I. Koren *et al.*, *Fault-tolerant systems*. Morgan Kaufmann, 2020.
[10] R. Santos, S. Venkataraman, A. Das, and A. Kumar, "Criticality-aware scrubbing mechanism for SRAM-based FPGAs," in *FPL*, 2014.
[11] M. B. Grabber and G. W. Kreutzberg, "Astrocytes increase in glial fibrillary acidic protein during retrograde changes of facial motor neurons," *Journal of Neurocytology*, vol. 15, no. 3, pp. 363–373, 1986.
[12] M. V. Sofroniew and H. V. Vinters, "Astrocytes: biology and pathology," *Acta Neuropathologica*, vol. 119, pp. 7–35, 2010.
[13] Z. Han, N. Luo, W. Ma, X. Liu, Y. Cai, J. Kou *et al.*, "AAV11 enables efficient retrograde targeting of projection neurons and enhances astrocyte-directed transduction," *Nature Communications*, 2023.
[14] M. Varshika, S. Johari *et al.*, "Design of a tunable astrocyte neuromorphic circuitry with adaptable fault tolerance," in *MWSCAS*, 2023.
[15] M. Isik, A. Paul, M. L. Varshika *et al.*, "A design methodology for fault-tolerant computing using astrocyte neural networks," in *CF*, 2022.
[16] A. P. Johnson, D. M. Halliday, A. G. Millard, A. M. Tyrrell, J. Timmis, J. Liu, J. Harkin, L. McDaid, and S. Karim, "An FPGA-based hardware-efficient fault-tolerant astrocyte-neuron network," in *SSCI*, 2016.
[17] M. Rastogi *et al.*, "On the self-repair role of astrocytes in STDP enabled unsupervised SNNs," *Frontiers in Neuroscience*, vol. 14, 2021.
[18] Z. Han, A. N. Islam, and A. Sengupta, "Astromorphic self-repair of neuromorphic hardware systems," in *AAAI*, 2023.
[19] J. Liu, L. J. McDaid, J. Harkin, S. Karim *et al.*, "Exploring self-repair in a coupled spiking astrocyte neural network," *IEEE TNNLS*, 2018.
[20] M. L. Varshika *et al.*, "Design of many-core big little μbrains for energy-efficient embedded neuromorphic computing," in *DATE*, 2022.
[21] A. Matinizadeh, Shadi, N. Pacik-Nelson, I. Polykretis, A. Mishra *et al.*, "A fully-configurable digital spiking neuromorphic hardware design with variable quantization and mixed precision," in *MWSCAS*, 2024.
[22] S. Matinizadeh and A. Das, "An open-source and extensible framework for fast prototyping and benchmarking of spiking neural network hardware," in *FPL*, 2024.
[23] A. Das, "Real-time scheduling of machine learning operations on heterogeneous neuromorphic SoC," in *MEMOCODE*, 2022.
[24] S. Song, A. Balaji, A. Das, N. Kandasamy *et al.*, "Compiling spiking neural networks to neuromorphic hardware," in *LCTES*, 2020.
[25] A. Balaji, A. Das, Y. Wu, K. Huynh, F. G. Dell'Anna, G. Indiveri, J. L. Krichmar, N. D. Dutt, S. Schaafsma, and F. Catthoor, "Mapping spiking neural networks to neuromorphic hardware," *IEEE Transactions on Very Large Scale Integration (VLSI) Systems*, vol. 28, no. 1, pp. 76–86, 2019.
[26] A. Das, "A design flow for scheduling spiking deep convolutional neural networks on heterogeneous neuromorphic system-on-chip," *ACM Transactions on Embedded Computing Systems*, 2023.
[27] J. Yik *et al.*, "Neurobench: Advancing neuromorphic computing through collaborative, fair and representative benchmarking," *arXiv*, 2023.
[28] J. K. Eshraghian, M Ward, E. O. Neftci *et al.*, "Training spiking neural networks using lessons from deep learning," *Proc. of the IEEE*, 2023.
[29] T. Spyrou, S. A. El-Sayed, E. Afacan, L. A. Camuñas-Mesa *et al.*, "Neuron fault tolerance in spiking neural networks," in *DATE*, 2021.

An AI-Assisted Connection Weight Prediction for Regression Testing of Integrated Circuits

Abishaan Ravikumar*, Xiaohan Yang[†], Rajendra Prasad[†], RajaNataraj Sivaraj[†], Alexander Rast*, and Abusaleh Jabir*

*School of ECM, Oxford Brookes University, Oxford, UK. E-mail: {19232440, arast, ajabir}@brookes.ac.uk
[†]Infineon Technologies, Bristol, U.K. Ltd. E-mail:{Xiaohan.Yang, Rajendra.Prasad, RajaNataraj.Sivaraj}@infineon.com

Abstract—Integrated Circuit (IC) verification, i.e. the process of ensuring that it performs according to the design specifications, is highly resource intensive. This often depends on an IC's complexity, e.g. the number of gates/transistors which translates into expressions, branches, blocks, etc in its high level descriptions. To reduce overall verification and hence design time, industries resort to "Regression Testing" where a very small test suite, with very high test coverage, is selected to verify any modified design block and its dependencies. One of the key steps in regression test-based verification is distributing the tests to the various interconnected blocks under tests based on their functionality and accessibility, which translates into a block's "connection strengths" among other parameters. The existing approaches currently define the connection strengths manually by the design experts which often lead to inconsistency in the test results. In this paper, we propose a Graph Neural Network (GNN) based approach to estimate the connection strengths of different interconnected blocks and evaluate its effectiveness with industrial designs in conjunction with a technique called "SMART Regression" compared to random and full regression testing.

Index Terms—IC verification, regression testing, graph neural network

I. INTRODUCTION

As integrated circuit (IC) designs grow in complexity, verifying whether they conform to their specification becomes increasingly challenging. Verifying a design requires subjecting it to millions of test scenarios depending on design complexity during the development stage. The complexity of an IC is governed by integration of many millions of transistors onto a single chip. Furthermore, bugs undetected in the early stage of design development can be significantly expensive [1], leading to increased development cost and delays in product delivery ultimately culminating into diminished competitive edges in industries. Hence, early and timely bug detection in IC verification—ensuring its functional correctness, quality, and design reliability—emerges as one of the most critical steps in its development cycle. The verification stage of an IC often consumes a significant proportion of its development resources [2]. Therefore, it is imperative to optimise this stage to reduce resource consumption (e.g. verification time) without compromising its effectiveness. To this end, IC manufacturers primarily rely on simulation-based verification among other

verification techniques as it offers robust framework for testing IC design against a wide range of scenarios [3].

To effectively deal with the rising complexities of very large IC design verification, industries adopt Regression Testing (Section II-B) whereby verification is carried out incrementally by focusing on that part of a circuit which has undergone modifications or design changes. To this end, this paper presents an application of Graph Neural Network (GNN) to assist with test selection for "SMART regression" [4] (Fig. 1). We use GNN for determining the connection weights among interconnected blocks in a design, within the knowledge base in Phase 1 (Fig. 1), to optimise test selection in the regression flow. Previously, design engineers manually estimated connection weights—a labour-intensive process consuming time and resources. This manual approach often resulted in inconsistencies in connection weights. However, this process has now been largely automated, streamlining the estimation process and reducing errors. Our experimental results show that the AI-assisted method is capable of producing better coverage compared to random testing techniques and similar coverage, but with more than two orders of magnitude fewer tests, compared to full regression.

II. BACKGROUND AND RELATED WORK

A. Formal and Simulation Based Verification

Traditional verification methods [5], including manual inspection and simulation, have become inadequate for achieving comprehensive coverage of all potential design scenarios [6]. Over the past few years IC design verification has evolved, leveraging advancements in hardware and software technologies [7]. Simulation-based verification techniques [8] are crucial for IC verification and are continually being enhanced to improve their effectiveness and efficiency. Formal verification [9] employs mathematical algorithms [10] to exhaustively verify the correctness of a design, providing guarantees that cannot be achieved through simulation alone. Meanwhile, emulation and hardware acceleration techniques [11] use specialized hardware platforms to accelerate the verification process, enabling quicker turnaround times and comprehensive testing of complex designs. In addition to technological advancements, the IC design verification ecosystem has been introduced with industrial standards, techniques, and best practices aiming towards improving verification productivity and effectiveness. Universal Verification Methodology (UVM) [12] is a standard technique that provides a common framework for developing verification environments, promoting re-usability and collaboration across design teams.

In [13] the author proposed an optimization technique for random test constraints in ICs assisted by machine learning algorithm. This method involves monitoring the flop pairs and the coverage information is extracted from the toggle matrix report. Remarkably, the approach yields promising results, achieving a pass rate of 98.5% after executing 6 iterations of 500 tests, encompassing 15.4 million cycles. However, it is noteworthy that this process is exceptionally exhaustive for verification purposes. Despite these advancements, there is a growing demand for innovative verification methodologies and tools to address emerging verification challenges.

B. Regression Testing

Since our work is on regression testing, we review current trends in more detail. Regression testing [14] [15] is carried out to systematically select a portion of previously conducted tests after implementing changes to the IC design. This ensures that any modifications made do not impact the existing functionality or introduce new bugs into the system. In contrast, full regression testing [16] is the process of exhaustively executing all the possible tests to achieve maximum coverage which is far more time/resource consuming compared to 'selective' regression testing.

In IC development, lack of high-quality regression tests can lead to bugs going unnoticed, which results in significant debugging efforts and engineering time due to multiple changes in the design. In complex IC designs, identifying the most appropriate tests for regression testing within a given test scenario is challenging because it is impractical to predict tests that are necessary solely based on the test scenario, especially considering the vast array of potential tests available (potentially many millions) depending on the design's complexity.

To this end, we consider SMART regression [4] a regression flow that leverages the Nearest Neighbors algorithm [17] to facilitate test selection and distribution by analysing the geographical closeness between modified and existing design blocks (Fig. 1). This flow identifies high-quality test suites for adaptive regression, ensuring efficient testing process. The SMART regression flow consist of 3 phases.

a) Knowledge Base Generation and Test Grading: Knowledge base generation and test grading are the two main sections of the first phase in SMART regression. It is important to establish the knowledge base that contains information about the design blocks in an IC and their connectivity while test grading process enables the evaluation of tests based on the coverage they achieve for the design blocks in the design.

b) Test Distribution Creation: In the second phase, the test scenario identifies the design blocks which needs to be subjected to regression testing. Distribution is established by mapping the changes across the geographical space.

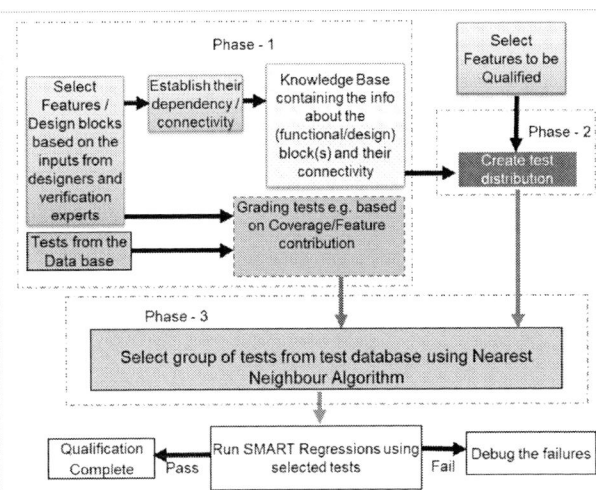

Fig. 1. Smart regression flow.

c) Test Scenario Based Test Selection: In the third phase, the tests are selected using the Nearest Neighbors algorithm [17].

1) Test Scenario: Fig. 2 illustrates a segment of a circuit design consisting of three design blocks labeled F1, F2 and F3. In this scenario, assuming that F1 has undergone modifications and, due to direct connections, both F2 and F3 may be affected as a result. Therefore, during regression testing, test cases will need to be distributed across all dependent design blocks with allocations influenced by the connection weights.

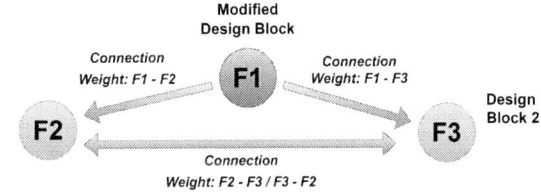

Fig. 2. Block F1 is modified with outgoing connections to F2 and F3. This is a Level 1 connection, with indirect connections at higher levels.

2) Connection Weights: The "connection weights", which lie between 0 and 1, represent the strength of the connection between two design blocks. These weights serve as reference inputs for optimizing test distribution strategies, ensuring a thorough coverage. A connection weight near 0 indicates a weak connection, suggesting the allocation of more test cases; whereas, a weight near 1 indicates a strong connection suggesting fewer tests. Previously design engineers provided an estimation of the connection weights, based on their knowledge of the designs, to facilitate the regression flow. Often there were inconsistencies in predicting the weights, which can also be time (resource) intensive.

III. PROPOSED APPROACH FOR WEIGHT PREDICTION

In this section, we present an AI based weight prediction for reflecting connection strengths between various design blocks.

This is to facilitate improved test distribution between the blocks.

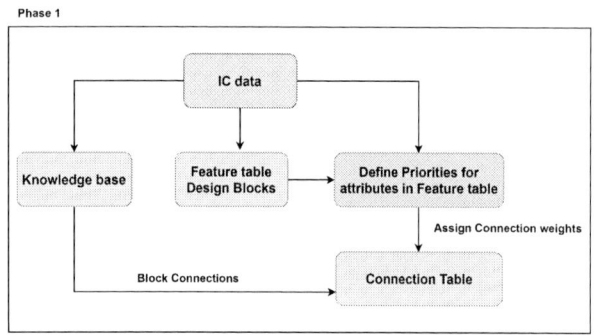

Fig. 3. Block diagram for data preparation.

Fig. 3 shows the block diagram of the data preparation stage for training the proposed approach. The primary objective is to extract relevant information from the database to create feature table and establish connection for the IC design under test. This is then used to train the GNN model as shown in Fig. 4, which predicts the connection weights for test selection.

A. Attributes for Connection Weight Prediction

A GNN was trained to predict the connection weights based on the design parameters of an IC which included:

- The number of incoming connections.
- The number of outgoing connections.
- The number of cover points, i.e. the critical points in a design that the tests are required to reach.
- Whether a block is self contained or not, i.e. it shows the existence of connection to itself.
- A block's intrinsic weight, indicating its priority with respect to the rest of the connected blocks

$$IW = SF + \frac{CP_b}{CP_T} + \frac{Out}{In + Out}. \quad (1)$$

where IW denotes intrinsic weight, SF denotes self-contained value, CP_b denotes cover points, CP_T denotes total cover points in a design, In and Out denotes the number of in coming and out going connections. By applying the given formula Eq. 1 a block's intrinsic weight can be derived.

We found out that these combination of attributes provided very good indication of a block's role, dependencies, and importance, facilitating efficient test allocation over entire regression cycles.

TABLE I
SAMPLE FEATURE TABLE FOR DESIGN BLOCKS

Blocks	Cover points	Self-Contained	In	Out	I.weight
F1	78	0	6	10	0.64
F2	450	1	4	6	0.50
F3	241	0	2	4	0.73

B. Graph Neural Network For Weight Prediction

Utilizing GNN to predict connection weights between design blocks introduces a novel approach that replaces the manual method of generating knowledge bases for regression testing. GNN operates on graph data structures, facilitating the exchange of information among nodes via edges that allow sharing of information among nodes through edges. By training on a given datasets, GNNs leverage neural networks applied to graphs to accomplish diverse tasks [18].

C. Data Analysis and Preparation

In this paper, we used industrial data consisting of comprehensive details regarding the design specifications and inter connectivity among design blocks. The proposed approach necessitates pre-processing the raw circuit data and converting them into graphical data structure to be compatible with the GNN model. To achieve this, the following steps are performed.

1) Refine and organise relevant information to construct a feature table (Table I) for the design blocks, each represented as nodes in GNN.
2) Extract relevant information from the database by correlating the IC design, thereby formulating a foundational connectivity table which included the source design block, target design block and the respective connection weights of the connections.
3) Once the base connectivity table is derived, reconstruct the connectivity table by including all connections between design blocks, accommodating all possible combinations and permutations. This comprehensive table reflects the full spectrum of potential connections between the design blocks present in an IC.

D. Constructing the Connection Table

To construct the connection table (Step 3 from previous section). We defined the priority of each attribute based on the impact of the attribute as shown in (Table II).

TABLE II
PRIORITY OF FEATURE TABLE ATTRIBUTES OUT OF 100%

Attributes	Cover points	Self-Contained	In	Out	I.weight
Priority %	20%	25%	10%	30%	15%

After defining the attribute's priorities (Table II), we determine the priority score PS of each block based on the following equation to the feature table.

$$PS = \frac{CP \cdot CP_P + SF \cdot SF_P}{100} + \frac{In \cdot In_P + Out \cdot Out_P + IW \cdot IW_P}{100}, \quad (2)$$

where the subscript P denotes the priority allocation of each attribute. We use the PS of both source and target blocks, along with connection levels indicating the presence of a connection, to compute the connection weight for the design block under test. Establishing a standard technique for

979-8-3503-6689-1/24 $31.00 © 2024 IEEE

deriving connection weights between design blocks is crucial for maintaining consistency (Table III). By applying the given formula connection weight, C_W, can be derived effectively as follows.

$$C_W = \frac{PS_{SB}}{L \cdot (PS_{SB} + PS_{TB})}. \quad (3)$$

Here, C_W denotes the connection weight between design blocks, PS_{SB} and PS_{TB} denote priority scores of the source and the target blocks of a connection, and L denotes the connection level indicating whether a connection is directly established or not. In Fig. 2 when F1 is under test, F2 is a direct connection i.e. $L = 1$ whereas F3 is connected directly and indirectly because F2 has a bi-directional connection with F3 i.e. $L = 2$.

TABLE III
CONNECTION TABLE

Source Block	Target Block	Level	Connection Weight
F1	F2	1	0.1740
F1	F3	1	0.2776
...	...	...	...
...	...	...	...
F2	F3	3	0.5517
F3	F2	3	0.9201

E. Preprocessing and Training

Before training the GNN model with the prepared data, it is essential to preprocess the data. A python framework based on PyTorch [19] called "Sentence Transformer" [20] is used for text embedding of non-numerical value in the dataset into a numerical embedding. Additionally, numerical values are reshaped to ensure compatibility with other tensors during matrix operations. To train the GNN, the preprocessed data is converted into data containers tailored for heterogeneous graphs using the PyTorch geometric library. This process involves assigning feature vectors to nodes i.e. source blocks and the connections and defining the edge connections with edge indices. This specifies the connection between the nodes in the graph. Finally, the connection weights in the connection table are converted to PyTorch tensor to be assigned as the edge labels for the connections.

For model evaluation, the data is randomly partitioned into training, validation and test sets with 80%, 10% and 10% of data allocated to each set, respectively. The proposed GNN model is simply based on two message passing network layers to encode and process the nodes in the graph. This model leverages the encoder-decoder architecture from PyTorch Geometric [19]. The encoder is one of the components in the model that is capable of retrieving accurate representations of the node under test from the connected blocks. This consists of two graph SAGE convolutional (SAGEConv) layers [21]. It functions as a message-passing neural network layer, aggregating features from neighboring nodes to enhance information gathering and representation for each node in the graph. These layers are organised in a way that the output of the first

message passing layer will be the input for the second layer, enabling the model to capture high level data representations and dependencies in the graph. Along with each message passing layer, a Rectified Linear Unit (ReLU) [22] activation function is applied after each SAGEConv layer to make the model non-linear. This allows the model to learn complex connections and patterns in the training data. The decoder concatenates the output from the encoder by passing it through two linear layers [23] along with ReLU activation to produce a predicted connection weight for various design block combinations.

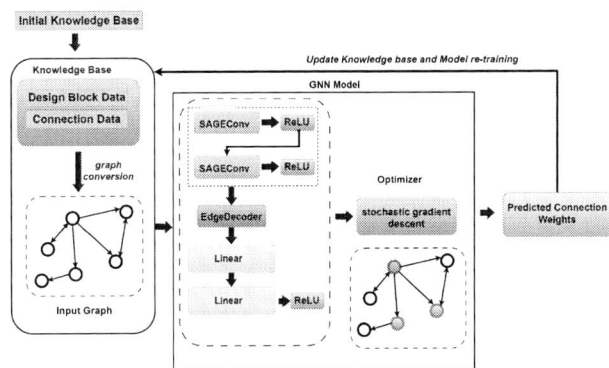

Fig. 4. The proposed GNN model architecture for weight prediction.

F. Training and Evaluation

Algorithm 1 GNN Training and Evaluation

1: **Phase 1: Training Setup**
2: Import necessary libraries and define training class
3: Set model to train model
4: Perform zero gradients
5: Compute predictions
6: Compute loss
7: Backpropagate gradients
8: Connection Prediction = model (Training data parameters)
9: **Phase 2: Train and Test the model**
10: **for** each $epoch$ **in** $range(1, n)$ **do**
11:     Perform training
12:     Compute train RMSE
13:     Compute validation RMSE
14: **end for**

For training the model, an optimizer is initialized that adjusts the parameters. Optimisation is achieved through Stochastic Gradient Descent [24], while updating parameters to minimize losses.

Algorithm 1 shows the GNN training setup and the training mechanism. In Phase 1, prediction, Root Mean Squared Error (RMSE) loss gradients are determined and the model parameters are updated. In Phase 2, the model is evaluated by executing iterative runs which were fine-tuned to be 300 iterations to avoid potential over fit. Fig. 5 shows that the model initially converges at 300 iterations but experiences a higher loss beyond that point. Each execution trains the model and computes the RMSE.

Once trained on the transformed data, the model will take the source and target blocks as input and predict the connection weights, outputs a value between 0 and 1 that represents the strength of the connection between these blocks. This prediction is based on the attributes defined during data preparation. If significant changes are made to the circuit that require adjustments to the connectivity between blocks, the model can be retrained using the updated connectivity data.

IV. RESULTS AND DISCUSSION

The technique was tested on a 535K μm^2 design with the 28nm technology node, which included an FPU, decoders, load/store units, fetch modules, bus interface units, and an arithmetic and logic unit among other features. Overall, 33K test cases were considered, which included good and bad tests. Out of these, 100 tests were selected both randomly and based on the proposed approach.

In Table IV, we compare the execution time/cycles, the number of tests required, and the coverage results produced by each testing technique. According to [13], execution time is measured by the number of test repetitions, represented as cycles. This method achieves 98% coverage with 3K tests; however, the complexity of the circuit design used remains unspecified. In contrast, full regression, random testing, SMART regression with random weights and weights generated with the proposed approach are all tested on the same IC design.

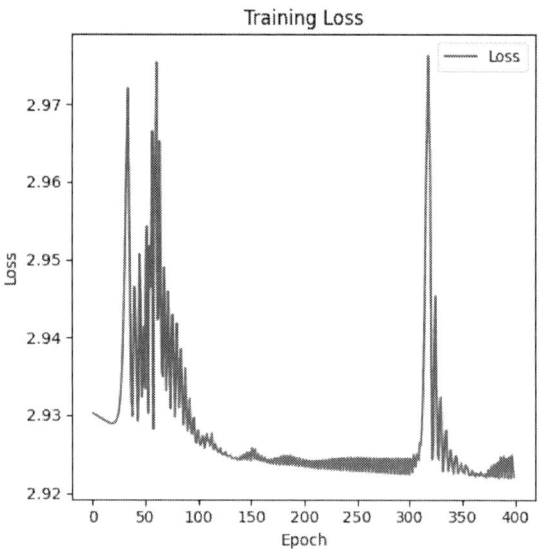

Fig. 5. Training loss plotted to find the ideal number of Epoch for Training.

Fig. 6 illustrates the RMSE on the against 300 connections from validation set, which fluctuates from 1.65 to 1.85 out of 10, indicative of good performance; and Fig. 7 illustrates the closeness accuracy of the predicted weights compared to actual weights which exhibited test accuracy ranging between 80%

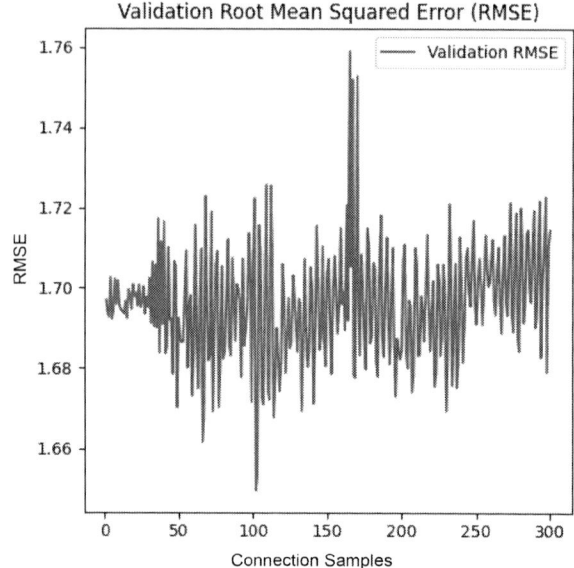

Fig. 6. RMSE for the validation dataset.

and 85%. Table V shows the expression coverage of the Bus Interface Unit (BIU) [25] design block, evaluating variables and sub-expressions in conditional statements, along with the overall coverage achieved from the selected tests. The overall coverage is calculated by considering the percentage values and cover points for block, branch, expression, and toggle coverages. The results are compared with full regression, three random regression tests, and the proposed approach.

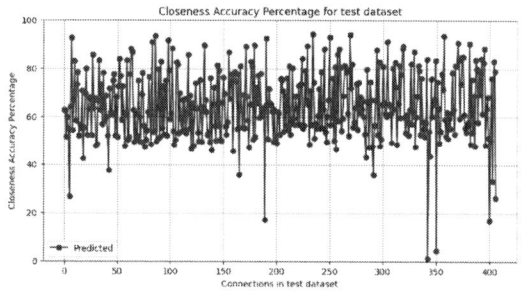

Fig. 7. Closeness Accuracy between predicted and target connection weights.

TABLE IV
EXECUTION TIME, NO.OF TESTS COMPARISON AND COVERAGE.

	[13]	Full Reg	Rand Test	SR. Rand W	SR. AI
T.Time or Cyc	15,400k	2 Days	50 mins	50 mins	50 mins
No Tests	3k	30k	100	100	100
Coverage %	98%	94%	80%	84%	91%

V. CONCLUSIONS

This paper presented an AI assisted weight prediction method applicable for regression test based IC verification.

TABLE V
COVERAGE COMPARISON: FULL REGRESSION (33K TESTS) VS RANDOM TESTING VS PROPOSED APPROACH (100 TESTS).

Code Coverage	BUS Interface Unit (BIU)				
	Full Regression	Rand1	Rand2	Rand3	Proposed
Expression	89.84%	66.10%	71.42%	69.55%	84.75%
Total	94.62%	80.23%	83.48%	79.95%	91.59%

Our approach is showing noticeable improvements over the random cases, which is quite popular for simulation based verification. Additionally, considerable accuracy improvement over a full regression test is also observed, where a full regression test required 27K tests (vs 100 tests) to achieve 89.52% expression coverage (vs 84.75%) over a 2-day run (vs about 50 mins).

In conclusion, our AI-assisted weight prediction for regression testing is proving to be very encouraging. Work is under way in automating various other stages of regression testing with AI-assisted technology.

REFERENCES

[1] R. N. Charette, "Toyota's sudden acceleration problems: What did it know and when did it know it?" *IEEE Spectrum*, 2010. [Online]. Available: https://spectrum.ieee.org/toyota-sudden-acceleration-problems-what-did-it-know-and-when-did-it-know-it-

[2] Y. Li, W. Wu, L. Hou, and H. Cheng, "A study on the assertion-based verification of digital ic," in *2009 Second International Conference on Information and Computing Science*, vol. 2, 2009, pp. 25–28.

[3] R. K. M. Vangara, B. Kakani, and S. Vuddanti, "An analytical study on machine learning approaches for simulation-based verification," in *2021 IEEE International Conference on Intelligent Systems, Smart and Green Technologies (ICISSGT)*, 2021, pp. 197–201.

[4] R. Prasad, R. Sivaraj, "Smart adaptive regression using nearest neighbours algorithm," US Patent 4 303 753, 2022, patent.

[5] H. D. Foster, "2018 fpga functional verification trends," in *2018 19th International Workshop on Microprocessor and SOC Test and Verification (MTV)*, 2018, pp. 40–45.

[6] D. Wang, J. Yan, and Y. Qiao, "Research on chip verification technology based on uvm," in *2021 6th International Symposium on Computer and Information Processing Technology (ISCIPT)*, 2021, pp. 117–120.

[7] B. Abir and M. Salah, "Towards formal verification of cryptographic circuits: A functional approach," in *2018 3rd International Conference on Pattern Analysis and Intelligent Systems (PAIS)*, 2018, pp. 1–6.

[8] L. Zhu and Z. Fang, "Modeling and simulation of mixed-signal integrated circuits based on verilog-ams," in *2013 2nd International Symposium on Instrumentation and Measurement, Sensor Network and Automation (IMSNA)*, 2013, pp. 1–4.

[9] T. Joshi, S. Chakraborty, S. Ambuskar, and A. K. Singh, "Automated formal verification methodology for digital circuits," in *2023 14th International Conference on Information & Communication Technology and System (ICTS)*, 2023, pp. 277–282.

[10] F. El-Licy and H. Abdel-Aty-Zohdy, "Formal verification and performance evaluation of logic integrated systems based on hierarchical analysis," in *Proceedings of the 44th IEEE 2001 Midwest Symposium on Circuits and Systems. MWSCAS 2001 (Cat. No.01CH37257)*, vol. 2, 2001, pp. 651–655 vol.2.

[11] Y. Serrestou, V. Beroulle, and C. Robach, "Impact of hardware emulation on the verification quality improvement," in *2007 IFIP International Conference on Very Large Scale Integration*, 2007, pp. 218–223.

[12] N. B. Harshitha, Y. G. Praveen Kumar, and M. Z. Kurian, "An introduction to universal verification methodology for the digital design of integrated circuits (ic's): A review," in *2021 International Conference on Artificial Intelligence and Smart Systems (ICAIS)*, 2021, pp. 1710–1713.

[13] S. Sokorac, "Optimizing random test constraints using machine learning algorithms," in *2017 design and verification conference and exhibition US(DVCon)*, 2017, pp. 583–588.

[14] M. Cieplucha, "Metric-driven verification methodology with regression management," *Journal of Electronic Testing*, vol. 35, 02 2019.

[15] G. Parthasarathy, A. Rushdi, P. Choudhary, S. Nanda, M. Evans, H. Gunasekara, and S. Rajakumar, "Rtl regression test selection using machine learning," in *2022 27th Asia and South Pacific Design Automation Conference (ASP-DAC)*, 2022, pp. 281–287.

[16] Y. Lee, J. Park, M. A. Elsayed, M. E. Gadallah, and M. Alimam, "Drc code coverage test a novel qa methodology," in *2018 International Conference on IC Design & Technology (ICICDT)*, 2018, pp. 93–96.

[17] T. Mladenova and I. Valova, "Comparative analysis between the traditional k-nearest neighbor and modifications with weight-calculation," in *2022 International Symposium on Multidisciplinary Studies and Innovative Technologies (ISMSIT)*, 2022, pp. 961–965.

[18] J. Liu, X. Xing, T. Wang, and Z. Jia, "A survey on graph neural networks method under social recommendation," in *2023 9th Annual International Conference on Network and Information Systems for Computers (ICNISC)*, 2023, pp. 583–588.

[19] PyTorch Contributors, "Pytorch," https://pytorch.org/, accessed: May 13, 2024.

[20] N. Reimers and I. Gurevych, "Sentence-bert: Sentence embeddings using siamese bert-networks," arXiv preprint arXiv:1908.10084, 2019. [Online]. Available: https://arxiv.org/abs/1908.10084

[21] J. Leskovec, "Graphsage: Inductive representation learning on large graphs," https://snap.stanford.edu/graphsage/, accessed: May 13, 2024.

[22] J. Brownlee, "Gentle introduction to the rectified linear unit (relu)," https://machinelearningmastery.com/rectified-linear-activation-function-for-deep-learning-neural-networks/, accessed: May 13, 2024.

[23] D.-S. Huang and S.-D. Ma, "Linear and nonlinear feedforward neural network classifiers: A comprehensive understanding," *Journal of Intelligent Systems*, vol. 9, no. 1, pp. 1–38, 1999. [Online]. Available: https://doi.org/10.1515/JISYS.1999.9.1.1

[24] Y. Tian, Y. Zhang, and H. Zhang, "Recent advances in stochastic gradient descent in deep learning," *Mathematics*, vol. 11, no. 3, 2023.

[25] Y. Xiao, X. Fan, and D. Yang, "The design of local bus interface unit based on pci9054," in *2012 2nd International Conference on Consumer Electronics, Communications and Networks (CECNet)*, 2012, pp. 2964–2967.

979-8-3503-6689-1/24 $31.00 © 2024 IEEE

BayWatch: Leveraging Bayesian Neural Networks for Hardware Fault Tolerance and Monitoring

Julian Hoefer, Matthias Stammler, Fabian Kreß, Tim Hotfilter, Tanja Harbaum, Juergen Becker

Karlsruhe Institute of Technology (KIT), Karlsruhe, Germany

Email: {julian.hoefer, stammler, fabian.kress, hotfilter, harbaum, becker}@kit.edu

Abstract—As Deep Neural Networks are increasingly being utilized in safety-critical domains, assessing the uncertainty of the models during inference will be a crucial component in enhancing the system's dependability. Furthermore, dependable AI systems need to provide measures against random hardware faults, since they can lead to false and possibly hazardous predictions.

To address this, we present a novel fault analysis methodology, specifically tailored for Bayesian Neural Networks (BayNNs). We utilize variational inference and dropout techniques to transform deterministic models into Bayesian models, thereby increasing their fault tolerance and enabling uncertainty monitoring. We conduct an extensive comparative analysis involving over 12 million fault injection experiments across two distinct networks, providing a comprehensive understanding of the network's uncertainty under fault conditions. Our results demonstrate that BayNNs can exhibit up to 2x greater fault tolerance compared to their deterministic counterparts. Furthermore, we show that uncertainty monitoring can serve as an effective tool for revealing fault-induced errors, since a bit-flip results in an error only in predictions with high uncertainty.

Index Terms—Dependable AI, Bayesian Neural Networks, Fault Injection, Fault Tolerance

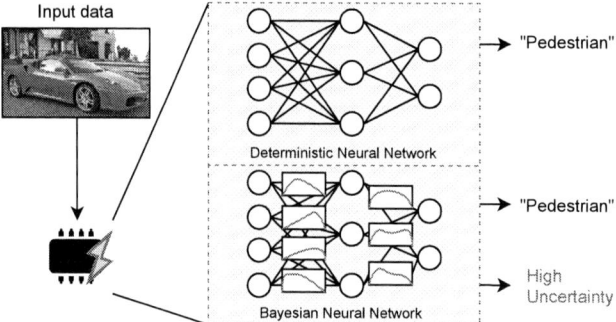

Fig. 1: In contrast to deterministic models, Bayesian models provide a measure of uncertainty in the model's output. Our results demonstrate that in the event of a hardware fault leading to an incorrect prediction, the associated uncertainty is significantly high.

I. INTRODUCTION

When deploying algorithms based on Artificial Intelligence (AI) into safety-critical systems, it is crucial to consider two distinct types of fault sources during operation. For road vehicles, these are explicitly defined and addressed by two safety standards [1], [2] in terms of error handling and mitigation.

On the one hand, a system must provide mechanisms against random hardware faults and towards *Functional Safety* [1]. On the other hand, faults may arise when the system encounters a state (scenario or environmental condition) for which it was not originally designed. These must be addressed to secure the *Safety of the intended functionality (SOTIF)* [2].

Requiring the system to check whether it still operates in its operational design domain (ODD), *SOTIF* distinguishes three cases: *1)* normal operation; *2)* an out-of-distribution (OOD) operation, e.g., when the class of the input is not present in a model's training dataset; and *3)* a distributional shift, altering the input through corruptions such as weather or light conditions or sensor noise or aging. One promising solution to detect such conditions is to quantify the uncertainty of predictions by using Bayesian Neural Network (BayNN)-based approaches [3], [4].

To enable the deployment of complex AI algorithms into embedded devices, new types of hardware accelerators and domain-specific architectures such as Neural Processing Units (NPUs) need to be used, satisfying the required computing capabilities. While these types of accelerators provide decent energy-efficiency, their high degree of data-reuse makes them susceptible to significant fault propagation [5]. Fulfilling *Functional Safety* requirements [1], this issue is often countered utilizing redundancy-based safety mechanisms, resulting in a significant area, energy, and runtime overhead.

To combine these two safety paradigms, in this work we show that a critical hardware fault leading to a false prediction will express itself in increased prediction uncertainty, as outlined in Figure 1. Critical faults occurring in an NPU are thereby covered by monitoring the uncertainty of a Bayesian model. By using uncertainty monitoring as a safety mechanism to detect OOD data or distributional shifts, as specified by *SOTIF* [2], we can also provide a safety mechanism against critical hardware faults, possibly reducing the burden of costly redundancy simultaneously.

In this paper, we make the following contributions:

- We introduce our novel fault analysis methodology specifically tailored for Bayesian Neural Networks utilizing variational inference and dropout techniques.
- We conduct a comparative analysis involving over 12 million fault injection experiments on two different networks.
- We demonstrate that BayNNs can exhibit up to 2x greater fault tolerance compared to their deterministic counterparts, and that hardware faults only alter the Bayesian model output when the prediction has high uncertainty.

979-8-3503-6689-1/24 $31.00 © 2024 IEEE

II. BACKGROUND

While there are existing works intended to detect anomalies or out-of-distribution samples in input images [6]–[8], these methodologies do not account for the potential occurrence of hardware faults during the inference process. Rather, they focus solely on aspects of the input image, i.e., whether it is represented in the training data or corrupted by external noise. On the other hand, there are methodologies that simulate faults in AI accelerators [9]. However, the combination of these fields, i.e., using BayNNs and uncertainty estimation for better fault tolerance and monitoring has not been explored yet.

A. Anomaly Detection

In the context of anomaly detection methods, *inference trace-based* approaches are closest related to hardware fault detection methodologies [6]–[8]. *Inference trace-based* methodologies use feature maps as inputs. These feature maps are extracted as the output of considered convolutional layers of the classification model.

In [6] and [8], the Mahalanobis Distance between a sample pool and the test sample is used to detect anomalies such as out-of-distribution samples or adversarial attacks. While they achieve very high accuracy, precision and recall, they nevertheless require a significant amount of memory and execution time [10].

Faster and similar in performance, [7] provides a methodology to detect anomalies by using a neural network-based classification. In this, the intermediate feature maps are used as the input to a convolutional neural network. The faster execution is offset by the usage of machine learning-based methodologies to detect anomalies in neural network-based inference. This only transfers the problem, but does not provide an inherently more reliable solution to monitor inference. With a faster execution time and the necessity to only store weights and no sample set, [7] is more suitable for on board execution [10].

Conceptually, these methods rely on the presence of examples from the normal class and the anomaly class, essentially solving a classification problem [6]–[8]. Modelling hardware faults, manifesting themselves in multiple ways, as a binary classification problem, proves to be unsuitable.

B. Bayesian Neural Networks

Traditional Deep Neural Networks (DNNs) rely on fixed, deterministic weights w, established during the training phase and unaltered during inference. Because of this, traditional DNNs represent point estimates for a Gaussian process. In contrast, BayNNs take a probabilistic approach and represent weights as a probability distribution employing probability density functions.

Before training, a prior distribution $p(y|x, w)$ is defined for the network's weights w [11]. During training, the network encounters new data from a training set $D = \{x, y\}$ with input data $x = \{x_1, ..., x_n\}$ and corresponding ground truth $y = \{y_1, ..., y_n\}$, where n defines the index of the sample and y_n is the ground truth for input data x_n.

Ultimately, the posterior $p(w|D)$ is determined using Bayes' rule:

$$p(w|D) = \frac{p(y|x, w) \cdot p(w)}{\int p(y|x, w) \cdot p(w) dw} \qquad (1)$$

As a result, the inference of trained BayNNs produces a stochastic distribution, suitable for an estimation of the network's uncertainty [3], [4], providing a significant advantage over deterministic networks, which provide a single prediction that can be overly confident. This effect is especially prevalent, when faced with unseen or corrupted out-of-distribution inputs.

However, the training of BayNNs is a challenging task, especially for complex networks, since the computation of the posterior quickly becomes intractable [12].

Overcoming this problem, the posterior $p(w|D)$ is approximated using Monte Carlo sampling, by conducting multiple *stochastic* forward passes on the same input x, rather than one single *deterministic* prediction p. Each pass uses the same input x, but results in a set of sample predictions $\{p_1, ..., p_{N_{MC}}\}$, where N_{MC} denotes the number of Monte Carlo samples. If the sample predictions agree on the input x, a low uncertainty is associated with the forward pass. Conversely, if the predictions do not agree, a high uncertainty is associated with the forward pass. Two common approaches for approximating BayNNs are *Variational Inference* [13] and *Monte Carlo (MC) Dropout* [14].

In mean field *Variational Inference*, each weight is represented by a Gaussian distribution, parametrized by its variational parameters μ and σ [11]:

$$q(w) = \mathcal{N}(w|\mu, \sigma), \qquad (2)$$

Approximating the posterior distribution $p(w|D)$ is done by updating the variational parameters for each weight during the training phase. For each Monte Carlo forward pass of the network, the weight values are drawn from their respective Gaussian distributions.

A computationally less complex method is *Monte Carlo (MC) Dropout* [14]. A dropout layer sets values of the input tensor to zero with probability p, defined as a hyperparameter. Dropout is a widely utilized regularization technique employed during the training of a DNN model to mitigate overfitting. In the case of *MC Dropout*, the Dropout Layer remains active even after training, which is a deviation from the usual practice. Performing multiple forward passes with *MC Dropout* approximates the posterior and provides an efficient and powerful mean for uncertainty estimation.

C. Uncertainty Metrics

To quantify and assess the uncertainty of fault-affected predictions, we rely on three different metrics, *variation ratio*, *predictive entropy* and *mutual information*, with nuanced differences in their interpretation.

Variation Ratio quantifies the deviation from the majority opinion of MC samples. Formally, it is defined as the proportion of MC samples whose highest softmax output does not correspond to the overall selected class, as determined by majority voting. In a scenario with a prediction set $P = \{p_1, ..., p_n\}$ with mode p_{mod}, the variation ratio is defined by $VR = \frac{1}{n}\sum_{i=1}^{n} \mathbb{1}_{p_i \neq p_{mod}}$. For example, the variation ratio of the prediction set $P = \{a, b, a, a, c\}$, with classes a, b and c is given by $0.4 = 40\%$.

Predictive entropy is a measure derived from the principles of information theory. It quantifies the average amount of information contained in the predictions of Monte Carlo (MC) samples. Formally, predictive entropy reaches its maximum value when all the MC samples assign equal probabilities to different class outputs, indicating a state of maximum uncertainty or randomness. Conversely, it attains its minimum value of 0 when all MC samples predict one class with a probability of 100%, and all the remaining classes with a probability of 0%, indicating a state of minimum uncertainty. Consequently, a high predictive entropy indicates high uncertainty of the prediction, and conversely, low predictive entropy indicates low uncertainty.

While both Variation ratio and predictive entropy are effective measures for capturing prediction uncertainty, the concept of *mutual information* provides a more nuanced perspective on the model uncertainty. Mutual information, in the context of machine learning models, is defined as the measure of the mutual dependence between the prediction and the posterior distribution of the model. This provides insights into how much the model learns from the data. Briefly expressed, high values of mutual information reflect high model uncertainty, whereas low mutual information reflects low model uncertainty.

III. FAULT ANALYSIS METHODOLOGY

In this paper, we demonstrate that it is highly beneficial to run BayNNs instead of traditional deterministic DNNs, since they prove to be more fault-tolerant, and since critical hardware faults that result in false classifications can be revealed by monitoring the uncertainty of the prediction. We demonstrate this using DNNs that we transform to BayNNs by inserting dropout layers, or by replacing Convolutional Neural Network (CNN) layers with variational layers, as depicted in Figure 2. In variational convolutional layers, a mean tensor μ_{weight} and a standard deviation tensor σ_{weight} are employed instead of a single point estimate weight tensor. This enables drawing new weight values from the respective Gaussian distribution in each forward pass.

A. Fault Propagation in DNN Accelerators

DNN inference usually has very regular data access patterns, since convolutional layers or linear fully connected layers can be represented as matrix multiplications. This makes the implementation and mapping on highly parallel architectures like systolic arrays efficient. An often used data flow in such a parallel architecture is the weight stationary data flow, which means that the DNN weights are loaded once, then kept locally in Processing Elements (PEs) while the input data is pushed through the computation array. This guarantees a high

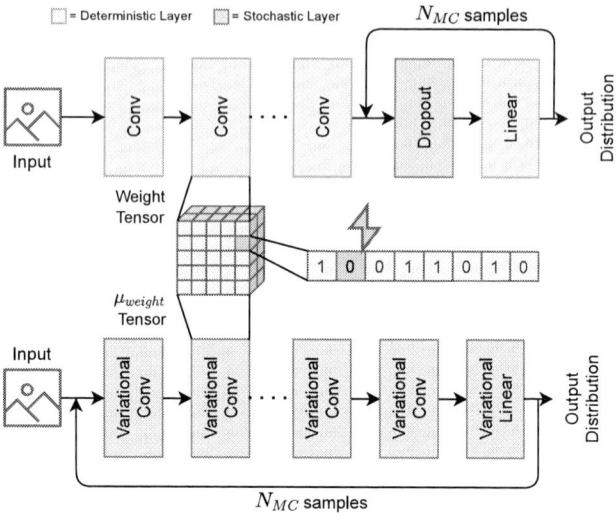

Fig. 2: Concept for the fault injection in BayNNs. For Dropout networks, faults are injected in the weight tensor of the non-stochastic conv layers (yellow). For Variational networks, the fault is analogously injected in the μ_{weight} tensor.

degree of data reuse and reduces the costs of off-chip memory transactions.

However, this approach makes such architectures also susceptible to fault propagation. Having a fault in a weight value will affect numerous operations and partial results of a convolutional layer. This erroneous layer is then passed to the following layers, which eventually can lead to a false classification in the final softmax layer. Due to this fault propagation, both in the hardware accelerator and the DNN layer structure, it is sufficient for a single bit to flip, for example due to a soft error, to cause a false classification [5] [9]. Thus far, to protect the system against such critical faults in the weight path, costly memory protection techniques and redundancy concepts have to be implemented [15]. This is primarily because standard DNNs lack a mechanism to detect hardware faults or other anomalies. Consequently, the classification score does not serve as an appropriate measure for the actual certainty of the prediction.

B. Fault Model and Fault Injection Framework

DNNs with complex architectures such as residual nets, often comprising millions of parameters and billions of operations per forward pass, present a significant challenge when it comes to fault analysis. Considering all possible fault combinations of both a network's operation and the underlying hardware accelerator register is computationally intractable. Therefore, we pursue a statistical fault injection approach that allows us to derive estimates of the fault tolerance of the Bayesian model, providing insights into the network's resilience against hardware faults.

In this paper, we primarily focus on faults occurring in the data path of an underlying DNN accelerator. We do not cover control faults as they are highly dependent on the specific hardware implementation, and thus, may vary widely across different systems. To encompass as many data fault locations in the hardware accelerator as possible, we model a fault as a memory bit-flip. This implies that the value is already corrupted before it is loaded into the accelerator, providing comprehensive coverage of potential fault scenarios. This approach guarantees a worst-case scenario, effectively simulating the effects of less critical faults that might occur in the data path or small buffers in the processing elements of the accelerator.

The concrete BayNN analysis procedure that we propose incorporates the fault injection mechanism, as detailed in Algorithm 1:

Algorithm 1 Fault Injection in BayNNs

1: **procedure** BAYNN($Dataset, Model, params_{FI}, N_{MC}$)
2:    **for** each $(Data)$ in $Dataset$ **do**
3:       $pred, pred_{FI} = []$
4:       $Model_{FI} \leftarrow FaultInj(Model, params_{FI})$
5:       **for** each run in 1 to N_{MC} **do**
6:          **for** each weight w_i in $Model$ **do**
7:             $w_i \sim \mathcal{N}(\mu_i, \sigma^2{}_i)$
8:          **end for**
9:          Compute out using $Model$ and $Data$
10:        Append out to $OutList$
11:        Compute out_{FI} using $Model_{FI}$ and $Data$
12:        Append out_{FI} to $pred_{FI}$
13:       **end for**
14:    **end for**
15:    Evaluate($pred, pred_{FI}$)
16: **end procedure**

The implementation of our approach necessitates a BayNN model, denoted as $Model$, which is subjected to testing. Additionally, a test dataset $Dataset$ is required to provide a statistical basis for the evaluation of the model. Likewise, the parameters for the fault injection can be defined in $params_{FI}$, specifically the layer where the bit-flips are injected and the bit position of the bit-flip (refer also to Figure 2). The number of MC samples is defined as a hyperparameter. Consequently, N_{MC} forward passes are carried out for each $Data$ input in total.

The fault injection is then performed in line 4. In our evaluation, we primarily focus on CNNs, whereby the weights are the most frequently reused operands. Given this, we inject a bit-flip into the respective weight or μ_{weight} tensor as depicted in Figure 2. This approach has the most significant impact on the predictions, aligning with our worst-case scenario assumption. The same fault is kept for all the N_{MC} forward passes, reflecting the typical fault behavior on a weight stationary hardware accelerator.

In lines 6-8, the random variables required for the BayNN are determined. The depicted case in Algorithm 1 is the case for variational inference, where all the weight values of the network are determined by drawing them from the normal distributions with the learned Gaussian variational parameters.

Error Probability	ResNet-18	ResNet-50
Deterministic (DNN)	1.63%	1.16%
Variational Inference (BayNN)	0.79%	0.64%
Dropout (BayNN)	1.30%	0.96%

TABLE I: Error Probability of BayNNs under faults

For the case of dropout, a dropout mask is created instead, that is then used for the computation of the respective outputs in lines 9 and 11. Once the dataset is completed, the uncertainty and fault tolerance statistics are evaluated, here represented in line 15. The process of determining the random variables before the computation of outputs, as illustrated in lines 9 and 11 of the algorithm, is a crucial step. This is because, if this step were to be performed during the inference instead, the second inference with a fault case would result in drawing different values from the stochastic distribution. Consequently, any variation in the predictions would be indistinguishable — it would be unclear whether the variation is attributable to the hardware fault or merely due to an unfortunate draw of weight values.

IV. EXPERIMENTS

For our experiments, we work on two pretrained DNNs of different complexity, ResNet-18 and ResNet-50 [16] as networks under test. We adopt two variants for transforming these DNNs into BayNNs, as depicted in Figure 2. On the one hand, we convert all the convolutional layers and the final linear layer into variational layers, using the MOPED-method of [11] and the framework *bayesian-torch* [17]. The MOPED-method allows deriving Gaussian parameter tensors directly from pretrained weights, by specifying an initial perturbation factor for the weights, which we set to a recommended value of 0.05.

On the other hand, we insert a dropout layer after the last convolutional layer of the network and set the dropout rate to 0.25. For the evaluation of both BayNN variants, we set the number of MC samples $N_{MC} = 10$ which has shown to be a good compromise to not excessively burden the hardware performance [18].

We apply post-training quantization as proposed in [19] such that the weight tensors are 8-bit signed values. As test dataset, we use a subset of the ImageNet validation dataset [20] with 10000 images, 10 per class. For every combination of network layer and bit position, the test dataset is applied once, thereby assuring uniform distribution of the bit-flips. With the 21 layers of ResNet-18 and 56 layers of ResNet-50 and 8 possible bit positions, our experiments total up to $(21 + 56) \cdot 10.000 \cdot 8 \approx 6.16M$ fault injections for each of the two BayNN variants, over $12M$ in total.

A. Fault Tolerance and Error Probability

First, we place a spotlight on the fault tolerance of BayNNs. Similarly to the authors in [9], we use the term *error* for a prediction, where the hardware fault has produced a false classification, i.e., a different overall chosen class (top-1) compared to the fault free prediction. Likewise, we use the

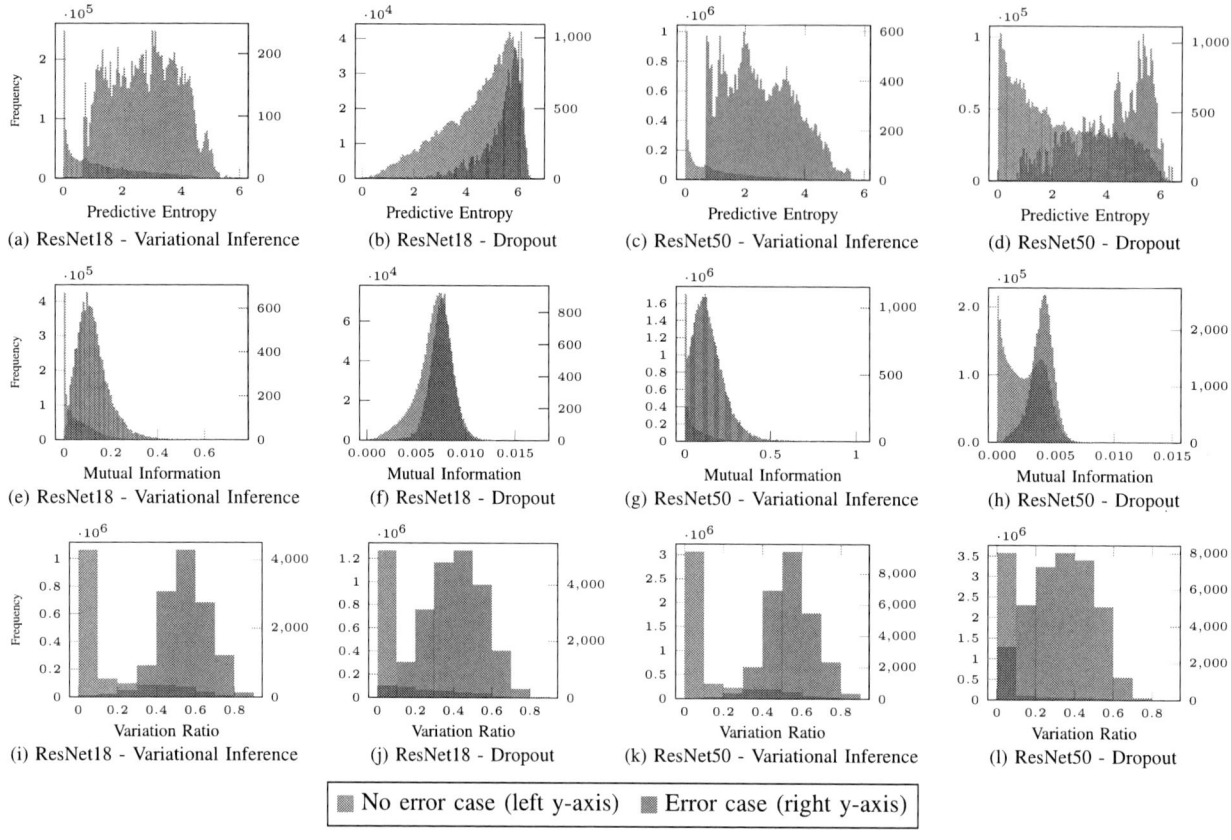

Fig. 3: Comparative analysis of fault injection experiments for ResNet-18 and ResNet-50 with two BayNN variants. The histograms illustrate the frequency of predictions from the test dataset over three uncertainty metrics. We categorize the outcomes into two scenarios: error cases (blue) where faults alter the original prediction, and non-error cases (green) where predictions remain consistent despite faults. A notable observation is the significantly elevated prediction uncertainty in error cases, as expressed by the predictive entropy and variation ratio, with a discernible boundary separating the two cases.

term *error probability* to denote the likelihood of a fault-induced false classification by the BayNN. The results are shown in Table I. First, it can be noted that both BayNNs variants provide significantly more fault-tolerance than their deterministic counterparts. Especially, variational inference is less likely to produce errors, cutting the error probability in half. If both networks are compared, an error in the larger ResNet-50 network is less significant, as the total number of weights and therefore the inherent redundancy of the network is higher overall.

B. Quantifying the Uncertainty of Error Predictions

Having demonstrated the superior fault tolerance of BayNNs compared to deterministic networks, we now aim to investigate the behavior of uncertainty in the presence of faults through a comparative analysis, summarized in Figure 3. We categorize the result of each fault injection experiment into two groups: cases with no error produced (green bars) and error cases (blue bars). In other words, we verify whether the induced error has led to an incorrect classification by the BayNN. The differences are visualized using histograms plotted over three uncertainty metrics: *predictive entropy*, *mutual information*, and *variation ratio*. It is important to note that the error cases, represented by the blue bars, employ a different scale compared to the no-error cases. This is done to enhance the clarity of the visualization, as the total quantity of errors is relatively small and would otherwise be indistinguishable. From the experiments, we can derive three major observations that we present in the following:

Errors only occur when the level of uncertainty is high.

In all combinations of networks and BayNN variants illustrated in Figure 3, it is noticeable that the uncertainty for error cases is higher than for no-error cases. While it might be intuitive to assume that the bit-flip has led to elevated uncertainty, our analysis suggests a different interpretation. We conclude that only in predictions where the uncertainty is already elevated, can a fault transform the prediction into an erroneous one, with the elevated level of uncertainty persisting. Conversely, if a prediction has low uncertainty, a single-bit weight fault is not significant enough for the network to change its prediction. This means that the inherent uncertainty in the model's prediction plays a crucial role in determining the impact of a fault. This observation also suggests that, depending on the uncertainty metric, a clear lower limit can be defined for which a false prediction due to a hardware fault is practically ruled out.

979-8-3503-6689-1/24 $31.00 © 2024 IEEE

Variational Inference generally outperforms Dropout, which requires careful, network-specific tuning.

When comparing both BayNN variants, it is evident that variational inference effectively separates error cases from the majority of no-error cases across both networks. Conversely, while Dropout performs well for the larger ResNet-50 network, significant overlaps are observed for the ResNet-18 network. This is likely due to the more pronounced impact of the dropout layer in the less complex ResNet-18 network. Consequently, configuring the dropout layer is network-dependent and requires careful attention to achieve results comparable to those of variational inference. Despite this, it is important to note that dropout can still be a viable choice for uncertainty estimation, given that the burden on hardware performance is much lower. In variational inference, every layer must be computed multiple times. This is not the case for dropout when it is placed before the final linear layer, as only this layer needs to be sampled accordingly.

Predictive Entropy and Variation Ratio work best for monitoring the uncertainty for fault detection.

When examining the different uncertainty metrics in our model's predictions, we observe that predictive entropy is close to zero for the majority of no-error predictions. A similar pattern is seen with variation ratio, where most predictions with no error produced have a variation ratio of zero due to identical MC sample outputs. This pattern, however, is not evident when looking at mutual information. For this metric, a significant number of error predictions exhibit low values, suggesting low uncertainty. We believe this is because mutual information, unlike predictive entropy and variation ratio, reflects model uncertainty rather than predictive uncertainty. Therefore, a hardware fault, even when introduced into the weight tensor, does not seem to affect the model's uncertainty. Instead, it appears to tip the balance in predictions that already exhibit high predictive uncertainty.

In summary, we observe that monitoring the uncertainty during runtime can serve as an effective measure for identifying hardware faults that occur during the inference process of neural networks. When a fault can only result in a false prediction when the uncertainty of the prediction is already elevated, monitoring the uncertainty can prevent of taking corrupted predictions into account, possibly reducing the need for expensive hardware redundancy. This opens up potential for future work, specifically in determining optimal uncertainty thresholds for fault detection.

V. CONCLUSION

In this paper, we addressed the need for dependable AI systems in safety-critical applications. Using variational inference and dropout techniques, we transformed deterministic models into Bayesian models, improving fault tolerance and enabling uncertainty monitoring.

We introduced a new fault analysis methodology, specifically tailored for BayNNs. Our extensive comparative analysis, with over 12 million fault injection experiments on two networks, demonstrated that BayNNs can achieve up to 2x greater fault tolerance compared to deterministic models. Additionally, we demonstrated that a bit-flip results in an error only in predictions characterized by elevated uncertainty, making uncertainty monitoring an effective measure for improved dependability.

ACKNOWLEDGMENT

This work was funded by the German Federal Ministry of Education and Research (BMBF) under grant number 16ME0096 (ZuSE-KI-mobil). The responsibility for the content of this publication lies with the authors.

REFERENCES

[1] "ISO 26262: Road vehicles — Functional safety," International Organization for Standardization, Geneva, CH, Standard, 2018.

[2] "ISO 21448: Road vehicles — Safety of the intended functionality," International Organization for Standardization, CH, Standard, 2022.

[3] M. Weiss and P. Tonella, "Uncertainty quantification for deep neural networks: An empirical comparison and usage guidelines," *Software Testing, Verification and Reliability*, vol. 33, no. 6, p. e1840, Sep. 2023.

[4] Y. Ovadia, E. Fertig, J. Ren, Z. Nado, D. Sculley, S. Nowozin, J. V. Dillon, B. Lakshminarayanan, and J. Snoek, "Can you trust your model's uncertainty," *Evaluating predictive uncertainty under dataset shift*, 2019.

[5] G. Li *et al.*, "Understanding error propagation in deep learning neural network (dnn) accelerators and applications," in *International Conference for High Performance Computing, Networking, Storage and Analysis*, 2017.

[6] K. Lee *et al.*, "A simple unified framework for detecting out-of-distribution samples and adversarial attacks," *Advances in neural information processing systems*, vol. 31, 2018.

[7] C. Schorn and L. Gauerhof, "Facer: A universal framework for detecting anomalous operation of deep neural networks," in *IEEE 23rd International Conference on Intelligent Transportation Systems (ITSC)*, 2020.

[8] Y.-C. Hsu *et al.*, "Generalized ODIN: Detecting out-of-distribution image without learning from out-of-distribution data," in *IEEE/CVF Conference on Computer Vision and Pattern Recognition (CVPR)*, 2020.

[9] J. Hoefer *et al.*, "SiFI-AI: A fast and flexible rtl fault simulation framework tailored for ai models and accelerators," in *Proceedings of the Great Lakes Symposium on VLSI 2023*, ser. GLSVLSI '23, 2023.

[10] M. Stammler *et al.*, "Effect: An end-to-end framework for evaluating strategies for parallel ai anomaly detection," *Procedia Computer Science*, vol. 222, pp. 499–508, 2023.

[11] R. Krishnan, M. Subedar, and O. Tickoo, "Efficient priors for scalable variational inference in bayesian deep neural networks," in *IEEE/CVF International Conference on Computer Vision Workshop (ICCVW)*, 2019.

[12] L. V. Jospin, H. Laga, F. Boussaid, W. Buntine, and M. Bennamoun, "Hands-on bayesian neural networks—a tutorial for deep learning users," *IEEE Computational Intelligence Magazine*, vol. 17, no. 2, 2022.

[13] A. Graves, "Practical variational inference for neural networks," in *Advances in Neural Information Processing Systems*, vol. 24. Curran Associates, Inc., 2011.

[14] Y. Gal and Z. Ghahramani, "Dropout as a bayesian approximation: Representing model uncertainty in deep learning," 2016.

[15] F. Kempf *et al.*, "The zuse-ki-mobil ai accelerator soc: Overview and a functional safety perspective," in *2023 Design, Automation & Test in Europe Conference & Exhibition (DATE)*, 2023, pp. 1–6.

[16] K. He, X. Zhang, S. Ren, and J. Sun, "Deep residual learning for image recognition," 2015.

[17] R. Krishnan, P. Esposito, and M. Subedar, "Bayesian-torch: Bayesian neural network layers for uncertainty estimation," https://github.com/IntelLabs/bayesian-torch, Jan. 2022.

[18] J. Hoefer *et al.*, "A hardware-aware sampling parameter search for efficient probabilistic object detection," in *Computer Vision Systems*. Cham: Springer Nature Switzerland, 2023, pp. 299–309.

[19] J.-L. Lin *et al.*, "Quantization for bayesian deep learning: Low-precision characterization and robustness," in *2023 IEEE International Symposium on Workload Characterization (IISWC)*. IEEE, 2023, pp. 180–192.

[20] J. Deng, W. Dong, R. Socher, L.-J. Li, K. Li, and L. Fei-Fei, "Imagenet: A large-scale hierarchical image database," in *2009 IEEE Conference on Computer Vision and Pattern Recognition*, 2009, pp. 248–255.

979-8-3503-6689-1/24 $31.00 © 2024 IEEE

Zero-Memory-Overhead Clipping-Based Fault Tolerance for LSTM Deep Neural Networks

Bahram Parchekani[1], Samira Nazari[1], Mohammad Hasan Ahmadilivani[2],
Ali Azarpeyvand[1,2], Jaan Raik[2], Tara Ghasempouri[2], and Masoud Daneshtalab[2,3]

[1]University of Zanjan, Zanjan, Iran
[2]Tallinn University of Technology, Tallinn, Estonia
[3]Mälardalen University, Västerås, Sweden

[1]{bahram.parchekani, samira.nazari, azarpeyvand}@znu.ac.ir
[2]{mohammad.ahmadilivani, ali.azarpeyvand, jaan.raik, tara.ghasempouri}@taltech.ee
[3]masoud.daneshtalab@mdu.se

Abstract—**Long Short-Term Memory (LSTM) Deep Neural Networks (DNNs) have shown superior accuracy in predicting and classifying time-series data. This has made them suitable for many applications, including safety-critical ones, such as healthcare, where fault tolerance is a major concern. Until now, fault resilience and mitigation in LSTMs have not been thoroughly explored, raising concerns about exploiting them in safety-critical use cases. This work, first, extensively explores the effect of faults on LSTM DNNs using fault injection into parameters. Moreover, the paper presents two effective zero-memory-overhead fault tolerance techniques for LSTM DNNs to protect them against random faults in their parameters. Experimental results indicate that the proposed techniques can improve fault tolerance of LSTM-based DNNs up to 278.6 times concerning unprotected ones.**

Index Terms—**Hardware Reliability, Fault Tolerance, Neural Networks, LSTMs, Healthcare.**

I. INTRODUCTION

Deep Learning (DL) is being increasingly employed in safety-critical applications, e.g., healthcare and automotive, due to their outstanding accuracy in classification, prediction, etc. [1], [2]. In this domain of applications, hardware reliability is a significant concern [3]–[5]. Hardware reliability is defined as the probability of hardware performing correctly with the presence of faults. Faults may occur due to temperature variation, aging, soft errors, etc., and flip the bits in logic or memory [3], [6], [7]. Such an effect may lead to catastrophic results in safety-critical applications. With transistor scaling, fault rates in memories have been increased, which threatens the hardware reliability significantly [8].

Healthcare applications exploit Deep Neural Networks (DNNs) extensively for various tasks such as diagnosis, treatment, and prediction of diseases and anomalies [9], [10] because of their outstanding strength in processing time-series data [11]. Long Short-Term Memory (LSTM) DNNs are a subset of Recursive Neural Networks (RNNs) that are remarkably effective in classifying and predicting time-series data. They retain long-term information through time via feedback loops making them highly desirable for disease prediction in healthcare applications [11].

Throughout the literature, several research works have thoroughly studied the reliability of DNNs in safety-critical applications [4], [6]. Fault injection at the software simulation level is the predominant method for analyzing and evaluating the reliability of DNNs due to their fast execution time [3]. The related studies indicate that faults in memory impacting the parameters of DNNs result in a substantial reduction of their accuracy [8], [12]. Therefore, various methods for improving their fault tolerance

are proposed. Model-level fault tolerance [13]–[15] significantly impacts their resilience against memory faults.

Nearly all existing works study the reliability of Convolutional Neural Networks (CNNs) for image classification and object detection [3]. Although LSTMs are widely deployed in safety-critical applications, their fault tolerance is not extensively explored [3]. In a recent paper, the effect of faults in the computational units of LSTMs for image classification in automotive is studied and a hardware-level fault tolerance approach is proposed [16]. In another prior work, a fine-grain and comprehensive resilience analysis for different sets of parameters in LSTMs is performed [17]. It is shown that their resilience can be remarkably improved by detecting the faulty weights and setting them to 0.

In this paper, we perform resilience analysis on StageNet [18] as a case study in healthcare applications for disease prediction. The contributions of this paper are as follows:

- Performing a comprehensive resilience analysis for various LSTM-based DNNs using fault injection in parameters, leading to observing the effect of different DNN structures and identification of critical bits;
- Proposing two fault-tolerant methods for LSTM-based DNNs, Weights Bit Clipping (WBC) and Activations Value Clipping (AVC), to effectively reduce the impact of faults in parameters on LSTM-based DNNs;
- Demonstrating the efficacy of the proposed methods in DNNs' resilience, leading up to 278.6 times less critical cases in the outputs caused by faulty parameters.

Section II provides a background on StageNet. Section III presents the proposed method for its resilience analysis and improvement. Section IV provides the results, and Section V concludes the paper.

II. PRELIMINARIES ON LSTMs AND STAGENET

StageNet [18] is an LSTM-based DNN designed for disease prediction in healthcare. It predicts the stage of a patient's disease according to the characteristics of the tests performed by the patient through time. Fig. 1 illustrates the overall structure of StageNet. It is composed of an LSTM layer for characterizing the disease stage over time, a convolutional (CONV) module, and an output Fully Connected (FC) layer to output the predicted disease condition. The CONV module contains one layer in parallel with two FC layers, and their outputs are multiplied point-wise.

StageNet receives time-series data as inputs according to the patient visits $(v_1, v_2, \cdots, v_t)$ which contain numerical clinical features at different times $(\Delta_1, \Delta_2, \cdots, \Delta_t)$. They pass through the LSTM layer with multiple cells inferring the variation of a

979-8-3503-6689-1/24 $31.00 © 2024 IEEE

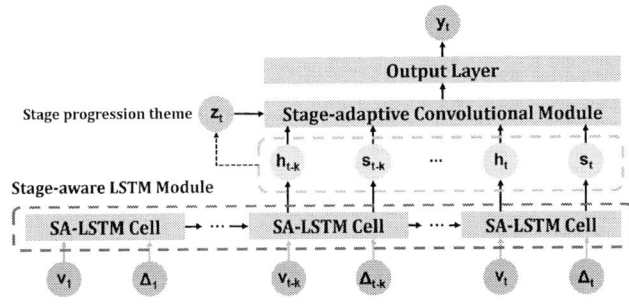

Fig. 1: Overal structure of StageNet [18].

patient's health stage considering their current status. The produced results are forwarded to the CONV module to learn patterns of the disease stages. Afterward, the output layer performs the classification for disease stage and risk prediction.

To measure the performance of StageNet, the following metrics are evaluated: **Accuracy:** This metric represents the percentage of correct predictions of StageNet compared to the expected outputs. **AUROC:** This metric refers to the area under the receiver operating characteristic curve illustrating the trade-off between true positive rate and false positive rate. It shows how a classifier can discriminate the positive and negative classes and it is extensively used when a dataset is imbalanced. Since the distribution of data in the dataset between different disease stages is imbalanced, AUROC is a more suitable metric to show the performance of StageNet. Therefore, for resilience analysis of this LSTM-based DNN, we consider AUROC drop along with accuracy drop.

III. FAULT PROPAGATION AND MITIGATION IN LSTM-BASED DNNS

A. Resilience Evaluation Using Fault Injection

To analyze the resilience of DNNs comprehensively, random bit flips are applied throughout the DNN's weights to assess the overall network's behavior in the presence of faults. To quantify the resilience analysis, once the average performance metrics (Accuracy and AUROC) for a DNN under test are obtained, their difference with the fault-free metrics is considered as accuracy drop and AUROC drop.

Furthermore, the output effect of faults is categorized into the following classes, to quantify the effect of faults on the DNNs' outputs: **Masked:** Outputs remain identical between faulty and fault-free executions. **Non-critical Silent Data Corruption (SDC):** Output values differ between faulty and fault-free executions, while the classification result remains consistent. **Critical SDC:** Both output values and classification results differ between faulty and fault-free executions. **Detected Unrecoverable Error (DUE):** The DNN generates "NaN" values in the output, indicative of a system exception.

To identify the critical bits in DNNs, we perform a bit-wise fault injection experiment throughout the DNNs. In such an experiment, one bit is considered as the target and it is flipped in all parameters and the inference is performed and the performance metrics are measured. The bit that has the highest impact on the performance metrics is identified as the most critical bit.

B. Resilience Improvement for LSTM-based DNNs

We propose two model-level fault tolerance techniques with zero memory overhead: 1) Weights Bit Clipping (WBC), and

2) Activations Value Clipping (AVC). In the WBC method, all weights of fault-free DNN models are profiled and their bit patterns are analyzed. As a result, a consistent bit pattern is revealed in the DNNs under study. Moreover, using fault injection, the most critical bit is identified. To this end, an extensive exploration of different bit flips is carried out and the resilience is measured for each bit. Therefore, the method suggests clipping the most critical bits to a certain value throughout the DNN, before an inference.

In the AVC method, first, the input values to each activation function of the LSTM cells as well as the CONV and FC layers in DNNs are profiled and their maximum and minimum values are obtained, during a fault-free forward pass with validation data. The obtained values are then utilized for detecting faults that is when an input to corresponding activation functions exceeds the determined value range. Once a fault is detected in a forward pass, the corresponding value is clipped. If an activation value falls below the minimum range value, it is set to that minimum threshold. Conversely, if an activation value exceeds the maximum profiled value, it is set to the maximum threshold.

Ultimately, the effectiveness of WBC and AVC is evaluated by fault injection to determine how they mitigate the fault impact and which technique offers superior fault tolerance in LSTM DNNs. By comparing these methods, we aim to identify the more robust approach for enhancing the reliability of LSTM neural networks in the presence of faults.

IV. EXPERIMENTS

A. Experimental Setup

To evaluate and improve the resilience of LSTM-based DNNs against faults in parameters, four variations of StageNet are experimented. Two variations represent the full StageNet model which includes CONV layer, with different numbers of LSTM cells (384 and 72), and the two variations that exclude the convolutional module, containing only LSTM layer, with 384 and 72 cells.

Test data is sourced from the Medical Information Mart for Intensive Care (MIMIC-III) dataset, which includes 17 physiological variables recorded at each visit. It is transformed into a 76-dimensional vector comprising numerical and one-hot encoded categorical clinical features for 33,678 patients.

Baseline metrics, including accuracy and AUROC in fault-free executions, are summarized in Table I, alongside the number of parameters for each model. All models were executed on a CPU supporting 32-bit floating-point IEEE-754 data representation.

Table I: Accuracy, AUROC, and the number of different weight sets of the LSTM-based ANNs in this work.

	Accuracy	AUROC	#weights in LSTM	#weights in CONV	#weights in FC
Stage-CONV-384	94.94%	79.21%	738618	442368	24960
Stage-CONV-72	83.75%	76.75%	48764	51840	1800
Stage-384	90.28%	79.29%	738618	0	384
Stage-72	77.28%	76.97%	48764	0	72

We conduct Fault Injection (FI) across all weights in the DNNs under study. The number of injected faults is determined using a Bit Error Rate (BER) ranging from 0.0001 to 0.01, covering a comprehensive range of potential errors. FI is repeated 1000 times to ensure an acceptable confidence level. For each iteration, a drop in accuracy and AUROC is obtained, and faults are classified according to Subsection III-A. Eventually, the average results over all iterations are reported in the paper.

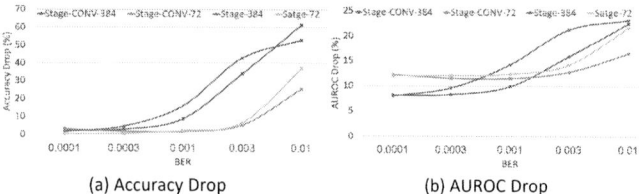

(a) Accuracy Drop   (b) AUROC Drop

Fig. 2: Model-wise FI results for DNNs.

All experiments are implemented and performed using PyTorch and executed on an Intel® Core™ i7-9700 CPU. Through these experiments, we aim to thoroughly assess the reliability of LSTM-based DNNs under faults in parameters and evaluate the effectiveness of the proposed protection techniques in preserving model performance.

B. Resilience Analysis Results

As illustrated in Fig. 2, the performance metrics for all models significantly drop under fault injection campaigns. In Fig 2a, as the BER increases, larger models (i.e., Stage-CONV-384 and Stage-384) demonstrate more accuracy drop than the other models, at the same BERs. However, the AUROC drop metric shows a different behavior. It is observed that Stage-CNN-72 and Stage-72 are remarkably sensitive to faults in terms of their AUROC, at the lowest BER. While at higher BERs, the AUROC drop is higher for larger DNNs.

Regarding Fig. 2b, as AUROC expresses the discrimination of classification over different thresholds, this observation shows that smaller DNNs are remarkably sensitive to faults to correctly distinguish the stage of patients' disease. According to the results, the AUROC metric for all DNNs falls below 60% when $BER = 0.01$. At such high BERs, although larger DNNs are shown to be more error-prone, in this safety-critical application, none of them are reliably functioning.

On the other hand, it is observed that DNNs possessing the CONV module are generally more resilient than the ones without the CONV module. It shows that the CONV module increases the capability of fault masking in LSTM-based DNNs and improves their inherent resilience.

According to Fig. 3, in all models, as the BER increases, the fault effects with masked and non-critical SDCs decrease, significantly leading to more critical SDCs and DUEs. This figure also evidences the fact that the DNNs with the CONV module are more resilient to faults than the ones without the CONV module. Obtained results indicate that when $BER = 0.01$, total critical SDC and DUE rate for StageNet-CONV-384, StageNet-CONV-72, StageNet-384 and StageNet-72 is 90.05%, 54.97%, 82.02% and 77.33%.

We perform a bit-level analysis of weights. First, we observe the average value of each bit throughout the weights in all DNNs, as illustrated in Fig. 4a. It is observed that bits 0 to 26 in 32-bit floating point data representation almost have a unified distribution between 0 and 1. while bits number 27, 28, and 29 are always '1' and bit number 30 is always '0'. It shows that we can protect the bits that are constant throughout the weights.

As depicted in Fig. 4b, the accuracy drop for bit 30 in all DNNs is significantly higher than bit-flip in other bits. Consequently, bit 30 is identified as the most critical bit in the DNNs, which is observed in Fig. 4a that its value is always '0'. As a protection mechanism, this bit is always set to '0', which ensures that the

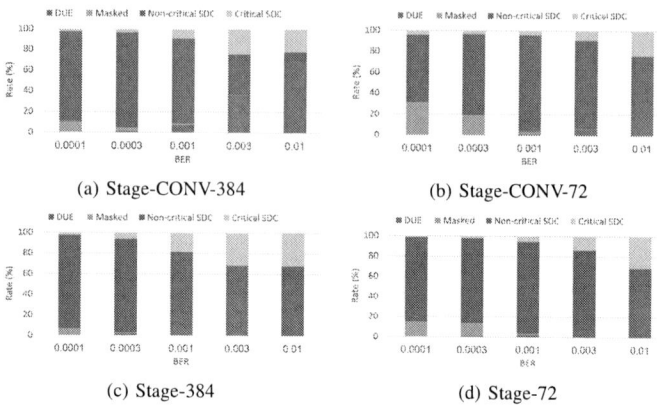

(a) Stage-CONV-384   (b) Stage-CONV-72

(c) Stage-384   (d) Stage-72

Fig. 3: Fault classification in model-wise FI for a) Stage-CONV-384, b) Stage-CONV-72, c) Stage-384, d) Stage-72.

most critical bit is consistently safeguarded, enhancing the overall reliability of the LSTM-based DNNs.

C. Fault Tolerance for LSTM-based DNNs

1) Weights Bit Clipping (WBC)

As shown in Fig. 5a and Fig. 5b, weights bit clipping lays a significant improvement on the reliability of LSTM-based DNNs. The accuracy drop is close to zero through all BERs for all protected models. On the other hand, the AUROC drop is remarkably improved for all DNNs. According to the results, the AUROC drop is reduced by up to 3.2x when $BER = 0.01$ and the total critical SDC and DUE rate across DNNs is reduced by up to 278.6x.

2) Activations Value Clipping (AVC)

Fig. 5c and Fig. 5d present how effectively activations value clipping protects the models against faults. According to the results, The accuracy drop and AUROC drop are reduced by up to 15.54x and 1.5x among the DNNs, respectively, when $BER = 0.01$. The fault classification results through the fault injection campaigns on protected DNNs by AVC indicate that this method is also capable of removing all DUE effects. As a result, the total DUE and critical SDC rate for StageNet-CONV-384, StageNet-CONV-72, StageNet-384, and StageNet-72 is 13.88%, 5.18%, 36.47%, and 16.86% resulting in up to 10.6x reduction across DNNs when $BER = 0.01$.

3) Comparison

Fig. 6 compares the proposed zero-memory overhead fault-tolerant techniques for LSTM-based DNNs. As observed, Weights Bit Clipping (WBC) generally demonstrates a more consistent protective effect across different DNNs. It effectively reduces accuracy drop and the incidence of critical faults across the board. However, Activations Value Clipping (AVC) slightly outperforms in a few cases. WBC achieves 2.36x, 1.19x and 2.26x less AUROC drop than AVC in Stage-CONV-384, Stage-CONV-72, and Stage-384, when $BER = 0.01$, whereas AVC provides 1.13x times less AUROC drop than WBC for Stage-72 at the same BER.

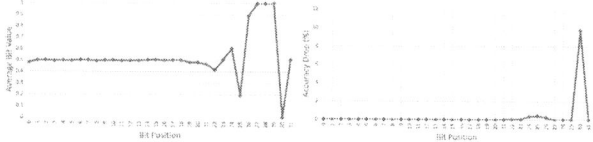

(a) Average value of different bits.   (b) Finding the most critical bit.

Fig. 4: Monitoring bit values

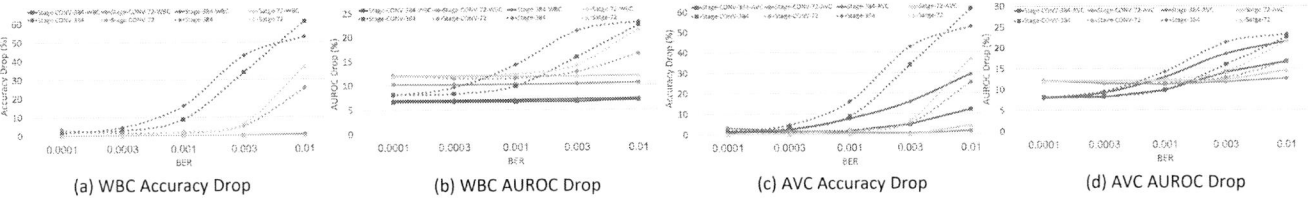

Fig. 5: Weights Bit Clipping protection and Activation Value Clipping and FI results.

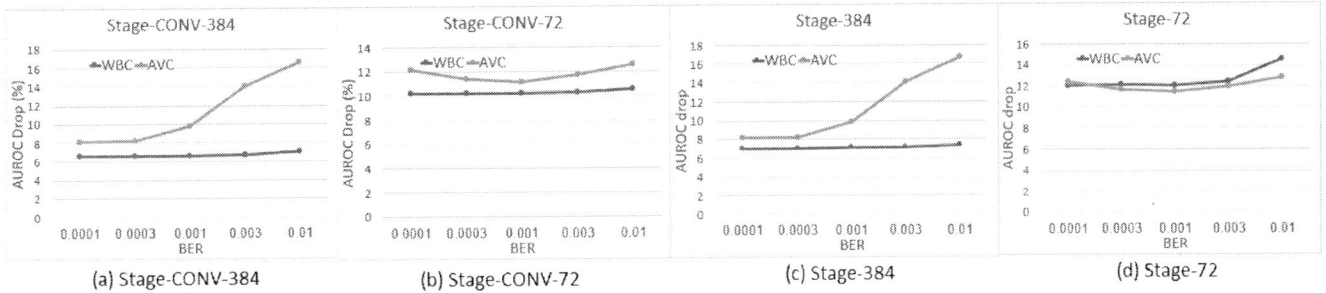

Fig. 6: Weights Bit Clipping (WBC) and Activations Value Clipping (AVC) comparison based on AUROC drop.

Nonetheless, each proposed fault-tolerance technique is applicable in different design scenarios. WBC applies directly to the memory and can be conducted to the bit values of the stored data before an inference. On the other hand, AVC is applied during the inference and prevents errors produced by faulty weights during the inference.

V. CONCLUSION

In this paper, we study the reliability of various LSTM-based DNNs (variants of StageNet) in healthcare as a case study with different structures and propose two zero-memory overhead fault-tolerance techniques for them. Results indicate that LSTM-based DNNs possessing convolutional layers demonstrate more resilience than the ones without convolutional layers. Furthermore, we performed bit-level analysis resulting in the identification of the most critical bits.

Moreover, two zero-overhead protection techniques to improve their fault tolerance are proposed: weights bit clipping and activations value clipping. It is shown that weights bit clipping can reduce the AUROC drop by up to 3.2x and DUE and critical faults by up to 278.6x compared to the unprotected DNNs at a high BER. Also, activations value clipping reduces the AUROC drop by 1.5x and DUE and critical SDCs by 10.6x, under the same conditions. Thus, the results demonstrate that the weights bit clipping method is extremely effective in mitigating the effect of faults occurring in the parameters of LSTM-based DNNs.

ACKNOWLEDGMENTS

This work is supported by PSG837 Estonian national funding.

REFERENCES

[1] M. Loni, H. Mousavi, M. Riazati, M. Daneshtalab, and M. Sjödin, "Tas: ternarized neural architecture search for resource-constrained edge devices," in *2022 Design, Automation & Test in Europe Conference & Exhibition (DATE)*. IEEE, 2022, pp. 1115–1118.

[2] B. Rokh, A. Azarpeyvand, and A. Khanteymoori, "A comprehensive survey on model quantization for deep neural networks in image classification," *ACM Transactions on Intelligent Systems and Technology*, vol. 14, no. 6, pp. 1–50, 2023.

[3] M. H. Ahmadilivani, M. Taheri, J. Raik, M. Daneshtalab, and M. Jenihhin, "A systematic literature review on hardware reliability assessment methods for deep neural networks," *ACM Computing Surveys*, vol. 56, no. 6, pp. 1–39, 2024.

[4] F. Su, C. Liu, and H.-G. Stratigopoulos, "Testability and dependability of ai hardware: Survey, trends, challenges, and perspectives," *IEEE Design & Test*, 2023.

[5] M. Taheri, N. Cherezova, S. Nazari, A. Rafiq, A. Azarpeyvand, T. Ghasempouri, M. Daneshtalab, J. Raik, and M. Jenihhin, "Adam: Adaptive fault-tolerant approximate multiplier for edge dnn accelerators," in *2024 IEEE European Test Symposium (ETS)*, 2024, pp. 1–4.

[6] Y. Ibrahim, H. Wang, J. Liu, J. Wei, L. Chen, P. Rech, K. Adam, and G. Guo, "Soft errors in dnn accelerators: A comprehensive review," *Microelectronics Reliability*, vol. 115, p. 113969, 2020.

[7] M. Nourazar, V. Rashtchi, F. Merrikh-Bayat, and A. Azarpeyvand, "Towards memristor-based approximate accelerator: Application to complex-valued fir filter bank," *Analog Integrated Circuits and Signal Processing*, vol. 96, no. 3, pp. 577–588, 2018.

[8] M. A. Neggaz et al., "Are cnns reliable enough for critical applications? an exploratory study," *IEEE Design & Test*, vol. 37, no. 2, pp. 76–83, 2019.

[9] R. Manne and S. C. Kantheti, "Application of artificial intelligence in healthcare: chances and challenges," *Current Journal of Applied Science and Technology*, vol. 40, no. 6, pp. 78–89, 2021.

[10] F. J. Harris et al., "A survey of human gait-based artificial intelligence applications," *Frontiers in Robotics and AI*, vol. 8, p. 749274, 2022.

[11] B. Lim and S. Zohren, "Time-series forecasting with deep learning: a survey," *Philosophical Transactions of the Royal Society A*, vol. 379, no. 2194, pp. 202–209, 2021.

[12] K. Adam et al., "A selective mitigation technique of soft errors for dnn models used in healthcare applications: Densenet201 case study," *IEEE Access*, vol. 9, pp. 65 803–65 823, 2021.

[13] L.-H. Hoang et al., "Ft-clipact: Resilience analysis of deep neural networks and improving their fault tolerance using clipped activation," in *DATE*, 2020, pp. 1241–1246.

[14] B. Ghavami et al., "Fitact: Error resilient deep neural networks via fine-grained post-trainable activation functions," in *2022 DATE*. IEEE, 2022, pp. 1239–1244.

[15] M. H. Ahmadilivani, S. Mousavi, J. Raik, M. Daneshtalab, and M. Jenihhin, "Cost-effective fault tolerance for cnns using parameter vulnerability based hardening and pruning," in *2024 IEEE IOLTS, inpress*, 2024, pp. 1–7.

[16] N. Nosrati and Z. Navabi, "Analysis and enhancement of resilience for lstm accelerators using residue-based ceds," *IEEE Access*, 2024.

[17] M. H. Ahmadilivani, J. Raik, M. Daneshtalab, and A. Kuusik, "Analysis and improvement of resilience for long short-term memory neural networks," in *2023 IEEE International Symposium on Defect and Fault Tolerance in VLSI and Nanotechnology Systems (DFT)*. IEEE, 2023, pp. 1–4.

[18] J. Gao, C. Xiao, Y. Wang, W. Tang, L. M. Glass, and J. Sun, "Stagenet: Stage-aware neural networks for health risk prediction," in *Proceedings of The Web Conference 2020*, 2020, pp. 530–540.

Inferred Fault Models for RISC-V and Arm: A Comparative Study

Ihab Alshaer*, Ahmed Al-kaf*,Valentin Egloff*, Vincent Beroulle*
*Univ. Grenoble Alpes, Grenoble INP, LCIS, 26000 Valence, France

Abstract—With the widespread adoption of embedded systems, security issues became a major concern. In particular, such systems are vulnerable to various kinds of physical attacks, and fault injection is one of the main physical attacks. Designers and developers require fault models to predict the effects of the fault injection, so that they can analyze possible vulnerabilities and develop countermeasures against such attacks. Thus, understanding the effects of fault injection is essential to provide realistic fault models. Moreover, many of the systems currently in use or planned for future deployment incorporate either Arm or RISC-V processors. In this paper, voltage glitch campaigns have been carried out on two microcontrollers that are widely used in the embedded system market. One embeds a RISC-V core, where the other embeds an Arm Cortex-M4 core. As a result, we provide comprehensive analysis for the obtained faulty behaviors using a set of inferred fault models. We show that the presented fault models are able to explain more than 99% of the observed faulty behaviors. We also show that some of these models are applicable to both cores. Furthermore, we illustrate that some of the presented models are also comparable to state-of-the-art models that are proposed as a result of clock glitch campaigns. The presented fault models enable better understating of the fault injection effects, and thus, easing the process of analyzing vulnerabilities, and developing cost-effective countermeasures against fault attacks.

Index Terms—fault injection attack, RISC-V, Arm Cortex-M, fault model

I. INTRODUCTION

The widespread use of embedded systems open the doors to various considerations that focus on performance, cost, complexity, and most importantly, security. Such systems are under questions when it comes to their immunity to physical attacks, with fault injection attacks being among the most significant.

In the context of hardware security, fault injection can be defined as a powerful active physical attack, where the attacker tries to perturb the normal behavior of a system by introducing faults. The potential faulty outcome is then examined for a possible vulnerability to be exploited. There are different techniques to perform the injection. The most known ones are: applying perturbations to the clock signal, which is known as clock glitch [1], applying variations to the power supply, which is known as voltage glitch [2], and exposing the device to electromagnetic (EM) pulses [3] or to laser beams [4].

For the sake of protecting embedded systems against fault attacks, a comprehensive understanding of the fault effect is required to provide fault models. Fault models are abstract representations of the physical fault effects. These models provide description for the effects of the faults at different levels of system abstraction. Hardware designers and software developers need such fault models, so that, they can predict the possible effects by applying these fault models. Thus, they will be able to identify possible vulnerabilities in software codes and hardware designs. Consequently, they will be be able to develop/design the most suitable countermeasures. Nevertheless, having inaccurate fault models would lead to proposing either excessive protections, which affects the cost/performance ratio, or insufficient countermeasures, which means potential vulnerabilities are still exploitable.

A. Related works

Numerous studies [2]–[9] have focused on analyzing the effects of fault injection at the instruction set architecture (ISA) level, leading to the proposal of various fault models. For example, instruction skip. [4], [6], [8], [9], instruction replay [4], [6], instruction corruption [2], [3], [7], [9], and register corruption [2], [5], [7]. These models appear to be quite generic, lacking precision in accurately portraying the actual consequences of fault injection. The term "corruption" lacks clarity in describing the fault's impact. Thus, there will be a difficulty in identifying vulnerabilities solely based on this information. As a result, this may lead to developing over- or under-protections.

In [1], [10], the authors provided a comprehensive analysis and rationale for the experimental findings, resulting from clock glitch fault injection campaigns on 32-bit microcontrollers embedding Arm Cortex-M3 and Arm Cortex-M4 cores. They demonstrated how the alignment of instructions in memory can influence the resulting faulty behaviors. Building on this insight, they introduced two fault models at the instruction encoding level: *Skip* specific number of bits and *Skip and repeat* specific number of bits. This number of bits is related to the flash memory access size or a cache line size, such as 32 bits or 64 bits. Additionally, they proposed another fault model, called *Partial update* [11], where some bits, while the data are propagated from memory to pipeline stages, will be updated correctly, and others will be updated in a faulty way. This faulty part received its update either from the previous value or from a precharge value. These fault models allowed explaining a large set of the observed faulty behaviors at ISA level. Nevertheless, the authors were not sure if such fault models would be applicable on different microcontrollers, or when employing an injection technique that is different from clock glitch. Table I summarizes the related works.

979-8-3503-6689-1/24 $31.00 © 2024 IEEE

TABLE I
SUMMARY OF RELATED WORKS ON FAULT INJECTION EFFECT
CHARACTERIZATION AND MODELING.

Reference	Injection type	Target	Fault model
[2]	voltage	ARMv7-A	instruction corruption
[3]	EM	Arm Cortex-M3	instruction replacement
[5]	EM	Arm Cortex-M4	register, xPSR corruption, opcode or operand substitution
[6]	EM	Arm Cortex-A9	register corruption, operand substitution, control-flow corruption
[7]	EM	Arm Cortex-A53, Intel i3-6100T	register corruption, instruction corruption
[4]	laser	Arm Cortex-M0+	instructions skip, skip and replay
[8]	laser	Arm Cortex-M4	multi-instruction skip
[9]	EM	RISC-V E31	instruction corruption, instruction skip
[1], [10], [11]	clock	Arm Cortex-M3, Arm Cortex-M4	Skip # of bits, Skip and repeat # of bits, Partial update
This work	voltage	RISC-V E31 Arm Cortex-M4	Skip # of bits, Skip & repeat # of bits, Non-sequential skip & repeat, Partial update, Combination, Skip with forwarding

B. Contributions

In this paper, we present a set of inferred fault models that are able to describe and explain more than **99 %** of the obtained faulty behaviours from voltage glitch fault injection campaigns on two different 32-bit microcontrollers embedding RISC-V and Arm Cortex-M4 processors. We show that some of these models are applicable to both devices. This includes *Skip*, *Skip & repeat*, and *Partial update* fault models, which are proposed in the literature for clock glitch campaigns on Arm targets. Furthermore, we illustrate *new* fault models that are strongly related to specific features supported by the target devices. This encompasses *Skip with forwarding* for RISC-V target, and *Non-sequential Skip & repeat* For Arm Cortex-M4 target. Other observed faulty behaviors required proposing another fault model: *Combination*. More details in Section IV.

C. Outline

The rest of this paper is organized as follows: Section II presents the followed inference methodology to explain the obtained results. Section III illustrates the experimental setup, then the definitions of the inferred models and the experimental results are reported and discussed in Section IV. The paper is concluded along with perspectives in Section V.

II. FOLLOWED APPROACH

To derive fault models capable of describing the effects of fault injection, we utilized a methodology akin to that of previous works in [3], [11], [12]. The essence of our analysis involves comparing the results of a comprehensive and tailored

set of programs executions, encompassing both simulations and physical fault injections, across different levels of digital system abstraction. In this study, we concentrate our analysis on the instruction execution level and the encoding level.

Fig. 1 illustrates the followed approach for the sake of inferring and confirming the use of fault models at encoding and execution levels. Initially, at step 1, physical fault injection campaigns are conducted on a target device running a set of target programs, comprising assembly instructions. Subsequently, in step 2, fault models are deduced from the results of these physical fault injections. The inferred fault models are then applied, as mutants to the encoding level, to simulate the execution of the same target program utilized in step 1 (step 3 in Fig. 1). The last step involves comparing the outcomes of physical fault injection with those of software executions to validate the inferred fault models, along with the ability to explain the obtained faulty behaviors from the physical fault injection (step 4 in Fig. 1). This comparison is conducted based on the output values of the processor's general-purpose registers, each of which has a predetermined initial value, allowing for the detection of any alterations as a result of the fault injection.

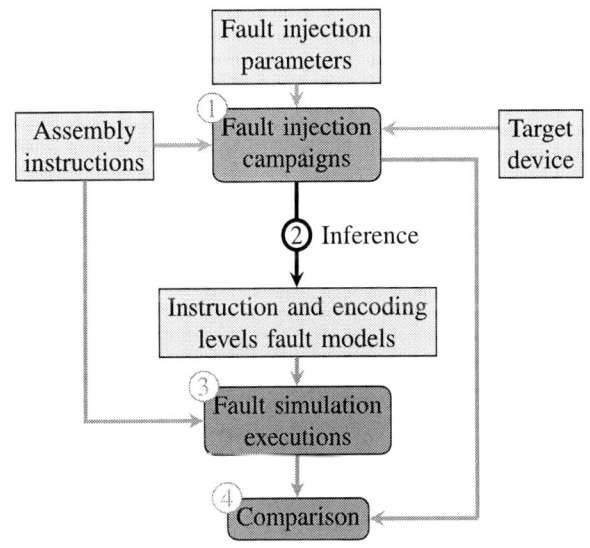

Fig. 1. Followed approach for inferring and confirming fault models

III. EXPERIMENTAL SETUP

Physical fault injection experiments were conducted to examine the effects of fault injection when targeting 32-bit microcontrollers embedding different cores. The objective is to establish reliable and realistic fault models for the observed faulty behaviors for each core. This endeavor aims to enhance the description of faulty behaviors at the ISA level, thereby offering a more comprehensive understanding. Furthermore, it is imperative to assess whether state-of-the-art fault models, typically applied when targeting Arm-based microcontrollers using clock glitch, remain applicable when targeting RISC-V and Arm-based microcontrollers using voltage glitch.

979-8-3503-6689-1/24 $31.00 © 2024 IEEE

The following subsections present the target devices we have used, the target programs, and the fault injection technique we have employed.

A. Target devices

Two different 32-bit microcontrollers have been employed as target devices. The *first device* is a SiFive 32-bit microcontroller (FE310-G002) that embeds an E31 RISC-V core. The E31 core has a 5-stage pipeline: fetch, decode and register fetch, execute, data memory access, and register writeback. It has 32 general-purpose 32-bit registers, X0 to X31. E31 core is based on RISC-V architecture and supports the RV32IMAC instruction set. Therefore, it supports the standard Multiply (M), Atomic (A), and Compressed (C) RISC-V extensions. Supporting the compressed extension makes RV32IMAC a variable-length instruction set that offers two encoding lengths: 16 and 32 bits. The instruction has a 32-bit encoding if the least significant two bits of a 32-bit word have 0b11 value. Otherwise, the least significant 16 bits belong to a 16-bit instruction [13]. Thanks to the observed faulty behaviors in Section IV, it seems that E31 core supports *operand forwarding*, which is used to resolve data hazards by bypassing data from one pipeline stage to another without writing to and reading from the register file.

This RISC-V device flash memory access size is 32 bits, allowing for the simultaneous retrieval of either one 32-bit instruction or two 16-bit instructions. Furthermore, since the instruction set encoding is variable in length, it permits fetching misaligned instructions in various configurations, as elaborated in [1]. For instance, during a clock cycle, the first half of a 32-bit instruction may be fetched, with the second half retrieved in the subsequent clock cycle.

The *second device* is an STM32L4 microcontroller. It is a 32-bit microcontroller that embeds an Arm Cortex-M4 core. Arm Cortex-M4 has a 3-stage pipeline: fetch, decode, and execute. It has 13 general-purpose 32-bit registers, R0 to R12. Arm Cortex-M4 is based on ARMv7-M architecture and supports the Thumb2 instruction set, consisting of variable-length instructions: 16-bit and 32-bit instructions. The instruction has a 32-bit encoding if the most significant five bits of a 32-bit word have either 0b11101, 0b11110, or 0b11111 value. Otherwise, the most significant 16 bits belong to a 16-bit instruction. This Arm Cortex-M4 device supports cache lines of 64 bits. Therefore, in this case, the flash memory access size is 64-bit wide, allowing fetching misaligned instructions in different configurations as detailed in [10].

B. Target programs

Listing 1 shows the RISC-V target program instructions along with their encoding in hexadecimal format. All these instructions are 32-bit instructions. Thus, a complete 32-bit instruction will be fetched at a given clock cycle, and hence, the code in this case is aligned.

Listing 2 presents the Arm Cortex-M4 target program instructions along with their encoding in hexadecimal format. These instructions are the same as the instructions used for

RISC-V, but clearly, they have different syntax and encoding as the supported instruction set is Thumb2. Moreover, we added a dummy 16-bit instruction at the beginning of the target program: MOVS R0, R0, whose encoding is 0x0000. This is done to make the code misaligned and to showcase that the provided fault models in this work apply to the observed faulty behaviors regardless of the alignment in memory.

```
1 ADDI x28, x28, 0x3b   // 0x03be0e13
2 ADDI x29, x29, -0xc   // 0xff4e8e93
3 ADDI x7,  x29, 0x27   // 0x027e8393
4 OR   x6,  x28, x7     // 0x007e6333
5 XORI x6,  x6,  0xf    // 0x00f34313
6 ADDI x31, x31, 0xd    // 0x00df8f93
```
Listing 1. RISC-V target program with its encoding in hex. format

```
1 MOVS R0, R0      // 0x0000f103
2 ADD r3, r3, 0x3b   // 0x033bf1a4
3 SUB r4, r4, 0xc    // 0x040cf104
4 ADD r2, r4, 0x27   // 0x0227ea43
5 ORR r1, r3, r2     // 0x0102f081
6 EOR r1, r1, 0xf    // 0x010ff106
7 ADD r6, r6, 0xd    // 0x060d0000
```
Listing 2. Arm Cortex-M4 target program with its encoding in hex. format

These instructions serve as examples in this paper, as the presented fault models hold regardless of the target instructions. Moreover, they are selected for different reasons. First of all, any software application has arithmetic and logical instructions. Additionally, they streamline the process of characterizing the fault injection effects, allowing identifying potential faulty behaviors. Thus, the tractability of a faulty behavior and the applicability of a specific fault model are all achievable with high probability. Upon completion of the regular execution, each register holds a distinct value compared to the others, thereby enhancing the detectability of any faults that may occur.

During the experiments, the processors undergo a predetermined setup before each fault injection, achieved through initialization instructions positioned ahead of the target instructions. Subsequently, following each execution, the data stored in the general-purpose registers is transmitted to a control computer via serial communication for analysis of the outcomes.

C. Voltage glitch fault injection

Introducing variations to the power supply that feeds an embedded system is an effective and low-cost fault injection technique. This method provides acceptable controllability in terms of timing accuracy. However, determining which part of the system is affected by the injection can be challenging.

In this work, ChipWhisperer [14] environment has been employed to perform the voltage glitch fault injection. In this setup, the perturbation to the power supply (Vcc) is performed by underpowering to ground while configuring three parameters as shown in Fig. 2:

979-8-3503-6689-1/24 $31.00 © 2024 IEEE

- **Width:** the period of time in which the variation on the power supply is applied.
- **Shift:** the offset of the starting point of the glitch to the rising edge of the targeted clock cycle.
- **Delay:** the duration between the rising edge of a trigger signal and the rising edge of the targeted clock cycle.

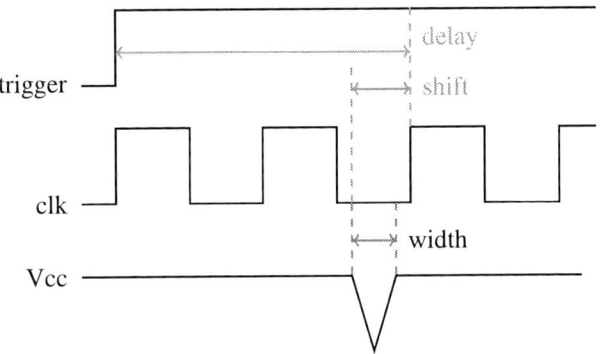

Fig. 2. Voltage glitch parameters

Table II shows the employed values for Shift and Width parameters. The values are expressed as a percentage of one clock period. The negative value for the shift means that the glitch is injected before the rising edge of the targeted clock cycle. Two different delay values are used for each injection campaign. The delay values depend on the number of initialization instructions before the target code and the start of the trigger signal within the code. For each combination of Shift, Width, and Delay, the experiments are repeated 20 times. Thus, the number of executions for each injection campaign is 20 000 (= *10 widths * 50 shifts * 2 delays * 20 repetitions*).

The Width values have been tuned to maximize the number of faults, as it has been observed that the probability to observe faults is much higher in this range. For Shift values, a wide range has been employed to make sure that most of the possible faults can be covered by the proposed models. Two different delay values has been used to target different locations within the target program, also to ensure the applicability of the proposed fault models. In summary, we wanted to ensure that our proposed fault models can cover the obtained faulty behaviours, regardless of the employed parameters. It should be mentioned that, by tuning the parameters, it was possible to maximize the observability of a specific faulty behavior, as we were able to do so.

TABLE II
PARAMETERS SETUP FOR BOTH DEVICES

Parameter	Values
Width	[40,49]
Shift	[-49,0]
repetitions	20
Total number of executions for each device	20 000

IV. EXPERIMENTAL RESULTS AND ANALYSIS

The outcome of a glitch injection in an execution leads to one of the following classes:
- **Silent:** the execution outcome is equivalent to a normal execution outcome without an injection.
- **Crash:** a reset or a crash occurs as a result of the injection.
- **Fault:** the execution outcome is different from the normal execution outcome.

The obtained classifications for both injection campaigns, *i.e.,* when targeting RISC-V device and Arm device are presented in Table III.

TABLE III
PERCENTAGE OF SILENT, CRASH & FAULT OVER THE TWO CAMPAIGNS.

Class	Target device	
	RISC-V	Arm Cortex-M4
Silent	96.71 %	92.405 %
Crash	0.005 %	0.925 %
Fault	3.285 %	6.67 %

The following subsections present the inferred fault models that allowed us to explain the observed faulty behaviours, along with examples for observed experimental results. Table IV shows the percentage of the observed faulty behaviors concerning each fault model over all obtained faulty behaviors for each target device. *Other* means that the observed faulty behavior cannot be modeled by the presented fault models. However, it is shown that *Other* only counts for less than 1 %.

TABLE IV
PERCENTAGE OF THE CLASSIFICATION OF THE OBSERVED FAULTS UNDER THE INFERRED FAULT MODELS FOR EACH TARGET DEVICE.

Fault model	Target device	
	RISC-V	Arm Cortex-M4
Skip	13.85 %	95.2 %
Skip & repeat	36.23 %	0.075 %
Non-sequential skip & repeat	0 %	0.45 %
Partial update	0.3 %	4.2 %
Skip with forwarding	1.98 %	0 %
Combination	47.03 %	0 %
Other	0.61 %	0.075 %

A. Skip n bits

A block of size *n bits* of instructions encoding data is skipped and the execution resumes from the next block. This *n* bits is related either to flash memory access size, cache line size, or internal register size. If the code is aligned, then the skipped block refers to complete instructions. However, the skipped block corresponds to different possibilities when the code is misaligned as explained in [10]. *Skip 32 bits* and *Skip 64 bits* are observed for Arm target device, while only *Skip 32 bits* is observed for RISC-V target.

979-8-3503-6689-1/24 $31.00 © 2024 IEEE

For RISC-V campaign and referring to Listing 1, this model led to a complete instruction skip, as the code is aligned and each line corresponds to a complete 32-bit instruction. Conversely, this model led to different faulty behaviors depending on the target lines in Listing 2.

For Arm target, an observed example of *Skip 64 bits* is shown in Listing 3. In this example, the first two lines in Listing 2 are skipped. As a result, the remaining half `0x040c` of `0xf1a4040c` instruction is executed as a *new* 16-bit instruction: `LSLS r4, r1, 0x10`.

```
1  LSLS r4, r1, 0x10   // 0x040c
2  ADD r2, r4, 0x27    // 0x0227ea43
3  ORR r1, r3, r2      // 0x0102f081
4  EOR r1, r1, 0xf     // 0x010ff106
5  ADD r6, r6, 0xd     // 0x060d0000
```

Listing 3. Observed execution as a result of skipping lines 1 and 2 (64 bits) in Listing 2

B. Skip & repeat n bits

A block of size *n bits* of instructions encoding data is skipped and the previous block with the same size has been repeated. As in the previous model, this *n* bits is related either to flash memory access size, cache line size, or internal register size. If the code is aligned, then the skipped and the repeated blocks refer to complete instructions. Yet, they correspond to different possibilities when the code is misaligned as in [10].

In this paper, *Skip & repeat 64 bits* is observed for Arm target, while *Skip & repeat 32 bits* is observed for RISC-V target. This is due to the flash memory access size as mentioned earlier.

For RISC-V campaign and referring to Listing 1, this model led to skipping a complete instruction and repeating the previous instruction.

For Arm campaign, the observed execution depends on the affected part of the code. An observed example is when skipping the encoding at lines 3 and 4 in Listing 2, and repeating the encoding at lines 1 and 2. Listing 4 shows the resulting execution of this example. In this example, `SUB r0, r4,0x0 (0xf1a40000)` has been executed instead of `SUB r4, r4, 0xc(0xf1a4040c)`. Additionally, `ADD r3, r3, 0x3b` has been repeated and `ADD r2, r4, 0x2b` has been skipped. Finally, `SUB r1, r4, 0x2(0xf1a40102)` has been executed instead of `ORR r1, r3, r2(0xea430102)`.

```
1  MOVS R0, R0         // 0x0000f103
2  ADD r3, r3, 0x3b    // 0x033bf1a4
3  SUB r0, r4,0x0      // 0x0000f103
4  ADD r3, r3, 0x3b    // 0x033bf1a4
5  SUB r1, r4, 0x2     // 0x0102f081
6  EOR r1, r1, 0xf     // 0x010ff106
7  ADD r6, r6, 0xd     // 0x060d0000
```

Listing 4. Observed execution as a result of skipping the encoding at lines 3 and 4 (64 bits) and repeating the encoding at lines 1 and 2 (64 bits) in Listing 2

C. Non-sequential skip & repeat 32 bits

This model is only observed for the Arm target. Referring to Listing 2, this model is defined as follow: 32 bits at line *i+2* are skipped, while the 32 bits at line *i* are repeated. Such faulty behavior may occur because of the supported cache lines. A cache line of size 64 bits can be seen as two chunks of 32 bits. Thus, each chunk may get its update separately from the other one at a given clock cycle. Consequently, an occurred fault prevents a chunk from being updated, resulting in repeating 32 bits and skipping another 32 bits.

An observed example of this is when skipping line 3 in Listing 2, and repeating line 1. The resulting execution is demonstrated in Listing 5. It is shown how the misalignment has a significant impact in this case. Nevertheless, the fault model works regardless of the alignment.

```
1  MOVS R0, R0         // 0x0000f103
2  ADD r3, r3, 0x3b    // 0x033bf1a4
3  SUB r0, r4, 0x0     // 0x0000f103
4  ADD r2, r3, 0x27    // 0x0227ea43
5  ORR r1, r3, r2      // 0x0102f081
6  EOR r1, r1, 0xf     // 0x010ff106
7  ADD r6, r6, 0xd     // 0x060d0000
```

Listing 5. Observed execution as a result of skipping the encoding at line 3 and repeating the encoding at line 1 in Listing 2

D. Partial update

While the instructions data are propagated through the microcontroller from flash memory to the pipeline stages, the injected glitch may affect some bits to be updated improbably in internal registers. As a result, the faulty bits get their values either from the *previous value*, the *precharge value*, or even from the *next* value. Update from next is only observed for RISC-V target, and in many cases, it is occurred over 16 bits, especially when the code is misaligned. Section IV-F presents an example for this model as a part of *Combination* on RISC-V. For examples on Arm target, we refer the reader to [11].

E. Skip with forwarding

This model is only observed for RISC-V target device. In this case, an instruction is skipped, however, its expected resulting value is forwarded to be used in another instruction. This can be explained as the instruction has finished its execute pipeline-stage, and the value has been forwarded, however, it has not been written back to the register file, thus, the corresponding register keeps its previous value. In [12], the authors noticed, using *only RTL fault simulation* on a RISC-V core, that bit flips could lead to forwarding faults, leading in certain scenarios, to break control flow integrity of programs.

An observed example for this model is when `ADDI x28, x28, 0x3b` instruction at line 1 in Listing 1 has been skipped, but the correct value of **x28** has been forwarded to be used in `OR x6, x28, x7`. As a result, **x6** had its golden value, however, **x28** kept its initial value.

It has been noticed that a value can be forwarded from an instruction at line i to dependent instructions at lines $i+1$, $i+2$, or $i+3$, but not further. This can be explained as the value can be forwarded, at max, from writeback pipeline-stage (5th stage) to decode and register fetch stage (2nd stage).

F. Combination

In this case, the observed faulty behavior is modeled by a combination of more than one fault model from the afore-mentioned fault models. This *Combination* fault model has been observed only for RISC-V target device. This might be explained due to the higher number of pipeline stages (5 stages) compared to Arm device (3 stages).

An observed example of this behavior is depicted in Listing 6. In this example, *Partial update* from previous value has occurred at line 3, so that **x29** got the result of: x29+0x27, instead of **x7**. However, due to forwarding, the correct value of **x7** has been correctly used for OR at line 4, but finally **x7** kept its initial value. Furthermore, XORI has been skipped.

```
1 ADDI x28, x28, 0x3b     // 0x03be0e13
2 ADDI x29, x29, -0xc     // 0xff4e8e93
3 ADDI x29, x29, 0x27     // 0x027e8e93
4 OR   x6, x28, x7        // 0x007e6333
5 XORI x6, x6, 0xf
6 ADDI x31, x31, 0xd      // 0x00df8f93
```

Listing 6. Observed execution example on RISC-V target for *Combination* fault model (*Partial update, forwarding, Skip*)

It might be noticed that predicting faulty behaviors, based on *Combination* and *Partial update* models, could be difficult. However, we have observed repetitive and reproducible patterns that show that some behaviors are more probable than others. This might be related to *Bit sensitivities* mentioned in [11]. Further investigation is required to confirm this.

V. CONCLUSION AND FUTURE WORK

In conclusion, voltage glitch campaigns have been performed on two different devices, embedding RISC-V and Arm Cortex-M4 cores. In order to explain the obtained faulty behaviors, different fault models have been inferred. These models allowed explaining more than **99%** of the obtained faults. Some of these models are applicable to both target devices: *Skip, Skip & repeat*, and *Partial update* fault models. These three models are also applicable to clock glitch results, as shown in the literature. Moreover, additional faulty behaviors have occurred due to some features. These features encompass variable-length encoding, cache lines, and forwarding. Therefore, new device features could lead to new possible exploitation. To deal with this, other fault models have been proposed: *Non-sequential skip & repeat, Skip with forwarding*, and *Combination*. The presented fault models enable better understanding of the fault effects. Thus, easing the process of vulnerability analyses, and hence, simplifying the thinking of cost-effective countermeasures.

As future work perspectives, countermeasure design at software and/or hardware levels would be very important. Also,

evaluating real-life security applications using the proposed fault models would be captivating. Finally, another interesting perspective is a deep investigation of the observed faulty behaviors at hardware level. This will help in determining the vulnerable registers, and thus, easing the design of protections.

ACKNOWLEDGMENT

This work has been partially supported by the LabEx PERSYVAL-Lab (ANR-11-LABX-0025-01) funded by the French program Investissements d'avenir, and by ARSENE project (PEPR PP7 ARSENE — ANR-22-PECY-0004).

REFERENCES

[1] I. Alshaer, B. Colombier, C. Deleuze, V. Beroulle, and P. Maistri, "Variable-length instruction set: Feature or bug?" in *25th Euromicro Conference on Digital System Design*. Maspalomas, Spain: IEEE, Aug. 2022, pp. 464–471.

[2] N. Timmers, A. Spruyt, and M. Witteman, "Controlling pc on arm using fault injection," in *2016 Workshop on Fault Diagnosis and Tolerance in Cryptography (FDTC)*, 2016, pp. 25–35.

[3] N. Moro, A. Dehbaoui, K. Heydemann, B. Robisson, and E. Encrenaz, "Electromagnetic fault injection: Towards a fault model on a 32-bit microcontroller," in *2013 Workshop on Fault Diagnosis and Tolerance in Cryptography, Los Alamitos, CA, USA, August 20, 2013*. IEEE Computer Society, 2013, pp. 77–88.

[4] V. Khuat, J.-L. Danger, and J.-M. Dutertre, "Laser fault injection in a 32-bit microcontroller: from the flash interface to the execution pipeline," in *2021 Workshop on Fault Detection and Tolerance in Cryptography (FDTC)*, 2021, pp. 74–85.

[5] O. Trabelsi, L. Sauvage, and J.-L. Danger, "Characterization of electro-magnetic fault injection on a 32-bit microcontroller instruction buffer," in *2020 Asian Hardware Oriented Security and Trust Symposium (Asian-HOST)*, 2020, pp. 1–6.

[6] J. Proy, K. Heydemann, A. Berzati, F. Majéric, and A. Cohen, "A first ISA-level characterization of EM pulse effects on superscalar microarchitectures: A secure software perspective," in *Proceedings of the 14th International Conference on Availability, Reliability and Security, ARES 2019, Canterbury, UK, August 26-29, 2019*. ACM, 2019, pp. 7:1–7:10.

[7] T. Trouchkine, G. Bouffard, and J. Clédière, "EM fault model character-ization on socs: From different architectures to the same fault model," in *2021 Workshop on Fault Detection and Tolerance in Cryptography (FDTC)*. IEEE, 2021, pp. 31–38.

[8] V. Werner, L. Maingault, and M. Potet, "An end-to-end approach for multi-fault attack vulnerability assessment," in *Workshop on Fault Detection and Tolerance in Cryptography*. Milan, Italy: IEEE, 2020, pp. 10–17.

[9] M. A. Elmohr, H. Liao, and C. H. Gebotys, "EM fault injection on ARM and RISC-V," in *2020 21st International Symposium on Quality Electronic Design (ISQED)*, 2020, pp. 206–212.

[10] I. Alshaer, G. Burghoorn, B. Colombier, C. Deleuze, V. Beroulle, and P. Maistri, "Cross-layer analysis of clock glitch fault injection while fetching variable-length instructions," *Journal of Cryptographic Engineering*, pp. 1–18, 2024.

[11] I. Alshaer, B. Colombier, C. Deleuze, V. Beroulle, and P. Maistri, "Mi-croarchitectural insights into unexplained behaviors under clock glitch fault injection," in *Smart Card Research and Advanced Applications*, ser. Lecture Notes in Computer Science. Springer Nature Switzerland, 2024, pp. 3–22.

[12] J. Laurent, C. Deleuze, F. Pebay-Peyroula, and V. Beroulle, "Bridging the gap between RTL and software fault injection," *ACM J. Emerg. Technol. Comput. Syst.*, vol. 17, no. 3, pp. 38:1–38:24, 2021.

[13] S. Inc. and B. EECS Department, University of California, "The RISC-V Instruction Set Manual Volume I: Unprivileged ISA," https://riscv.org/wp-content/uploads/2019/12/riscv-spec-20191213.pdf, [Accessed: February 16, 2024].

[14] C. O'Flynn and Z. D. Chen, "Chipwhisperer: An open-source platform for hardware embedded security research," in *International Workshop on Constructive Side-Channel Analysis and Secure Design*, ser. Lecture Notes in Computer Science, E. Prouff, Ed., vol. 8622. Paris, France: Springer, 2014, pp. 243–260.

Image Degradation due to Interacting Hot Pixels and SEUs

Glenn H. Chapman, Alireza Farahmandpour,
Amit Chakma
School of Engineering Science
Simon Fraser University
Burnaby, B.C., Canada, V5A 1S6
glennc@cs.sfu.ca , alireza_farahmandpour@sfu.ca ,
amit.chakma@stud.hshl.de

Israel Koren, Zahava Koren
Dept. of Electrical and Computer Engineering
University of Massachusetts
Amherst, MA, 01003
koren@ecs.umass.edu

Abstract— **Hot pixels (induced by cosmic rays) in digital imaging sensors accumulate with camera age, impacting the quality of all images produced by the camera. During its lifetime, the imager also experiences Single Event Upsets (SEUs) that create transient defects affecting a single image. The SEUs occur at a much higher rate (events/second) compared to 10-100s additional permanent hot pixels that occur per year. We explore in this paper how an SEU that occurs within a close proximity (within a 5x5 pixel area) of an existing Hot Pixel may distort the image. We also consider situations where an SEU impacts two or more adjacent pixels. When such events happen, the currently employed color demosaicing and JPEG image compression algorithms spread the damage into a 16x16 pixel area, creating significant color changes and resulting in noticeable image degradation. We use formulas developed for a two Hot Pixel interaction to estimate the probability of a Hot Pixel-SEU interaction, and predict its increase with camera age.**

Keywords: imager hot pixel defect growth, active pixel sensors, Single Event Upsets, imager degradation with time, interactions of Hot pixel and SEUs, adjacent defect interactions.

I. INTRODUCTION

The trend of digital camera manufacturers, for both stand alone cameras and embedded imagers in other products, is to shrink pixel size and increase Megapixel count in order to attract users, shrink camera areas and increase apparent resolution. However, as we have shown in the past [1], as pixel sizes shrink, cameras experience a substantial increase in the development of defects over time. Defects generated by cosmic rays induce two types of in-field errors: First, permanent "hot pixel" defects (created by high energy particles), which accumulate continuously over the lifetime of the sensor. Second, SEUs which result from lower energy cosmic particles and are transitory in nature. Importantly, unlike regular ICs where a single permanent fault may end the circuit's life, imagers have the advantage that the appearance of defects (hot pixels or SEUs) only degrades the image quality. It is often assumed that with current cameras the defects accumulation rate is so small that the damage will not be noticeable. However, we have previously shown that within 1-2 years, the accumulating permanent hot pixels (at current pixel sizes) may result in interacting adjacent defects (within a 5x5 pixel area). When this occurs, the currently

used image manipulation algorithms create a noticeable image degradation. In this work we show that the transitory SEUs may interact with the already present hot pixels causing further image degradation. Furthermore, some SEUs hit two or more adjacent pixels resulting in a similar degradation. As SEUs occur at substantially higher rates than the permanent hot pixels, their impact on image quality may be considerable. This indicates that the low energy SEUs may cause a more noticeable image damage than was previously assumed.

In our previous work [1-6] we demonstrated that Hot Pixels (HPs) (producing a signal independent of their light exposure, see Fig. 1) develop as a camera ages. Since they are generated by cosmic ray hits [1-3,7,8], they cannot be prevented. We have developed an empirical power law formula estimating the HP defect density rate D (defects per year per mm² of sensor area) as a function of the pixel size S (in microns) raised to the inverse ~1/3 power and the square root of sensor gain (ISO) [9]. The important point is that when pixel sizes decrease by a factor of 2, the defect density D grows by 8X, while doubling of ISO (gain) increases D by 1.4X. Yet for practical camera applications, shrinking pixels and higher sensitivity (ISO) are the current trends.

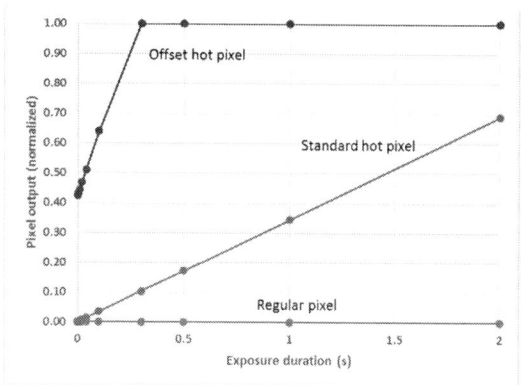

Figure 1: Comparing the dark response of imager pixels: a good regular pixel, a standard hot pixel, and an offset hot pixel.

Although the current literature suggests that even hundreds of defects among megapixels are almost undetectable, we have shown in [16] that accumulation of only 90-200 hot pixels over time can cause significant (noticeable) image degradation. Analysis and experiments have shown that when two defective pixels are sufficiently close (within a 5x5 pixel area), the color demosaicing and JPEG compression algorithms that cameras use to produce the final image, generate interactions between the two hot pixels thus spreading the image damage to a much larger area, which tends to be quite visible. We developed empirical formulas showing that it takes a surprisingly small number of hot pixels to cause noticeable damage for megapixel cameras, which occurs in about 1-2 years.

In addition, we have shown that besides the high energy cosmic rays which create permanent HP defects, the much larger number of low energy particles create transient, bright pixels that appear in a single picture but cause no lasting damage. These are called *soft errors* or *SEUs*. Prior studies have treated these two effects separately. What we explore in this paper is the combined impact of HPs and SEUs. The background level of permanent HPs creates a matrix of defects with which the SEUs interact via the demosaicing and JPEG compression processes. Moreover, while the HP defect growth rate D adds about 20-50 HPs per year, the SEU event rate is 0.2-0.7 per second, or 15 million times greater. When the number of accumulated HP defects is sufficient to generate a 4% probability of two interacting hot pixels, we can expect a significant chance of SEU hits occurring close enough to an existing HP and thus creating a large area damage in a single image. Furthermore, about 2-5% of SEUs hit several adjacent pixels, thus further degrading the image quality.

II. Hot Pixels

Hot pixel defects generate a non-zero output even under no incoming illumination (i.e., dark field) and this output increases with the exposure time T_e as shown in Figure 1 (output 1 is saturation). Undamaged pixels output (under no illumination) grows only with thermal noise. There are two types of hot pixels [5]: standard HPs and offset HPs that have a dark current R_{dark} (photodiode junction leakage current). For both types, the output current grows linearly with T_e. The output of the offset HP starts at a value $a \times b$ (where a is the amplification from the ISO setting) and is more likely to reach saturation. Under no illumination, the output of an HP is given by

$$I_{hp} = a * [R_{dark}T_e + b] \qquad (1)$$

Generated by the bombardment of cosmic rays, HP defects are randomly distributed over the imager [1-6,10]. We have monitored for many years the HP generation process for 32 cameras ranging from 4-7 µm pixels (DSLRs/Mirrorless) with large sized sensors (864 and 364mm^2) down to cell phones with small 24 mm^2 imagers and 1.34 µm pixels, for ISOs (gains) from 100 to 102,400. Based on these measurements, we have developed in [9,11] the following empirical power law formula to estimate the defect density rate D (defects per year per mm^2 of sensor area) as a function of the pixel size S (in µm) and sensor gain (ISO):

$$D = 10^{-0.98}S^{-3.03}ISO^{0.506} \qquad (2)$$

This expression indicates that the defect density increases drastically when the pixel size goes below 2µm and reaches 24 defects/year×mm^2 at ISO 102,400 (already used in many high-end cameras). This has important implications as the current trend in cameras is to shrink the pixel size (4 µm in DSLRs and 1.2µm in cell phones) to increase the megapixel resolution. Therefore, cosmic ray generated defects may present important limits to camera system lifetimes.

As we have shown previously, the important effect of the defect rate acceleration is only seen by injecting the defect in the camera raw image data before it is processed and compressed. The hot pixel's damage spreads by two important camera processing algorithms operating on the defect before the image is displayed: color demosaicing and JPEG compression (see Figure 2). As we have shown in [13-16], Color Demosaicing generates the Red (R), Green (G) and Blue (B) values for each pixel using interpolations from its neighbors, while the compression algorithm in JPEG involves a still larger 8x8 pixel area when creating the final image. The impact of these is to spread the damage of a defect from a single point to several neighboring pixels, see Fig. 2 (see [5,16] for the detailed analysis).

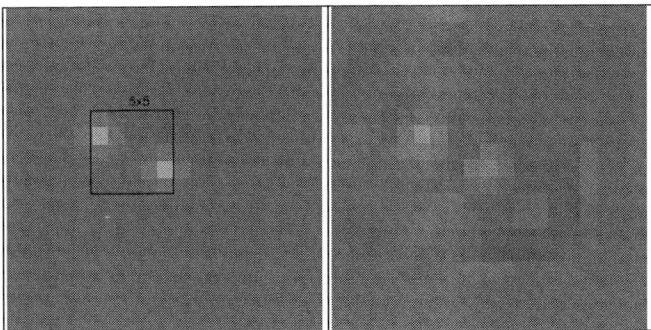

Figure 2: Impact of 2 hot pixels within 5x5 square (left) after demosaicing (right) after adding JPEG (median)

III. Two Hot pixels interaction Probability

In [13,16] we have shown that two hot pixels within a 5x5 square (Fig. 2) create up to a 16x16 disturbed area when we combine the demosaicing algorithms and JPEG compression. Using Monte Carlo simulation, we developed a formula to predict the probability of this occurring, using as a start the solution to the birthday problem that asks: Given n people in a room, what is the probability P_{BD} of at least two people having the same birthday within an m=365 day year. Using the generalized formula [12], if n HPs occur on a sensor with m 5x5 pixel squares (m=total number of pixels/25) then the probability P_{BD} of having at least two HPs occurring in a 5x5 square is:

$$P_{BD} = 1 - exp\left[\frac{-(n-1)n}{2m}\right] \qquad (3)$$

The Monte Carlo (MC) simulations in [16] showed the need to modify the above probability to reflect that there are many ways that neighboring 5x5 pixel squares can create interacting pixel situation. Fortunately, the actual probability P_{2hp}, derived from the MC simulations, exhibits a simple second order relation with the probability P_{BD}

$$\frac{P_{2hp}}{P_{BD}}(P_{BD}) = 4.651 - 7.5095 * P_{BD} + 4.11 * P_{BD}^2 \qquad (4)$$

Therefore, to calculate the number n of hot pixels for a given probability, we first calculate P_{BD} from (3) and then use (4) to get P_{2hp}. Solving for the number n of hot pixels needed for a given m and probability P_{BD} yields (for large m)

$$n = \sqrt{-2m * ln(1 - P_{BD})} + 0.5 \tag{5}$$

Therefore, we can compute, for a given sensor size, the number of hot pixels which will cause a certain percentage of cameras to experience the problem of two hot pixels within a 5x5 square. For example, for a 10 Mpix camera with a 4.3 µm pixels and 864 mm² size (mid range DSLR camera), to get a P_{2hp} =4.57% (1 in 22 cameras) we need a P_{BD}=1% from (4) which occurs, from equation (5), with only 90 defects, or only 0.0009% of the pixels. Doubling the probability to 8.9% (1 in 11 cameras) only requires an increase in the number of defects by ~1.414 in equation (5), i.e., only 128 hot pixels.

IV. SEUs IN DIGITAL IMAGERS

Digital camera sensors allow us to investigate SEUs behavior by using dark-frame (no illumination) photographs, where SEUs can be identified in both location and intensity. In dark frames, an SEU appears as a transient bright pixel that shows up in a single image (within a series of images), but not in the other images, and causes no lasting damage [2,3]. As noted in the literature [3,4,5] all these events appear to be caused by cosmic ray particles striking the sensor at random times and locations. The HP defects are caused by the strongest of the particles. SEUs create 3 types of defects: Most (95%) are single pixel events (see Fig. 3(a)) and the rest are either small adjacent clusters or long streaks (Fig. 3(b)). For radiation type events, a common assumption is that their number follows a Poisson process, where λ denotes the event rate (per sec×cm².) and λt is the expected number of events in t seconds per cm². Our measurements [17,18] have shown typical values of λ between 0.1 and 0.2 SEUs/sec×cm² which for a 862 mm² camera sensor amounts to 0.8-1.7 events per second.

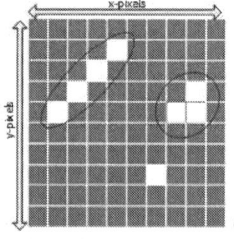 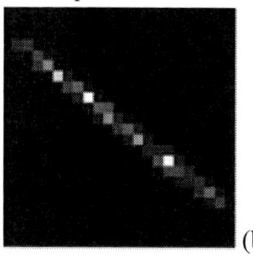

Figure 3: (a) SEU types – simple SEU spot (bottom) and SEU streak (top left), and cluster (top right) (b) 23x23 pixels of 6.5 µm (raw pixel image)

V. HOT PIXEL AND SEU INTERACTIONS

When considering HP:SEU interactions it is important to note that when determining the impact on the resulting image, it does not matter whether the defects are HPs or SEUs, only their position in the imager array and intensity are relevant. Let us analyze first the effect of a single pixel SEU interacting with a hot pixel. Consider an imager that already has n HPs and then an SEU hit occurs. The probability of any two interacting damaged pixels (either two HPs or an HP and an SEU) can be calculated using equations (3)-(5) by replacing n with $n+1$. Therefore, the probability of an HP:SEU interaction is the difference between the interaction probability of $n+1$ damaged pixels and that of n damaged pixels, i.e.,

$$P_{HP:SEU}(n) = \left\{ exp\left[\frac{-n(n-1)}{2m}\right] - exp\left[\frac{-(n+1)n}{2m}\right] \right\} \frac{P_{2hp}}{P_{BD}} \tag{6}$$

Due to the high rate of SEUs, many of them will occur while the number of HPs will remain **n**. Denote by w the number of such SEUs. Each SEU hit is independent of the others, so we can use the Binomial distribution. Thus, the probability of at least one HP:SEU interaction when w SEU hits occur is

$$P_{HP:SEU,w} = 1 - [1 - P_{HP:SEU}(n)]^w \tag{7}$$

Let us consider a test case of a 10 megapixel sensor for a digital DSLR of size 862 mm², like those we have tested for two HP interactions in the past [16]. We would expect that at a modest gain (ISO 1600) we would see between 90-127 HPs after two years, and up to 200 HPs after four years. From equations (3)-(5) for 10 Mpix, m is 4×10^5 and for 90 HPs we would get a 4.5% (1 in 25) probability of having a two HP interaction. Figure 4 shows the probability of an HP:SEU interaction as a function of the number n of hot pixels already present, for $w=1$.

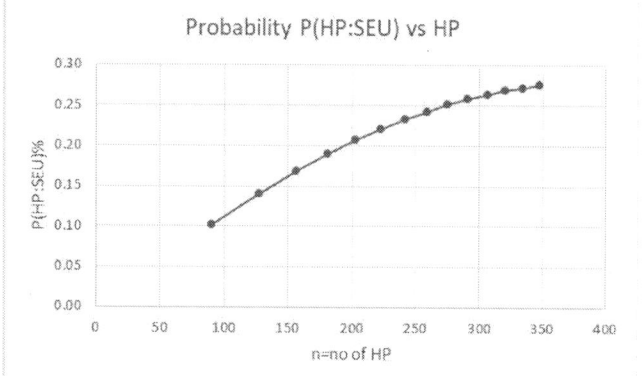

Figure 4: Probability (%) of a HP:SEU interaction vs the number n of HPs for $w=1$

Figure 5 shows how the probability of an HP:SEU interaction varies with the number w of SEUs for several values of n (the number of existing HPs). Note that we would expect that one SEU will occur almost every second.

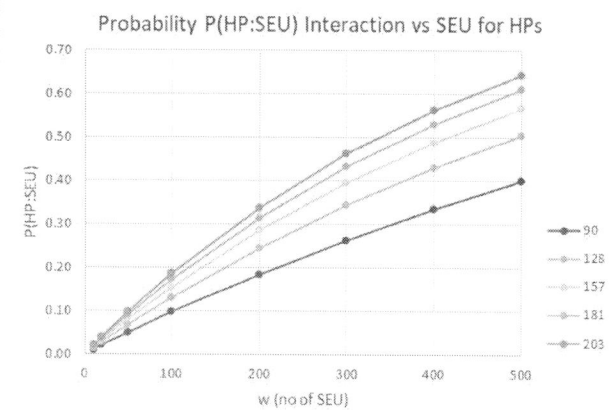

Figure 5: Probability of an HP:SEU interaction as a function of the number w of SEUs for different values of n (the number of HPs)

From Figure 5 we can estimate the cumulative exposure time of images that is needed for HP:SEU interactions to occur. Note that for an HP:SEU interaction to occur (given n existing HPs) it does not matter if the sensor already has two interacting HPs or not, the SEU interaction probability stays the same.

The above estimated HP:SEU interaction probability is just one factor in estimating the impact of SEUs as it only looks at the impact of a cosmic ray hitting a single pixel. As we noted in section IV, about 5% of SEUs are streaks or clusters (see Fig. 3). Small clusters often involve two or more adjacent bright pixels, Inherently, the demosaicing and JPEG steps will extend that into a large defect in the image as per Fig. 2. Streaks are even worse as they involve a line of bright spots, and thus the camera algorithms will spread the damage to a much larger area. In our experiments we have identified [17,18] many streaks that are more than 10 pixels long. Streaks that are at an angle to the pixel array row/column (see Fig 3(b)) will create large area defects in the final image. Our measurements show that the fraction of SEUs that are streaks or clusters is f_{streak}= 0.02-0.05.

From the point of view of the SEU and HP interactions, we can, as a first approximation, treat streaks/clusters just like all other SEUs. Thus, we can use the probability formula of equation (6) but the interaction distance may not be the 5x5 square that was used for two HPs.

Thus, for a sensor with HPs and SEUs we need to add to equation (7) the probability of the streak/cluster generated defects (i.e., the streak fraction f_{streak}) and subtract the probability of both types of interactions yielding

$$P_{HP:SEU,w} = \{1 - [1 - P_{HP:SEU}(n)]^w\} + f_{streak} \\ - \{1 - [1 - P_{HP:SEU}(n)]^w\} \times f_{streak}$$
(8)

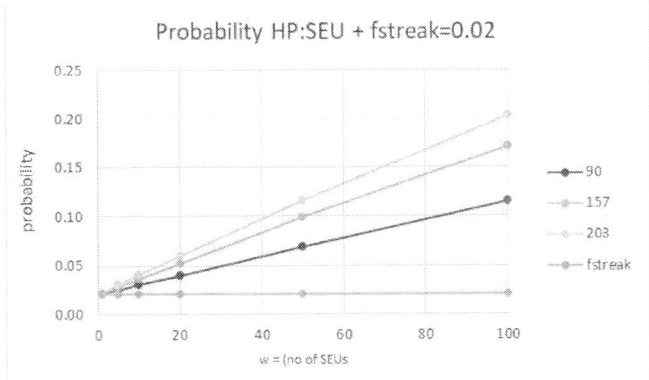

Figure 6: Probability of HP:SEU interactions + SEU streaks creating image defects vs the number of SEU events (w) for various HP levels

Figure 6 shows that for the current measured SEU rates and streak fractions, the streak/cluster effects almost overwhelm the HP:SEU interactions for our 10 Mpix camera and for low values of w. First note that the f_{streak} line, which is the probability contribution due to the streaks, is constant for all values of w (number of SEUs) as indicated in equation (8). Initially the streak/cluster effects dominate, forming a background defect level independent of the number of HPs. As w increases, the contribution of the HP:SEU interactions starts to dominate. The curve in Figure 5, which shows only the contribution of the HP:SEU interactions, looks very similar to the curve in Figure 6 for $w>50$. At about $w=10$ the probability of the HP:SEU interactions equals that of the streaks. For larger values of n (the number of HPs), the contribution of the HP:SEU interactions starts to dominate earlier and greatly increases the probability of image degradation.

The growth in hot pixels increases the probability of two HP interactions as shown in Figure 7. For a given imaging sensor, as the camera ages, the number of HPs increases roughly linearly. However, the probability of two HP interactions grows more rapidly (Fig. 7). Note that what Figure 7 shows is the fraction of cameras that from that point in time forward have every picture showing a defect from the two HP interaction. For the 10 Mpix camera example used, the n=90 HPs point is reached at about 2 years.

Equation (8) and Figure 6 show that at the same point in time, the probability of HP:SEU interactions grows significantly above the background level provided by the streaks (a few %) to the same probability as the 2 HP defects.

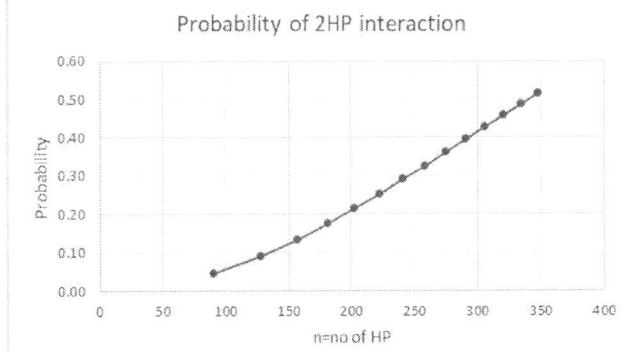

Figure 7: Probability of two HP interactions vs the number n of HPs (10 Mpix imager, 862 mm²)

Consider now the practical implications of the above. As cameras age and the HPs accumulate linearly, a growing fraction of the imagers that have two HP interactions will create one large defect in all images from that point on. In our example, that occurs for about 4.5% of imagers at the two years point for typical gain (ISO) level images. As time goes on, that fraction grows more rapidly as per Fig. 7.

Even cameras that do not have two HP interactions are not free of defects. In our example, those 95.5% of cameras are subject to many SEUs causing a growing probability of having one image in a series showing a large degraded area. In the example. at about 50 SEUs, there is a 10% chance that one image will show a large area defect, and in 100 SEUs a 15% chance. As the cameras age further, gaining more HPs, two things happen. First. the fraction of cameras exhibiting permanent defects grows more rapidly. To reach n=180 HPs it takes double the time (4 years), but the fraction of cameras with two HP interactions rises to 17.4%. Moreover, the remaining 82.6% of cameras have now increased their occurrence of HP:SEU interactions to 20% of images for 100 SEUs. Note that cameras with two HP interactions also experience HP:SEU interactions at that same rate, and the produced images now have two large area defects.

We wish now to estimate the impact that the above has on the total number of images taken by the camera. This depends on the exposure time T_e used by the camera for the images it is collecting. Denote by N_d the number of images taken until the first SEU. Given the SEU rate λ (in sec⁻¹cm⁻²), the sensor area A (in cm²) and the exposure time T_e (in sec) we obtain

$$N_d = \frac{1}{\lambda A T_e} \qquad (9)$$

For a DSLR with A=8.6 cm², λ=0.01 and typical exposure times (say 1/50 of a second), the result would not be initially worrisome, for 50 SEUs it would be a 10% chance of a defect in 2500 pictures. However, for long exposure (e.g., 1 sec) night shots, it would be 10% in only 50 pictures. Moreover, for video, roughly 50% of each second of "film" is camera exposure time. Hence, a two second video run would experience one defective image with a 10% chance. Note that the 10 Mpix camera is close to what 4K video uses. However, at twice the camera age, the HP count doubles, and the HP:SEU interaction probability rises to 15% and the number of cameras showing defects in all images increases to 17% .

VI. STREAK DAMAGE

Initially we assumed that streaks would behave similarly to single pixel SEU's in creating damage when interacting with HPs but did not take into account the fact that streaks are large. We have then done more simulations of streaks interacting with HPs and found that the image damage is much more extensive and the probabilities of interactions are considerably higher. To this end, we simulated streaks of various lengths (5 to 15 pixels) at various distances from a hot pixel and at various angles to the color array rows ranging from 0° (horizontal/vertical) to 30°-45°

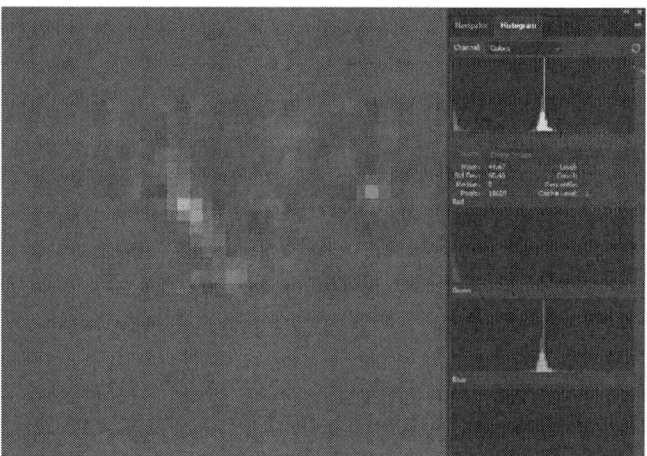

Figure 8: Streak length of 5 pix at 45° with single HP at a distance of dx=15 from the streak center on 50% green background with demosaicing and Jpeg damaging 215 pixels

Figure 8 shows a length 5 streak at 45° on a green background (50%) which is similar to the picture in Fig. 2 with two HP at 5 pixels apart. Note that due to the JPEG compression, the point where the distance at which a damage is created expands from just 5 pixels apart to about 15 pixels from the streak center. This creates a defect area of 215 pixels, more than twice that of two adjacent HPs and a much larger color spread. If the hot pixel is adjacent to either the streak beginning or end it does not have to be very close to create damage, it is sufficient to be at about 10-12 pixels away from the streak body. Other angles in the range of 10-80 degrees give similar results. However, if the streak is in line with the pixel rows or columns the interaction distance reduces to 10 like for streak ends. Changes in angles do generate different color distributions

because the streak passes through all 3 colors at every angle except 45° and 0/90°. For a particular 45° diagonal, where the streak passes through green pixels only, the damage is reduced, but this is a low probability event.

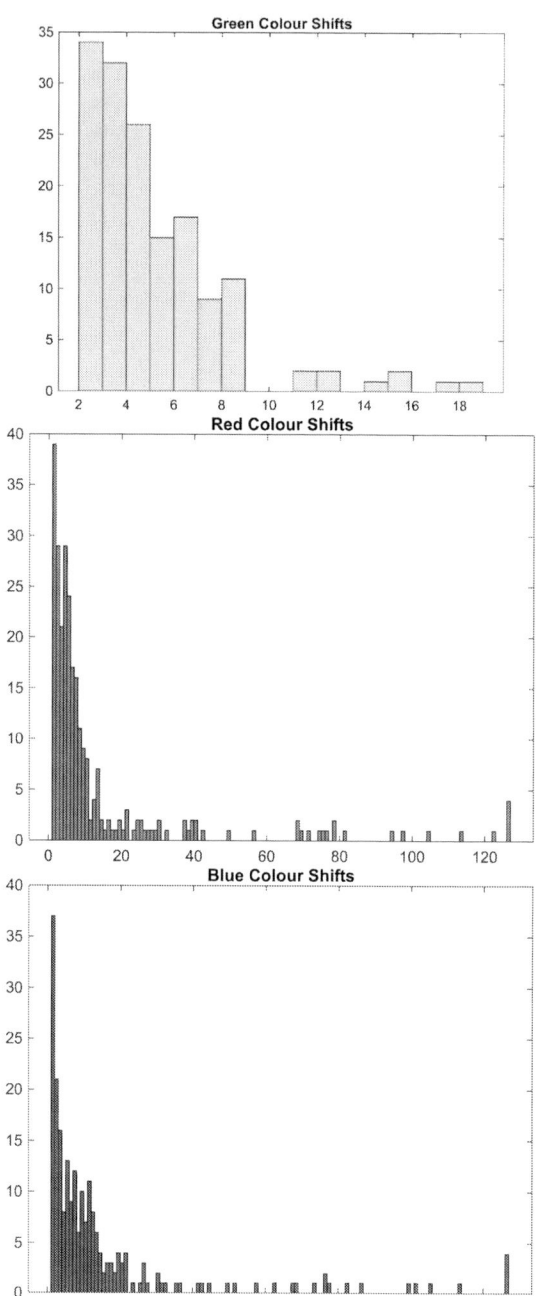

Figure 9: Color shift for 5 pix Streak at 45° angle interacting with a single HP at a distance of dx=15 from the streak center on a 50% green background with demosaicing and Jpeg damaging: Green shist is absolute change from background (125 value), Red/Blue shifts are from 0.

In our analysis we have only considered interactions with a single SEU or a streak. Using the HP SEU formula (8) but with an interaction space of 15 (as we found for streaks in a 10 Mpix camera with 90 HPs) the probability of a streak interacting with a HP is 0.75% while at 157 HP it rises to 0.85%. A preliminary Monte Carlo simulation with streaks at many angles and lengths produced similar probabilities.

The above shows that the damage due to a streak is much larger than that due to two adjacent hot pixels. Previously [13] we found that human observers could immediately identify and locate two interacting hot pixels in complex random images due to the distortion in the local scene. It is also notable that the higher color shift in the affected area resulted in a higher detectability of the image distortion by human observers and streaks inherently involve more color shifts and intensity because they excite more pixels. Figure 9 shows the color shift for G, R, B from the background of 50% green for the streak+HP of Figure 8. We did not include those within only a small shift (2 or less) from the background, and for green shift we use the absolute value of the change. Fig. 9 indicates that the majority of the color changes are significantly larger than the minimal value of 2 which is at the edge of detectability. We plan to explore in our future work possible image damage metrics that will depend on the relative color and intensity changes of the affected pixels, to characterize how bad the local damage is.

VII. CONCLUSIONS

In this paper we have extended our previous work that showed how two hot pixels occurring in close proximity (within a 5x5 pixel square) interact, creating noticeable image degradation due to the color demosaicing and Jpeg algorithms that create the color images. We explored how lower energy cosmic particles created SEUs interact with these hot pixels to further increase the damage to the images. We presented an expression for the probability of HP:SEU interactions and estimated how often such interactions would occur for a given imager and how would this effect grow with time. The main point is that the damage due to the accumulating hot pixels significantly worsens by the occurrence of SEUs experienced by cameras. For SEU streaks, more distanced HPs will interact with them and generate much larger and more noticeable, image degradation. In future work we will explore how SEUs/streaks plus HP could cause problems for AI image detection in auto drive systems as we had demonstrate for just HP in [18].

REFERENCES

[1] J. Dudas, L. M. Wu, C. Jung, G.H. Chapman, Z. Koren, and I. Koren, "Identification of in-field defect development in digital image sensors," *Proc. Elect. Imaging, Digital Photo III*, v6502, 65020Y1-0Y12, San Jose, Jan. 2007.

[2] J. Leung, G. H. Chapman, I. Koren, and Z. Koren, "Statistical Identification and Analysis of Defect Development in Digital Imagers," *Proc. SPIE Electronic Imaging, Digital Photography V*, v7250, 742903-1 − 03-12, Jan. 2009.

[3] J. Leung, G. H. Chapman, I. Koren, and Z. Koren, "Automatic Detection of In-field Defect Growth in Image Sensors," *Proc. of the IEEE Intern. Symposium on Defect and Fault Tolerance in VLSI Systems*, 220-228, Oct. 2008.

[4] J. Leung, G. H. Chapman, I. Koren, Z. Koren, "Tradeoffs in imager design with respect to pixel defect rates," *Proc. of the IEEE Intern. Symposium on Defect and Fault Tolerance in VLSI*, pp. 231-239, Oct. 2010.

[5] J. Leung, J. Dudas, G. H. Chapman, I. Koren, Z. Koren, "Quantitative Analysis of In-Field Defects in Image Sensor Arrays," *Proc. Intern. Sym on Defect and Fault Tolerance in VLSI*, pp. 526-534, Sept. 2007.

[6] J. Leung, G. H. Chapman, Y. H. Choi, R. Thomson, I. Koren, and Z. Koren, "Tradeoffs in imager design parameters for sensor reliability," *Proc., Elect Imaging, Sensors, Cameras, and Systems for Industrial/Scientific Applications XI*, v 7875, 78750I1-0I12, Jan. 2011.

[7] A.J.P. Theuwissen, "Influence of terrestrial cosmic rays on the reliability of CCD image sensors. Part 1: experiments at room temperature," *IEEE Transactions on Electron Devices*, Vol. 54 (12), 3260-6, 2007.

[8] A.J.P. Theuwissen, "Influence of terrestrial cosmic rays on the reliability of CCD image sensors. Part 2: experiments at elevated temperature," *IEEE Transactions on Electron Devices*, Vol. 55 (9), 2324-8, 2008.

[9] G. H. Chapman, R. Thomas, I. Koren, and Z. Koren, "Empirical formula for rates of hot pixel defects based on pixel size, sensor area and ISO," *Proc. Electronic Imaging, Sensors, Cameras, and Systems for Industrial/Scientific Applications XIII*, v8659, 86590C-1-C-11, Jan. 2013.

[10] J. Leung, "Measurement and Analysis of Defect Development in Digital Imagers," MASc thesis, Simon Fraser University, Canada, 2011.

[11] G. H. Chapman, R. Thomas, K.J. Meneses, P. Purbakht, I. Koren, and Z. Koren, "Exploring Hot Pixel Characteristics for 7 to 1.3 micron Pixels," *Proc. Electronic Imaging: Image Sensors and Imaging Systems*, Feb. 2018.

[12] J. E. Hill, "Birthday Paradox Calculations and Approximation," https://www.untruth.org/~josh/beta/birthdayparadoxapproximation.pdf (2015).

[13] K. J. Meneses, "What Creates Noticeable Defects on Digital Imagers?" MASc (Eng) Thesis, Simon Fraser Univ, 2021.

[14] W.T. Freeman, "Median Filter for Reconstructing Missing Color Samples," U.S Patent, 2724395, 1988.

[15] R. Kimmel, "Demosaicing: Image Reconstruction from CCD Samples," *Proc. Trans. Imaging Progressing*, vol. 8, pp. 1221-1228, Sept. 1999.

[16] G. H. Chapman, K. J. Silvia Menese, I. Koren, and Z. Koren, "Image Degradation due to Interacting Adjacent Hot Pixels," *Proc. IEEE Int. Symp. on Defect and Fault Tolerance in VLSI (DFT)*, pp 1-6, Oct. 2022.

[17] G. H. Chapman, P. Purbakht, P. Le, I. Koren, and Z. Koren, "Exploring Soft Errors (SEUs) with Digital Imager Pixels ranging from 7 to 1.3 μm," *Proc. IEEE Int. Symp. on Defect and Fault Tolerance in VLSI*, pp. 139-142, Oct. 2017.

[18] G. H. Chapman, R. Thomas, K. J. Meneses, I. Koren, and Z. Koren, "Using digital imagers to characterize the dependence of energy and area distributions of SEUs on elevation," *Proc. IEEE Int. Sym on Defect and Fault Tolerance in VLSI (DFT) 2020*, pp 1-4, Oct. 2020.

A Novel Digitisation Method for Pulse Switchable Memristive Chemical Sensors

Meenakshi Devi*, Saurabh Khandelwal*, Marek Vidiš†, Tomas Plecenik†, and Abusaleh Jabir*

*School of ECM, Oxford Brookes University, UK. Email: {19186574, skhandelwal, ajabir}@brookes.ac.uk
†Faculty of Mathematics, Physics and Informatics, Comenius University Bratislava, Slovakia.
Email: {marek.vidis, tomas.plecenik}@fmph.uniba.sk

Abstract—Memristors, typically considered for their non-volatile resistive memory for high-density memory designs, have shown very good sensitivity to various chemicals. As such, these devices can also be fabricated as chemical sensors with intrinsic memory. When fabricated for sensing chemicals, the switching state of the devices, depending on the amount of the applied bias voltage/current, also changes in the presence of the chemicals, compared to when the chemicals are not present. We have observed that this property can be combined with the device's intrinsic memory to directly digitise sensed information. To this end, in this paper, we propose an innovative technique to directly digitise the sensor readings, e.g. gas concentrations, simply by pulsing the devices to exploit the memory behaviour in the presence of a chemical to change the state of the device. Essentially, we are relying on the sensors to tell us when a certain property of a chemical is sensed by switching its state while this information is digitised. This method obviates the need to use separate Analogue-to-Digital converters (ADC), thereby significantly simplifying the sensor architecture in terms of power consumption and circuit complexity. Additionally, owing to the observed high non-linearity of the fabricated devices, this digitisation method is also highly nonlinear, which can provide an added layer of security in the sensed information.

Index Terms—Memristor, Sensors, Resistive switching, Encoding and Decoding.

I. INTRODUCTION

The ubiquitous and reliable chemical sensing is becoming increasingly important, for example, in factories and manufacturing facilities, to identify poisonous gas leaks such as carbon monoxide (CO) or to prevent the spread of harmful gases to organic life. The gas sensing properties of metal oxide semiconductors (MOS) have been widely studied over the years [1]–[5]. With respect to other sensor technologies, they are simple, inexpensive, miniaturizable, and have good sensitivity [6]. Despite their simple operation, they are scarcely selective and may require operating at high temperatures [7]. However, several advances have been made in some of these areas. For example, it has recently been demonstrated that nanomaterials MOS gas sensors can perform optimally at

M. Devi is sponsored by the Indian Government under the award letter no. $K - 11015/56/2021 - SCD - V(NOS)$.
M. Vidiš and T. Plecenik are supported by the Slovak Research and Development Agency under contract No. APVV-21-0053 and the Scientific Grant Agency of the Slovak Ministry of Education, Sciences, Research and Sport (Grant No. VEGA 1/0062/22).

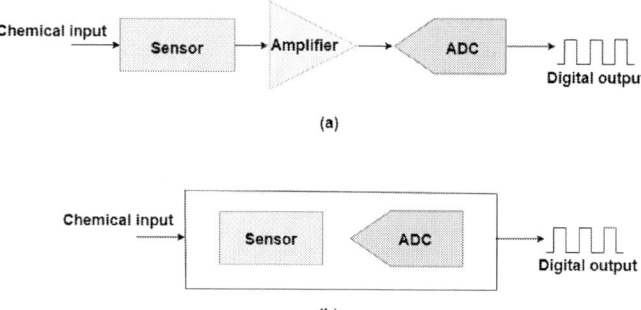

Fig. 1: (a) Traditional sensing method to obtain digital output [9]; (b) Proposed Memristive sensor approach for directly digitising sensed information.

room temperature [8]. Unlike other technologies, sensors with memristive hysteresis properties offer dual capability for sensing and storage. However, fabricating, characterising and modelling memristor-based gas sensors that hold the desired characteristics for efficient sensing of reducing and oxidising gasses remain mostly unexplored.

Traditionally, memristive devices find applications in memory systems [10], [11], logic circuits [12], [13], reconfigurable computing [14] and neuromorphic systems [15]–[17]. Recent research suggests that the metal oxide semiconductor-based memristors also have applications in chemical sensing, owing to their ability to react to chemicals when exposed to them [18]–[25]. The reaction can manifest in various measurable ways, such as a change in the device's resistive behaviour depending on the chemical's nature, composition, and concentration. To this end, much evidence of fabricated practical memristive sensors exists. For example, in [23], memristive hydrogen gas sensors highly suitable for fuel cells and hydrogen safety applications were presented. As another example, memristive sensors for detecting liquid glucose concentrations suitable for implementation in bio-sensing applications were also fabricated in [21], [22]. With this added sensing property, traditional architectures for sensing often rely on analogue-to-digital converters (ADC) for processing [9], [26] and to convert the analogue signal from sensors into digital format. ADC converters employ a resistance ramp line for every pulse [27]–[31], indicating the completion of the conversion process once the progressive resistance matches

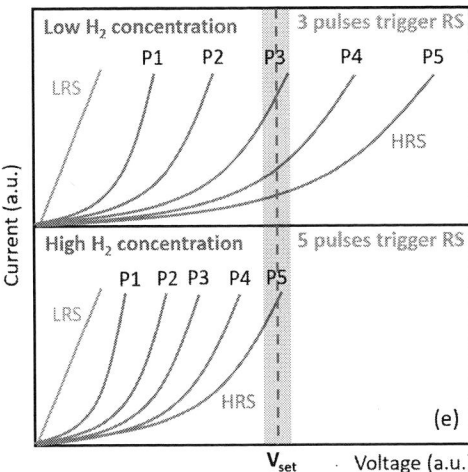

Fig. 2: No. of pulses required for certain gas concentrations.

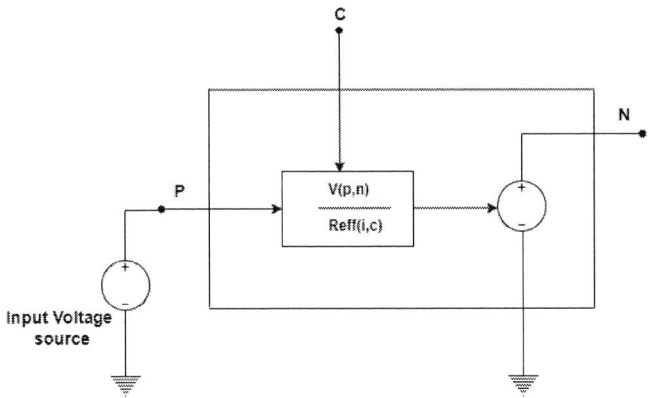

Fig. 3: Voltage driven Memristive model.

the reference resistance line. However, the cost and power consumption of ADCs can be significant challenges. To this end, we propose an innovative solution to address these issues by directly converting gas sensor readings into digital format. This consideration holds the potential to simplify the overall architecture and obviates the necessity for complex analogue-to-digital conversion circuits. Overall, the proposed architecture aims to reduce overheads on memory oriented sensing or processing, where sensing, storage, and processing of chemical information may be carried out within the same structure. The paper is structured as follows: Section. II elaborates the proposed approach. Section. III presents the results and discussion. Finally, Section. IV provides the conclusion of the paper.

II. PROPOSED ARCHITECTURE

This paper proposes an architecture for measuring gas concentrations using pulse-switchable memristive sensors. Leveraging their high sensitivity [23] and selectivity [32], [33], memristive sensors possess the capability to directly encode sensed information into pulse counts, capitalising on the distinctive characteristics of memristors, including their non-volatile nature and resistance modulation without requiring external processing. Essentially, we have integrated the functions of both sensors and ADC converter into a single, streamlined unit. This technique simplifies sensor architecture and reduces power consumption due to its inherent simplicity. Fig. 1 illustrates the comparison between the conventional sensor system and the pulse-based proposed sensing approach. In this system (Fig. 1b), the sensor plays a critical role in signalling the completion of the digitisation process by measuring gas concentrations and switching the memristive sensor state once the threshold is exceeded. We can accurately determine when the conversion process is complete by monitoring the sensor's status. In [34], it was demonstrated that multilevel resistive switching (RS) can be achieved by applying short voltage pulses while biasing the device with a constant current. However, the resistance levels of the memristive sensor device

vary with different gas concentrations. The device operation can be adjusted to reach the threshold / set voltage (V_{set}) after a concentration-dependent number of voltage pulses. By counting the number of pulses required to switch the device, digitised information about the specific gas concentration can be obtained, as illustrated in Fig. 2. Currently, in our work, the memristive sensor device [23] is considered a single unit for detecting hydrogen gas, but the architecture can be modified to detect a wide range of chemicals. Our proposed architecture is based on a voltage-driven memristive gas sensor model, as depicted in Fig. 3. To ensure an accurate representation of the device model for our proposed architecture, we specifically consider the device presented in [23] and utilise the current-driven memristive model presented in [35] as the basis for our voltage-driven memristive sensor model. This proposed voltage-driven model provides a precise depiction of the sensing behaviour of the memristive device compared to other existing models [18], [36]–[40].

A. Voltage Driven Memristive Model

In this work, we focus on driving a memristive sensor using voltage. This involves deriving a voltage model from a current-driven model developed in [35], which accurately describes the behaviour of the gas-sensitive memristors considered in this study under varying gas concentrations and applied current. The internal state of the memristor, represented by the state variable $X(t)$, evolves based on the current $I(t)$ and gas concentration C.

$$R_{\text{eff(i,c)}} = XtoR(X(t)) + R_{\text{diff}}(i(t), C) \qquad (1)$$

In the voltage-driven memristive sensor model, the function $XtoR$ is essential for converting $X(t)$ back to $M(t)$ (memristance of the device) during the modelling process. This transformation is supported by $R_{\text{diff}}(i(t), C)$, which accounts for resistance changes due to variations in gas concentrations. Consequently, we considered Eq. (1) and Eq. (2) to derive a voltage-driven current model for a memristive sensor, as depicted in Fig. 3. The iterative modelling process involves updating the state variable $X(t)$ and determining the next

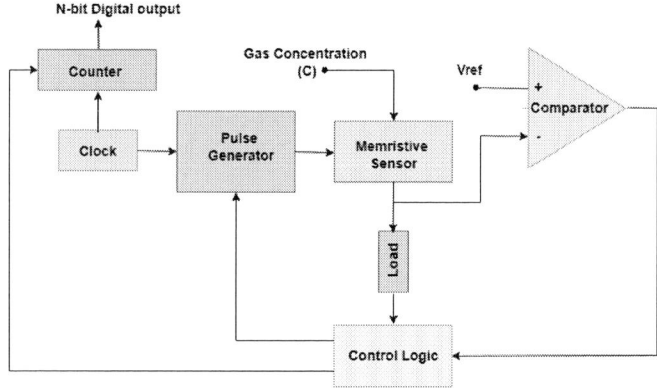

Fig. 4: Proposed architecture for direct digitisation of chemical readings from a gas switchable memristor device.

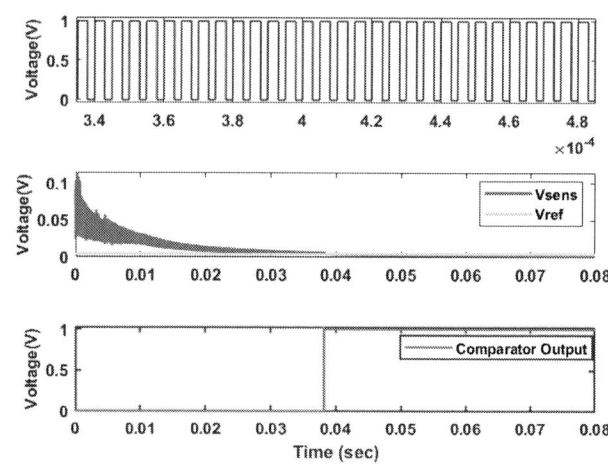

Fig. 5: Simulation Results of the memristive sensor with applied pulses (Zoomed); Sensor voltage (V_{sens}) compared to a reference voltage (V_{ref}) and Comparator output.

memristance $(M(t) + \Delta M(t))$ value based on the effective resistance $R_{eff}(i,c)$ and gas concentration C, starting with the initial current obtained by dividing the terminal voltage of the memristor $V(p,n)$, by the initial resistance R_{on}.

$$Gmem = \frac{V(p,n)}{(R_{eff(i,c)})} \qquad (2)$$

Here, *Gmem* represents the current passing to the memristive sensor during iterations. This transformation enables the model to shift from a current-driven to a voltage-driven perspective.

B. Direct Digitisation Method of Sensor Reading

This section presents an architecture for directly digitising sensor readings, with a specific emphasis on gas concentrations and the exploitation of device memory behaviour in response to chemicals. Illustrated in Fig. 4, the architecture employs a pulsed technique wherein gas concentration critically influences the switching of the memristive sensor state upon surpassing the threshold is exceeded, thereby indicating completion of digitisation alongside the pulse count required to sense specific concentrations by the counter. The architecture predominately consists of a reference voltage, denoted as V_{ref}, a gas-sensitive memristor, and load resistors, configuring the voltage dividers. Additionally, it integrates a pulse generator that is responsible for delivering a series of pulses to the memristive sensor. Initialising the memristance of the sensor device, represented as M_{sens}, to R_{off} (a resistance value in the high resistance state of the device), a known programming voltage, designated as V_{write}, is applied to the sensor. Programming the device involves adjusting the V_{write} to a sufficiently high level ($V_{write} > V_{hrs}$) to appear across the memristive sensor, even with the load, while V_{ref} is set to a low resistive state V_{lrs} ($V_{ref} < V_{hrs}$). Similarly, it is important to adjust the V_{hold} to a sufficient value, but it should be between $V_{on} \geq V_{hold} \leq V_{off}$, allowing the V_{read} to appear across the memristive sensor.

During the programming period (T_{prg}), V_{write} initiates the shifting of the barrier of the memristive sensor from the R_{off} region towards the R_{on} region (i.e. a low resistance state of the device), albeit non linearly, causing the voltage across the sensor V_{sens} to decline. Subsequently, the comparator compares the voltages during the hold period (T_{hold}), generating a logic 0 that directs the control logic to pass input pulses through the pulse generator until V_{sens} exceeds V_{ref}. As a result, V_{sens} gradually decreases during each clock cycle until it equals V_{ref}. Upon V_{sens} exceeding V_{ref}, the sensor switches its state for the specific gas concentration applied, triggering the control logic to produce a logic 1. Subsequently, based on this output, the pulse generator is deactivated by the control logic, ceasing the passage of input pulses, indicative of the completion of the conversion. Concurrently, the n-bit counter counts the pulses required to measure the specific gas concentration. The counter's output represents the encoded value, which, being non-linear, serves as both an encoder and decoder for programming the device and retrieving its sensed information. Our architecture generates non-linear encoding, which may offer inherent security. To reset the sensor from the low resistive state (LRS) to the high resistive state (HRS), a high *CLR* pulse is applied to the negative terminal of the memristive sensor. During CLR pulse, the comparator is disabled via the control signal and produces a constant zero.

III. RESULTS AND DISCUSSION

The voltage-driven model and memristive sensor architecture presented in this paper were coded and tested via simulations in LTSpice. Fig. 3 and Fig. 4 shows the schematic diagram of the voltage-driven memristive sensor model, discussed in Section II-A, and its utilisation in designing the architecture for digitisation, respectively. Fig. 6 illustrates the Spice implementation of this architecture, which incorporates a memristive sensor, comparators, logic gates, and an n-bit counter. The design of the pulse-based direct digitisation circuit begins with an input pulse generator that triggers an AND gate (G1),

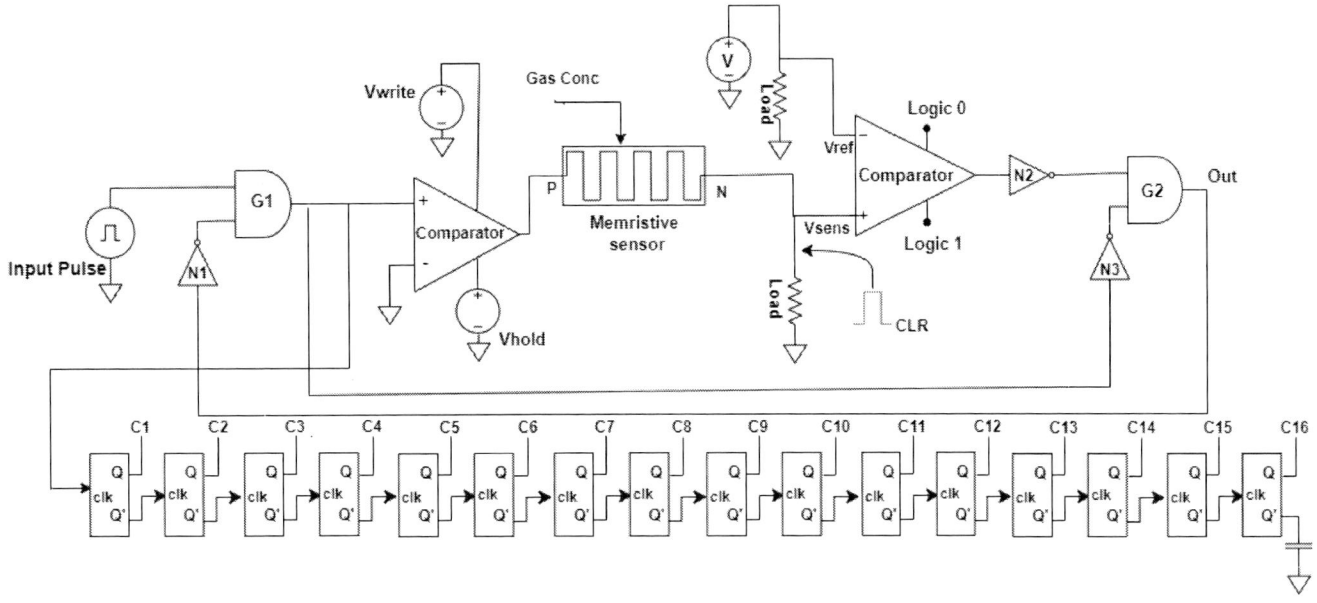

Fig. 6: Spice implementation of the architecture shown in Fig. 4.

TABLE I: No. of Pulses required to switch the device at different H_2 Gas concentration.

Gas conc.(H_2 ppm)	Pulse count	Switching Time(ms)
0	13069	38.9555
1000	13097	38.9946
2000	13104	39.005
3000	13119	39.0191
4000	13183	39.0541
5000	13202	39.1155
6000	13261	39.1442
7000	13273	39.1924
8000	13355	39.2036
9000	13389	39.3083
10000	13447	39.3370

subsequently driving the first comparator. This comparator modulates the voltage V_{write} applied to the memristive sensor, whose resistance varies with gas concentration. The output of the second comparator is a digital logic signal (either logic 0 or logic 1) based on the comparison between the sensor voltage V_{sens} and a reference voltage V_{ref}. This logic signal is further processed by inverters (N2 and N3) and a logic AND gate (G2) to produce the final output. The process continues until a logic 1 is generated, which in return produces a logic 0 on G1 via inverter (N1), halting the input pulse. To digitise the sensed data, a series of counters (C1 to C16) is employed, with each counter representing a digitised state ($Q \approx 1 or\ Q' \approx 0$). Our simulation results demonstrate that the number of pulses

required to trigger the device varies with the concentration of gas present, indicating that the device can measure gas concentration effectively. Specifically, we observed that the device required more pulses to switch resistive states when H_2 gas was present, compared to when there was no gas. Notably, higher gas concentrations necessitate more pulses to switch the device from one state to another, whereas at low concentrations, the device switches earlier.

To design the architecture and perform simulations, our assumptions included $V_{\mathrm{write}} \approx 1.5V$, $V_{\mathrm{ref}} \approx 4.5mV$, $V_{\mathrm{hold}} \approx 0.8V$, $T_{\mathrm{prg}} \approx 2.5\mu s$, $V_{\mathrm{on}} \approx V_{\mathrm{lrs}} \approx -1.69V$, and $V_{\mathrm{off}} \approx V_{\mathrm{hrs}} \approx 0.852V$. The complete digitisation process at a gas concentration of 7000 ppm (parts per million) is illustrated in Fig. 5. We created a calibration curve by calibrating the number of pulses against the known gas concentrations exposed to the sensor, as presented in Table I. This table demonstrates a non-linear correlation between the number of pulses and the gas concentration. For instance, if the number of pulses is 13,273, we know from this table that the gas concentration is 7000 ppm.

Our architecture generates non-linear digital codes that indicate the number of pulses required to sense a specific gas concentration, making it a pulse-based direct digitisation method for pulse-switchable memristive chemical sensors. This is owing to firstly the devices non-linear characteristics, which is amplified by the proposed architecture during the digitisation process. This added complexity ensures that any unauthorised attempts to decode the information will likely be difficult, as the intricate interplay of voltage, chemical properties, and resistance values makes it challenging to interpret the encoded data without a comprehensive understanding of the parameters involved.

In our current work, the memristive sensor device [23]

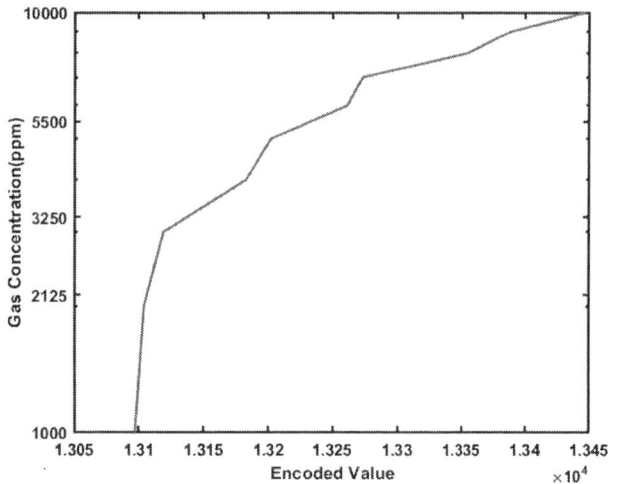

Fig. 7: Calibration curve of H_2 gas concentration with corresponding non-linear Encoded Values.

is considered as a single unit for detecting hydrogen gas. However, the devices can also be considered as a sensor and a memristor in parallel [33]. Therefore, it is possible to separate the sensor and memristor and by placing any sensor in parallel with a memristor we can detect any chemical/chemical properties to which the sensor is sensitive to. Hence, the architecture can be used to detect and digitise a wide range of chemicals, thus expanding its capabilities.

IV. CONCLUSIONS

In this paper, we introduce a novel digitisation method for pulse-switchable memristive chemical sensors that leverages the intrinsic memory properties of memristive devices. This method allows for the direct digitisation of sensor data, eliminating the need for traditional analog-to-digital converters (ADCs), thereby simplifying sensor architecture and reducing power consumption and complexity. Through simulations, we demonstrate that the number of pulses required to switch the device state varies with the gas concentration, thus providing a direct digitally encoded value of sensed data. By preparing a calibration curve, we can precisely measure the gas concentration by simply counting the number of pulses. Our proposed architecture enhances security through non-linear digital codes, making unauthorised access to the information difficult without comprehensive knowledge of the system parameters. Although the circuit has been simulated with H_2 gas concentration, the method can be modified to detect a wide range of chemicals. A parallel arrangement of the sensor and memristor can be implemented to detect other greenhouse gases, thus expanding the device's capabilities beyond H_2 gas.

REFERENCES

[1] V. Bochenkov and G. Sergeev, "Sensitivity, selectivity, and stability of gas-sensitive metal-oxide nanostructures," *Metal Oxide Nanostruct. Appl.*, vol. 3, pp. 31–52, 2010.

[2] N. Yamazoe, "New approaches for improving semiconductor gas sensors," *Sens. Actuators B: Chem.*, vol. 5, no. 1, pp. 7–19, 1991.

[3] D. E. Williams, "Semiconducting oxides as gas-sensitive resistors," *Sens. Actuators B: Chem.*, vol. 57, no. 1, pp. 1–16, 1999.

[4] P. Clifford and D. Tuma, "Characteristics of semiconductor gas sensors i. steady state gas response," *Sens. Actuators*, vol. 3, pp. 233–254, 1982.

[5] M. Devi, S. Khandelwal, and A. Jabir, "Memristor materials, fabrication, and sensing applications," in *Nanoscale Memristor Device and Circuits Design*. Elsevier, 2024, pp. 209–227.

[6] G. F. Fine, "Metal oxide semi-conductor gas sensors in environmental monitoring," *Sensors*, vol. 10, no. 6, pp. 5469–5502, 2010.

[7] A. Damico and C. D. Natale, "A contribution on some basic definitions of sensors properties," *IEEE Sens. J.*, vol. 1, no. 3, pp. 183–190, 2001.

[8] S. Vallejos, I. Grácia, O. Chmela, E. Figueras, J. Hubálek, and C. Cané, "Chemoresistive micromachined gas sensors based on functionalized metal oxide nanowires: Performance and reliability," *Sensors and Actuators B: Chemical*, vol. 235, pp. 525–534, 2016.

[9] J. Horstmann, M. Ramsbeck, and S. Bosse, "Analog sensor signal processing and analog-to-digital conversion," *Material-Integrated Intelligent Systems-Technology and Applications: Technology and Applications*, pp. 257–280, 2018.

[10] D. B. Strukov, G. S. Snider, D. R. Stewart, and R. S. Williams, "The missing memristor found," *nature*, vol. 453, no. 7191, pp. 80–83, May 2008.

[11] A. Adedotun, S. Khandelwal, and A. Jabir, "Techniques for crossbar array read operation," in *Nanoscale Memristor Device and Circuits Design*. Elsevier, 2024, pp. 181–207.

[12] X. Yang, A. A. Adeyemo, A. Jabir, and J. Mathew, "High-performance single-cycle memristive multifunction logic architecture," *Electron Lett.*, vol. 52, no. 11, pp. 906–907, May 2016.

[13] X. Yang, S. Khandelwal, and A. Jabir, "Novel memristive physical unclonable function," in *Nanoscale Memristor Device and Circuits Design*. Elsevier, 2024, pp. 59–89.

[14] J. Cong and Bingjun Xiao, "mrfpga: A novel fpga architecture with memristor-based reconfiguration," in *2011 IEEE/ACM International Symposium on Nanoscale Architectures*, June 2011, pp. 1–8.

[15] A. Thomas, "Memristor-based neural networks," *Journal of Physics D: Applied Physics*, vol. 46, no. 9, p. 093001, Feb. 2013.

[16] A. Bala, S. Khandelwal, A. Jabir, and M. Ottavi, "Yield evaluation of faulty memristive crossbar array-based neural networks with repairability," in *2022 IEEE 28th International Symposium on On-Line Testing and Robust System Design (IOLTS)*. IEEE, 2022, pp. 1–5.

[17] A. Bala, X. Yang, A. Adedotun, S. Khandelwal, and A. Jabir, "Memristor crossbar-based learning method for ex situ training in neural networks," in *Nanoscale Memristor Device and Circuits Design*. Elsevier, 2024, pp. 91–109.

[18] S. Carrara, D. Sacchetto, M. A. Doucey, C. B. Rossi, G. D. Micheli, and Y. Leblebici, "Memristive-biosensors: A new detection method by using nanofabricated memristors," *Sensors and Actuators B: Chemical*, vol. 171-172, pp. 449 – 457, Sep. 2012.

[19] A. Adeyemo, J. Mathew, A. Jabir, C. D. Natale, E. Martinelli, and M. Ottavi, "Efficient sensing approaches for high-density memristor sensor array," *J. Comput. Electron.*, vol. 17, no. 3, pp. 1285–1296, Sep. 2018.

[20] A. A. Haidry, A. Ebach-Stahl, and B. Saruhan, "Effect of pt/tio2 interface on room temperature hydrogen sensing performance of memristor type pt/tio2/pt structure," *Sens. Actuators B Chem.*, vol. 253, pp. 1043–1054, Dec. 2017.

[21] N. S. M. Hadis, A. A. Manaf, and S. H. Herman, "Comparison on tio2 thin film deposition method for fluidic based glucose memristor sensor," in *IEEE International Circuits and Systems Symposium (ICSyS)*, Sep. 2015, pp. 36–39.

[22] N. S. M. Hadis, A. A. Manaf, S. H. Herman, and S. H. Ngalim, "High roff/ron ratio liquid based memristor sensor using sol gel spin coating technique," in *IEEE SENSORS*, Nov. 2015, pp. 1–4.

[23] M. Vidis, T. Plecenik, M. Mosko, S. Tomasec, T. Roch, L. Satrapinskyy, B. Grancic, and A. Plecenik, "Gasistor: A memristor based gas-triggered switch and gas sensor with memory," *Appl. Phys. Lett.*, vol. 115, no. 9, p. 093504, Aug. 2019.

[24] S. Khandelwal, M. Ottavi, E. Martinelli, and A. Jabir, "Low power memristive gas sensor architectures with improved sensing accuracy," *Journal of Computational Electronics*, vol. 21, no. 4, pp. 1005–1016, 2022.

[25] M. Ottavi, V. Gupta, S. Khandelwal, S. Kvatinsky, J. Mathew, E. Martinelli, and A. Jabir, "The missing applications found: Robust design techniques and novel uses of memristors," in *IEEE 25th International*

Symposium on On-Line Testing and Robust System Design (IOLTS), July 2019, pp. 159–164.

[26] T. Rengachari, "A 10 bit algorithmic a/d converter for a biosensor," 2004.

[27] A. Adeyemo, X. Yang, A. Bala, J. Mathew, and A. Jabir, "Analytic models for crossbar read operation," in *IEEE 22nd International Symposium on On-Line Testing and Robust System Design (IOLTS)*, 2016, pp. 3–4.

[28] A. Adeyemo, A. Jabir, J. Mathew, E. Martinelli, C. Di Natale, and M. Ottavi, "Reliable gas sensing with memristive array," in *IEEE 23rd International Symposium on On-Line Testing and Robust System Design (IOLTS)*, July 2017, pp. 244–246.

[29] A. Adeyemo, X. Yang, A. Bala, and A. Jabir, "Analytic models for crossbar write operation," in *Sixth International Symposium on Embedded Computing and System Design (ISED)*, 2016, pp. 313–317.

[30] X. Yang, S. Khandelwal, and A. Jabir, "Secure memristor replicator architecture with physical uncloneability," *Electron Lett.*, vol. 55, no. 24, pp. 1275–1277, Dec. 2019.

[31] X. Yang, S. Khandelwal, A. Jiang, and A. Jabir, "A modelling attack resistant low overhead memristive physical unclonable function," in *2020 IEEE International Symposium on Defect and Fault Tolerance in VLSI and Nanotechnology Systems (DFT)*. IEEE, 2020, pp. 1–4.

[32] M. Vidiš, I. O. Shpetnyi, T. Roch, L. Satrapinskyy, M. Patrnčiak, A. Plecenik, and T. Plecenik, "Flexible hydrogen gas sensor based on a capacitor-like pt/tio2/pt structure on polyimide foil," *International Journal of Hydrogen Energy*, vol. 46, no. 36, pp. 19 217–19 228, 2021.

[33] M. Vidiš, M. Patrnčiak, M. Moško, A. Plecenik, L. Satrapinskyy, T. Roch, P. Ďurina, and T. Plecenik, "Gas-triggered resistive switching and chemiresistive gas sensor with intrinsic memristive memory," *Sensors and Actuators B: Chemical*, vol. 389, p. 133878, 2023.

[34] S. Stathopoulos, A. Khiat, M. Trapatseli, S. Cortese, A. Serb, I. Valov, and T. Prodromakis, "Multibit memory operation of metal-oxide bi-layer memristors," *Scientific reports*, vol. 7, no. 1, p. 17532, 2017.

[35] M. Devi, S. Khandelwal, M. Vidiš, T. Plecenik, and A. Jabir, "A gas-sensitive current-driven memristor: Characterisation and modelling," in *2024 IEEE International Conference on Interdisciplinary Approaches in Technology and Management for Social Innovation (IATMSI)*, vol. 2. IEEE, 2024, pp. 1–6.

[36] S. Kvatinsky, E. G. Friedman, A. Kolodny, and U. C. Weiser, "Team: Threshold adaptive memristor model," *IEEE Transactions on Circuits and Systems I: Regular Papers*, vol. 60, no. 1, pp. 211–221, 2013.

[37] S. Kvatinsky, M. Ramadan, E. G. Friedman, and A. Kolodny, "Vteam: A general model for voltage-controlled memristors," *IEEE Trans. Circuits Syst., II, Exp. Briefs*, vol. 62, no. 8, pp. 786–790, Aug. 2015.

[38] D. Sacchetto, M. A. Doucey, G. D. Micheli, Y. Leblebici, and S. Carrara, "New insight on bio-sensing by nano-fabricated memristors," *BioNanoScience*, vol. 1, pp. 1–3, Apr. 2011.

[39] S. Khandelwal, A. Bala, V. Gupta, M. Ottavi, E. Martinelli, and A. Jabir, "Fault modeling and simulation of memristor based gas sensors," in *2019 IEEE 25th International Symposium on On-Line Testing and Robust System Design (IOLTS)*, July 2019, pp. 58–59.

[40] V. Gupta, D. Pellegrini, S. Khandelwal, A. Jabir, S. Kvatinsky, E. Martinelli, C. Di Natale, and M. Ottavi, "Sensing with memristive complementary resistive switch: Modelling and simulations," in *2020 IEEE International Symposium on Defect and Fault Tolerance in VLSI and Nanotechnology Systems (DFT)*. IEEE, 2020, pp. 1–6.

Implementation and Reliability Evaluation of a ChaCha20 Stream Cipher Hardware Accelerator

Wesley Grignani*, Khalil G. Q. Santana*, Douglas A. Santos[†], Luigi Dilillo[†], and Douglas R. Melo*

*LEDS, University of Vale do Itajaí, Itajaí, Brazil
[†]IES, University of Montpellier, CNRS, Montpellier, France
{wesley.grignani, khalil.santana}@edu.univali.br, {douglas.santos, luigi.dilillo}@umontpellier.fr, drm@univali.br

Abstract—Cryptography is fundamental to ensuring the security and privacy of space communications, enabling the reliable exchange of sensitive data between spacecraft, ground stations, and other components of space infrastructure. However, using encryption can be computationally costly, resulting in lower throughput or higher latencies. One strategy to mitigate the costs imposed by cryptography is using accelerators. In addition, these systems that operate in space are susceptible to faults due to adverse conditions and require the implementation of protection techniques to mitigate these faults and ensure correct operation. In this context, this paper presents a fault-tolerant and high-performance encryption accelerator for the ChaCha20 stream cipher. We present a low-cost optimized implementation and a fault-tolerant version using Hamming Error Correction Code (ECC) and Triple Modular Redundancy (TMR). Results show that the optimized and hardened versions accelerated the application by 23× and 17× compared to the software solution. The optimized solution can process 281.25 MB/s, presenting a higher throughput than state-of-the-art works that rely on FPGA implementation. The hardened solution can process 187.50 MB/s, leading to an overhead of 1.52× more Look-Up Tables (LUTs) and 1.03× more Flip-Flops (FFs) compared to the optimized solution of this work.

Index Terms—Cryptography, RFC 8439, Hardware Acceleration, Fault Tolerance.

I. INTRODUCTION

Encryption is widely used in various applications to enhance security and meet compliance requirements. However, implementing encryption can be challenging in scenarios with limited resources, such as embedded devices, due to potential overheads [1]. In space systems, data cryptography is vital for securing sensitive data transmissions and ensuring data privacy and security. The encryption protects satellites, subsystems, mission data, and the entire space-ground system [2].

Cryptography categorizes ciphers into several subdivisions, including symmetric, asymmetric, block-based, or stream-based. Stream cryptography involves generating a key stream from an initial key and combining it with cleartext to produce ciphertext [3].

This work was supported in part by the Foundation for Support of Research and Innovation, Santa Catarina (FAPESC-2021TR001907), the Brazilian National Council for Scientific and Technological Development (CNPq - process 50794/2023-5), the Brazilian National Coordination of Superior Level Staff Improvement (CAPES/PROSUC), the EU project RADNEXT - Horizon 2020 (grant 101008126), and Project HARV in the framework of the action "Accelerateur d'innovation" from the University of Montpellier.

To mitigate encryption/decryption overheads, cryptographic accelerators and lightweight encryption algorithms can be employed. The study [4] shows that ChaCha and ASCON offer faster performance and lower energy consumption than traditional AES ciphering. However, the same study concludes that ASCON has a higher cost in some of its functions, resulting in ChaCha being the best choice for most use cases.

ChaCha, a stream cipher proposed by Daniel J. Bernstein in 2008, is used in various software applications like WireGuard, OpenSSH, and TLS. It was standardized in RFC 8439, describing its key features and parameters [5]–[7]. Its adaptability to resource-constrained environments makes it well-suited for space systems.

Some works presented the development of hardware cryptographic accelerators for the ChaCha cipher. The work [8] introduced a scalable hardware implementation capable of achieving high throughput across different configurations for ChaCha8/12/20. Semenov [9] emphasized the potential of hardware acceleration, achieving a 3.61× speedup over software by optimizing logic resource usage even with certain operations, like adding internal states and XOR with cleartext, being performed in software. Similarly, the work [10] implemented the ChaCha20 cipher in FPGA, providing insights into resource utilization.

This work introduces a fault-tolerant ChaCha20 hardware accelerator aimed at high performance with low resource utilization. The accelerator was designed with a combination of spatial and information redundancies to improve the accelerator reliability, including the Hamming Error Correcting Code (ECC) and the Triple Modular Redundancy (TMR).

II. HARDWARE IMPLEMENTATION

The accelerator was initially designed in a standard form (STD), utilizing a single instance of the `Qround` component and additional registers for storing intermediate results, which were periodically written back to the main state matrix register.

Subsequently, we developed an optimized design (OPT) featuring two sets of four instances of the `Qround` component, facilitating the computation of parallel column- and diagonal-wise rounds. These sets are interconnected via internal wiring to map each output of the first set to its corresponding input in the subsequent set. This configuration enables the computation of a full `DoubleRound` at once, simplifying the controller and enhancing throughput.

979-8-3503-6689-1/24 $31.00 © 2024 IEEE

Considering the architecture design of standard and optimized solutions, we developed a fault-tolerant (FT) solution from the optimized version. We chose the optimized version because it uses fewer memory elements, which leads to fewer Single-Event Upsets (SEUs) in the design.

A. Hardware Architecture

The architecture of the hardened accelerator is illustrated in Fig. 1. In this optimized configuration, two sets of four elements compute the `Qround` function connected in series. The first set concurrently computes the four column-wise rounds using an instance of the `QuarterRound`. Subsequently, the outputs of these blocks are linked to the inputs of the subsequent set of `QuarterRound` blocks, which concurrently compute the diagonal-wise rounds. This approach eliminates the need for a write-back register to store intermediate results, as each clock cycle computes an entire `DoubleRound`.

The `Concat` component merges various cipher inputs, including the current block counter (`i_BLK_COUNT`), the key (`i_KEY`), and the nonce (`i_NONCE`). These inputs constitute 3/4 of the initial state matrix, while the remaining 128 bits correspond to the constant "expand 32-byte k". The `MUX` switches between the initial state (`i_DATA_FROM_CONCAT`) and the state of the previous round (`i_DATA_FROM_LAST_ROUND`), feeding into the `RegPP` component.

The `QRound` function takes four 32 bit inputs as a 128 bit vector, which undergo various logic and arithmetic operations, including 4 additions, 4 XORs, and 4 rotations. The `DoubleRound` component encompasses two sets of four instances of the `QRound` component (totaling 8), arranged to execute all column-wise operations in parallel, followed by simultaneous computation of diagonal-wise rounds.

The `RoundCounter` signals through its output (`o_ALL_ROUNDS_FINISHED`) when all 20 rounds are done. The `SMAdder` module adds each of the 16 elements of the current state matrix (`RegPP`) to the initial state matrix (`Concat`), without carrying bits between matrix elements.

After all rounds, a bitwise XOR operation is performed between the cleartext and the generated One-Time Pad (OTP), generating the ciphertext in the output `o_CIPHER_TEXT`.

B. Fault Tolerance Implementation

We applied Hamming ECC to harden the memory elements of the datapath (`RegPP` and `Round Counter`). From Fig. 1, the `ENC` and `DEC` components represent the encoding and decoding logic from Hamming and generate the ECC that is stored with its respective value. The controller was hardened by using TMR with a bitwise majority voter system, sufficient to protect against SEUs affecting one of its triplicated modules, mitigating most of the errors in memory elements for radiation environments, as seen in [11].

C. Fault Injection

We performed a simulation-based fault injection to evaluate the reliability of the accelerator. The solution presented in [12] was used to perform SEU injections affecting a single bit into the registers and was customized to operate on the designed accelerator. This fault injection technique uses built-in commands of the ModelSim simulator.

The fault injection strategy for each iteration consists of initially simulating without fault injections to obtain a golden run. The next stage consists of listing all the registers in the circuit and randomly choosing one bit to perform the fault injection. Next, a time is randomly calculated to perform the fault injection within the simulation execution time. In

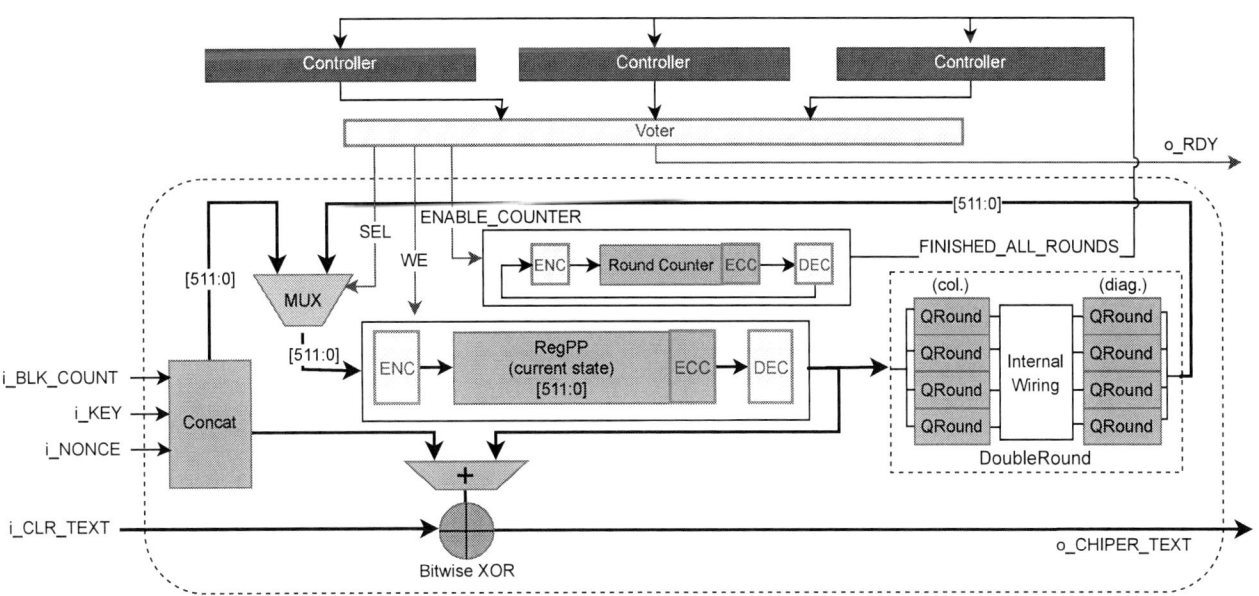

Fig. 1. Block Diagram of the hardened ChaCha20 Accelerator

979-8-3503-6689-1/24 $31.00 © 2024 IEEE

each simulation, a single fault is injected by inverting a bit within the value of the target signal. Whether the output of any external port differed from the golden run at the end of the simulation, it is assumed that the fault resulted in an error in the ciphertext. We performed 1000 simulations, each configured to encrypt 1MB of data.

III. RESULTS

We used Xilinx Vivado 2020.1 to synthesize the hardware described in VHDL and the Mentor ModelSim 20.1 for design verification and reliability analysis. We used the Xilinx Zedboard development kit (Zynq-7000) for prototyping and the OpenSSL [13] as a software implementation reference.

A. ChaCha20 Accelerator

Table I summarizes the usage of Look-Up Tables (LUTs) and Flip-Flops (FFs), with the maximum operating frequency and estimated power dissipation for each implementation.

TABLE I
CHACHA20 ACCELERATOR SYNTHESIS RESULTS.

Implementation	LUTs	FFs	Fmax (MHz)	Power (mW)
ChaCha20 STD	1,562	1,044	81.80	167
ChaCha20 OPT	2,676	518	52.64	202
ChaCha20 FT	4,086	532	35.09	212

The optimized (OPT) solution requires about $1.63\times$ more LUTs and $2\times$ less FFs than the standard (STD) solution, and presents a lower maximum frequency because the circuit modifications increased the critical path. The hardened (FT) version requires around 52% more LUTs and 3% more FFs than the optimized version, resulting in the hardened version having a 33% reduction in the maximum frequency.

Considering the number of cycles required to encrypt 512 bits of data, we estimated the throughput and energy consumed to process 1MB in each version, as shown in Table II.

TABLE II
CHACHA20 ACCELERATOR PERFORMANCE.

Implementation	Cycles	Throughput (MB/s)	Energy* (mJ)
ChaCha20 STD	103	51.25	3.26
ChaCha20 OPT	12	281.25	0.72
ChaCha20 FT	12	187.50	1.13

* Estimated energy consumed to process 1MB.

Even with a lower frequency than the standard solution, the optimized solution achieves a $5.4\times$ higher throughput while consuming $4.6\times$ less energy. The throughput of the hardened version is lower than the optimized version due to the reduction in frequency, but it is still $3.6\times$ higher than the standard version, consuming $2.9\times$ less energy.

B. ChaCha20 SoC

The accelerator was integrated into an SoC to compare the processing time with the software implementation on an ARM processor (Table III). We used the AXI4-Stream interface with a Direct Memory Access (DMA) controller for integration.

TABLE III
CHACHA20 SOC SYNTHESIS RESULTS.

Implementation	LUTs	FFs	Fmax (MHz)	Power (W)
ChaCha20 SoC STD	3,109	4,391	72.84	1.696
ChaCha20 SoC OPT	3,914	3,871	44.60	1.702
ChaCha20 SoC FT	5,416	3,895	31.78	1.711

The introduction of DMA caused an increase of $2\times$ in LUTs and $4.2\times$ in FFs for the standard solution, and by $1.5\times$ in LUTs and $7.4\times$ in FFs for the optimized and hardened versions. Including DMA with the AXI4-Stream interface slightly reduced the maximum frequency compared to the standalone accelerator.

We compared the SoC versions of the accelerator to the software solution running at 667 MHz on the ARM processor. The system was initially configured to perform the cryptography only by the ARM processor. Then, the ARM processor configures the DMA to send the 512-bit blocks to the accelerator, which controls the cryptography process. Table IV presents the execution time for encrypting 1MB of data.

TABLE IV
EXECUTION TIME TO CIPHER 1MBYTE.

Implementation	Execution time (ms)	Acceleration
Software (ARM)	140.13	-
ChaCha20 STD	25.28	5.54
ChaCha20 OPT	5.87	23.87
ChaCha20 FT	8.02	17.47

The standard and optimized solutions show an acceleration of approximately $5.5\times$ and $23\times$ compared to the application entirely in software, and the hardened version shows an acceleration of $17\times$. The acceleration differs due to the lower maximum frequency reported in the hardened version compared to the optimized one.

C. Reliability analysis

The reliability evaluation consisted of ModelSim simulations executed on a computer with an Intel Core i7-12700 processor and 16 GB of RAM, running a Linux operating system. We conducted 1000 simulations for the optimized and hardened solutions, consuming about 62 hours to validate both configurations. Table V shows the simulation results for the fault injection campaign.

TABLE V
RELIABILITY RESULTS.

Implementation	Runs	Errors	Sim. Time
ChaCha20 OPT	1000	918	9h17min
ChaCha20 FT	1000	0	52h35min

The simulation results from the optimized version showed 918 errors in 1000 performed simulations. We observed that most of the errors were caused by faults injected into the RegPP component, which represents the most significant part of the memory elements in the circuit.

TABLE VI
COST AND PERFORMANCE RESULTS OF THE ACCELERATOR IN COMPARISON WITH RELATED WORKS.

Work	Implementation	Device	LUTs	FFs	Fmax (MHz)	Power (mW)	Throughput (MB/s)
[8]	-	Virtex-7 XC7VX485T	2,369	2,152	362.50	-	270.00
[9]	Verilog	Cyclone V	1,440	1,094	50.00	-	126.25
[10]	-	Virtex-7 XC7VX485T	2,288	1,050	-	-	-
This work STD	VHDL	Zynq-7000	1,562	1,044	81.80	167	51.25
This work OPT	VHDL	Zynq-7000	2,676	522	52.64	202	281.25
This work FT	VHDL	Zynq-7000	4,086	538	35.09	212	187.50

In the hardened version, the implementation detected and corrected all the faults. This demonstrates that the applied Hamming ECC and TMR techniques can protect the circuit against SEUs consisting of single-bit flips.

D. Discussion

The comparison with related works is presented in Table VI. Compared to [8], our optimized solution achieves higher throughput, utilizing only 8% more LUTs and consuming $4\times$ fewer FFs, despite operating at a frequency $7\times$ lower.

When compared to [9], our optimized version achieves a throughput $2.2\times$ higher while using half the FFs. Additionally, the authors emphasize that two accelerator units were required to achieve the reported throughput, and [9] employs software for a portion of the cipher, resulting in a lower LUT count.

The work [10] utilizes a comparable number of LUTs and roughly double the FFs compared to our optimized version, but they do not provide any performance metrics. It is important to mention that the related works do not present hardened solutions.

Despite a decrease in throughput compared to the optimized version, our hardened solution achieves a $1.5\times$ higher throughput than [9] even with a lower frequency. In addition, it is important to mention that the related works do not present hardened solutions.

Regarding resource utilization, our hardened implementation uses $4\times$ fewer FFs than [8] and half the FFs compared to [9], [10]. This resource reduction in FFs represents a valuable result regarding reliability facing SEUs, since memory elements are considered one of the most sensitive parts of circuits in the space environment [14]. Moreover, unlike Single Event Transients (SETs), in FFs, the bit-flips remain stored and potentially create problems up to the FF update.

Our optimized implementation achieves the highest throughput compared to related works implemented in FPGA, even when utilizing a mid-range FPGA device. Additionally, the lower resource utilization of FFs in optimized and hardened solutions sets our approach apart from all other related works.

IV. CONCLUSION

This paper presented the implementation of a standard and an optimized encryption accelerator for the ChaCha20 stream cipher. We also applied reliability techniques such as Hamming ECC and TMR to generate a fault-tolerant accelerator version.

We performed a fault injection campaign to evaluate the reliability of the accelerator. We observed that the techniques applied were sufficient to protect from SEUs affecting a single bit. We also compared the cost and performance achieved with the related works. Our optimized solution presented higher throughput with a good trade-off between resource utilization.

For future work, we plan to perform a reliability analysis of the accelerator through particle accelerator tests and integrate it into a reliable SoC solution for space applications. Finally, implementing the Poly1305 message authentication code (MAC) would increase the applicability of the accelerator.

REFERENCES

[1] P. Panahi, C. Bayılmış, U. Çavuşoğlu, and S. Kaçar, "Performance evaluation of lightweight encryption algorithms for iot-based applications," *Arabian Journal for Science and Engineering*, vol. 46, pp. 4015–4037, 2021.

[2] Y. Zhang, S. Zhao, J. He, Y. Zhang, Y. Shen, X. Jiang *et al.*, "A survey of secure communications internet based on cryptography and physical layer security," *IET Information Security*, vol. 2023, 2023.

[3] L. G. de Alvarenga, *Criptografia Clássica e Moderna*. Clube de Autores, 2011.

[4] L. E. Kane, J. J. Chen, R. Thomas, V. Liu, and M. Mckague, "Security and performance in iot: A balancing act," *IEEE access*, vol. 8, pp. 121 969–121 986, 2020.

[5] D. J. Bernstein *et al.*, "Chacha, a variant of salsa20," in *Workshop record of SASC*, vol. 8, no. 1, 2008, pp. 3–5.

[6] A. L. Yoav Nir, "Chacha20 and poly1305 for ietf protocols," Internet Requests for Comments, IRTF, RFC 4180, 03 2018. [Online]. Available: https://datatracker.ietf.org/doc/html/rfc8439

[7] V. Krasnov, "It takes two to chacha (poly)," *Cloudflare*, 2016. [Online]. Available: https://blog.cloudflare.com/it-takes-two-to-chacha-poly/

[8] J. Pfau, M. Reuter, T. Harbaum, K. Hofmann, and J. Becker, "A hardware perspective on the chacha ciphers: Scalable chacha8/12/20 implementations ranging from 476 slices to bitrates of 175 gbit/s," in *2019 32nd IEEE International System on-Chip Conference (SOCC)*. IEEE, 2019, pp. 294–299.

[9] I. Semenov, "An implementation of chacha20 stream cypher in all-programmable socs," Ph.D. dissertation, The University of Alabama in Huntsville, 2020.

[10] R. Serrano, C. Duran, T.-T. Hoang, M. Sarmiento, A. Tsukamoto, K. Suzaki, and C.-K. Pham, "Chacha20-poly1305 crypto core compatible with transport layer security 1.3," in *2021 18th International SoC Design Conference (ISOCC)*. IEEE, 2021, pp. 17–18.

[11] D. A. Santos, A. M. P. Mattos, D. R. Melo, and L. Dilillo, "Characterization of a fault-tolerant RISC-V system-on-chip for space environments," in *2023 IEEE International Symposium on Defect and Fault Tolerance in VLSI and Nanotechnology Systems (DFT)*, 2023, pp. 1–6.

[12] D. R. Melo, C. A. Zeferino, L. Dilillo, and E. A. Bezerra, "Maximizing the inner resilience of a network-on-chip through router controllers design," *Sensors*, vol. 19, no. 24, p. 5416, 2019.

[13] The OpenSSL Project Authors, "Openssl," 2022. [Online]. Available: https://github.com/openssl/openssl/tree/master/crypto/chacha

[14] D. J. Sorin, *Fault Tolerant Computer Architecture*. Morgan and Claypool Publishers, 2009.

Digital generation of RF phase-modulated test stimuli: application to BPSK modulation scheme

K. Tahraoui
LIRMM, Univ. Montpellier, CNRS
Montpellier, France
kamilia.tahraoui@lirmm.fr

R. Burelle
LIRMM, Univ. Montpellier, CNRS
Montpellier, France
richard.burelle@lirmm.fr

T. Vayssade
LIRMM, Univ. Montpellier, CNRS
Montpellier, France
thibault.vayssade@lirmm.fr

F. Lefevre
NXP Semiconductors
Caen, France
francois.lefevre@nxp.com

L. Latorre
LIRMM, Univ. Montpellier, CNRS
Montpellier, France
richard.burelle@lirmm.fr

F. Azaïs
LIRMM, Univ. Montpellier, CNRS
Montpellier, France
florence.azais@lirmm.fr

Abstract— **This paper presents an original strategy for low-cost generation of Radio-Frequency (RF) phase-modulated test stimuli using a standard digital Automated Test Equipment (ATE). The main idea is to generate a modulated digital signal at relatively low-frequency and exploit one of its harmonic replicas to get a signal at higher frequency. Given the specificity of a digital ATE, which manipulates data in the discrete-time domain, one of the cornerstones of the technique is to identify favorable sampling conditions that preserve the spectral content of the generated signal around the targeted harmonic replica. To this end, a corruption estimator is defined based on an analytical expression of a sampled-and-held digital carrier. The approach is illustrated in this paper using the Binary Phase Shift Keying (BPSK) modulation scheme with the objective of generating a 2.4GHz signal, assuming a maximum ATE sampling rate of 1.6Gbps. Simulation results and hardware measurements are presented, validating the proposed solution.**

Keywords— *Analog/RF test, test stimulus generation; digital ATE; sampling theory; BPSK modulation*

I. INTRODUCTION

Modern applications of RF devices include consumer appliances, wireless sensor networks and IoT, which use standard narrow-band communication protocols such as Wi-Fi, Bluetooth or Zigbee. Related markets are very competitive, urging the Integrated circuits (IC) industry to lower production expenses. A significant part of the expenses is devoted to production test, which is responsible for device quality. Testing transmit (TX) channels consists in a full characterization of the transmitted signal (spectrum, Error Vector Magnitude EVM), whereas testing Receive (RX) channels is essentially a functional test of the correct reception of a known message, at various transmitted power levels. These tests are carried out on ATE equipped with dedicated RF instrumentation that can generate or analyze RF modulated signals. This specific instrumentation is extremely expensive and constitutes a major contributor to RF testing costs.

To reduce analog/RF testing costs, a number of works were developed targeting the development of digital solutions with the aim to avoid the use of expensive analog/RF test resources. These solutions include the use of Built-In-Self Test (BIST) within the circuit [1], modification of the ATE architecture [2], or dedicated analysis of analog/RF test responses using standard digital test resources [3-5]. Some works focus more specifically on the generation of analog test stimuli using digital resources, either pure sine-wave signals [6,7] or multi-tone signals [8,9]. Only a limited number of works deal with the generation of RF modulated signals. An exploratory study is presented in [10] for the generation of Offset-Quadrature Phase Shift Keying OQPSK test stimuli, based on the exploitation of a harmonic replica of a digital modulated signal. Following the same approach, a theoretical analysis is presented in [11] considering the case of elementary analog modulation scheme, namely Frequency Modulation (FM) or Phase Modulation (PM). This analysis was validated only in simulation. In this paper, we extend the latter work to a practical case of a digital modulation scheme with the generation of BPSK test stimuli in the 2.4GHz ISM band, and we validate the approach through hardware measurements.

II. DIGITAL STRATEGY FOR TEST STIMULUS GENERATION

A. Principle

Our strategy for reducing the testing costs of RF communicating devices is to develop new solutions that replaces RF instruments with a standard digital ATE. Significant test cost reduction is expected, taking into account that the cost of an RF channel is generally 50 times higher than that of a digital one.

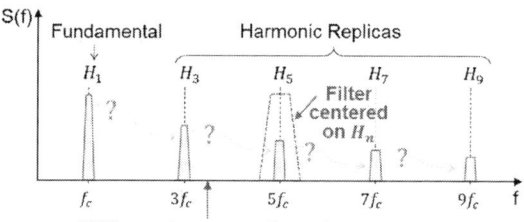

(b) Harmonic filtering of the digital signal

Fig. 1. Strategy for RF test stimulus generation using ATE digital channel

Obviously, the electrical signal generated by a digital channel strongly differs from that generated by an RF channel. One characteristic of a digital signal is that it exhibits harmonic replicas located at multiples of the fundamental frequency (odd multiples only, in case of a digital signal with 50% duty cycle). Interestingly, this characteristic can be exploited in the context of RF testing for today's communication devices, which use narrow-band modulation schemes. Indeed in this case, the harmonic replicas have a limited bandwidth and are well separated from each other. The idea is then to filter one of the harmonic replicas of the digital signal to obtain the analog signal that will be applied to the receiver input as test stimulus, as illustrated in Figure 1. The filter can be placed on the load board that provides the interface between the ATE and the device under test (DUT), or may be already present within the DUT itself.

A key feature of this approach is that the RF signal is generated from a digital signal at lower frequency. This is particularly important taking into account that digital ATEs are limited in terms of sampling capabilities, typically with a maximum sampling rate of 1.6Gbps for standard digital channels. By exploiting a harmonic replica of the digital signal, it is possible to generate a signal at a frequency higher than the ATE's maximum sampling frequency, and therefore to address the classical 868MHz, 915MHz, and 2.4GHz ISM frequency bands, which are used by most RF communication devices.

Two main challenges have to be solved for the practical implementation of this strategy. The first one is to understand how the amplitude and the spectral content of the harmonic replicas are related to the amplitude and the spectral content of the digital signal spectrum around its fundamental frequency (hereafter referred to as the baseband digital signal spectrum). This understanding is a necessary step to encode appropriate information into the digital signal such that the targeted harmonic replica exhibits the desired characteristics according to the modulation scheme considered.

The second challenge is related to the hardware architecture of a digital tester, which manipulates data in the discrete-time domain. Figure 2 gives a simplified diagram of the hardware resources involved in signal generation (drive resources); It comprises a test processor, a vector memory, a waveform composer associated with a timing formatter and a driver associated with a level formatter. The content of the vector memory is sequentially transferred to the waveform composer which formats it into a logical waveform according to the programmed timing information, the duration of each bit being set by the tester period $T_s = 1/f_s$. The binary waveform is then supplied to the driver which formats it into a voltage signal according to the programmed level information, with V_{IH} for a high level and V_{IL} for a low level. The whole process is controlled by the test processor.

Digital channel (Drive resources)

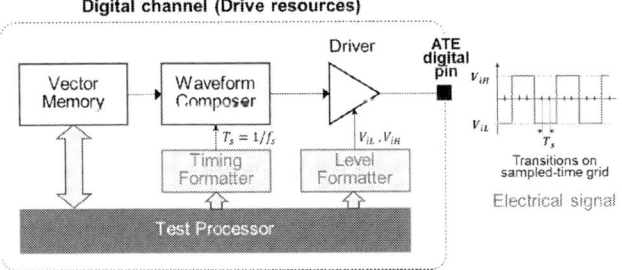

Fig. 2. Simplified diagram of ATE digital channel (Drive ressources)

The electrical signal delivered on the ATE digital pin is therefore a binary signal whose transitions occur on a discrete-time grid determined by the ATE sampling frequency f_s and with an amplitude $A = (V_{IH} - V_{IL})/2$ (in following, we assume a symmetrical signal with $V_{IL} = -V_{IH}$). The fact that transitions occur on a discrete-time grid has an impact on the spectral content of the generated signal and is susceptible to alter the desired characteristics around the targeted harmonic replica. The second challenge is therefore to identify appropriate sampling conditions to limit spectral corruption around the targeted harmonic replica.

B. Mathematical modeling

In order to understand the specific properties of a modulated digital signal generated by means of an ATE digital channel, a mathematical model has been defined. This model, depicted in Figure 3, represents how the signal generated by the ATE digital channel can be derived from a conventional modulated analog signal on the basis of three successive mathematical transformations:

– 1-bit quantization, i.e. Zero-Crossing (ZC):
The first operation transforms the original analog voltage signal into a binary signal with only two possible levels; the continuous-time nature of the signal is preserved.

– Time discretization, i.e. Sampling (S):
The second operation converts the continuous-time binary signal into a discrete sequence of samples; each sample directly corresponds to the binary value that will be stored in the ATE vector memory.

– Zero-Order Hold (ZOH):
Finally, the last operation is a zero-order hold operation in order to restore a continuous-time signal; the signal still has only two possible levels, but the transitions are now synchronous with the sampling clock.

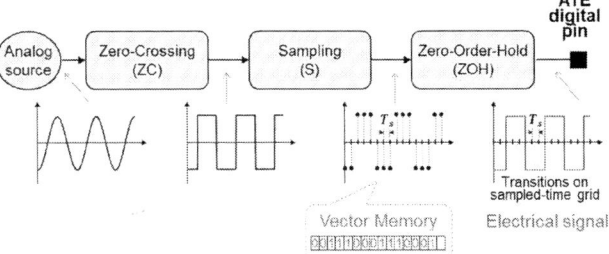

Fig. 3. Mathematical model for signal generation using ATE digital channel

Theoretical developments presented in this paper are based on this conceptual model.

C. Theoretical basis on analog modulation scheme (FM/PM)

The main outcomes of the theoretical analysis conducted on the generation of FM/PM test stimuli with an ATE digital channel [11] are discussed in this section, as they constitute the foundation of the proposed strategy.

In its general form, an angle-modulated signal can be written as:

$$y(t) = A\cos\bigl(\Omega(t)\bigr) = A\cos(\omega_c t + \phi(t)) \tag{1}$$

where $\Omega(t)$ is the instantaneous phase of the modulated signal, A is the amplitude of the carrier signal, ω_c is the angular frequency of the carrier signal, and $\phi(t)$ is the instantaneous phase deviation.

In case of single-tone modulation, the message signal is a sinusoidal signal. Eq.1 can then be expressed as:

$$y(t) = A\cos(\omega_c t + \beta \sin(\omega_m t)) \tag{2}$$

where ω_m is the angular frequency of the message signal, and β is the modulation index.

1) Effect of zero-crossing

Referring to the mathematical model of Figure 3, the first transformation applied on the analog signal is a zero-crossing operation. The digital signal resulting from a zero-crossing operation applied on a modulated analog signal can be expressed as an infinite sum of modulated analog signals:

$$y(t) = \frac{4}{\pi}\left(A\cos(\Omega(t)) + \frac{A}{3}\cos(3\Omega(t)) + \frac{A}{5}\cos(5\Omega(t)) + \cdots\right) \quad (3)$$

where A is the amplitude of the modulated digital signal. The first term corresponds to the baseband signal while the following terms correspond to harmonic replicas located at odd multiples of the baseband signal.

Inserting $\Omega(t) = \omega_c t + \beta\sin(\omega_m t)$ in Eq.3, the expression of the modulated digital signal is then given by:

$$y(t) = \frac{4}{\pi}\sum_i \frac{A}{i}\cos(i\omega_c t + i\beta\sin(\omega_m t)) \quad for\ i = 1, 3, 5, \dots \quad (4)$$

This expression indicates that the modulated digital signal resulting from a zero-crossing operation can be considered as a sum of modulated analog signals with a modification of both the amplitude and the modulation index for each individual modulated analog signal. The harmonic order comes then as multiplier for the modulation index and as a divider for the amplitude. This equation is a key element of the proposed strategy because it establishes the link between the spectral content and the amplitude of a given replica and the parameters used for the modulation of the baseband digital signal. It therefore permits to choose appropriate settings of the baseband signal to reach desired characteristics around a given replica. Practically, to obtain a modulated signal with carrier frequency $f_{c_{target}}$, amplitude A_{target} and modulation index β_{target}, the baseband digital signal should be generated with $f_c = f_{c_{target}}/i$, $A = \pi i A_{target}/4$ and $\beta = \beta_{target}/i$, where i is the order of the selected harmonic replica.

2) Effect of sample-and-hold

Referring again to the mathematical model of Figure 3, the following transformations applied after zero-crossing are sampling and zero-order hold operations. Sampling is the process of converting a continuous-time signal into a discrete sequence of samples. In the frequency domain, it introduces a periodization of the spectrum, with the creation of copies of the original spectrum (called images) shifted by multiples of the sampling frequency. Zero-order hold is the process converting a sampled signal to the continuous-time domain by maintaining each sample until the next one. In the frequency domain, it introduces a global shaping of the spectrum by the sin_c function.

The Nyquist theorem states that a band-limited signal can be fully represented if sampled at a frequency f_s which is greater than twice the maximum frequency component f_M in the signal: $f_s > 2 * f_M$. Indeed, when this criterion is satisfied, there is no overlap between the baseband spectrum and the images created by the sampling process. An extension of the Nyquist theorem concerns narrowband signals, i.e. signals that have a limited bandwidth around a given frequency and that do not extent to DC. In this case, it is possible to sample the signal below the Nyquist rate while still obtaining a perfect signal representation in the Nyquist band, upon specific conditions on the sampling frequency. Such process is called undersampling, harmonic sampling or bandpass sampling.

In our context, the modulated digital signal obtained from zero-crossing has an infinite number of replicas with decreasing amplitudes, each replica being a narrow-band signal. So, depending on the value of the sampling frequency, some replicas will satisfy the Nyquist criterion while others

will be undersampled. The resulting spectrum is therefore complex, as illustrated in Figure 4 which shows the spectrum of the sampled-and-held modulated digital signal for a sampling frequency $f_s = 6.3 * f_c$. The global shaping introduced by the sin_c function is clearly visible, with a cancellation of all components close to the local zeros present at every multiple of the sampling frequency. The expected harmonic replicas are present but they are mixed with other components of similar or even higher amplitude. The components of high amplitude actually correspond to high-frequency images of the harmonic tones located below the Nyquist frequency, while the other additional components correspond to low-frequency images of the harmonic tones located above the Nyquist frequency. Although the additional components are distributed all over the spectrum, components of significant magnitude do not necessarily fall in the vicinity of a given harmonic replica, depending on the chosen value of the sampling frequency.

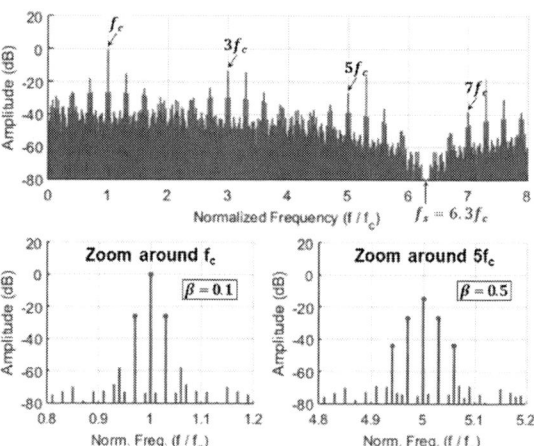

Fig. 4. Spectrum of the sampled-and-held modulated digital signal for a sampling frequency $f_s = 6.3 * f_c$

For instance, with the particular value $f_s = 6.3 * f_c$, additional components are indeed present around the 5th-order harmonic replica as illustrated in the close-up view, but their amplitude is much lower than that of the main components related to the modulation (denoted with a marker). This example shows that despite the hairy aspect of the overall spectrum, the modulation-related spectral content can be preserved within a given limited region, upon an appropriate choice of the sampling frequency. Note also that compared to the baseband spectrum, the number of sidebands with significant power is increased, corresponding to the foreseen increase in the modulation index.

3) Conditions for non-destructive sampling

The main challenge is to be able to identify the values of the sampling frequency that allow non-destructive sampling for a given harmonic replica, i.e. no overlap in the signal bandwidth between the targeted replica and images of other replicas with significant magnitude. To this end, a corruption estimator has been defined, which is based on the expression established in [11] for a sampled-and-held digital carrier:

$$\bar{X}_c(f) = A\,e^{-i\pi T_s f}\sum_{k=-\infty}^{+\infty}\sum_{l=-\infty}^{+\infty}\alpha_{k,l}\;\delta(f - (kx + ly)\Delta f) \quad (5)$$

$$with\ \alpha_{k,l} = sin_c\left(\frac{k}{2}\right) * sin_c\left(\frac{k}{NSPP} + l\right)\ and\ \Delta f = gcd(f_c, f_s)$$

where $NSPP$ stands for Number of Samples Per Period, defined as the ratio between the sampling frequency and the signal frequency $NSPP = f_s/f_c$.

This expression reveals that, because of the sampling process, a sampled-and-held digital carrier exhibits unwanted components located at $if_c \pm m\Delta f$ in addition to the expected harmonic tone located at if_c, as illustrated in Figure 5. The amplitude of the harmonic tone as well as unwanted components depends on the $NSPP$ value.

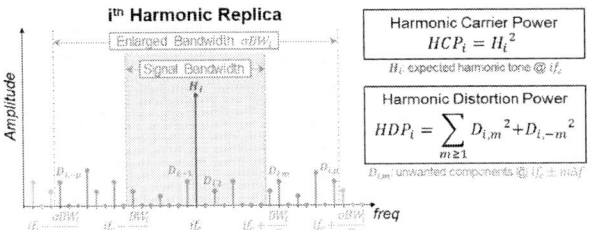

Fig. 5. Performance metrics defined on sampled-and-held digital carrier

Two performance metrics have been defined to evaluate how the sampled-and-held digital carrier differs from a regular square-wave signal.

- HCP_i: the Harmonic Carrier Power of a replica i is defined as the power contained in the harmonic tone located at if_c:

$$HCP_i = H_i^2 \qquad (6)$$

The value expected under ideal conditions is derived from the amplitude of the ith-order component of a square-wave signal shaped by the sin_c function:

$$HCP_i^{expected} = \left(\frac{4A}{\pi i} * sin_c\left(\frac{i}{NSPP}\right)\right)^2 \qquad (7)$$

A deviation from the expected value is representative of images superimposed onto the harmonic tone, images that will alter the spectral content in case of a modulated signal.

- HDP_i: the Harmonic Distortion Power of a replica i is defined as the power contained by the unwanted components in an enlarged bandwidth σBW_i:

$$HDP_i = \sum_{m \geq 1}^{p} \left(D_{i,m}^2 + D_{i,-m}^2\right) \qquad (8)$$

The amplitude $D_{i,m}$ of unwanted components is evaluated using Eq.5, considering all frequency components except the central one that fall in the interval $\left[if_c - \frac{\sigma BW_i}{2}, if_c + \frac{\sigma BW_i}{2}\right]$, where σ is a factor higher than 1 that permits to choose to size of the enlarged bandwidth. Under ideal conditions, no image of harmonic tones should be present in this region; the expected value is therefore zero. Any deviation from zero is representative of the presence of unwanted components likely to alter the spectral content in case of a modulated signal.

Based on these metrics, the corruption estimator associated with a given harmonic replica is defined by:

$$Corr\_Est_i = \frac{\left|HCP_i - HCP_i^{expected}\right| + HDP_i}{HCP_i^{expected}} \qquad (9)$$

The numerator returns the power deviation while the denominator quantifies it according to the expected one. This estimator permits then to have a quantitative evaluation of the quality of the signal generated by the ATE around a given

harmonic replica, for any potential sampling frequency. The closer this estimator is to zero, the better the quality of the generated signal.

III. APPLICATION TO BPSK MODULATION

The theoretical foundations of the proposed strategy were developed considering an elementary analog modulation scheme. The objective in this paper is to demonstrate the applicability of the solution for actual RF communication devices that employ digital modulation schemes.

A. Case study

As a case study, we select the IEEE 802.11 standard which operates in the 2.4GHz ISM band using BPSK with a stream data rate of 1Mbps and a bandwidth of 22MHz.

BPSK is a digital modulation scheme where the 0's and 1's of a binary message are transmitted by changing the phase of a constant frequency carrier by $180°$. It is a widely used modulation scheme due to its high robustness to noise and interference.

A classical expression in the time domain is given by:

$$y(t) = A\cos(\omega_c t + \phi(t)) \text{ with } \phi(t) = \begin{cases} 0 \text{ for binary "0"} \\ \pi \text{ for binary "1"} \end{cases} \quad (10)$$

Using $b(t) = \sum b(lT_b) \Pi_{Tb}(t - lT_b)$, with $b(t)$ being the continuous-time expression of the discrete bit sequence that composes the message, T_b the bit duration and $\Pi_{Tb}(t)$ the rectangular function of duration T_b, Eq.10 can be expressed:

$$y(t) = A\cos\left(2\pi f_c t + \Delta\phi \sum_{l=0}^{N-1} b(lT_b)\Pi_{Tb}(t - lT_b)\right) \quad (11)$$

where $\Delta\phi = \pi$ is the peak phase deviation.

The expression in the frequency domain is given by [12]:

$$Y(f) = \frac{A}{2}\frac{sin[\pi(f - f_c)T_b]}{\pi(f - f_c)\sqrt{T_b}} = \frac{A\sqrt{T_b}}{2} sin_c[(f - f_c)T_b] \quad (12)$$

The spectrum of a BPSK signal is illustrated in Figure 6. It has zeros at every $f - f_c = \pm iT_b$ except for $i = 0$. The central lobe has therefore a width of $2/T_b$, while secondary lobes have a width of $1/T_b$. The amplitude of secondary lobes decreases by $(f - f_c)^2$ as f moves away from f_c.

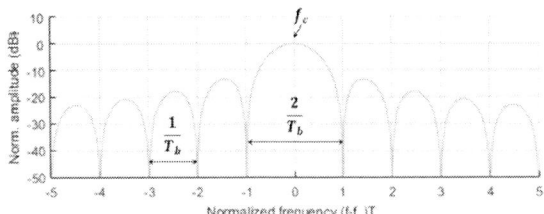

Fig. 6. Theoretical spectrum of a BPSK signal (Eq.12)

Our objective is to generate a 2.4GHz BPSK signal using a digital signal delivered by standard ATE digital channel, which has a maximum sampling rate of 1.6Gbps (e.g., Advantest's V93000 Pin Scale 1600 digital channel card).

B. Definition of the baseband digital signal

In order to define the baseband digital signal, the first step is to select the harmonic replica exploited for the generation of the BPSK signal. The main requirement is that the fundamental frequency of the baseband digital signal should comply with the Nyquist criterion, taking into account the maximum sampling rate of the test equipment. In other words, the carrier frequency of the baseband signal $f_{c_{BB}} = f_{c_{target}}/i$ has to be lower than half the maximum sampling frequency:

$$\frac{f_{c_{target}}}{i} < \frac{f_{s_{max}}}{2} \tag{13}$$

For the considered case study with $f_{c_{target}} = 2.4GHz$ and $f_{s_{max}} = 1.6GHz$, the smallest odd integer that satisfies this requirement is $i = 5$. This is the choice used in this work, which gives a fundamental frequency of the baseband digital signal fixed at $f_{c_{BB}} = 480MHz$.

The second step is to determine the content of the binary sequence that will be stored in the ATE vector memory by defining appropriate modulation encoding of the baseband digital signal. The theoretical developments conducted on analog angle modulation have established that the baseband signal should be encoded with a modulation index $\beta = \beta_{target}/i$. In case of a digital phase modulation like BPSK, the modulation index corresponds to the peak phase deviation $\Delta\phi = \pi$. Encoding of the baseband digital signal should therefore be done using a peak phase deviation $\Delta\phi_{BB} = \pi/i$:

$$y_{BB}(t) = A\,cos\left(2\pi\frac{f_{c_{target}}}{5}t + \frac{\pi}{5}\sum_{l=0}^{N-1}b(lT_b)\,\Pi_{Tb}(t-lT_b)\right) \tag{14}$$

Practically, a dedicated processing algorithm is used to determine the content of the binary sequence that will be stored in the ATE vector memory. As illustrated in figure 7, it involves three main steps:

- computation of the baseband analog modulated signal $y_{BB}(t)$ using Eq.14 and the message input bits $b(lT_b)$,
- computation of the digital modulated signal $\widetilde{y_{BB}}(t)$ by applying zero-crossing on the analog modulated signal,
- computation of the ATE binary sequence bits $b_{ATE}(kT_s)$ by sampling the digital modulated signal at a given f_s.

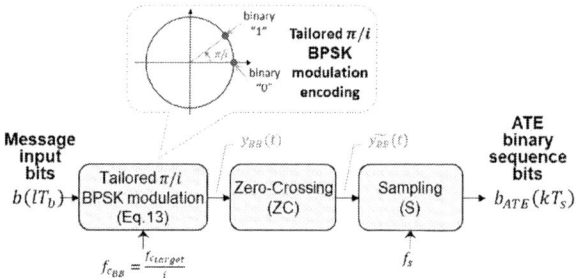

Fig. 7. Processing algorithm for the determination of ATE binary sequence

C. Identification of non-destructive sampling conditions

The corruption estimator defined in section II.C.3 is the central element for the identification of favorable sampling conditions. Computation of this estimator was performed on the considered case study for different values of $NSPP$ varied between 2 and 3.3 by step of 0.05, where the lowest $NSPP$ value corresponds to the lowest sampling frequency satisfying the Nyquist criterion and the highest $NSPP$ value to the highest sampling frequency compatible with the capabilities of the test equipment. The enlarged bandwidth was set to twice to bandwidth specified by the IEEE 802.11 standard.

Results are summarized in Figure 8 which reports the corruption estimator $Corr\_Est_5$ for the different $NSPP$

values. It can be observed that 16 of the 27 explored solutions have a low corruption below 5%; these solutions are therefore a priori valid solutions. The best solution (the one with minimal corruption), is obtained for $NSPP = 3.15$, corresponding to a sampling frequency of $1.512GHz$ and a corruption of less than 0.5%.

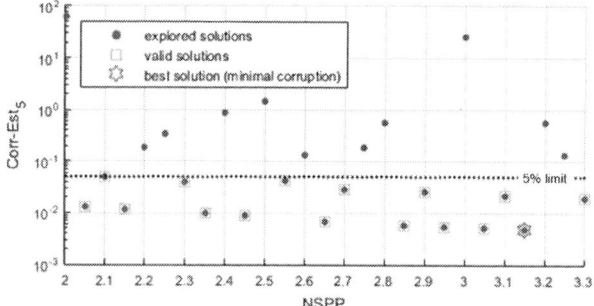

Fig. 8. Corruption estimator vs. $NSPP$ for the 5th-order harmonic replica

IV. EXPERIMENTAL VALIDATION

A. Hardware setup

A laboratory test bench has been developed to validate the proposed solution through hardware measurements. As shown in Figure 9, the operation of an ATE digital channel is emulated using a Universal Software Radio Peripheral (USRP) associated with a latched comparator. The USRP is programmed to deliver the analog modulated signal $y_{BB}(t)$ ($70MHz - 6GHz$) with the tailored $\pi/5$ modulation encoding. This analog signal is then converted into a digital signal by passing through the latched comparator (ADCMP572) clocked at f_s rate. The signal at the output of the comparator thus corresponds to the electrical signal that would be delivered by an ATE digital channel. Acquisition of this signal is performed by a digital storage oscilloscope at 25Gsps and the captured transient data are transferred to a PC for further processing. In particular, FFT will be applied to calculate the experimental spectrum and verify that its spectral content around the selected harmonic replica matches with the expected one. The digital signal delivered by the comparator is also passed through a bandpass filter centered on the targeted carrier frequency and the resulting signal is used as test stimulus for a BPSK receiver implemented in a second USRP. Demodulated data will be compared to the original input bits to ascertain the quality of the generated test stimulus.

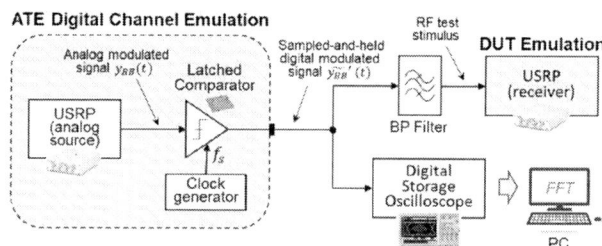

Fig. 9. Experimental setup for lab experiments

B. Hardware measurements

Hardware measurements have been carried out on the considered case study. Note that the fundamental frequency of the targeted RF test stimulus has been set to $f_{c_{target}} = 2.442GHz$ to adapt with the center frequency of the bandpass filter. The carrier frequency of the analog modulated signal is therefore adjusted to $f_{c_{BB}} = 488.4 MHz$.

The solution identified as the best one has been selected for the first experiment. i.e. the clock frequency of the latched comparator was set to $f_s = 3.15 * f_{c_{BB}} = 1.53846 GHz$. The signal was generated for a duration of $250 \mu s$, using a test sequence composed of 250 random input bits. Figure 10 shows the global view of the spectrum computed using an FFT on the captured transient data, as well as the close-up view around the 5th-order harmonic replica. Despite the hairy aspect of the global spectrum, the spectral content around the targeted RF signal frequency exhibits the desired characteristics and presents a good agreement in all the signal bandwidth with the theoretical spectrum of a BPSK spectrum.

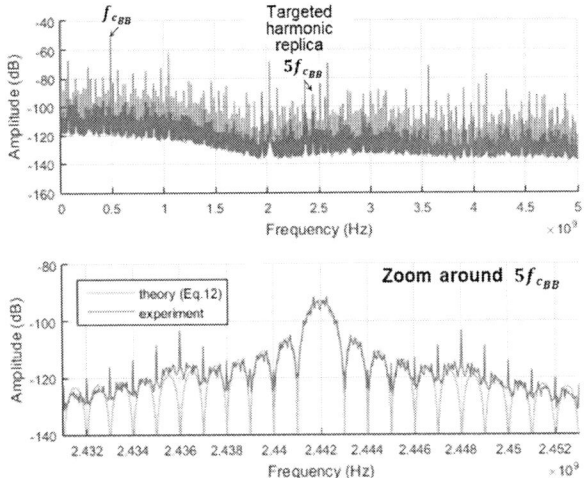

Fig. 10. Experimental spectrum of the modulated digital signal using $f_{c_{BB}} = 488.4 MHz$ and $f_s = 1.53846 GHz$ (solution with lowest corruption)

For demonstration purposes, a solution that was not retained by the corruption estimator has also been implemented, with $f_s = 2.40 * f_{c_{BB}} = 1.17216 GHz$. The spectrum observed around the targeted RF signal frequency is shown in Figure 11. Compared to the previous experiment, the alteration of the spectral content in the signal bandwidth is clearly visible, with only the first lobes that are in good agreement with the theoretical spectrum.

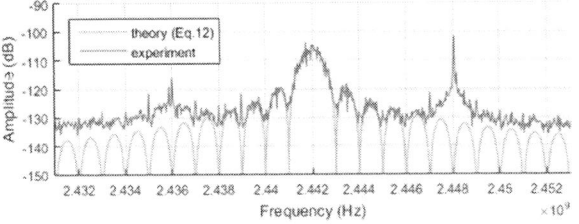

Fig. 11. Illustration of a non-favorable sampling condition: $f_s = 1.172 GHz$

Finally, to further assess the effectiveness of the corruption estimator in identifying favorable sampling conditions, the 16 valid solutions have been implemented on the test bench. A test stimulus of $100 ms$ duration was generated, corresponding to 10^5 bits. In all cases, demodulation of the RF test stimulus by the receiver implemented on the USRP was successful, with a perfect match between the input bits and the demodulated data.

Overall, these hardware experiments demonstrate the

validity of the proposed approach for obtaining a modulated signal with the desired characteristics from a digital signal. They also confirm the feasibility of generating test stimuli in the classical frequency bands used by RF communication devices, despite the limited sampling capabilities of an ATE standard digital channel.

V. CONCLUSION

In this paper, we have presented a low-cost solution for the generation of BPSK test stimuli in the 2.4GHz ISM band using a standard ATE digital channel with a maximum sampling rate of 1.6Gbps. The approach relies on the generation of a baseband digital signal and the exploitation of one of its harmonic replicas to get a signal at higher frequency. Three main requirements for obtaining a good quality BPSK test stimulus have been identified and discussed. Firstly, the fundamental frequency of the baseband digital signal is chosen as the smallest odd submultiple of the targeted test stimulus frequency that satisfies the Nyquist criterion, taking account the ATE sampling capabilities. Then, the baseband digital signal is generated with a tailored π/i modulation encoding such that it exhibits typical BPSK spectral characteristics around the selected harmonic replica. Finally, value of the sampling frequency that minimizes spectral corruption around the selected harmonic replica is chosen, based on a corruption estimation computed using the analytical expression of a sampled-and-held digital carrier. Hardware measurements using off-the-shelf components that emulate the operation of an ATE digital channel have demonstrated the validity of the proposed solution. Future work will target the practical implementation on an actual ATE, as well as extension to other digital modulation formats.

REFERENCES

[1] C. H. Peng et al., "A novel RF self test for a combo SoC on digital ATE with multi-site applications," Proc. Int'l Test Conf. , pp. 1-8, 2014.

[2] M. Ishida K. Ichiyama, "An ATE System for Testing RF Digital Communication Devices With QAM Signal Interfaces," IEEE Design & Test, vol. 33, no. 6, pp. 15-22, 2016.

[3] N. Pous et al., J.,"A Level-Crossing Approach for the Analysis of RF Modulated Signals using only Digital Test Resources", J. of Electronic Testing: Theory and App. (JETTA), vol. 27, no. 3, pp 289-303, 2011

[4] S. David-Grignot et al., "Low-cost phase noise testing of complex RF ICs using standard digital ATE," Proc. Int'l Test Conf. (ITC), pp. 1-9, 2014.

[5] T. Vayssade et al., "EVM measurement of RF ZigBee transceivers using standard digital ATE", Proc. Design Automation & Test in Europe Conference (DATE), pp. 396-401, 2021.

[6] H. Malloug et al., "Mostly digital design of sinusoidal signal generators for mixed-signal bist applications using harmonic cancellation," Proc. Int'l Mixed-Signal Testing Workshop (IMSTW), pp. 1–6, 2016.

[7] S. David-Grignot et al., "Analytical study of on-chip generations of analog sine-wave based on combined digital signals," Proc. Int'l Mixed Signals Testing Workshop (IMSTW), pp. 1-5, 2017.

[8] A. Banerjee et al., "Optimized digital compatible pulse sequences for testing of RF front end modules," IEEE 16th Int'l Mixed-Signals, Sensors and Systems Test Workshop (IMS3TW), pp. 1-6, 2010.

[9] M. A. Zeidan et al., "Phase-Aware Multitone Digital Signal Based Test for RF Receivers," in IEEE Trans on Circuits and Systems I (TCAS), vol. 59, no. 9, pp. 2097-2110, Sept. 2012.

[10] T. Vayssade et al., "Exploration of a digital-based solution for the generation of 2.4GHz OQPSK test stimuli," Proc. European Test Symp. (ETS), pp. 1-6, 2021.

[11] K. Tahraoui, F. Azaïs, L. Latorre, F. Lefevre and T. Vayssade, "Digital generation of single-tone FM/PM test stimuli: a theoretical analysis," Proc. IEEE Latin-American Test Symp. (LATS), pp. 1-6, 2024.

F. Amoroso, "The bandwidth of digital data signal," in IEEE Communications Magazine, vol. 18, no. 6, pp. 13-24, Nov. 1980.

A Flexible FPGA-based Test Equipment for Enabling Out-of-Production Manufacturing Test Flow of Digital Systems

Nicola di Gruttola Giardino, Francesco Angione,
Paolo Bernardi, Tommaso Foscale
Dip. di Automatica e Informatica,
Politecnico di Torino Turin, Italy
name.surname at polito.it

Claudia Bertani, Vincenzo Tancorre
STMicroelectronics, Italy
name.surname at st.com

Abstract—In the past years, the complexity of Automotive System-on-Chips (SoCs) has risen dramatically, mainly dictated by the increasing application requirements. As the complexity increases, safety standards, such as *ISO 26262*, impose quality requirements on manufactured devices. Manufactured SoCs go through a manufacturing test flow where, at every step, a set of tests is conducted to verify that the SoCs are manufactured correctly. Although the manufacturing test flow has been stabilized among companies, with the rising complexity of SoCs, the test pace is hard to keep. Consequently, in the last decade, many companies introduced the so-called *System-Level Test* (SLT) as an additional test step. SLT could be combined with other test steps in order to screen faulty SoCs effectively.

Industrial semiconductor testers are available in the market. However, to the best of our knowledge, in the state-of-the-art FPGA-based semiconductor tester, no tester can combine different advanced test strategies, such as advanced functional applications (SLT) and structural tests (scan-based). The proposed FPGA-based Test Equipment aims to overcome the limitations in Research Labs (at universities or in companies), for failure analysis and test-flow prototyping; since an industrial tester, called Automatic Test Equipment (ATE) is expensive and requires physical space.

Experimental results show how the hardware is carried out on an FPGA platform with its performance and utilization factor. Meanwhile, the execution time of the software is profiled. Using an industrial Automotive SoC, it shows how the proposed semiconductor tester enables the development of different test strategies.

Index Terms—FPGA, digital tester, manufacturing test flow, hardware testing, patterns application, debugger, System-Level Test, test equipment, structural test.

I. INTRODUCTION

The complexity of Automotive System-on-Chips (SoCs) has surged due to rising application demands [1]. Safety standards like ISO 26262 [2] mandate high quality for these devices. In the semiconductor industry, particularly for safety-critical uses, manufacturers must ensure SoCs are correctly produced, screening out faulty and early-failure devices.

In the manufacturing test flow, different test phases are present [3]. The tests in the different phases can be grouped according to two macro-categories for the sake of this work:

- Structural tests, they are tests that resort to the execution of scan-based patterns [4], or internal Built-In Self Tests (BISTs) [5], [6], to test the SoCs. Scan-based patterns are

patterns applied when the SoC is transformed, thanks to the Design For Test approaches [7], into a shift register, i.e., all the internal FFs are connected in order to create a shift register in which patterns are loaded and applied to the combinational logic. Afterward, their results are shifted out and compared to the expected output. The FFs can be connected to create a single chain of FFs or be divided into different chains with their inputs and outputs on the SoC pins. On the other hand, BISTs are custom hardware modules in charge of testing a single module in the SoC.

- Functional tests are tests that resort to the execution of a program in order to verify the behaviour of SoCs.

Although the manufacturing test flow has been stabilized among companies, in the last decade, some companies started to introduce an additional test step in the manufacturing test flow [8]–[10], the so-called *System-Level Test* (SLT), supported by Automatic Test Equipment (ATE) providers [11].

As depicted in [11], SLT has emerged as an essential phase in the semiconductor production cycle due to the increasing Quality Requirements from safety standards. There are tremendous efforts in keeping the pace of test coverage following the exponential scaling density.

SLT mainly consists of running functional applications that mimic the in-field behavior of the device, as well as the workload and the environment. SLT could be an effective and significantly quicker solution compared to structural tests (scan-based patterns) for further screening faulty devices, avoiding field-return and overtesting [12].

Moreover, companies are looking for flexible System-Level Testers capable of combining capabilities for executing functional test programs and structural tests [13]. Combining different test steps into a single step requires a combination of efforts from tester providers, test engineers, and technologies, which is not negligible and cheap. Therefore, it is challenging to develop and manufacture Test Equipment that combines two different and complementary test steps.

Industrial semiconductor testers available in the market serve this purpose. However, to the best of our knowledge, in the state-of-the-art FPGA-based tester for digital systems, no tester can combine different advanced test strategies, such as advanced functional applications (SLT) and structural tests (scan-based). The proposed FPGA-based flexible Test Equipment aims to overcome the limitations in Research Labs (at universities or in companies); since an industrial tester, called Automatic test equipment (ATE) is expensive and requires physical space in the lab.

The proposed tester enables researchers in the direction of SLT and how to extract useful information and apply different test strategies effectively. More in detail, the proposed flexible tester enables:

- To Execute manufacturing tests (scan-based test patterns).
- To Command the BIST engines.
- To Execute Functional test programs.
- To Extract information from executed functional test programs, e.g., instruction traces.
- To Allow tester-SoC functional communications, based on communication protocols.
- To Gather information for manufacturing process refinement.

Its modular design allows the introduction of additional IPs for functional hardware testing, for example, communication peripherals, or different configurations for structural tests, for example, single or multiple scan chains for structural tests.

Experimental results show the performance, FPGA utilization, and the profiled execution of the software stack. The proposed tester is used to test with structural and functional stimulus an Automotive SoC produced by STMicroelectronics.

Moreover, the proposed hardware architecture and software stack are ported to a different FPGA family to show the design's portability.

The paper provides in Section II some background about manufacturing test flow and an overview of the state-of-the-art for FPGA-based testers. Section III explains the proposed hardware architecture, while Section IV describes the software framework, and Section VI shows the experimental results obtained from the case study and evaluates the design portability, while Section VII draws some conclusions.

II. BACKGROUND

A. Manufacturing test flow

As transistors undergo scaling, the density of transistors per unit area rises, as noted in the work by Campbell et al. (2010) [1]. This significant advancement has resulted in a substantial increase in the complexity of integrated circuits, consequently amplifying the challenges in the testing domain [1].

To address the escalating testing complexity, structural tests, particularly those based on scan techniques, have been implemented. These approaches aim to streamline the testing process and enhance the automation of test pattern generation for specific integrated circuits, as highlighted in [7].

The testing process is segmented into distinct stages, each targeting specific defects with the goal of identifying and discarding faulty devices, as discussed in works by Chen et al. (2020) [23] and Polian et al. (2020) [3]. These stages include:

- Wafer Test: Conducted at the wafer level, this stage assesses the primary electrical functionalities of the chip and executes structural test patterns.
- Package Test: Performed at the package level, this stage measures and tests essential electrical characteristics of the pins.
- Burn-In: Primarily employed for automotive and safety-critical devices, the Burn-In phase aims to exacerbate latent defects [24], [25]. Devices manufactured with weaknesses can be captured in subsequent test steps after Burn-In.
- Final Test: This stage is designed to detect faults by applying a combination of structural and functional tests [26].
- System-Level Test: Added as an additional final test for safety-critical and automotive devices [27], it verifies the correctness of devices using complex functional programs.

With the growing complexity of modern System-on-Chips (SoCs), defects may manifest in the field, leading to a field-return scenario where structural tests might need to be re-executed by the manufacturer, potentially resulting in a No-Trouble-Found (NTF) situation [3].

B. State-of-the-art for FPGA-based tester

In order to provide placement of the proposed tester into the state-of-the-art FGPA-based tester, Table I is presented. The Comparison is done from the perspective of FPGA-based tester for digital systems. Moreover, in the literature there is a lack of FPGA-based testers, advanced and non, in the last years. Industrial ATEs were not taken into account because the purpose of this work is not to compete with ATEs on

TABLE I: Comparison between different FPGA-based tester.

Tester	Description	Host PC	Hardware requirements	Structural capabilities			Functional capabilities			
				Max Pattern feeding speed	Pattern Memory requirements	Single/Multi chain	Executable flashing	Executable debugging	Instruction Trace capabilities	Protocol-based communication
[14]	FPGA-based with data compression	Required	FPGA design	50MHz	5MB	NA	yes	NA	NA	NA
[15]	FPGA-based with on-board power supplies	Required	FPGA custom board	NA	NA	NA	NA	NA	NA	NA
[16]	FPGA-based applying simulation-extracted patterns	Required	FPGA design	6 MHz	576KB	NA	NA	NA	NA	NA
[17]	FPGA-based for multi-IC, on-chip pattern generation	Required	FPGA design and multiple hw connections	1 MHz	NA	NA	NA	NA	NA	NA
[18], [19]	FPGA-based functional protocol	Required	FPGA design	NA	NA	NA	NA	NA	NA	Yes
[20]	FPGA-based functional GPIO test propagation delay and power consumption test	Required	FPGA design Power measurements HW	200 MHz (funct.)	32KB	NA	NA	NA	NA	Yes
[21]	FPGA-based with precomputed scan pattern, IOs stimulus	No	FPGA design Socket-Specific test board	50 GHz	16GB	Multi	NA	NA	NA	NA
[22]	FPGA-based with on-chip pattern generation (SW) precomputed scan pattern sequencer	Required	FPGA-design	6.25 MHz	Max 2GB	Single	NA	NA	NA	NA
Proposed Tester	FPGA-based with on-chip pattern generation (HW) precomputed scan pattern sequencer and full functional capabilities	No	FPGA-design SW development	100 MHz[1]	Max 2GB	Single/Multi	yes	yes	yes	yes

[1] The DUT was capable of reaching only 20MHz due to the device's I/O limitation on the maximum frequency.

979-8-3503-6689-1/24 $31.00 © 2024 IEEE

the market, but to allow research centres with low budgets to perform their activities on in-production devices, at affordable prices. Table I compares different FPGA-based testers from a structural and functional perspective.

In [14], a tester capable of compressing test programs is presented, but it has no enhanced feature for functional test programs. Furthermore, in [15]–[17], different testers are presented for applying test patterns on the IOs of SoCs, by either generating on-the-fly or applying precomputed patterns.

In [18], an FPGA-based tester is presented for only functionally verifying a communication. Meanwhile, in [20], it focuses on testing GPIO and verifying propagation delay and power consumption.

On the other hand, authors in [21] presents a tester capable of feeding precomputed test patterns (in multi-chain configuration) at 50GHz, thanks to a high-speed interface between a Socket-specific test board and custom IPs. However, it does not provide any functional capabilities.

The proposed tester aims to enhance a preliminary version (with only limited structural capabilities) presented in [22]. However, none of the aforementioned FPGA-Based Testers provide a combination of structural and advanced functional capabilities as well as design modularity, functional test program information extraction, and protocol-based communication.

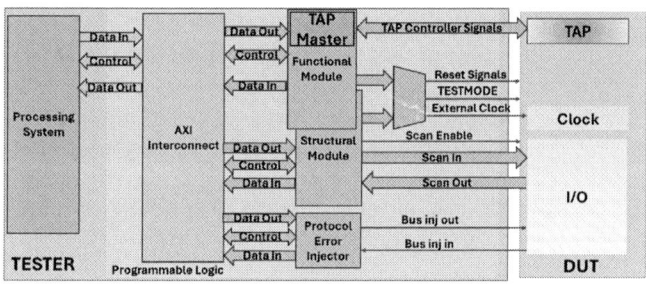

Fig. 1: FPGA-based Tester Architecture.

III. PROPOSED HARDWARE ARCHITECTURE

It is challenging for research labs, in companies or universities, to develop new test strategies since semiconductor ATE may not be affordable, and unnecessary. This greatly limits the ability of research labs.

The hereinafter proposed flexible FPGA-based tester aims at overcoming the aforementioned limitations. It requires an affordable cost and space from research labs, compared to industrial ATE. The hardware architecture of the proposed tester is shown in Figure 1, it is a flexible FPGA-based tester tailored for SLT, with the additional capabilities to structurally test Devices Under Test (DUTs), addressing issues of adaptability, modularity, and cost-effectiveness found in conventional testers. This has been achieved using the capabilities of a System on Chip which hosts both a Processing System and a Programmable Logic, in which the digital design has been synthesized, capable of communicating with each other.

In the following subsections, the two main hardware blocks of the proposed tester are presented. The first one focuses on functional testing by exploiting the JTAG Standard [7]. The

structural module is capable of executing the reconfiguration of the circuit for structural tests.

A. Functional Module

The IEEE 1149.1 JTAG standard defines the Test Access Port (TAP) [7] as the interface to access and control the device in debug mode. All standardized devices integrate a slave TAP Controller which, through the five required pads (TCK, TDI, TDO, TMS and TRST) can be controlled via a master implemented in the tester.

Following this principle, the functional module for the proposed tester, in Figure 2, was developed as a block composed of a Master TAP Controller, which provides full access to the internal TAP FSM of the DUT. To access the registers and the overall debugging features, an IP has been designed around this module.

Fig. 2: Functional Module Block Design

The module can be controlled through an AXI bus by the PS. It is attached to two double port Block RAMs, dedicated to store the data exchanged on the JTAG TDI and TDO lines. The address to read and write are provided to the module through two GPIOs, and can be accessed, not concurrently, by the PS.

A third GPIO is used to activate the module and wait for the completion of the execution, acting as a hardware mutex, avoiding concurrent access from different users. In Figure 1 the main structure can be observed.

The operations to be executed by the functional module, in compliance with the JTAG standard, are provided through an instruction encoded to accept the following information:

- Reset status of the TAP.
- A bit to reach the Test Logic Reset state.
- A bit to decide whether a write operation needs to be performed on an Instruction or Data Register.
- Number of bits to shift out through TDI.
- One bit encoding whether the read operation needs to be performed on an Instruction or Data Register.
- Number of bits to shift in through TDO.
- Bits controlling the status of the Reset signals of the DUT (Power-On Reset and External Reset pads).

B. Structural Module

The structural block is designed starting from the previously designed functional module. The functional module can be used to execute the Test Mode Entry, by just adding minor features. Furthermore, the tester requires the ability to sequence patterns into a scan chain.

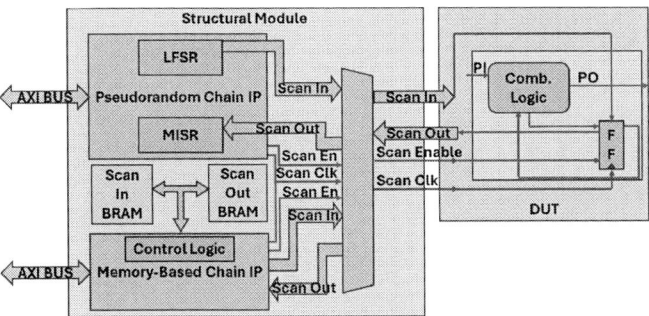

Fig. 3: Structural Module Block Design

To do this, a module has been designed (in Figure 3) to take as input the patterns and output them to the scan in pins, while reading the scan outs. The design has two Block RAMs, one for the patterns to scan in, and the other with the data read from the chain. To keep it modular, the number of scan chains can be set up before synthesis thanks to a configurable IP. It works by reading the first word of the BRAM, which contains the number of bits to shift in the chain, then reading N words (with N being the number of chains) of 32 bits, shifting them in, and repeating the last two steps until the number of bits requested has been scanned into the chain. Reading the chain is done the same way. Additionally, the module is able to automatically generate patterns through a Pseudo-Random Pattern Generator (PRPG), and collecting the output signature, a condensed representation of a circuit's output response, through a Multiple Input Shift Register (MISR), which compresses the data read from the chains.

IV. PROPOSED SOFTWARE FRAMEWORK

The Hardware Architecture needs to be addressed from the Processing System via software, to exploit as much as possible the Hardware to create complex functions. To make sure that the tester's software kept both the highest modularity possible and a user-friendly design, scripting via the Python language is used, via linux drivers to access the design on the FPGA. All modules in the PL are accessible and controllable by the

Functional Debug Script	Trace Download Script
Fault Injection Script	Test-Mode Entry Script
TCU managing script	Multiple Chain Pattern Sequencer
STIL Parser	LBIST customize and launch script
Pseudo-Random Pattern Sequencer	Single Chain Pattern Sequencer
PYNQ Framework	APP
Custom HALs	
Operating System	
Hardware Design	

Fig. 4: Software stack

PS through the AXI. On the latter, a customized version of Linux runs, the Pynq OS [28]. The usage of an operating system comes with all the features it supports, most notably the *mmap* library [29], which was used to create a Hardware Abstraction Layer for the peripherals connected to the AXI. This way, the hardware keeps just the basic features to access the DfT (scan chain, LBIST engine) and debug infrastructure

of the device, while the software is customizable.

On top of the HALs, a software stack resides (Fig. 4), which, through the aforementioned instruction to be passed to the hardware module, gives the tester the capabilities to perform the following tasks:

- Access to the on-chip Test Access Port (TAP) controller to extract basic device information.
- Extract information on registers and memories.
- Set on-chip breakpoints.
- Modify registers and memories.
- Inject errors in the memory words and their relative Error Correction Code (ECC).
- Execute Instruction trace download in both single and multi-core applications.
- Flash erase, programming and verify.
- Perform Test-Mode Entry for structural testing for single or multiple scan chains.
- Structural Testing for different fault models.
- Diagnostics of DUTs through the data retrieved by applying precomputed stimuli through a STIL file [30].

All the above capabilities are device specific, and are to be coded for each Device Under Test (DUT) used. The generic HALs to interface with the debugger, as well as the hardware design, are generic and do not necessitate modifications, unless required. The usage of PYNQ framework allows to easily read, understand and script the operations.

V. ENHANCING MODULARITY

By using a System-On-Chip featuring an FPGA, the proposed tester gains an inherent modularity, both on the hardware and software level, which gives the user the ability to create and integrate specialized modules directly on the FPGA fabric. These tailored modules seamlessly integrate with the core design, enabling targeted and comprehensive testing scenarios. This approach offers distinct advantages, including:

- Scalability: Easily add or remove modules, customizing the design for specific tests;
- Flexibility: Effortlessly customize to match requirements;
- Cost-effectiveness: Eliminate the need for expensive dedicated testing hardware, reducing costs.

An application example can be the necessity to functionally test the CAN Module of the DUT. SLT are used to functionally test the DUT, and peripherals are a fundamental block in modern safety-critical systems. In this case, a generic communication peripheral can be added in the Programmable Logic to communicate with the chip. Moreover, if the user needs a protocol error injector to test the answer of the communication peripheral to errors within the frame, a design can be integrated as part of the tester [31].

VI. EXPERIMENTAL RESULTS

In the following subsections, a description of the experimental setup is presented, with the tester connected to an Automotive SoC. Moreover, a description of the performances and utilization of the design are given.

979-8-3503-6689-1/24 $31.00 © 2024 IEEE

TABLE II: Experimental results for Hardware Design[1].

[1]Comparison with State-of-the-art testers in Table I is missing due to lack of comparison data.

Version	Modules	Eval. Board	Area			Timing			Power Consumption (W)	
			LUT(%)	FF(%)	BRAM(%)	Maximum Clock (MHz)	WNS (ns)	WHS (ns)	Static	Dynamic
V1	Functional	ZCU104	13,411 (5.82)	15,132 (3.28)	15 (4.81)	0.5	9.42	0	0.714	2.691
V2	Functional	ZCU104	8,841 (3.84)	9,013 (1.96)	4 (1.28)	100	2	0.017	0.715	2.737
V3	Functional + Structural Single Chain	ZCU104	14,324 (6.22)	16,158 (3.51)	8 (2.56)	100	0.037	0.01	0.714	2.643
V3	Functional + Structural Multi-Chain	PYNQ Z2	7236 (13.6)	8827 (8.30)	8 (5.71)	100	0.168	0.02	0.141	1.362

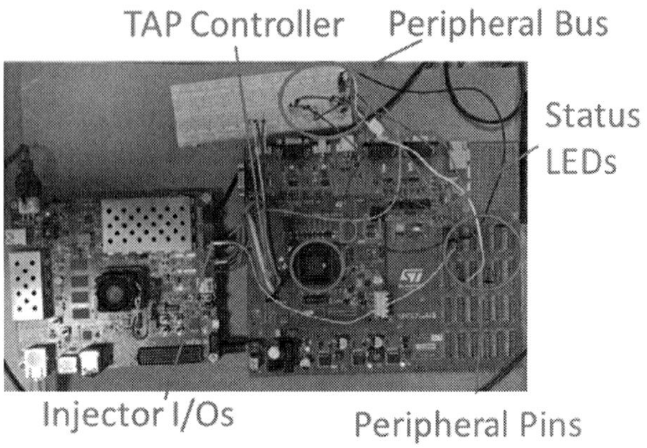

Fig. 5: Experimental setup with a ZCU104 Evaluation Board.

A. Experimental setup

The tester has been synthesized for an AMD-Xilinx Zynq Ultrascale+ MPSoC ZCU104 Evaluation Kit [32]. It includes a ZU7EV device is equipped with a quad-core ARM Cortex-A53 applications processor, dual-core Cortex-R5 real-time processor, Mali-400 MP2 graphics processing unit. The FPGA fabric has 504k System Logic Cells and 38Mb of memory.

The Device Under Test (DUT), in Figure 5 is an automotive SoC produced by STMicroelectronics, ISO26262 ASIL-D compliant. The SoC has a multicore architecture with three 32-bit cores using the PowerPC Variable-Length Encoding (VLE) instruction set. It has 6Mbyte of Flash memory and 128Kbyte of general-purpose SRAM.

In Figure 5, the tester is connected to the TAP Controller of the DUT for debugging purposes, and to the CAN Bus, to perform more in-depth functional testing through error injections.

B. FPGA Utilization and Performance evaluation

Three versions of the design have been evaluated. The first was a beta version, with a heavier design based on a core-like design (divided into a Datapath and a Control Unit), and was capable of reaching max 500kHz of clock speed. This design has a high utilization of LUTs due to its poor design choices. In its second version, the tester was completely re-designed as explained in Section III, making it lighter and able to reach up to 100MHz, halving the required logic elements. The final version just adds the capabilities to enable structural testing in single chain configuration, increasing the utilization. In Table II the utilization, timing and power are reported.

Speeds higher than 20MHz were tested with logic analysers and signal generators because the DUT's evaluation board I/Os could not manage higher speeds.

C. Software Performance Evaluation

Thanks to the ditching of Pynq [28] in favour of the more performing *mmap* Linux library with custom HALs based on Python, the performances of the debugger grew 10x, going from needing $10ms$ to read the IDCODE to $1ms$ at 500kHz. A significant improvement can be brought by using compiled code instead of interpreted. In Table III, an overview of the results for the software stack is given, showing the execution time for different operations performed by the system. For these benchmarks, the final version of the design is used, with a clock frequency of 5MHz.

TABLE III: Software Stack Profiling using V3[2].

[2]Comparison with State-of-the-art testers in Table I is missing due to lack of comparison data..

Operation	Benchmark Size	STIL Size	Execution Time (s)
Flash Erase, Program, Verify	290kB	N/A	103.1
Program Trace Multicore	1,383,003 instructions	N/A	93,420
Pre-computed pattern stimuli, Single Chain, $\sim 700k$ FFs	N/A	1 pattern	1.57
Pre-computed pattern stimuli, 8 Chains, ~ 900 FFs per chain	N/A	136 patterns	26.61

D. Portability Evaluation

While the ZCU104 served as a suitable initial platform, its high cost and limitations in I/O expansion without additional hardware misaligned with the design goals. Recognizing this, the design was successfully ported to the Pynq Z2 [33], a significantly more affordable board offering over 30 readily available I/Os and more than enough logic elements, as shown in Table II, giving us the possibility of structurally testing the DUT through its multiple scan chains. This not only addresses our current resource constraints but also provides the flexibility to accommodate future improvements and add-on modules, making the Pynq Z2 an ideal long-term solution for our evolving design. The usage of this board does not bring any downgrade in performances, the design is still capable of functional and structural testing up to 100MHz, and multiple scan chain testing could be performed thanks to the additional I/Os on the board, as shown in II. In order to port the design to a different FPGA (of the same family) it is necessary to adjust only the pinout.

VII. CONCLUSIONS & FUTURE WORKS

Companies need System-Level Testers who can seamlessly blend functional and structural testing [13]. However, integrating these test steps demands significant investment in collaboration, expertise, and tools. Finding Test Equipment that combines these diverse test methods remains a complex

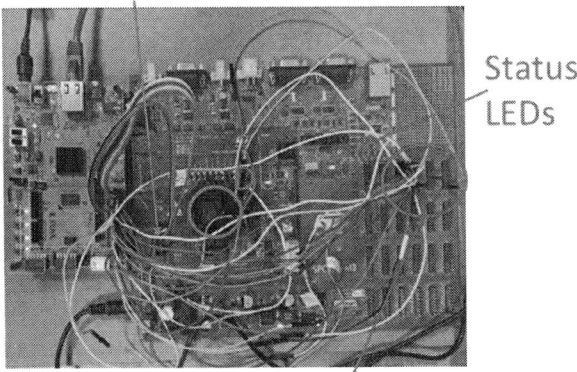

Fig. 6: Experimental Setup with a Pynq Z2 Evaluation Board.

and costly hurdle. While industrial semiconductor testers exist, no FPGA-based tester on the market unifies advanced SLT and structural testing.

The proposed tester aims to help solve these problems by using a standard architecture, with modular capabilities. It aims to give the users the ability to run multiple typologies of tests, ranging from functional to structural tests, leaving the space to enhance its features without additional external hardware requirements (or minimal).

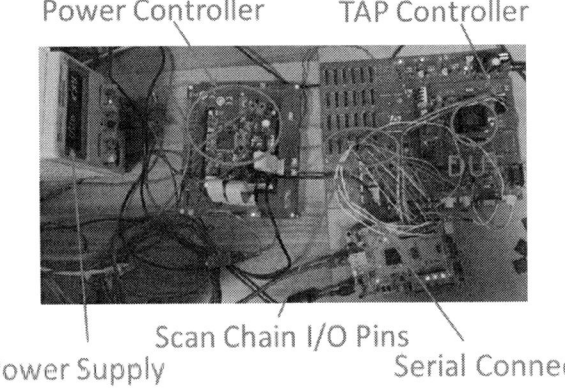

Fig. 7: Experimental setup with power controller board.

The proposed approach enables, at low costs, researchers to perform:

- General device debugging.
- Program tracing as in [34], [35].
- Load programs into the DUT and test communications with external peripherals [31].
- Execute Test-Mode Entry and use of scan chains (single/multiple) using pre-computed or pseudo random patterns.
- Customization and launch of Logic BISTs engine.
- Diagnosis of DUTs through DfT.

The tester's abilities can be enhanced by adding the capabilities to control the input voltage of the DUT, with an

experimental setup like the one shown in Figure 7. The idea is to execute diagnostics (functional and structural) and tests, forcing different voltages for characterizing faults due to process variations, at the cost of an additional custom board for voltage regulation.

REFERENCES

[1] M. Campbell, "Plenary presentations: Keynote: The product complexity and test — how product complexity impacts test industry," in *IEEE ETS*, 2010, pp. 9–9.

[2] "Iso 26262-[1-10], road vehicles – functional safety," 2011.

[3] I. Polian *et al.*, "Exploring the mysteries of system-level test," in *IEEE ATS*, 2020.

[4] G. Iaria *et al.*, "A novel pattern selection algorithm to reduce the test cost of large automotive systems-on-chip," in *IEEE LATS*, Sep. 2022.

[5] A. Lotfi *et al.*, "Configurable architecture for memory bist," in *IEEE EWDTS*, Sep. 2011.

[6] D. Tille *et al.*, "Towards an automated flow for implementation of dedicated lbist scan chains for functional safety," *TUZ*, 2020.

[7] "Ieee draft standard test access port and boundary scan architecture," *IEEE P1149.1/D2012.e27, September 2012*, pp. 1–434, 2012.

[8] D. Appello *et al.*, "System-level test: State of the art and challenges," in *IEEE IOLTS*, 2021.

[9] D. K. R. Tipparthi and K. K. Kumar, "Concurrent system level test (cslt) methodology for complex system-on-chip," in *IEEE EPTC*, 2014.

[10] S. Biswas and B. Cory, "An industrial study of system-level test," *IEEE Design Test of Computers*, vol. 29, no. 1, pp. 19–27, 2012.

[11] P. Reichert, "System Level Test," *Teradyne*.

[12] P. Bernardi *et al.*, "Applicative system level test introduction to increase confidence on screening quality," in *IEEE DDECS*, 2020.

[13] F. Almeida et al., "Effective screening of automotive socs by combining burn-in and system level test," in *IEEE DDECS*, 2019.

[14] L. Ciganda *et al.*, "An enhanced fpga-based low-cost tester platform exploiting effective test data compression for socs," in *IEEE DDECS*, 2009.

[15] A. Patel *et al.*, "Fpga based low-cost portable tester with on-board supplies," in *IEEE ICECA*, 2020.

[16] L. Mostardini *et al.*, "Fpga-based low-cost system for automatic tests on digital circuits," in *IEEE ICECS*, 2007.

[17] B. Rabakavi and S. Siddamal, "Design of high speed, reconfigurable multiple ics tester using fpga platform," in *IEEE ICEECCOT*, 2018.

[18] T. Lyons *et al.*, "The implementation and application of a protocol aware architecture," in *IEEE ITC*, 2013.

[19] Y. Fan and Z. Zilic, "Ber testing of communication interfaces," *IEEE Transactions on Instrumentation and Measurement*, vol. 57, 2008.

[20] A. A. Bayrakci, "Elate: Embedded low cost automatic test equipment for fpga based testing of digital circuits," in *IEEE ELECO*, 2017.

[21] M. de Carvalho *et al.*, "A flexible stand-alone fpga-based ate for asic manufacturing tests," in *IEEE LATS*, 2018, pp. 1–6.

[22] F. Angione *et al.*, "Test, Reliability and Functional Safety trends for Automotive System-on-Chip." European Test Symposium, 2022.

[23] C. He and Y Yu, "Wafer level stress: Enabling zero defect quality for automotive microcontrollers without package burn-in," *IEEE ITC*, 2020.

[24] A. Benso *et al.*, "Atpg for dynamic burn-in test in full-scan circuits," in *2006 15th ATS*, 2006, pp. 75–82.

[25] F. Angione *et al.*, "A toolchain to quantify burn-in stress effectiveness on large automotive system-on-chips," *IEEE Access*, pp. 1–1, 2023.

[26] A. Birolini, "Reliability engineering theory and practice," *Springer*, 2017.

[27] H. H. Chen, "Beyond structural test, the rising need for system-level test," in *IEEE VLSI-DAT*, 2018.

[28] Xilinx, "Pynq," https://github.com/xilinx/pynq.

[29] L. Manual, "mmap," https://man7.org/linux/man-pages/man2/mmap.2. html.

[30] "Ieee standard test interface language (stil) for digital test vector data," *IEEE Std 1450-1999*, pp. 1–140, Sep. 1999.

[31] F. Angione *et al.*, "A system-level test methodology for communication peripherals in system-on-chip," in *submitted to IEEE Transactions on Computers,2024*.

[32] AMD-Xilinx, "Zynq ultrascale+ mpsoc zcu104 evaluation kit," https://www.xilinx.com/products/boards-and-kits/zcu104.html.

[33] ——, "Zynq ultrascale+ mpsoc zcu104 evaluation kit," https://www.tulembedded.com/fpga/ProductsPYNQ-Z2.html).

[34] F. Angione *et al.*, "An innovative strategy to quickly grade functional test program," in *IEEE ITC*, 2022.

[35] ——, "A guided debugger-based fault injection methodology for assessing functional test programs," in *IEEE VTS*, 2023.

Large Language Model-based Optimization for System-Level Test Program Generation

Denis Schwachhofer[1,3] ⓞ, Peter Domanski[2] ⓞ, Steffen Becker[3],
Stefan Wagner[3,4], Matthias Sauer[5], Dirk Pflüger[2], Ilia Polian[1]

[1]Institute of Computer Engineering and Computer Architecture, University of Stuttgart, Stuttgart, Germany
[2]Institute for Parallel and Distributed Systems, University of Stuttgart, Stuttgart, Germany
[3]Institute of Software Engineering, University of Stuttgart, Stuttgart, Germany
[4]Technical University of Munich, Heilbronn, Germany
[5]Advantest Europe, Boeblingen, Germany

Abstract—**System-Level Test (SLT) is essential for testing integrated circuits, focusing on functional and non-functional properties of the Device under Test (DUT). Traditionally, test engineers manually create tests with commercial software to simulate the DUT's end-user environment. This process is both time-consuming and offers limited control over non-functional properties. This paper proposes Large Language Models (LLMs) enhanced by Structural Chain of Thought (SCoT) prompting, a temperature schedule, and a pool of previously generated snippets to generate high-quality code snippets for SLT. We repeatedly query the LLM for a better snippet using previously generated snippets as examples, thus creating an iterative optimization loop. This approach can automatically generate snippets for SLT that target specific non-functional properties, reducing time and effort. Our findings show that this approach improves the quality of the generated snippets compared to unstructured prompts containing only a task description.**

Index Terms—**Large Language Models, Test Generation, System-Level Test, Optimization, Functional Test**

I. INTRODUCTION

System-Level Test (SLT) is increasingly used to improve the quality assurance of complex Systems-on-Chip (SoC). In SLT, the Device under Test (DUT) is placed into an environment that emulates its end-user environment as closely as possible. Typically, SLT entails executing off-the-shelf software in the DUT's operational mode [1]. Presently, test engineers rely on the manual construction of the test suite, drawing from field returns and personal insights. During SLT, the DUT is treated as a black box, mainly because Intellectual Property (IP) cores within the SoC may only be accessible as black boxes. External entities lacking DUT-specific knowledge may also engage in SLT, such as integrators.

It is expected that SLT programs will explore execution paths and transactions that are not typically covered by structural tests, thus revealing defects that would otherwise remain undetected [2]. Nevertheless, certain defects may remain latent, requiring specific conditions for activation or detection, such as particular temperature ranges or gradients within the DUT. Consequently, SLT workloads targeting non-functional properties of the DUT hold significance. However, this endeavor presents challenges, mainly due to missing structural insights into the DUT.

Large Language Models (LLMs), such as transformer-based neural networks like Code Llama [3], are commonly employed for code synthesis tasks. Their efficacy has been evidenced across various coding tasks, underscoring their potential utility in software development. Wang *et al.* [4] highlight the prospects for LLMs in software testing, encompassing unit test case generation and test oracle creation. Leveraging methodologies effective in software test generation is a promising foundation for generating SLT programs from natural language prompts.

Previous research showed that while LLMs effectively generate SLT programs controlling non-functional properties, there is room for improvement even after fine-tuning. In [5], we introduced a Reinforcement Learning-based framework with basic prompts to tune and evaluate LLMs, demonstrating significant improvements. However, we concluded that optimizing prompt design is another aspect that can improve performance.

This work investigates the effectiveness of using LLMs as an iterative optimizer by combining state-of-the-art prompt engineering and optimization techniques. Therefore, we run the following experiments: First, we use basic and Structural Chain of Thought (SCoT) prompts without applying any optimization loop. Then, we run optimization twice, starting with a basic prompt. In the second run, we replace the basic prompts with SCoT prompts. Both prompt types contain multiple examples and ask the LLM to return a better code snippet. We compare the best results from each run against each other and relevant statistics, such as pass@k metrics, which evaluate the percentage of snippets that compile and run.

II. RELATED WORK

Yang *et al.* [6] introduce Optimization by PROmpting (OPRO), an iterative methodology that uses the LLM as an optimizer. Their model incorporates two inputs: previously generated outputs with associated scores and the problem description, enabling the formulation of new prompts in each optimization iteration, thereby enhancing test precision. This novel approach showcases versatility in various applications, making it promising for SLT program synthesis. Thus, it has already found application in unit test generation [7]. Since its introduction, the OPRO concept has been extended to various optimization settings, such as Gradient Descent, Hill Climbing, or Black-Box Optimization [8]. Additionally, multiple works proposed employing it in frequently

979-8-3503-6689-1/24 $31.00 © 2024 IEEE

used optimization frameworks, e.g., for genetic or evolutionary algorithms, replacing traditional selection, crossover, or mutation operations [9]–[11].

Liu *et al.* [10] first introduced a self-adaptation mechanism that dynamically adjusts the LLM's temperature, demonstrating superior performance over OPRO, especially w.r.t. exploration capabilities. While most works focus on single-objective optimization, Liu *et al.* [12] expand their work [10] in a multi-objective optimization framework, achieving comparable performance to traditional multi-objective evolutionary algorithms.

Chain-of-Thought (CoT) prompting [13] has gained attention in natural language processing. It describes a prompting design that consists of a series of intermediate reasoning steps and has improved LLM performance, especially in complex reasoning tasks. Since the first introduction of CoT, many variants have been proposed that can be divided into three categories: Manual CoT, automatic CoT, and semi-automatic CoT [14]. Moreover, several works introduce modifications to the original chain structure, such as a tree structure [15] or graph structure [16].

In previous work, we demonstrated that fine-tuning an LLM using Reinforcement Learning increases the quality of snippets for SLT compared to querying it off-the-shelf [5]. For this purpose, we have fine-tuned an LLM to generate snippets with high Instructions per Cycle (IPC). The results were promising; nonetheless, we noted the need for improved prompt engineering. This work focuses on novel prompting techniques and extended conversations in a multi-step structure to receive more reasonable answers off the shelf and thus without fine-tuning. Furthermore, we do not rely solely on IPC but leverage multiple metrics to optimize SLT programs. Therefore, we propose an iterative optimization framework inspired by Simulated Annealing (SA) for multi-objective SLT program generation, which in the future can be combined with the approach in [5] to fine-tune the optimization framework for SLT program generation.

III. PROPOSED APPROACH

This section details our LLM-based optimization framework for SLT program generation. Fig. 1 provides an overview of the proposed optimization framework. We start with a pool of N_{Pool} random candidates from which we randomly sample n_{examples} solutions that we include in the second step (prompt generation). Afterward, the LLM generates an output snippet, and we evaluate its score based on a given objective. Depending on a selection criterion, e.g., the score value, the initial pool of candidates is optionally updated with the new solution. Finally, a stopping condition, e.g., the maximum number of iterations, decides whether to continue the iterative optimization process, including an optional LLM temperature adjustment or to stop and return the best-found candidate solution. In the subsequent sections, we describe the individual modules and their use in optimizing SLT program generation.

A. Structured Chain-of-Thought Prompting

Chain-of-thought (CoT) prompting is a novel technique that has been successful in various reasoning tasks [14], [17]. It consists of multiple intermediate natural language reasoning steps, providing a structured framework to explore thought and uncover deeper layers of understanding. After several intermediate steps, the output is generated, e.g., a code snippet.

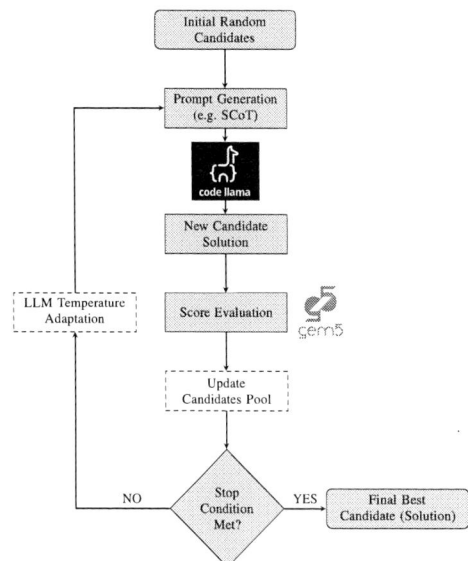

Figure 1: Schematic of our LLM-based optimization framework for SLT program generation.

However, CoT was designed for natural language generation tasks and thus often performs poorly in code generation tasks, such as SLT program generation. Therefore, our approach builds upon Structured CoT (SCoT) [18], which uses structural information (program structures) in code snippets, such as sequences, loops, and branches, to build CoTs specifically for code generation tasks. SCoT is language-agnostic, making it a suitable method for our approach to generating C code snippets. Moreover, it is robust to different examples in the prompts, and its two-step pipeline reduces the risk of error accumulation. Nonetheless, the performance of an LLM largely depends on the prompt design [19]. Thus, an effective prompt strategy for SLT program generation needs further refinements. Section IV-A provides more details on our specific SCoT prompt design for SLT program generation. In the following section, we describe how we include LLMs and SCoT prompts within an iterative optimization framework inspired by temperature schedules of Simulated Annealing (SA) and evolutionary algorithms, e.g., a pool of candidate solutions that is regularly updated.

B. LLM-based Optimization

LLMs have shown promising results in non-linguistic tasks involving numerical optimization, e.g., black-box optimization tasks. This work introduces LLMs as multi-objective optimizers for system-level test program generation. Therefore, we use two recently introduced approaches: OPRO [6] and LLM-driven evolutionary algorithm (LEMA) [10] or its multi-objective extension (MOEA) [12], respectively. Furthermore, we propose a temperature schedule similar to Liu *et al.* [10], which first introduced a self-adaptation mechanism, dynamically adjusting the temperature of LLMs to effectively address the exploration versus exploitation trade-off in optimization problems.

We propose the following stop conditions, but other user-defined stopping criteria are possible:

- $n_{snippets}$ snippets have been evaluated
- n_{evals} total number of evaluations, with $n_{evals} > n_{snippets}$

We introduce the second condition because we observed that LLMs can start to hallucinate (nonsense answers) or return no answers, possibly getting stuck in either state.

1) Pool Update Strategies: We use a pool of examples to increase the diversity of the generated snippets. In every iteration, we sample $n_{examples}$ snippets where $n_{examples}$ is smaller than the total pool size. Two strategies for sampling can be employed: always pick the top $n_{examples}$ snippets or randomly sample them from the pool. In the latter case, the probabilities can be weighted by their score. The advantage of random sampling is that the diversity of examples presented to the LLM is higher than in the former case. This is primarily because, over time, the diversity of the top $n_{examples}$ snippets will decrease while the LLM is refining existing ones further.

The pool is updated when a new snippet fulfills specific requirements, e.g., has a Levenshtein distance above a certain threshold or a better score than any other snippet. Optionally, the snippet with the lowest score can be removed to increase the pool's average score. This is beneficial when sampling randomly, as the LLM will use the examples as a reference for its output.

2) Temperature Schedule: The temperature of an LLM defines the randomness in the output, or, in other words, the "creativity" of the LLM. Higher temperatures increase output randomness, thus allowing more diverse answers, which can eventually become nonsensical. In contrast, lower temperatures give more conservative responses until they become deterministic (always predict words with the highest output probability), increasing the risk of repetitive outputs. More formally, the temperature is a parameter that controls the probabilities of randomly sampling highly probable candidate words different from the most likely word at the final prediction layer.

In our work, we aim to use both properties to our advantage in the iterative optimization process. At first, we set the temperature to a higher value to generate diverse snippets, putting the focus on exploration. However, with an increasing number of iterations, the temperature should decrease so that the LLM can refine promising snippets instead, exploiting already gained knowledge. We allow the temperature to increase if the LLM gets stuck in a local optimum to avoid overly extensive exploitation, potentially resulting in sub-optimal solutions.

Therefore, we introduce a temperature schedule that defines how temperature behaves with increasing iterations. This schedule increases or decreases the LLM's temperature depending on the chosen parameters, e.g., Levenshtein distance to previous or top n snippets or a score based on feedback metrics. Moreover, we implement a restart mechanism, i.e., set the temperature to its initial value if specific conditions are fulfilled, such as, for example, the last n snippets are similar, their Levenshtein distance is below a threshold, or we reached the lower bound of the temperature schedule. The goal of the schedule is to guide the LLM from the local optima to the global optimum.

IV. EXPERIMENTAL SETUP

In this section, we introduce our experimental settings. First, we describe the setting of the LLM we used and show our task-specific prompt design. Afterward, we illustrate the evaluation process and highlight the challenges of the SLT program generation task using LLM-based approaches. Tab. I shows the four scenarios we define for our experiment.

Table I: Scenarios investigated in this work

Name	Description
Basic	Basic prompt without optimization
BasicOpt	Basic prompt with optimization
SCoT	SCoT prompt without optimization
SCoTOpt	SCoT prompt with optimization

```
You are a C code generator that optimizes the instructions
↪   per cycle of a target processor.
In the following I will describe your task and give details
↪   about the target processor.
Your task is to write a single C program that aims for a
↪   high number of instructions per cycle.
The processor you target has a super-scalar pipeline with
↪   seven stages, out-of-order execution and three issues.
The processor has 4GB of RAM available.
All three issues contain an ALU, one contains a
↪   Multiplier/Divider, one contains the FPU and one
↪   contains an Address Generation Unit.
There is no operating system running on the target
↪   processor.
The code must contain an infinite loop, that contains
↪   instructions or function calls.
Please wrap the output code in ```.`
```

Listing 1: Prompt used in Basic and BasicOpt scenarios

A. Prompts

The basic prompt shown in Listing 1 introduces the context and the task to Code Llama and is used in the Basic and BasicOpt scenarios. For the BasicOpt scenario, we add examples in the format shown in Listing 2. If we additionally use the proposed pool update strategy, then the examples are always updated to contain the best $n_{examples}$ snippets found in the optimization run so far.

Listing 3 shows the structure of the first SCoT prompt used in the experiments. The first prompt is used to generate the SCoT, in the form of a rough solving process in natural language, for the second prompt. We include $n_{examples}$ examples so the LLM can understand the expected structure of the SCoT prompts. We keep the task description as simple as possible because we observed that if more details, e.g., about the processor, are present, the LLM focuses more on these and does not generate the desired structure.

Listing 4 shows the format of the second SCoT prompt to generate code. We also include $n_{examples}$ examples here and update the examples to contain the best $n_{examples}$ if we use the pool update strategy. The two-step process helps the LLM to generate the desired output, i.e., C code that fulfills the task. Furthermore, we provide additional feedback to the LLM in the SCoTOpt scenario by including the metric values for each example in the SCoT prompt.

```
Example 1:
C code
Scores: {IPC, NumFPInsts, NumALUInsts, NumMemInsts}

Example 2:
C code
Scores: {IPC, NumFPInsts, NumALUInsts, NumMemInsts}
...
Example n_examples:
C code
Scores: {IPC, NumFPInsts, NumALUInsts, NumMemInsts}

Please look at the above examples and return a new, but
↪   completely different code snippet that achieves overall
↪   higher scores.
```

Listing 2: Examples presented in BasicOpt scenario

```
Example 1:
int main(void) {
    /*
     * <task description>
     */
}
Please understand the requirement and write a rough solving
↪ process.
You should use three basic structures to build the solving
↪ process, including sequences, branches and loops.
The necessary details should be written in natural language.
<SCoT>
More examples ...
Task:
You are an expert in software development.
Use the same structure as in the examples to solve the
↪ following task:
int main(void) {
    /*
     * Write a C main function that maximizes the
     * number of instructions per cycle and the number of
     ↪ executed instructions of a high-performance
     ↪ processor.
     * The code must contain an infinite loop, that contains
     ↪ instructions or function calls.
     */
}
```

Listing 3: Prompt to generate SCoT prompts

```
Example 1, Scores: Instructions per Cycle = <IPC>, Number of
↪ Floating-Point Instructions = <#FPInsts>, Number of
↪ Integer Instructions = <#IntegerInsts>, Number of Memory
↪ Instructions = <#MemInsts>:
<header files>
int main() {
    /*
     * <Example SCoT>
     */
    // Please check the above solving process and write a
    ↪ code based on it.
    // Note that the solving process may contain errors.
    <Code>
}
More examples ...
Task:
int main(void) {
    /*
     * <SCoT provided by LLM>
     */
}
//Please check the above solving process and write C code
↪ based on it.
//Note that the solving process may contain errors.
```

Listing 4: SCoT Prompt to generate code

B. Temperature Schedule

To adapt the temperature, we use an adaptive schedule based on the Levenshtein distance of the last snippet to previous snippets and the score of the last snippet. We include the distance because a high temperature will increase the diversity of the generated snippets and, thus, the distance. We can reduce the amount of repeated or similar snippets by increasing the temperature whenever the distance falls below a threshold d_t. We propose the formula given in Eq. (1) to calculate the new temperature. T^i is always between T_{low} and T_{init}. i is the current iteration. We divide the sum of the deltas by factor C to scale the temperature changes. This factor can be adjusted, making the schedule move faster or slower. Additionally, we change the weight of each delta per iteration. Initially, we prioritize generating diverse snippets, so the temperature is primarily adjusted based on the distance. Over time, the score becomes more important than the distance; thus, we shift the weight towards it.

$$T^{(i+1)} = T^{(i)} + \frac{(1 - \frac{i*2}{n_{\text{snippets}}})\Delta T^{(i)}_{distance} + \frac{i*2}{n_{\text{snippets}}}\Delta T^{(i)}_{score}}{C} \quad (1)$$

$$\Delta T^{(i)}_{distance} = \sigma^{-1}(d^{(i)} - d_t, \frac{d_t}{5}) - 0.5$$
$$\Delta T^{(i)}_{score} = \sigma^{-1}(s^{(i)} - s_t, \frac{s_t}{5}) - 0.5 \quad (2)$$

σ^{-1} used in Eq. (2) is the complementary cumulative distribution function of the logistic distribution. The second parameter for σ^{-1} denotes the scale. We chose the first parameter and the second parameter such that $d_{mid} = \frac{d_t}{2}$ and $s_{mid} = \frac{s_t}{2}$ respectively are the values for which σ^{-1} becomes 0.5. s_t should represent the expected mean score throughout the optimization process. d_t is a "soft" lower bound for the minimum expected distance for new snippets. We subtract 0.5 in Eq. (2) such that values below half of the threshold values are positive and those above are negative. When the distance falls below d_{mid}, it implies that the current snippet is too similar to existing ones. Similarly, scores below s_{mid} indicate that the current snippet's quality is insufficient; therefore, we need to increase the temperature to find new, more diverse, and better snippets. Otherwise, we reduce the temperature so the LLM focuses more on improving existing, known-good snippets.

Further, we implemented a restart condition in our temperature schedule: If the temperature stays at T_{low} for n_{examples} snippets, we set it back to T_{init}.

C. Objective Metrics Evaluation

We use Gem5 [20], a microarchitectural simulator that simulates multiple different ISAs (x86, ARM, RISC-V, and more) and processor architectures. Our simulation involves a super-scalar, out-of-order RISC-V processor featuring three execution units modeled after the BOOM core [21]. The simulation is carried out over 1×10^9 ticks, equivalent to one simulated millisecond. During this process, Gem5 gathers numerous performance metrics, such as the IPC, which serve as valuable feedback.

In our experiments, we chose the following metrics as feedback: Instructions per Cycle (IPC), number of floating-point instructions, number of integer instructions, and number of load/store instructions executed. Eq. (3) shows the formula for the score, where $s(x) = \sigma(x) - \sigma(0)$ and σ denotes the logistical function with scale parameter $s = 220000$ and location parameter $\mu = 550000$.

$$F = IPC + s(FP) + s(ALU) + s(MEM) \quad (3)$$

Furthermore, the output of $s(x)$ ranges from 0.0 to around 1.0, bringing it close to expected IPC values. Consequently, the magnitude of each metric in the score is equal. The definition of $s(x)$ is chosen so that zero will be added if zero instructions of a specific type are executed.

We chose IPC as a feedback value because we assume that a higher IPC relates to a higher switching activity and, therefore, a higher power consumption [22]. However, IPC itself is insufficient to generate code snippets that load the functional units of a processor. Previous experiments have revealed that the LLM favors an empty infinite loop as the optimal solution to maximize IPC in simulations. Thus, to prevent this, we also include the number of executed instructions of different types. As we will demonstrate, this causes the LLM to generate infinite loops with instructions utilizing functional units.

Another significant consideration is that LLMs exploit already-known solutions rather than explore the solution space. This behavior likely stems from their tendency to derive new outputs based on provided examples. As these examples persist into subsequent iterations, they diminish the diversity of generated snippets, resulting in very similar or identical outputs.

To tackle this challenge, we compute the Levenshtein distance between the most recent snippet and the best n_{examples} snippets, opting for the snippet with the minimal distance. If the distance between the new and the chosen snippet is below a predefined threshold d_t, we assess the new snippet's quality against the selected snippet; otherwise, we store the new snippet. If the new snippet proves superior, we replace the selected snippet with it; if not, we discard the new one.

V. RESULTS

Our experiments use Code Llama with 34 billion parameters based on Llama 2 [3]. We run it on an NVIDIA A100 GPU with 40 GB VRAM and apply 4-bit quantization to reduce the model's memory footprint. Tab. II shows the parameters of our experiments. If not stated otherwise, we let Code Llama generate 100 snippets per run and execute five runs per scenario.

Table II: Parameters used in the experiment

Parameter	Value	Parameter	Value
T_{init}	1.0	C	10
T_{low}	0.1	n_{examples}	3
d_t	200	n_{snippets}	100
s_t	1.7	n_{evals}	1000

A. Basic vs. SCoT without Optimization

To evaluate the performance of the Basic and SCoT scenarios, we compare the following metrics: Score of the best snippet, mean score, and pass@k. Pass@k is the probability that at least one of the top k generated responses is a pass [23]. We defined pass as "compiles and finishes the simulation." Tab. III shows results of different metrics for the Basic and SCoT scenarios.

Table III: Pass@k metrics and score values for Basic and SCoT scenario (average of 5 runs)

	Pass@1	Pass@5	Best Score	Mean Score
Basic	30.20%± 1.55%	83.57%± 1.87%	2.236 ± 0.064	1.719 ± 0.135
SCoT	41.71%± 3.98%	93.17%± 2.14%	2.423 ± 0.084	1.791 ± 0.304

Tab. III clearly shows the superiority of SCoT prompts compared to the basic prompt design. First, we observe that the pass@k metrics show a significant improvement in the SCoT scenario, indicating that the two-step prompting structure with a description of the solving process (Section IV-A) increases the percentage to generate SLT programs that compile and run on the target processor. This implies that the time spent to generate

a set amount of snippets is lower for SCoT than for Basic. Moreover, SCoT prompting achieved a notable improvement in the best achievable score, underlining the benefits of SCoT, especially in code generation tasks. Even though the mean scores of both scenarios are comparable, we see a larger standard deviation in the SCoT scenario, indicating additional exploration of this technique. Thus, a larger variety of solutions is tried, improving the optimization of SLT program generation, especially regarding the best achievable score.

B. Basic vs. SCoT with Optimization

Again, we use the same performance metrics to evaluate the BasicOpt and SCoTOpt scenarios.

Tab. IV lists the results of the different metrics for the optimization scenarios. Similar to the results in Section V-A, we observe an improvement in the pass@k metrics and the best achievable score for SCoT prompting, also showing the superiority of SCoT in optimization tasks that involve generating code snippets. Nevertheless, the pass@k metrics for SCoT and SCoTOpt are very similar, whereas BasicOpt is distinctly worse than Basic (cf. Tab. III), especially in the pass@5 metric. This difference indicates a negative impact on the ability of Code Llama to return passing code snippets, most likely due to the nature of the optimization prompt. At the same time, SCoT prompting seems to be robust against changing the prompt. Interestingly, we observe that BasicOpt has a higher tendency for exploitation and thus often gets stuck in local optima, leading to repeated snippets. Therefore, Tab. IV shows a better mean score value for BasicOpt, even if compared to Basic. Nonetheless, the best score and the higher variance in the mean score indicate that SCoTOpt performs more exploration, decreasing the mean score slightly but often leading to better solutions regarding the best achievable score. SCoTOpt also seems to explore more than only SCoT, as the latter has a higher mean score and lower variance. However, SCoTOpt additionally found the best snippet of all scenarios.

Table IV: Pass@k metrics and score values for BasicOpt and SCoTOpt scenario (average of 5 runs)

	Pass@1	Pass@5	Best Score	Mean Score
BasicOpt	28.04%± 12.31%	74.20%± 24.93%	2.141 ± 0.132	1.823 ± 0.294
SCoTOpt	42.31%± 7.24%	92.85%± 3.82%	2.514 ± 0.083	1.536 ± 0.522

Another interesting observation is the impact of the pool updating strategy. Our findings indicate that Code Llama can not benefit from bootstrapping. Instead of improving the LLM's performance in the optimization framework, it often leads to more hallucinations and sub-optimal results. Fig. 2 visualizes the score distributions of five independent runs for the different scenarios in Tab. I.

Interestingly, SCoT prompting performs best within the optimization framework and boosts the performance without optimization, even outperforming BasicOpt w.r.t. the best score, showing the strength of SCoT prompting in code generation tasks. If we compare the shape of the distributions, we see that SCoT, without optimization, has difficulties improving the

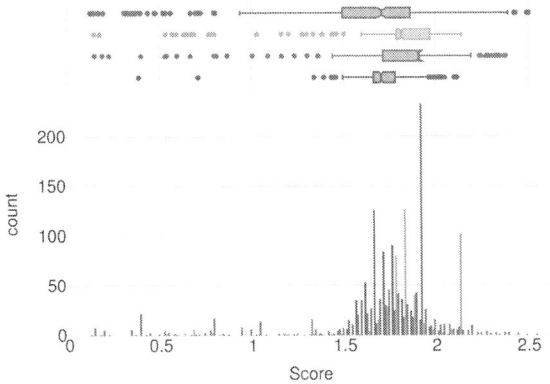

Figure 2: Distribution of score values for the different scenarios Basic (blue), SCoT (red), BasicOpt (green), and SCoTOpt (purple).

solution above a score of around 2.0. Thus, the mean value of SCoT is larger than that of SCoTOpt. However, the distribution (red) is narrower and shows more outliers, indicating that our optimization framework helps to explore more candidates and to exploit already known, well-performing solutions.

To study the impacts of the pool updating strategy and the temperature schedule of our optimization framework, we introduce three additional scenarios: SCoTOpt-Pool, SCoTOpt-Temperature, and SCoTOpt-Plain, which uses neither the pool update strategy nor the temperature schedule.

1) Pool Update Strategy (SCoTOpt-Pool): In this scenario, we keep the temperature constant while updating the pool of candidates see Section III-B1. Tab. V shows the results for the SCoTOpt-Pool scenario. The optimization performance with a constant temperature is comparable to the performance of our adaptive schedule, shown in the SCoTOpt scenario in Tab. IV. However, the pass@k metrics indicate that the LLM shows more hallucinations than SCoTOpt-Plain, leading to longer runtimes.

2) Temperature Schedule (SCoTOpt-Temperature): This scenario uses our temperature schedule (see Section III-B2) while leaving the initial candidate pool unchanged. Tab. V shows the results for the SCoTOpt-Temperature scenario. Compared to the SCoTOpt scenario, the performance is similar, underlining the observation that bootstrapping shows no benefits.

Table V: Pass@k metrics and score values for SCoTOpt-Pool and SCoTOpt-Temperature scenario (average of 5 runs)

	Pass@1	Pass@5	Best Score	Mean Score
SCoTOpt-Plain	29.85%± 0.92%	83.21%± 1.11%	2.389 ± 0.134	1.505 ± 0.055
SCoTOpt-Pool	19.22%± 2.66%	65.41%± 5.30%	2.505 ± 0.094	1.538 ± 0.531
SCoTOpt-Temperature	38.30%± 3.80%	90.92%± 2.91%	2.558 ± 0.069	1.635 ± 0.526

To summarize, the temperature seems more important for the performance of our optimization framework than the pool updating strategy. Whereas the pool updates improve the best score but seem to decrease the pass@k metric scores, a temperature schedule enhances the pass@k metrics and the best score compared to SCoTOpt-Plain. Even though the mean score only changes slightly, the standard deviations show that both pool update and temperature schedule improve the exploration capabilities of our optimization framework.

VI. CONCLUSION

In this work, we have introduced an LLM-based optimization framework for SLT program generation. Using SCoT prompting, which we adapted for our task and a multi-objective task formulation, we successfully automatized SLT program generation while achieving the best-performing C code snippets regarding non-function properties of the given DUT. Compared to the Basic scenario, we achieved an improvement of around 8% in the best score and 10% in pass@k metrics. Our optimization framework, including the temperature schedule and pool updates, improves the best achievable score of pre-trained LLMs by 12%. It allows the balance of exploration and exploitation to generate high-quality code snippets for SLT optimized w.r.t. non-functional DUT properties.

ACKNOWLEDGEMENTS

This work was supported by Advantest as part of the Graduate School "Intelligent Methods for Test and Reliability" (GS-IMTR) at the University of Stuttgart.

REFERENCES

[1] H. H. Chen, "Beyond structural test, the rising need for system-level test," in *VLSI-DAT*, ISSN: 2472-9124, IEEE, Apr. 2018.

[2] I. Polian, J. Anders, S. Becker, *et al.*, "Exploring the mysteries of system-level test," in *2020 IEEE 29th ATS*, ISSN: 2377-5386, IEEE, Nov. 2020.

[3] B. Roziere, J. Gehring, F. Gloeckle, *et al.*, "Code llama: Open foundation models for code," *arXiv preprint arXiv:2308.12950*, 2023.

[4] J. Wang, Y. Huang, C. Chen, Z. Liu, S. Wang, and Q. Wang, *Software testing with large language model: Survey, landscape, and vision*, 2023.

[5] D. Schwachhofer, P. Domanski, S. Becker, *et al.*, "Training large language models for system-level test program generation targeting non-functional properties," in *2024 IEEE European Test Symposium (ETS)*, IEEE, 2024.

[6] C. Yang, X. Wang, Y. Lu, *et al.*, "Large language models as optimizers," *arXiv preprint arXiv:2309.03409*, 2023.

[7] M. Schäfer, S. Nadi, A. Eghbali, and F. Tip, "An empirical evaluation of using large language models for automated unit test generation," *arXiv preprint arXiv:2302.06527*, 2023.

[8] P.-F. Guo, Y.-H. Chen, Y.-D. Tsai, and S.-D. Lin, "Towards optimizing with large language models," *arXiv preprint arXiv:2310.05204*, 2023.

[9] E. Meyerson, M. J. Nelson, H. Bradley, *et al.*, "Language model crossover: Variation through few-shot prompting," *arXiv preprint arXiv:2302.12170*, 2023.

[10] S. Liu, C. Chen, X. Qu, K. Tang, and Y.-S. Ong, "Large language models as evolutionary optimizers," *arXiv preprint arXiv:2310.19046*, 2023.

[11] S. Brahmachary, S. M. Joshi, A. Panda, *et al.*, *Large language model-based evolutionary optimizer: Reasoning with elitism*, 2024.

[12] F. Liu, X. Lin, Z. Wang, *et al.*, "Large language model for multi-objective evolutionary optimization," *arXiv preprint arXiv:2310.12541*, 2023.

[13] J. Wei, X. Wang, D. Schuurmans, *et al.*, "Chain-of-thought prompting elicits reasoning in large language models," *NeurIPS*, vol. 35, 2022.

[14] Z. Chu, J. Chen, Q. Chen, *et al.*, "A survey of chain of thought reasoning: Advances, frontiers and future," *arXiv preprint arXiv:2309.15402*, 2023.

[15] S. Yao, D. Yu, J. Zhao, *et al.*, "Tree of thoughts: Deliberate problem solving with large language models, 2023," *URL https://arxiv. org/pdf/2305.10601. pdf*, 2023.

[16] M. Besta, N. Blach, A. Kubicek, *et al.*, "Graph of thoughts: Solving elaborate problems with large language models," in *AAAI Proceedings*, vol. 38, 2024.

[17] H. Cai, S. Liu, and R. Song, "Is knowledge all large language models needed for causal reasoning?" *arXiv preprint arXiv:2401.00139*, 2023.

[18] J. Li, G. Li, Y. Li, and Z. Jin, "Enabling programming thinking in large language models toward code generation," *arXiv preprint arXiv:2305.06599*, 2023.

[19] Z. Zhao, E. Wallace, S. Feng, D. Klein, and S. Singh, "Calibrate before use: Improving few-shot performance of language models," in *ICML*, PMLR, 2021.

[20] J. Lowe-Power, A. M. Ahmad, A. Akram, *et al.*, *The gem5 simulator: Version 20.0+*, 2020.

[21] J. Zhao, B. Korpan, A. Gonzalez, and K. Asanovic, "SonicBOOM: The 3rd Generation Berkeley Out-of-Order Machine," en, *Fourth Workshop on Computer Architecture Research with RISC-V*, May 2020.

[22] Z. Hadjilambrou, S. Das, P. N. Whatmough, D. M. Bull, and Y. Sazeides, "Gest: An automatic framework for generating CPU stress-tests," in *ISPASS 2019, Madison, WI, USA, March 24-26, 2019*, IEEE, 2019.

[23] M. Chen, J. Tworek, H. Jun, *et al.*, "Evaluating large language models trained on code," *arXiv preprint arXiv:2107.03374*, 2021.

An Automated and Effective Approach for SBST Generation Targeting RISC-V CPUs

Endri Kaja*†, Nicolas Gerlin*†, Jad Al Halabi*†, Ares Tahiraga*, Sebastian Prebeck*,
Dominik Stoffel†, Wolfgang Kunz†, Wolfgang Ecker*‡

*Infineon Technologies AG, Germany
†Rheinland-Pfälzische Technische Universität Kaiserslautern-Landau, Germany
‡Technische Universität München, Germany

Abstract—The trend toward scaling and more complex fabrication techniques in digital circuit design often leads to numerous faults within Integrated Circuits (ICs). To address these vulnerabilities, various Design-for-Test (DFT) techniques have been developed for thorough testing during IC manufacturing. However, integrating these DFT infrastructures introduces significant overhead in both area and performance. To overcome these challenges, especially for testing processor cores, Software-Based Self Test (SBST) has emerged as a promising alternative. This paper proposes a novel, automated, and efficient approach for generating SBST tailored for RISC-V processor cores. By combining formal verification with fault simulation, the number of required assertions is reduced by eliminating faults detected by existing test patterns. In addition to SBST generation, our approach introduces a novel Program Flow Checking (PFC) technique, ensuring adherence to ISO 26262 standards and providing a high Fault Detection Rate (FDR). Experimental results show that various RISC-V processor components achieve Fault Coverage (FC) greater than 91%, with the PFC providing a 100% FDR for most components. The methodology is fully automated, utilizing Model-Driven Architecture (MDA) principles.

Index Terms—SBST, PFC, Fault Simulation, Model-Driven, RISC-V.

I. INTRODUCTION AND MOTIVATION

Digital circuit designs are becoming more sophisticated, with continuous scaling and intricate manufacturing techniques increasing the risk of faults. These faults, if not identified and corrected, can lead to severe functional failures. ISO 26262 mandates rigorous testing and verification of safety-critical systems in the automotive sector. Over time, Design-for-Test (DFT) methodologies have been developed for comprehensive testing of integrated circuits (ICs), including System-on-Chip (SoC) designs that integrate additional testing circuits. While DFT infrastructures streamline testing, they also impose additional area and performance penalties and increase Automated Test Equipment (ATE) [1] costs due to extensive test vectors. To address these issues in processor-centric designs, Software-Based Self Test (SBST) has gained traction. SBST enables testing of processor cores using existing instructions, facilitating tests during normal operation without needing design modifications or additional DFT infrastructure [2].

The main challenge of SBST techniques is the generation of effective test patterns that ensure extensive fault coverage. To overcome this hurdle, different test pattern generation methods are utilized. Psarakis et al. [3] categorize these methods into distinct groups: (i) functional, (ii) structural, hierarchical with precomputed stimuli, (iii) structural, hierarchical using constrained test generation, and (iv) structural, RTL. All SBST techniques face similar challenges such as: achieving a high fault coverage, reducing the complexity of test generation, scalability to complex designs. and high degree of automation.

In this paper, we present an automated and optimized technique for generating SBST designed for RISC-V processor cores, addressing the previously mentioned challenges. The contributions of this paper are summarized as follows:

- We present an SBST generation technique that takes advantage of formal methods to generate test patterns to achieve a high fault coverage. This is proven by numerous experimental results. Formal verification is combined with fault simulation to further reduce the complexity.
- We provide an engineering driven method to strive for SBIST as generic commercial tools are applied. The SBST generation is fully automated following a model-driven approach, requiring minimal manual effort to configure the required parameters.
- We introduce a novel hardware- and software-based PFC technique that ensures high fault detection rates, aligned with ISO 26262 standard.

The paper is organized as follows: Section II summarizes related works, Section III provides an overview of the SBST flow, Section IV explains the test pattern generation flow, Section V discusses the PFC mechanism, Section VI presents experimental results, and Section VII concludes the paper.

II. RELATED WORK

Chen et al. [2] introduce an SBST technique that embeds a software-based tester within the processor's memory for structural testing. This approach uses software to generate and adjust pseudo-random test patterns, enhancing fault coverage. Initially, it creates test patterns for individual CPU components and uses process-level instructions to test them. Paschalis et al. [4] present a deterministic SBST strategy, focusing on the processor's functional units. They create test routines using the processor's arithmetic capabilities, particularly for multiplier-accumulator units, achieving high fault coverage with repetitive patterns. Kranitis et al. [1] introduce a high-level SBST technique aimed at maximizing fault coverage while minimizing test development costs. In contrast to conventional SBST methods targeting programmer-visible elements like the ALU, Gizopoulos et al. [5] focus on the processor's pipelining logic, including hazard-detection and address-calculation components. Riefert et al. [6] and Faller et al. [7] apply formal verification techniques to create test programs for medium-sized processor units. Using SAT-solvers to verify structural testability, Riefert et al. introduce a Validity Checker Module (VCM) to aid SBST generation, integrating functional constraints within the DUT. Faller et al. [7] adapt the VCM for different processor families, enhancing versatility. These approaches have been validated to provide high fault coverage and improve processor testability.

Differentiation from related work: This paper presents a fully automated and structured SBST technique. Unlike previous techniques that demanded manual input and deep processor knowledge, as required in [1], [2], [5], [8], this new method automates the entire process, requiring less human intervention and processor expertise. Similar to the SBST methodologies presented in [1], [4], [6], [7], the proposed approach is deterministic. The approach developed in the paper shares similarities with those introduced by Riefert at al. [6]

979-8-3503-6689-1/24 $31.00 © 2024 IEEE

and Faller et al. [7], but it also has key distinctions. While the aforementioned works incorporated testing constraints directly into the hardware description language (HDL) of the system, our technique uses property-based constraints to create test patterns, which lessen the amount of additional hardware needed. Moreover, traditional methodologies generally allow memory content observation after tests have been performed, which can delay fault detection. The approach in this paper, however, integrates the PFC module that allows for immediate fault detection as soon as errors reach certain outputs—eliminating the necessity for external memory to store and analyze test results later.

III. OVERVIEW OF THE SBST

SBST generation is in general a complex, time-consuming, and error-prone task. To address these issues and to facilitate the SBST generation, we utilized extensively in-house automation frameworks such as MetaRTL [9], MetaProp [10], and the fault injection framework known as MetaFI [11].

A. Automation frameworks

MetaRTL is a model-driven RTL generation framework that follows the MDA [12] principles. The main idea of MetaRTL is to encourage a design-centric development of RTL. MetaRTL is composed of three different layers. The framework supports currently VHDL, Verilog, and SystemVerilog. Additionally, the MDA principles have been adopted to generate formal verification properties via MetaProp. This framework follows the same three-layered structural approach as MetaRTL. The currently supported languages are SystemVerilog Assertions (SVA) and Interval Temporal Logic (ITL). MetaFI is a model-driven fault injection framework that enables fault injection on design models with mixed gate-level/RTL granularity by leveraging MetaRTL. The design is further transformed by inserting *fault injectors* that enable injection of various fault models. Since fault injection is performed directly on the design, any RTL simulator tool can be used to perform fault simulation.

B. SBST generation overview

The SBST generation flow is composed of two major tasks such as *test pattern generation* and *PFC*. Fig. 1 displays an overview of the SBST generation flow. The DUT, i.e. CPU, undergoes the fault injection transformation via MetaFI. The component that will be tested is represented via a gate level granularity while the rest of the design is kept on the original RTL. The complete fault list is automatically generated during this process. Once the DUT has been transformed into the Fault Injection DUT (FI-DUT), a miter circuit is automatically created. This circuit combines the original DUT with FI-DUT using identical set of inputss. The next phase involves injecting faults into the design using formal properties and checking whether the fault affects the design. The main idea is to use a single property that checks that the miter circuit's outputs remain the same despite the presence of the fault. If there is a mismatch, i.e. the fault affects the design, the formal verification tool generates a counterexample. This counterexample represents the input pattern needed for *fault sensitization* and *fault propagation*. A script is then used to retrieve input values from the counterexample, creating a test pattern capable of propagating the fault to the primary outputs. With the test pattern for a single fault created, fault simulation is performed for all other faults using this pattern as a stimulus. If another fault is sensitized and propagated by the same pattern, *fault dropping* is applied, excluding this fault from future pattern generation. Formal properties are reapplied to remaining faults until all are either detected by test patterns or deemed undetectable, having no impact on the design. Individual test patterns are then combined into a complete test

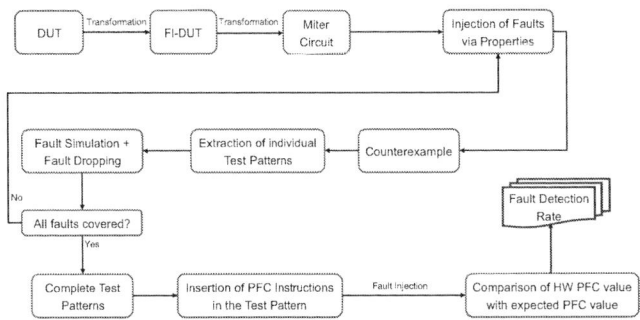

Fig. 1. SBST generation flow

pattern. To reset the CPU, a custom Control Status Register (CSR) instruction is added at the start of each test pattern. Two unique custom CSR instructions, *PFC Start* and *PFC End*, are integrated. *PFC Start* initiates the PFC hardware module to begin hashing instructions, while *PFC End* concludes the hashing process. The hash value (see Section V) is stored in the internal PFC CSR. The final step involves comparing the hash value in the PFC CSR with a previously calculated expected hash value. Any mismatch indicates a detected fault.

IV. TEST PATTERN GENERATION

The core concept of the test pattern generation method outlined in this paper involves employing formal properties to check whether a fault impacts the CPU (DUT). If a fault does indeed impact the CPU, the flow then determines the sequence of inputs, i.e. instructions, necessary to sensitize and propagate the effects of the fault to the primary outputs of the design. A fault that would modify only the state of the CPU is always equivalent to a fault that eventually has an effect on the primary output. Fig. 2 displays the overall setup of DUT and properties. The DUT, i.e. the RISC-V CPU, is initially generated via MetaRTL. The main element of the RTL generation framework is the metamodel that defines the CPU specifications. Some of the main features of RISC-V architecture are the extensability and customizability, thus a custom CSR instruction has been created that can reset the CPU's Register File. This reset instruction is particularly important for testing the CPU. It is inserted at the start of each test pattern to avoid data dependencies between individual test patterns. Essentially, each test pattern begins with a 'clean state', aligning with formal properties that presume the CPU starts from a reset state. An illustrative sequence of a complete test pattern set (CTP) that includes multiple test patterns looks like in the following: CTP = {custom CSR, TP_1, custom CSR, TP_2,..., custom CSR, TP_n}, where n represents the amount of individual test patterns (TP). Once the CPU is

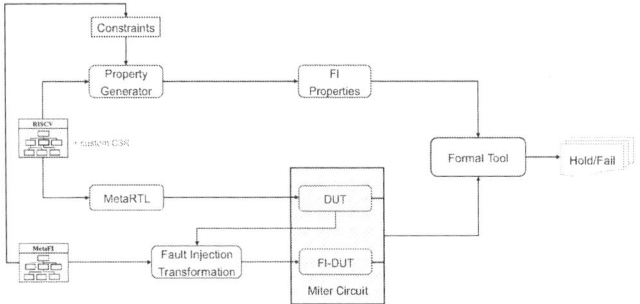

Fig. 2. Setup of the DUT and properties

generated, fault injection transformation is performed where specific components, as specified by the user, are equipped with fault injection capabilities. A miter circuit is automatically generated, combining two versions of the CPU: the original unmodified CPU (the original DUT) and the modified

CPU with fault injection features enabled. Simultaneously, the RISC-V metamodel serves as the blueprint for property generation. Functional constraints, known as 'assume' type properties, are generated according to the model specifications for the RISC-V CPU. These constraints act as rules to prevent formal verification tools from considering scenarios that violate permitted operations, such as illegal operation codes. Additionally, a separate set of test constraints is automatically integrated into the property generator to ensure the Self-Test Built-In Self-Test (SBST) operates independently, avoiding dependency issues during testing. These constraints ensure robust and isolated testing conditions.

- External interrupts are disabled.
- Control flow instructions are disabled, thus SBST is independent of its memory location.
- The PC is constrained to increase linearly.
- Only a single fault is allowed to be injected per property.
- All CSR instructions are disabled except custom CSR to reset the Register File and PFC CSR. This is done such that the SBST does not change CSR states of the original program. Testing of the CSRs is eventually straightforward by using CSR-type instructions. As an example, the instructions write a value to the register and read back the register to check it.

In the approach presented in this paper, rather than embedding constraints in a hardware module as mentioned in [7], these constraints are encoded using 'assume' properties. Once these constraints are established, an 'assert' property is then formulated. This property's role is to test the impact of a fault on the miter circuit. The procedure for this test, which is described in a pseudocode following the SystemVerilog Assertion (SVA) standard, can be found illustrated in Fig. 3.

property *generate_test_pattern;*
@ (posedge clock) disable iff (reset)
  fault_injection_macro |− >
  DUT.Outputs == FI_DUT.Outputs;
endproperty

Fig. 3. Fault Injection Property

The purpose of this property is to determine if the design with fault injection (FI-DUT) and the original design remain equivalent when a single fault is injected. A permanent 'stuck-at' fault is injected using the *fault_injection_macro* for the entire duration of the property. If the fault affects the FI-DUT, the property is violated, and the formal tool provides a counterexample with instructions that sensitizes and propagates the fault. If the property holds, it indicates the fault is undetectable. Timing faults can also be applicable.

Generating test patterns for many faults using only formal properties can be very time-consuming. This paper employs a hybrid strategy combining fault simulation with formal properties to accelerate test pattern generation while ensuring comprehensive coverage. The procedure begins by injecting the first fault using a predefined macro. The formal tool, with a timeout feature, determines whether the fault impacts the design. If the property does not fail, the fault is deemed undetectable. If the property fails, the formal tool generates a counterexample with a test pattern, saved as a '.hex' file. A fault simulation testbench is generated via the MetaFI framework to sequentially inject all single faults. A TCL script extracts the input sequence from the counterexample file and saves it into *test_pattern_file.hex*, starting with a custom CSR instruction. The script also adjusts the testbench to use these sequences as stimuli for the CPU inputs. Another script activates the simulator to perform fault simulation, injecting other faults using the generated pattern. Detectable faults are removed from the fault list (fault dropping). After simulating all faults, the next untested fault is injected via formal properties. Successive patterns from counterexamples are appended to *test_pattern_file.hex*. This iterative process continues until all faults are addressed, resulting in a comprehensive *test_pattern_file.hex* containing all test patterns.

V. PROGRAM FLOW CHECKING

Faults in the CPU can lead to anomalies in program execution, such as out-of-sequence or illegal instructions, and misaligned memory addresses. To detect these faults, Program Flow Checking (PFC) is used [13]. PFC embeds unique identifiers, or signatures, into the program during compilation and verifies these signatures during execution using a hashing function like Cyclic Redundancy Check (CRC).The PFC process involves generating an initial signature, updating it continuously by hashing the current instruction with the preceding signature, and verifying the signature by comparing the runtime computed signature with the compile-time signature. Any mismatch indicates a fault. While software-based runtime signature calculation is possible, it can degrade system performance. This paper proposes a dedicated hardware module for PFC to efficiently compute the runtime signature, enhancing fault detection. This module, integrated within the CPU, updates the signatures for each instruction and is expected to improve fault detection compared to software-only solutions. State-of-the-art PFC methods typically focus on instruction signatures derived from disassembled code. However, this paper extends traditional techniques by also comparing outputs from the Arithmetic Logic Unit (ALU) and the Register File, thereby ensuring both control flow and data flow integrity. Internal signals like ALU outputs and Register File outputs, not available in the disassembly file, are captured into a separate file, and signatures are generated.

A. PFC Hardware

The PFC hardware is automatically integrated with the CPU core, placed within the same pipeline stage as the signal it monitors. It uses a CRC hashing function to generate signatures for ensuring data integrity, with parameters like the CRC polynomial specified within the RISC-V CPU generation framework. Illustrated in Fig. 4, the PFC hardware includes a special state register for storing the runtime signature. This state register is connected to the RISC-V CSR interface, taking advantage of the architecture's extensibility to allow new CSR addresses for custom operations. Specifically, a unique CSR address is reserved for the PFC module.

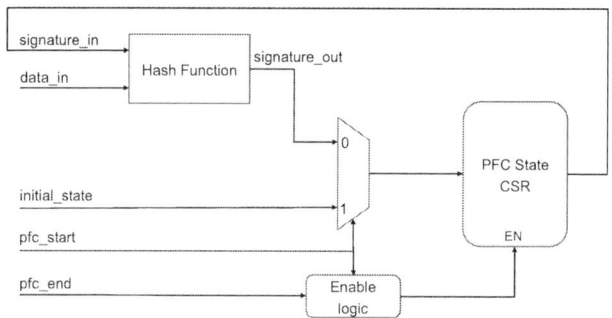

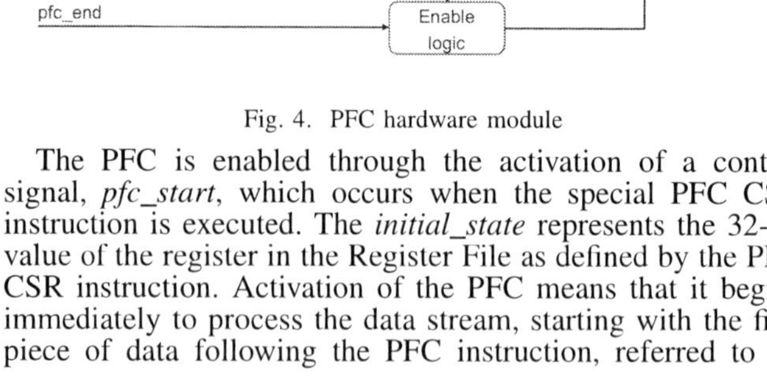

Fig. 4. PFC hardware module

The PFC is enabled through the activation of a control signal, *pfc_start*, which occurs when the special PFC CSR instruction is executed. The *initial_state* represents the 32-bit value of the register in the Register File as defined by the PFC CSR instruction. Activation of the PFC means that it begins immediately to process the data stream, starting with the first piece of data following the PFC instruction, referred to by *data_in*. This *data_in* could be the current instruction, ALU outputs, or Register File outputs, depending on what is chosen to be monitored. The hash function, which is a hardware-implemented CRC, computes a new signature *signature_out* by hashing the incoming data *data_in* with the existing value in the PFC State register. The PFC can be turned off, halting the data stream processing, through an instruction that sets the *pfc_end* signal. This instruction is the final input to the hash function, after which the PFC State's value remains unchanged and can be read out as the final signature.

B. Fault Detection Flow

The PFC hardware flow is designed for fast runtime calculation of signatures, which aids in detecting CPU faults. The core principle is comparing the runtime signature generated by the hardware with a precomputed signature. A mismatch indicates a potential hardware fault. Activation of the PFC module is triggered by a custom PFC CSR instruction, which initiates data hashing. This data includes test pattern instructions defined in the test pattern generation flow. These instructions are stored in a specific memory location and fetched and executed by the CPU immediately after the PFC CSR instruction.

After the CPU executes all test instructions, a custom PFC CSR instruction is issued to stop the PFC module from hashing. This final instruction is the last processed for signature generation, after which the signature remains unchanged and is saved in the PFC CSR register for later retrieval. Simultaneously, an expected signature value is computed using software tools. A Python script mimics the PFC hashing process by applying the same CRC polynomial used in the hardware to the test pattern. Under fault-free conditions, the hardware-generated signature should match this precomputed value. Any mismatch signals a detected fault. After hashing concludes, the runtime signature is read from the hardware using a PFC CSR read instruction. A subsequent comparison instruction is used to compare this value with the static signature. Additional instructions are added into the test pattern to facilitate this comparison after the PFC stop instruction.

VI. EXPERIMENTAL RESULTS

The generation methodology and the fault detection mechanism were tested on various components of a RV32IMC RISC-V CPU to determine the overall effectiveness of the technique. The tested components were chosen to include both combinational components, i.e. ALU, Hazard Detection Unit (HDU) and Decoder, and sequential components such as Register File and the Memory-Writeback pipeline register (MEM-WB). Table I shows the results of the SBST flow and gives figures for the number of faults, number of patterns, fault coverage (FC), PFC fault detection rate (FDR), and the total test generation time.

TABLE I
TEST PATTERN GENERATION FLOW RESULTS

Component	Results				
	# Fault	# Patterns	FC(%)	FDR(%)	Time (h)
HDU	111	19	91.89	100	0.61
ALU	4732	313	97.63	100	13.6
Register File	11747	492	96.86	100	40.5
Decoder	3224	104	76.3	92.19	38.87
MEM-WB	410	10	54.8	100	8.5

The **HDU** is a small component with 111 total faults. The process generated 19 patterns to detect 104 stuck-at faults in about 0.61 hours, with undetected faults classified as undetectable within a 2-minute timeout. For the **ALU**, 313 patterns identified 4620 faults in 13.6 hours, achieving a fault coverage of 97.63%. All detectable faults were found, as the PFC module connected to the ALU output ensures any faulty result causes a signature mismatch. The **Register File**,

a large sequential component with 11747 faults, generated 492 patterns to identify 111379 faults in approximately 40.5 hours with a 5-minute property checking timeout. This interval can be increased but we saw no difference in the behavior. The fault coverage was 96.87%, with the PFC units catching all faults due to their direct connection to the Register File outputs. The **Decoder** had around 22.02% redundancy due to a generation framework flaw. The process took about 38.87 hours, resulting in a fault coverage of 76.3%. The fault detection mechanism found 92.19% of these faults. The remaining faults are unlikely to affect CPU outputs as they did not influence the Register File or ALU.

VII. CONCLUSION AND FUTURE WORK

In this paper we introduced an efficient and automated technique for generating customized SBST for RISC-V processors. The proposed technique makes use of formal verification techniques to generate deterministic test patterns, thus leading to a high fault coverage, i.e. greater than 91% for most of the tested components. Through a hybrid approach by mixing formal verification and fault simulation, fault dropping is performed to reduce the number of required properties. Furthermore, the PFC module is added to the SBST generation flow. This fault detection mechanism guarantees a high fault detection rate of 100% for most of the components, thus aligning with various ASILs of ISO 26262 standard. The technique is fully automated by utilizing MDA principles.

Future work: We intend to apply the SBST technique beyond RISC-V processors to get insights on the adaptability of the technique across different processor architectures. Additionally we plan to activate CSR and branch instructions.

VIII. ACKNOWLEDGEMENTS

Part of the work described herein is funded by the German Federal Ministry of Education and Research (BMBF) as part of the research project Scale4Edge (16ME0122K). Part of the work has also been performed in the project ISOLDE under grant agreement No 101112274.

REFERENCES

[1] N. Kranitis, A. Paschalis, D. Gizopoulos, and G. Xenoulis, "Software-based self-testing of embedded processors," *IEEE Transactions on Computers*, vol. 54, no. 4, pp. 461–475, 2005.

[2] L. Chen and S. Dey, "Software-based self-testing methodology for processor cores," *IEEE Transactions on Computer-Aided Design of Integrated Circuits and Systems*, vol. 20, no. 3, pp. 369–380, 2001.

[3] M. Psarakis, D. Gizopoulos, E. Sanchez, and M. Sonza Reorda, "Microprocessor software-based self-testing," *IEEE Design & Test of Computers*, vol. 27, no. 3, pp. 4–19, 2010.

[4] A. Paschalis, D. Gizopoulos, N. Kranitis, M. Psarakis, and Y. Zorian, "Deterministic software-based self-testing of embedded processor cores," in *DATE 2001*, pp. 92–96, 2001.

[5] D. Gizopoulos, M. Psarakis, M. Hatzimihail, M. Maniatakos, A. Paschalis, A. Raghunathan, and S. Ravi, "Systematic software-based self-test for pipelined processors," *IEEE TVLSI*, vol. 16, no. 11, pp. 1441–1453, 2008.

[6] A. Riefert, R. Cantoro, M. Sauer, M. Sonza Reorda, and B. Becker, "A flexible framework for the automatic generation of sbst programs," *IEEE TVLSI*, vol. 24, no. 10, pp. 3055–3066, 2016.

[7] T. Faller, N. I. Deligiannis, M. Schwörer, M. S. Reorda, and B. Becker, "Constraint-based automatic sbst generation for risc-v processor families," in *IEEE ETS 2023*, pp. 1–6, 2023.

[8] L. Chen, S. Ravi, A. Raghunathan, and S. Dey, "A scalable software-based self-test methodology for programmable processors," in *DAC 2003*, pp. 548–553, 2003.

[9] J. Schreiner, R. Findenig, and W. Ecker, "Design centric modeling of digital hardware," in *IEEE HLDVT 2016*, pp. 46–52, 2016.

[10] K. Devarajegowda and W. Ecker, "Meta-model based automation of properties for pre-silicon verification," in *IFIP/IEEE VLSI-SoC 2018*, pp. 231–236, 2018.

[11] E. Kaja, N. Gerlin, M. Bora, K. Devarajegowda, D. Stoffel, W. Kunz, and W. Ecker, "Metafs: Model-driven fault simulation framework," in *IEEE DFT 2022*, pp. 1–4, 2022.

[12] "Omg group." https://www.omg.org/mda/. Accessed: 2024-05-07.

[13] D. Arora, S. Ravi, A. Raghunathan, and N. K. Jha, "Hardware-assisted run-time monitoring for secure program execution on embedded processors," *IEEE TVLSI*, vol. 14, no. 12, pp. 1295–1308, 2006.

A Low Power Oriented Multiple Target Test Generation Method for 2-Cycle Gate-Exhaustive Faults

Toshinori Hosokawa
College of Industrial Technology
Nihon University
Chiba, JAPAN
hosokawa.toshinori@nihon-u.ac.jp

Momona Mizota
Graduate School of Industrial Technology
Nihon University
Chiba, JAPAN
cimo22001@g.nihon-u.ac.jp

Masayoshi Yoshimura
Faculty of Information Science and Engineering
Kyoto Sangyo University
Kyoto, JAPAN
yoshimura.masayoshi@cc.kyoto-su.ac.jp

Masayuki Arai
College of Industrial Technology
Nihon University
Chiba, JAPAN
arai.masayuki@nihon-u.ac.jp

Abstract— **In recent years, the progress of VLSIs has caused an increase in the number of cell-internal defects in addition to signal line defects. In this paper, we focus on a fault where the transition on the output signal line of a cell is delayed in each input pattern combination of the gate due to cell-internal defects when the transition occurs on the output signal line of the cell. This fault model is called as a 2-cycle gate-exhaustive fault. In VLSI testing, high power consumption might occur because many faults are tested by a test vector from the view point of test cost reduction. However, high power consumption causes erroneous tests due to excessive IR drop, resulting in yield loss. Thus, this paper proposes a low power oriented multiple target test generation method for 2-cycle gate-exhaustive faults using pseudo Boolean optimization.**

Keywords— cell-internal defects; 2 cycle gate-exhaustive faults; pseudo Boolean optimization; WSA; multiple target test generation;

I. INTRODUCTION

In recent years, with the downsizing process technologies, density and complexity of Very Large-Scale Integrated circuits (VLSI) are rapidly progressing [1]. In addition, defective VLSIs have physical defects in the cells and signal lines, and there are various faults such as logical faults that change to another logic function and timing faults that cause incorrect operation when tested at-speed operation [1]. In VLSI testing, tests are often performed for stuck-at faults, open faults, bridge faults, transition faults, etc. that model defects on signal lines. However, for VLSIs built in mission-critical products related to human life, such as automotive and medical products, testing with higher quality is required. In such cases, it is necessary to consider not only defects on signal lines but also cell-internal defects. In this paper, we focus on cell-internal defects. By performing a layout analysis, it can determine that which input patterns or input sequences for each cell can excite the cell-internal defects at the output signal lines of the cells [2]. Cell-internal defects can be excited with only specific cell input

patterns or sequences and might not be detected by the test sets for classical fault models such as stuck-at faults and transition faults [3]. Therefore, it might cause test escapes that judge defective VLSI as good VLSI [3].

In this paper, we focus on faults where a delay occurs when the value of a cell's output signal line changes due to cell-internal defects, especially 2-cycle gate-exhaustive fault model [4]. In this paper, logic gates are referred to as cells or gates. 2-cycle gate-exhaustive fault model is defined as a fault where the transition of values on the output signal line is delayed with a specific input sequence for a gate. The length of the input sequence is 2. Since the number of 2-cycle gate-exhaustive faults depends on the number of gates and the number of gate inputs, the number of faults and test vectors might increase when the circuit size becomes larger.

In [4], in order to reduce the number of faults to be tested, test vectors are generated only for 2-cycle gate-exhaustive faults that affect circuit functions. However, since the functional states are determined using a random input sequence, there is a possibility that the number of identified functional states become insufficient. Therefore, the faults that affect circuit functions might be included in the undetected 2-cycle gate-exhaustive faults. In this paper, we target all 2-cycle gate-exhaustive faults and aim to generate test vectors that achieve the 100% fault efficiency. Our test generation method aims to reduce the number of test vectors and to make the power consumption during testing be smaller than the given threshold values.

Dynamic test compaction [5-8] is important to reduce the number of test vectors. Multiple Target Test Generation (MTTG) [6-8] has been proposed as one of the dynamic test compaction methods. MTTG is a method to generate test vectors for multiple target faults. In this paper, we use MTTG.

979-8-3503-6689-1/24 $31.00 © 2024 IEEE

The Launch-on-capture (LoC) approach [9] is widely used in at-speed scan testing [10]. In this paper, we use the LoC approach. In at-speed scan testing using the LoC approach, it is very important to reduce the number of signal line transitions (Launch Switching Activity: LSA) during capture operations. In the test set using the LoC approach, test vectors that do not exceed the power consumption threshold are called capture-safe test vectors [11], and test vectors that exceed the power consumption threshold are called capture-unsafe test vectors [11]. Many test generation methods have been proposed to reduce the number of capture unsafe test vectors in the LoC approach for transition faults [11-16]. Most of these methods identify (pseudo) primary input values unrelated to fault detection as don't cares for test sets that do not consider power consumption, and assign logical values to these don't cares in order to reduce the LSA.

In this paper, we propose a low power oriented multiple target test generation method for two-cycle gate-exhaustive faults. The proposed method is a pseudo-Boolean optimization (PBO) [17] based test generation method. In the proposed method, one fault is selected from the undetected faults of each gate, a target fault set with fewer elements than the given upper limit is generated, and PBO-based MTTG is performed for the target fault set. The PBO is formulated so that the number of target faults detected is maximum and smaller than the power consumption threshold.

The rest of this paper is organized as follows: Sect. II explains defects in cells and 2-cycle gate-exhaustive fault model. Sect. III proposes a low power oriented multiple target test generation method for 2-cycle gate-exhaustive faults using PBO. Sect. IV shows our experimental results, and Sect. concludes our paper.

II. 2-CYCLE GATE-EXHAUSTIVE FAULTS

A. Cell-Internal Defects

Fig. 1 shows an example of a defect in CMOS transistors modeling a 2-input NAND gate. The voltage level at the time 1 and the voltage level at the time 2 are shown near the input/output signal lines in Fig. 1. The ON and OFF states of each transistor at the time 2 are shown in Fig.1. Assume that there is a resistive open defect on the wiring between the PMOS transistor connected to the input B, and the output Y. Since the defect is not completely disconnected, the voltage level is propagated, but it is propagated later than in normal conditions. At the time 1, since the voltage levels of inputs A and B are both high, both PMOS transistors are turned off. Therefore, Y is not affected by the defect, and the voltage level becomes low as in normal conditions. Since the voltage level of B is low at the time 2, the PMOS transistor connected to B is turned on, but the voltage level of Y at the time 2 becomes low, the same level as the time 1, due to the defect. The low voltage is propagated to the next wiring. After the delay due to the defect, the voltage level of Y becomes high. Therefore, due to the delay, a voltage level different from the normal state might be captured at flip-flops (FFs). As mentioned above, in the case of a cell-internal defect, the input sequence of the cell which a

fault can be excited at the output of the cell depends on the type and location of the defect. In this paper, we will deal with 2-cycle gate-exhaustive fault model [4] which is a delay fault model occurred due to cell-internal defects. The model does not require layout analysis.

B. 2-Cycle Gate-Exhaustive Fault Model

2-cycle gate-exhaustive fault [4] is defined as a fault such that when the value of the output signal line at each gate transitions, the transition is delayed with each input sequence of that gate. By detecting all 2-cycle gate-exhaustive faults, it is possible to detect all cell-internal-defects which can be detected by 2-pattern testing.

TABLE I shows a 2-cycle gate-exhaustive fault set for a 2-input AND gate. When the number of gate inputs is n (n > 1), the number of 2-cycle gate-exhaustive faults is $2 \times (2^n - 1)$. The number of 2-cycle gate-exhaustive faults shown in TABLE I is $2 \times (2^2 - 1) = 6$. f_1, f_2, and f_3 are faults such that the rising transition of the output signal line is delayed, and f_4, f_5, and f_6 are faults such that the falling transition of the output signal line is delayed. For a two-input XOR (XNOR) gate, the number of two-cycle gate-exhaustive faults is 8. A 2-cycle gate-exhaustive fault set for each gate is an independent fault set which is not detected by the same test vector [18].

III. A LOW POWER ORIENTED TEST GENERATION METHOD FOR 2-CYCLE GATE-EXHAUSTIVE FAULTS

In this section, we propose a low power-oriented multiple target test generation method for two-cycle gate-exhaustive faults using PBO. The proposed method targets all two-cycle gate-exhaustive faults, reduces the number of test vectors, and guarantees that WSAs [19] of all generated test vectors are below a threshold. PBO is used to satisfy these conditions.

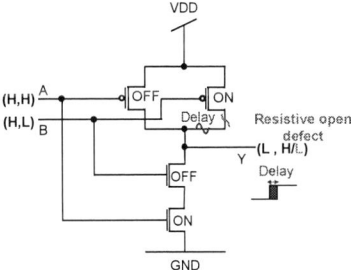

Fig 1: Example of a Cell-Internal defect in 2-Input NAND Gate Cell

TABLE I
A 2-CYCLE GATE-EXHAUSTIVE FAULT SET OF 2-INPUT AND GATE CELL-

Faults	Inputs				Output	
	Time 1		Time 2		Time 1	Time 2
	A	B	A	B	C	C
f_1	0	0	1	1	0	1/0
f_2	0	1	1	1		
f_3	1	0	1	1		
f_4	1	1	0	0	1	0/1
f_5	1	1	0	1		
f_6	1	1	1	0		

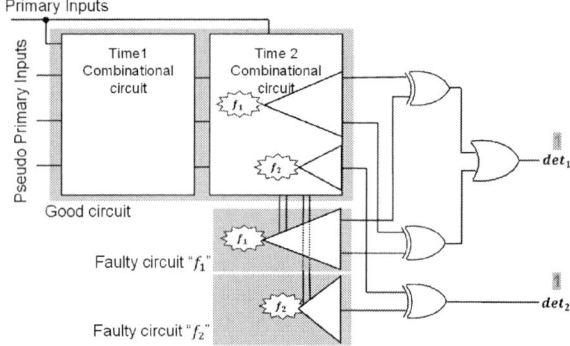

Fig 2: Example of Multiple Target Test Generation Model

A. Test Generation Model

Generally, in PBO-based ATPG as with SAT-based ATPG [20], a test generation model consisting of a good circuit and a faulty circuit is constructed for the circuit under test (CUT). Faulty circuits corresponding to target faults are prepared. Fault detection is determined by comparing the values of primary outputs (POs) and pseudo primary outputs (PPOs) for a good circuit and faulty circuits. Therefore, we construct a model that performs XOR operations for the POs and PPOs included in the faulty circuits and the POs and PPOs of the corresponding good circuit as the inputs.

Fig. 2 shows a test generation model with two target faults. Since the faulty circuit for the fault f_1 has two (pseudo) primary outputs, perform an XOR operation at each (pseudo) primary output. It is sufficient that the fault effects can be propagated to any one PO or PPO. Therefore, the test vector can be found by OR-operating the outputs of the two XOR operations and determining the primary input values and the pseudo primary input values such that the output variable det_1 becomes 1.

B. Circuit Behavior Constraints

Since PBO uses an optimization function and constraints as inputs to solve problems, it is necessary to express the test generation model using Boolean constraints. Circuit behavior constraints are equations which express the test generation model such as a good circuit and faulty circuits shown in Fig. 2. Circuit behavior constraints must be satisfied.

C. Fault Detection Constraints and Optimization Function

Not all faults in the target fault set can be detected with the same test vector. Therefore, fault detection constraints are formulated using relaxation variables. By using relaxation variables, it is possible to describe the optimization function such that simultaneously detects as many faults as possible. An example of the fault excitation constraints is shown in Equation (1). Equation (1) is the example of a two-input AND gate with the input signal lines A and B and the output signal line C in a circuit. It is constraints to excite the fault which causes that the falling transition at C is delayed when (the values of A and B at the time 1, the values of A and B at the time 2) = (11,01) is applied.

$$r + C_1 \geq 1,$$
$$r + {\sim}C_2 \geq 1,$$
$$r + F \geq 1,$$
$$r + A_1 \geq 1,$$
$$r + {\sim}A_2 \geq 1,$$
$$r + B_1 \geq 1,$$
$$r + B_2 \geq 1$$
$$r, F, C_1, C_2, A_1, A_2, B_1, B_2 \in \{0,1\} \quad (1)$$

In equation (1), A_1 (B_1, C_1) is the value of A (B, C) at the time 1 in the good circuit, and A_2 (B_2, C_2) is the value of A (B, C) at the time 2 in the good circuit. F is the value of C at the time 2 in the faulty circuit. r is a relaxation variable.

In addition, to detect the fault, it is necessary to set the output signal line of the XOR gate or OR gate to 1, which is used to compare the pseudo primary output values of the good circuit and the faulty circuit. An example of the fault detection constraint is shown in Equation (2).

$$r + det \geq 1$$
$$r \in \{0,1\} \quad (2)$$

r is the relaxation variable and det is the fault detection variable. If multiple target faults cannot be detected simultaneously, the constraints are satisfied by setting the relaxation variables in Equations (1) and (2) to 1.

In addition, relaxation variables r_i are used for each target fault f_i. Therefore, there are as many relaxation variables as the number of target faults. By introducing relaxation variables, there is a possibility to satisfy constraints without detecting the target faults during test generation. Therefore, the problem is solved by using an optimization function. The optimization function is shown in Equation (3).

$$minimize = \sum_{i=1}^{N} r_i \quad (3)$$

In Equation (3), N is the number of relaxation variables corresponding to target faults. Equation (3) allows each relaxation variable r_i which corresponds to the target fault f_i to be assigned 0 as much as possible, making it possible to maximize the number of detected faults for the target faults.

D. Transition Suppression Constraints

To keep the capture power consumption below the threshold value, it is necessary to suppress the number of transitions on signal lines. Constraints to suppress transitions on each signal line are given. However, the most important thing in test generation is to detect target faults. By assigning a transition suppression constraint to each signal line, it becomes difficult to detect the given target faults. Therefore, we formulate transition suppression constraints using relaxation variables. Equation (4) shows the transition suppression constraint for the signal line x.

$$sw + x_1 \cdot x_2 + \overline{x_1} \cdot \overline{x_2} \geq 1$$
$$sw, x_1, x_2 \in \{0,1\} \quad (4)$$

In Equation (4), x_1 is a variable representing the value of x at the time 1, x_2 is a variable representing the value of x at the time 2, and sw is a relaxation variable corresponding to x. Equation (4) is satisfied when $(x_1, x_2) = (0,0)$ or $(x_1, x_2) = (1,1)$ is assigned. On the other hand, when $(x_1, x_2) = (0,1)$ or $(x_1, x_2) = (1,0)$ is assigned, Equation (4) is satisfied by assigning 1 to sw.

By constructing the transition suppression constraint for each signal line in a good circuit and finding the sum of sw corresponding to each signal line, it is possible to calculate the WSA [19] value, which is one for estimating power consumption in VLSI. In this paper, we guarantee that only capture-safe test vectors are generated by adding a constraint that the WSA value is below a threshold value in the test generation. The power consumption constraint is shown in Equation (5).

$$Const_{WSA} \equiv \sum_{j=1}^{M} sw_j \leq WSA_{th} \quad (5)$$

In Equation (5), M is the number of relaxation variables for transition suppression constraints which is equal to the total number of signal lines, sw_j is the relaxation variable for the transition suppression constraint corresponding to the signal line j, and WSA_{th} is the WSA threshold value.

In addition to circuit behavior constraints, Equations (1), (2), (4), and (5) are connected using an AND operation and given to the PBO solver. Equation (6) shows the overall formulation. Equation (3) is an optimization function.

$$\phi_c \cdot \bigwedge_{i=1}^{N} (\phi_{f_i} \cdot \phi_{act_i} \cdot \phi_{det_i}) \cdot \bigwedge_{j=1}^{M} (\phi_{WSA_j}) \cdot Const_{WSA} \quad (6)$$

In Equation (6), N is the number of target faults and M is the number of signal lines. ϕ_c is the constraint for a good circuit behavior, ϕ_{f_i} is the constraint for a faulty circuit behavior corresponding to the target fault f_i, ϕ_{act_i} is the fault excitation constraint corresponding to f_i, ϕ_{det_i} is the fault detection constraint corresponding to f_i, ϕ_{WSA_j} is the transition suppression constraint corresponding to the signal line j, and $Const_{WSA}$ is the constraint for the WSA threshold value.

E. Whole Algorithm

Algorithm	Low-Power Oriented Multiple Target Test Generation for 2-Cycle Gate-Exhaustive Faults
Input	Circuit C, WSA WSA_{th}
Output	Test set T_{set}, Untestable fault set UTF_{set}, High-power Fault set HPF_{set}, Detected fault set DF_{set}

1. $T_{set} = \emptyset$, $DF_{set} = \emptyset$
2. $F_{set} = $ 2cycle_gate_exhaustive_fault_set(C);
3. $UTF_{set} = $ Untestable_Delete(C, F_{set});
4. $F_{set} = F_{set} - UTF_{set}$;
5. $HPF_{set} = $ High_Power_Fault_Delete(C, F_{set}, WSA_{th});
6. $F_{set} = F_{set} - HPF_{set}$;
7. **while**($F_{set} \neq \varphi$)then
8. $TF = $ Fault_Selection(C, F_{set});
9. $tv = $ PBO_Test_Generation(C, TF, WSA_{th});
10. $(F_{set}, DF_{TVset}) = $ Fault_Simulation(C, tv, F_{set});
11. $T_{set} = T_{set} \cup tv$;
12. $DF_{set} = DF_{set} \cup DF_{TVset}$;
13. **endwhile**
14. return(T_{set}, UTF_{set}, HPF_{set}, DF_{set});
15. end

Fig 3: Whole algorithm of Test Generation

Fig. 3 shows a low-power oriented multiple target test generation algorithm for 2-cycle gate-exhaustive faults using PBO. The input is the circuit C and the WSA threshold value WSA_{th}. The outputs are the test set T_{set}, the untestable fault set UTF_{set}, the unsafe fault set HPF_{set}, and the detected fault set DF_{set}. An unsafe fault [11] is a fault which is detected only by capture-unsafe test vectors [11]. Let F_{set} be a fault set, TF be a target fault set, tv be a test vector, and DF_{TVset} be the fault set detected by tv. First, T_{set} and DF_{set} are initialized to empty (line 1). Next, F_{set} is generated from C (line 2). Next, whether each fault is an untestable or not is determined and the determined untestable faults are substituted for UTF_{set} (line 3). Next, the faults in UTF_{set} are deleted from F_{set} (line 4). Next, HPF_{set} is generated and unsafe faults in HPF_{set} are deleted from F_{set} (lines 5 and 6). When no test vectors below WSA_{th} that detect a single target fault in F_{set} are generated, the fault is added to HPF_{set} as an unsafe fault. Next, the processing from lines 8 to 12 is iterated until F_{set} becomes empty and test generation is performed (line 7). First, the target faults from F_{set} are selected and TF is generated (line 8). Next, low power-oriented test generation for TF considering WSA_{th} is performed and tv is generated (line 9). Next, fault simulation is performed for F_{set} with tv, DF_{TVset} is generated from the faults detected by tv, and the faults in DF_{TVset} are deleted from F_{set} (line 10). Next, tv is added into T_{set} (line 11). Next, the faults in DF_{TVset} are added into DF_{set} (line 12). Finally, T_{set}, UTF_{set}, HPF_{set}, and DF_{set} are returned (line 14).

IV. EXPERIMENTAL RESULTS

The proposed method was implemented in C language, and experiments were conducted on ISCAS'89 benchmark circuits using a computer with a Core i7-13700 and 16GB memory. In this paper, Clasp [21] 3.3.4 was used as PBO solver. In this experiment, the time limit on PBO per 1 MTTG was set to 120 seconds which was enough time to avoid being classified to aborted faults. Also, to reduce the test generation time, the number of target fault per 1 MTTG was set to 100 or less.

TABLE II
EXPERIMENTAL RESULTS FOR FAULT COVERAGE

Circuits	FLT	DET	UNTST	FC(%)	FE(%)
s5378	13,502	6,556	6,946	48.56	100
s9234	23,102	13,533	9,569	58.58	100
s13207	33,306	17,119	16,187	51.40	100
s15850	37,856	16,749	21,107	44.24	100
s35932	80,946	35,901	45,045	44.35	100
s38417	89,458	71,748	17,710	80.20	100
s38584	106,194	43,456	62,738	40.92	100

TABLE III
EXPERIMENTAL RESULTS FOR DYNAMIC TEST COMPACTION

Circuits	#TV			ATPG time	
	STTG	MTTG	RR(%)	STTG	MTTG
s5378	1,752	672	61.64	1.00	20.88
s9234	3,128	1,365	56.36	1.00	4.94
s13207	3,144	1,165	62.95	1.00	27.17
s15850	2,167	304	85.97	1.00	18.06
s35932	1,420	191	86.55	1.00	1.00
s38417	14,929	1,643	88.99	1.00	0.03
s38584	11,892	1,186	90.03	1.00	0.48

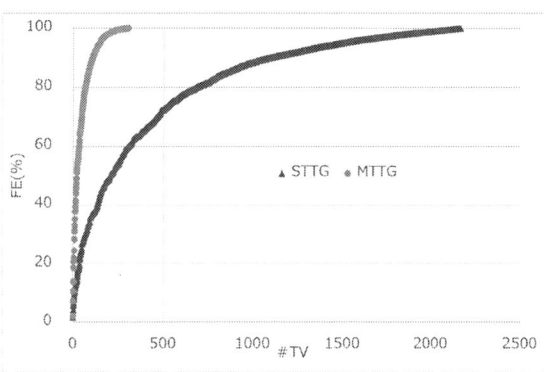

Fig 4: Cumulative Fault Efficiency of s15850

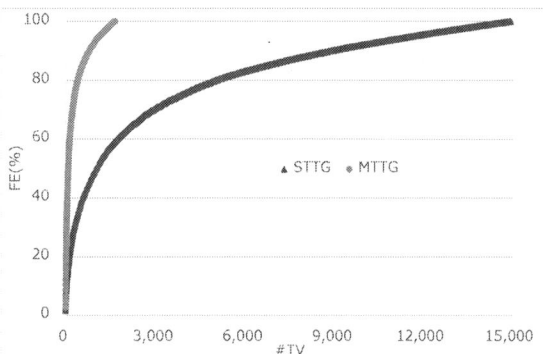

Fig 5: Cumulative Fault Efficiency of s38417

TABLE IV

EXPERIMENTAL RESULTS FOR WSA

Circuits	#TV		#STV		#USTV		ATPG time	
	OFF	ON	OFF	ON	OFF	ON	OFF	ON
s5378	555	672	0	672	555	0	1.00	38.21
s9234	1,014	1,363	53	1,363	961	0	1.00	33.15
s13207	931	1,165	1	1,165	930	0	1.00	96.44
s15850	239	304	78	304	161	0	1.00	44.96
s35932	64	191	0	191	64	0	1.00	6.97
s38417	600	1,643	0	1,643	600	0	1.00	0.64
s38584	1,110	1,186	618	1,186	492	0	1.00	3.09

A. Fault Coverage

TABLE II shows the experimental results for fault coverage. In TABLE II, "Circuits" denotes the name of circuits, "FLT" denotes the total number of 2-cycle gate-exhaustive faults, "DET" denotes the number of detected faults, "UNTST" denotes the number of untestable faults, "FC" denotes the fault coverage, and "FE" denotes the fault efficiency. Our proposed method could achieve 100 % fault efficiencies for all circuits. On the other hand, the fault coverage was 52.61 % on average. This means that the ratio of untestable faults was very high since the number of the necessary assignments to detect a 2-cycle gate-exhaustive fault was generally large and the possibility of value conflicts was increased. In addition, as the result of applying the fifth line in Fig. 3, the number of unsafe faults was 0 for all circuits.

B. Number of Test Vectors

TABLE III shows the experimental results for dynamic test compaction. In TABLE III, "Circuits" denotes the name of circuits, "#TV" denotes the number of test vectors, and "ATPG time" denotes the normalized test generation time. In addition, "STTG" denotes the experimental results of the test generation whose number of target faults is 1, "MTTG" denotes the experimental results of the test generation whose number of target faults is 100 or less, and "RR" denotes the reduction ratio for the number of test vectors in MTTG to that in STTG. In STTG and MTTG, test vectors were generated to satisfy the given WSA threshold value constraint. The WSA threshold was set to 20% of the total number of signal lines.

In MTTG, 100 gates with undetected faults were randomly selected. When the number of gates with undetected faults was smaller than 100, all the gates with undetected faults were selected. Then, one undetected fault was randomly selected from each selected gate. 2 or more undetected faults for a gate were never selected as target faults since all faults for a gate were independent faults [18]. Our proposed MTTG could reduce the number of test vectors by 76.07 % on average compared with STTG. The reduction ratio in the number of test vectors for s5378, s9234, and s13207 (relatively small circuits) was 56 to 62%. On the other hand, the reduction ratio in the number of test vectors for s15850, s35932, s38417, and s38584 (relatively large circuits) was 85 to 90%.

Next, we compare the test generation times of STTG and MTTG. The MTTG test generation time in Table II was normalized to the STTG test generation time of 1. For three relatively small circuits, the MTTG test generation time increased by 4.9 to 27.1 times compared to STTG since the reduction ratio for the number of test vectors was relatively low. For four relatively large circuits except for s38417, the MTTG test generation time increased by 3.0 to 44.9 times compared to STTG since the reduction ratio for the number of test vectors was relatively high. For s38417, the test generation time of MTTG was shorter than that of STTG. Fig.4 and Fig. 5 show the cumulative fault efficiency for s15850 and s38417, respectively. In Fig..4 and Fig. 5, the vertical axis represents cumulative fault efficiency, the horizontal axis represents test vector numbers arranged in the generated order, the blue curve is for STTG, and the orange curve is for MTTG. The reduction ratio for the number of test vectors was 85.97% and 88.99% for s15850 and s38417, respectively. The test generation time was 18.06 times and 0.03 times compared with STTG for s15850 and s38417, respectively. Although the reduction ratio for the number of test vectors is about the same, the test generation time is significantly different. For s15850, when we focus on the latter 30% of the test vectors in Fig. 4, we can see that the number of detected faults was saturated for both STTG and MTTG. On the other hand, for s38417, when we focus on the latter 30% of the test vectors in Fig. 5, we can see that the number of detected faults was saturated for STTG. However, the number of detected faults was increasing linearly for MTTG. Therefore, for s38417, we consider that the test generation time of MTTG was faster than that of STTG.

C. WSA

TABLE IV shows the experimental results for transition suppression constraints. In TABLE IV, "Circuits" denotes the name of circuits, "#TV" denotes the number of test vectors, "#STV" denotes the number of capture-safe test vectors, and "#USTV" denotes the number of capture-unsafe test vectors. In addition, "ON" denotes the experimental results for the test generation with the WSA threshold values, and "OFF" denotes the experimental results for the test generation without the WSA threshold values. Thus, in the test generation for OFF, $\bigwedge_{j=1}^{M}(\phi_{WSA_j}) \cdot Const_{WSA}$ was deleted from Equation (6).

The number of test vectors for ON increased by 77.64% on average compared with that for OFF. Also, the test generation time for ON increased by 31.92 times on average. ON had a large number of transition suppression constraints given to the PBO solver in order to satisfy the WSA threshold value constraints, which increased the difficulty of finding a solution and increased the test generation time. For s38417, the test generation time of ON was 0.64 times faster than that of OFF. However, 84.39% of the test vectors generated by OFF were capture-unsafe test vectors on average. On the other hand, since test vectors generated by ON were guaranteed to satisfy the WSA threshold value constraints, all test vectors generated by ON were capture-safe test vectors.

V. CONCLUSION

In this paper, we proposed a low power oriented multiple target test generation method for 2-cycle gate-exhaustive faults. Our proposed test generation method targeted all faults in circuits. Experimental results on ISCAS'89 benchmark circuits showed that our proposed method could achieve the 100 % fault efficiencies for all circuits. Moreover, our proposed method could reduce the number of test vectors by 76.07 % on average compared with the single target test generation method while guaranteeing that all the generated test vectors were capture-safe test vectors.

Our future work includes accelerating our proposed test generation method for application to larger circuits.

Acknowledgments

This work was supported in part by Japan Society for the Promotion of Science (JSPS) under Grants-in-Aid for Science Research C (No.21K11817).

REFERENCES

[1] H. Fujiwara, Logic Testing and Design for Testability, MIT Press, 1985.

[2] F. Hapke, W. Redemund, A. Glowatz, J. Rajski, M. Reese, M. Hustave, M. Keim, J. Schloeffel, and A. Fast, "Cell-Aware Test," IEEE Trans. on Computer-Aided Design, Vol. 33, No. 9, pp.1396–1409, Sept. 2014.

[3] K.Y. Cho, S. Mitra, and E.J. McClusky, "Gate Exhaustive Testing," IEEE International Conference on Test, pp.1–7, Austin, USA, Nov. 2005.

[4] I. Pomeranz, "Functional Constraints in the Selection of Two-Cycle Gate-Exhaustive Faults for Test Generation," IEEE Transactions on Computer-Aided Design of Integrated Circuits and Systems, vol.29, no.7, pp. 1500–1504, July 2021.

[5] P. Goel and B. C. Rosales, "Test generation and dynamic compaction of tests," in Proc. of the International Test Conference, pp.189–192, 1979.

[6] G. Tromp, "Minimal Test Sets for Combinational Circuits," IEEE International Conference on Test, no.7.3, pp.204–209, Nashville, USA, Oct.1991.

[7] J.S. Chang, and C.S. Lin, "Test set compaction for combinational circuits," Browse Journals & Magazines, vol.14, Issue.11, pp.1370–1378, Nov. 1995.

[8] S. Eggersglüß, K. Schmitz, R. Krenz-Bååth, and R. Drechsler, "On Optimization-Based ATPG and Its Application for Highly Compacted Test Sets," IEEE Trans. on Computer-Aided Design, Vol. 35, No. 12, pp.2104–2117, Dec. 2016.

[9] J. Savir, and S. Patil, "On Broad-Side Delay Test," IEEE Transactions on Very Large Scale Integration (VLSI) Systems, vol. 2, no. 3, pp.368–372, Sep. 1994.

[10] L. T. Wang, C. W. Wu, and X. Wen, VLSI Test Principles and Architectures: Design for Testability, Morgan Kaufmann, 2006.

[11] X. Wen, K. Miyase, S. Kajihara, H. Furukawa, Y. Yamato, A. Takashima, K. Noda, H. Ito, K. Hatayama, T. Aikyo and K. K. Saluja, "A Capture-Safe Test Generation Scheme for At-Speed Scan Testing," European Test Symposium, pp. 55–60, Verbania, Itary, July 2008.

[12] X. Wen, K. Miyase, T. Suzuki, Y. Yamato, S. Kajihara, L.-T. Wang and K. K. Saluja, "A Highly-Guided X-Filling Method for Effective Low-Capture-Power Scan Test Generation," Proc. ICCD, pp. 251–258, 2006.

[13] S. Remersaro, X. Lin, Z. Zhang, S. M. Reddy, I. Pomeranz and J. Rajski, "Preferred Fill: A Scalable Method to Reduce Capture Power for Scan Based Designs," Proc. ITC, paper 32.2, 2006.

[14] Y.-H. Li, W.-C. Lien, I.-C. Lin, and K.-J. Lee, "Capture-Power-Safe Test Pattern Determination for At-Speed Scan-Based Testing," IEEE Trans. Comput. Aided Design Int. Circuits & Syst., vol. 33, no. 1, pp. 127–138, 2014.

[15] M. Yoshimura, Y. Takahashi, H. Yamazaki, and T. Hosokawa, "A Don't Care Filling Method for Low Capture Power based on Correlation of FF Transitions Using SAT," IEICE Trans. Fundamentals, Vol. E100-A, no. 12, pp. 2824–2833, 2017.

[16] T. Hosokawa, K. Misawa, Y. Hirama, H. Yamazaki, M. Yoshimura, and M. Arai, "A Low Capture Power Oriented X-filling Method Using Partial MaxSAT Iteratively," Proc. DFTS, S2-4, 2019.

[17] V. Manquinho, R. Martins, and I. Lynce, "Improving Unsatisfiability-Based Algorithms for Boolean Optimization," International Conference on Theory and Applications of Satisfiability Testing, pp.181–193, Berlin, 2010.

[18] S. B. Akers, C. Joseph, and B. Krishnamurthy, "On the Role of Independent Fault Sets in the Generation of Minimal Test Sets," in Proc. Intl. Test Conf, pp.1100–1107, 1987.

[19] S. Jayanthy , and M.C. Bhuvaneswari, Test Generation of Crosstalk Delay Faults in VLSI Circuits, Springer Singapore, 2019.

[20] T. Larrabee, "Test pattern generation using Boolean satisfiability," IEEE Trans. Comput. Aided Design Int. Circuits & Syst., vol. 11, no. 1, pp. 4–15, 1992.

[21] M. Gebser, B. Kaufmann, A. Neumann, and T. Schaub, "Conflict-Driven answer set solving," Artificial Intelligence, vol.187-188, pp.52–89, Aug. 2012.

A Structural Testing Approach for SRAM Address Decoders using Cell-Aware Methodology

X. Xhafa[1], E. Faehn[2], P. Girard[1], A. Virazel[1]

[1]LIRMM - University of Montpellier/CNRS
Montpellier, France
xxhafa, girard, virazel@lirmm.fr

[2]STMicroelectronics
Crolles, France
eric.faehn@st.com

Abstract—Testing memory circuits is crucial to ensure the quality of a System on Chip (SoC) as technology nodes shrink, making circuits more prone to defects and reliability issues at nanometer scales. This paper presents an efficient testing flow based on an adaptation of the Cell-Aware (CA) test concept for the testing of memory address decoders. With the use of an Automatic Test Pattern Generator (ATPG), two different decoder architectures are tested for stuck and transition faults, addressing both intra and inter-cell defects. In this work, we show that this methodology results in a 100% intra and inter-cell defect coverage for both Transition Faults (TFs) and Stuck-At Faults (SAFs). To compare our results with existing solutions, the MATS++ algorithm has been used. A 51% improvement in TFs and 8% improvement in SAFs test coverages have been obtained through our Cell-Aware methodology.

Index Terms—Memory testing, Structural testing, SRAM address decoders, Cell-Aware models, ATPG

I. INTRODUCTION

Recent applications of Integrated Circuits (ICs) require an extensive amount of data to be stored and processed. Therefore, memory blocks now occupy a significant proportion, often up to 90%, of the System-on-Chip (SoC) area [1]. As the technology node shrinks, the memory density has increased significantly. However, with smaller transistor sizes and shorter proximities in interconnects due to higher densities, memories are now increasingly susceptible to defects and reliability challenges [2]. Thus, the testing of advanced and newly emerging memories has become a critical step in ensuring the quality and functional safety of manufactured ICs.

The most prominent memory testing approach is based on functional testing (i.e., March algorithms) [3]. However, with the increasing complexity of circuit designs, including emerging memory technologies such as MRAM and RRAM, functional testing becomes impractical due to high development costs and insufficient defect coverage to meet quality standards [4]. To address this, a shift from functional to structural testing is proposed in this paper. Structural testing relies on precise defect location and its impact on the input-output relationship, reducing fault modeling abstraction from the functional to transistor level, resulting in fewer test escapes

and higher defect coverage. One implementation of structural testing is the Cell-Aware (CA) methodology, which addresses intra-cell defects in standard cells to reduce test escapes not caused by interconnect defects [5]. An initial flow for adapting CA to memory testing has been presented in [6] and applied to an SRAM bit cell, tailored for the memory array but insufficient for ensuring maximum coverage of defects in blocks outside the array. The SRAM memory periphery, particularly address decoders that comprise a large number of transistors and interconnects, is highly prone to defects thus contributing to overall test escapes [7], [8].

In this paper, we address the limitations of the aforementioned flow and propose an improved version that is adapted to testing address decoders. We introduce a systematic and comprehensive approach for SRAM address decoder testing that is based on the CA methodology and aims to achieve maximum defect coverage. The defects considered are located on interconnections and at the transistor level in each standard cell that composes the address decoders. A 4x4 array SRAM architecture, including its periphery, has been used as case study. The results are compared and validated with a functional approach, the MATS++ algorithm that has been developed to target address decoder faults with minimal complexity. Our methodology has demonstrated a notable improvement in defect coverage, achieving a 4% increase in SAFs and a 49% improvement in TFs fault coverages.

The paper is organized as follows. In Section II, the modified flow for address decoder testing using the CA methodology is introduced. The structural testing of address decoders is described in Section III. The obtained results and validation are given in Section IV. Finally, conclusions and perspectives are drawn in Section V.

II. CELL-AWARE METHODOLOGY FOR ADDRESS DECODER TESTING

The flow presented in [6] introduces a preliminary adaptation of the Cell-Aware (CA) methodology for memory testing but is limited to detecting intra-cell defects in the SRAM bit-cell, excluding the memory periphery. Here, we expand this flow to test address decoders in an SRAM architecture (see Fig. 1).

979-8-3503-6689-1/24 $31.00 © 2024 IEEE

The CA methodology, initially designed for digital IC testing, involves generating CA models for each standard cell within a digital circuit. This process starts with identifying potential intra-cell defects, such as transistor-level shorts and opens, and layout-based defects. Defects are injected into the circuit, and SPICE simulations are run to observe deviations in output signals. Detected defects are associated with input patterns, and this information is compiled into a CA model, which includes both statically and dynamically detected defects. These models are then used by an Automatic Test Pattern Generator (ATPG) to produce patterns for testing inter-cell and intra-cell defects in the overall IC.

A 4x4 memory array designed using a 28nm FDSOI technology that includes the memory array, address decoders, sense amplifier, write driver and pre-charge circuits is used as a case study. Following the initial flow, the memory is described at the gate level (structural Verilog) to generate CA models. These models, along with the gate-level structure, are used by the ATPG to generate patterns for detecting interconnect and cell-aware defects, considering SAF for static defects and TF for dynamic defects (1st ATPG run, see Fig. 1).

The address decoders are purely composed of digital gates. Hence, when the decoder is tested on its own, a test coverage of 100% is achieved. However, when used in the full memory architecture a considerable percentage of the defects located in the decoder are classified as UnDetectable by the ATPG. The UD defects are analyzed and we have observed that for these defects dynamic patterns that address more than one cell in the memory array are necessary. To address this issue, the proposed solution is to map the effects of the UD defects into the memory array (Decoder to Memory Array defect mapping, see Fig. 1). Since the row decoder controls the activation of Word Lines (WLs), only one column of the memory array is sufficient for this task. In order for the ATPG to comprehend the information regarding the switching of WLs that allows two cells to be addressed, a new CA model is generated for the first column (CLM0) of the array. The idea is to emulate the detection conditions so that the ATPG can generate structural patterns. This step will be further detailed in Section III. To obtain the additional patterns after the CLM0 CA model generation, another ATPG run is necessary (2nd ATPG run, see Fig. 1). Finally, the flow ends by merging all the test patterns and calculating the test coverage. The test coverage is a ratio of detected defects over the total number of defects.

By using the CA methodology as a structural test approach for address decoders we therefore inherit the advantages that come with CA testing. Through CA models, the exact location information of intra-cell defects is available, hence also contributing to the diagnosis process [5].

III. STRUCTURAL TESTING OF ADDRESS DECODERS

A. CA model generation for row decoder cells

CA models need to be generated for each module type in the decoder. In this case, the standard cells from the 28nm technology library have been used for the design. A CA model is generated for NOR and AND gates, using the steps

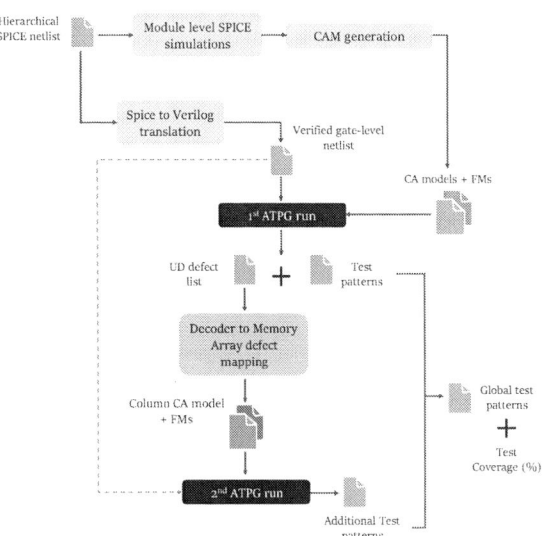

Fig. 1: Address Decoder Test Flow using a CA methodology

mentioned in the CA model generation flow (c.f. Section II). The CA model files contain all the information on the defect type and its location within the cell, as well as the detection tables with the necessary patterns to detect them. Defects that have equivalent detection conditions are classified as equivalent. When generating CA models, non-resistive open and short defects are considered, although the CA test can also address resistive defects [9]. Note that each defect is considered separately when running the SPICE simulations.

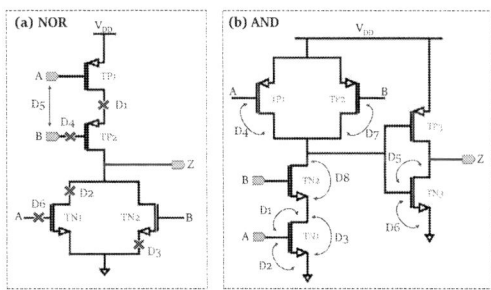

Fig. 2: The (a) NOR and (b) AND schematics including a subset of the injected defects

Depending on their detection conditions, the defects are classified and organized in static and dynamic tables. For demonstration purposes, we will show only one example for each of the tables. The location and type of the given defects in Tables I and II in the NOR gate and the AND gate are shown in Fig. 2. The static table of the AND CA model is shown in Table I. As seen under each defect name, the notation "1" is used to indicate that the pattern detects the defect, and a "0" indicates that the defect is not detected. Note that, in the static table, each pattern corresponds to at most one stimulus.

The dynamic table of the NOR CA model is shown in Table II. In the latter, the 'R' notation indicates a rising transition

	Inputs		Output	Defects							
	A	**B**	**Z**	**D1**	**D2**	**D3**	**D4**	**D5**	**D6**	**D7**	**D8**
Pt.1	0	0	0	0	0	0	1	1	1	1	0
Pt.2	0	1	0	1	0	1	1	1	1	0	0
Pt.3	1	0	0	0	0	0	0	1	1	1	1
Pt.4	1	1	1	1	1	0	1	1	0	1	0

TABLE I: Static detection table of AND

from logic '0' to logic '1'. On the contrary, the 'F' notation indicates a falling transition from logic '1' to '0'. To better understand Table II we can analyze pattern 1. The '0R-F' input-output pattern can detect defects D3 to D5. This means that a logic '0' on input A and a rising transition on input B of the NOR gate can cause the output to deviate from its golden value, in the presence of D3, D4, or D5.

	Inputs		Output	Defects					
	A	**B**	**Z**	**D1**	**D2**	**D3**	**D4**	**D5**	**D6**
Pt.1	0	R	F	0	0	1	1	1	0
Pt.2	R	0	F	0	1	0	0	1	1
Pt.3	0	F	R	1	0	0	0	0	0

TABLE II: Dynamic detection table of NOR

B. Decoder to Memory Array Defect Mapping

In a memory circuit, the effect of each inter and intra-cell defect needs to be propagated from the defective net/s located in the decoder to the output of the memory, which in this case is the output of the Sense Amplifier (SA) circuit. It is, therefore, possible that the propagation of certain defects can be hindered by other memory blocks. We have observed that a part of the open defects, predominantly in the NOR gates, are not detected using only the information provided by the detection tables. This is due to the fact that while one defect can cause one world line to have a falling TF, the access from the targeted memory cell to the output is blocked. Therefore, the defect is not propagated.

To better understand this issue, we can analyze defect D3, which is an open defect located at the source net of transistor TN2 of a NOR gate (see Fig. 2). After a SPICE simulation with the injected defect, we observe the following detection conditions:

- W0 in address <A0, A1> = <0, 0> : WL0 is active
- W1 in address <0, 1>: WL2 is active. Due to a falling transition delay caused by D3, WL0 remains active during this W1 operation
- Read in address <0, 0> : '0' is expected yet '1' is read.

The (A, B) inputs of the defective NOR gate correspond to the row decoder inputs <A0, A1>. For D3 detection, as in the SPICE simulation, the decoder address switches from <0, 0> to <0, 1>. In this memory architecture, the delay in WL0's falling transition is blocked by the SRAM bit cell's access transistor primitive, preventing defect propagation (see Fig. 3). To address undetectable (UD) defects, these defects are mapped into a new CA model allowing to control multiple WLs. Since the row decoder activates the WLs, only one

memory column (CLM0) is needed for this CA model. The new CA model is generated without SPICE simulations, using existing CA model information for NOR gates.

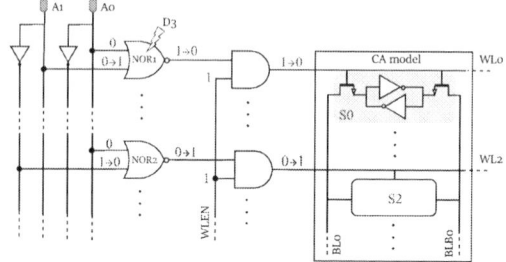

Fig. 3: Propagation of D3

An example of the CLM0 CA model dynamic table is given in Table III. In this table the pattern that detects the previously analyzed D3 in NOR1 is shown. Since the defect is synthesized by two consecutive write operations, the observable outputs are the inner nodes of the bit cells of CLM0 that contain the written value in the cell. This pattern detects all the defects detected by Pt.1 (see Table II) in the NOR dynamic table. The same procedure is followed for all UD defects in all of the gates of the row decoder.

Inputs						Outputs				Defects
WL0	**WL1**	**WL2**	**WL3**	**BL0**	**BLB0**	**S0**	**S1**	**S2**	**S4**	**D3,D4,D5**
F	0	R	0	R	F	0	0	R	0	1

TABLE III: Dynamic table of the CLM0 CA model

IV. RESULTS AND VALIDATION

The CA models obtained are utilized by ATPG to generate test patterns for inter and intra-cell defects (SAF and TF models) in the row decoder of a 4x4 SRAM architecture.

A. Static Faults

The SAF detection process in the decoder involves three steps. First, the ATPG generates circuit-level patterns to detect SAFs in interconnections. These patterns are saved, and then, after adding Static Cell-Aware (SCA) defects, a fault simulation is run using the initial patterns to check defect coverage. If coverage is below 100%, another ATPG run generates patterns targeting undetected intra-cell defects. For the SAF model, 112 inter and intra-cell defects are considered. After both ATPG runs, 12 patterns are generated: 8 for W1R1 operation on CLM0 and 4 for detecting D8 of the AND cell, which causes an SAF1 on each WL detected by keeping the WLEN signal at level '0'. The ATPG generates four read patterns while the decoder is inactive, in the presence of D8, the SA reads an unexpected value. This structural pattern is noted as R_EN0.

After the two ATPG runs, the defect coverage is 96%, with four undetected D3 defects on each AND gate. This defect, detected with a '01' input (see Table I), results in an inactive output for each corresponding WL. Similar to an open defect, D3 is mapped to the CLM0 CA model. Following this third step, the SAF defect coverage reaches 100%.

B. Transition Faults

For the TF model, 92 inter and intra-cell defects are considered. The pattern generation process for TF detection is similar to that for SAFs. First, inter-cell defects are targeted using 4 static patterns to write values in each cell of CLM0. The initial ATPG run produces 14 transition patterns (28 stimuli). Figure 4 (a) shows these patterns and corresponding memory operations in terms of row decoder inputs <A0, A1>. Next, intra-cell defects from the Dynamic Cell-Aware (DCA) table are added, and a fault simulation is run using these patterns. In the second run, no additional patterns are generated and a defect coverage of 60% is achieved. As discussed in Section III, a part of NOR intra-cell defects that correspond to open defects causing a falling TF at the WL-s, are not detected. For this reason, the defects are mapped in the CA model of CLM0. The procedure explained in Section III-B is followed and using a third ATPG run the defect coverage is increased to 100%. In Fig. 4 (c), the lined arrow indicates the W0W1 operations from one address (e.g. '00') to the other (e.g. '01'). The curved lines indicate the read operation in the initial address (e.g. '00'). This step requires an additional 20 stimuli, resulting in a total of 54 stimuli needed for all TF detection. In Fig. 4, the obtained defect coverage cumulative percentage is indicated after each ATPG run. By combining the three graphs (a), (b) and (c) in Fig. 4, we observe that the obtained transition patterns cover the entire graph.

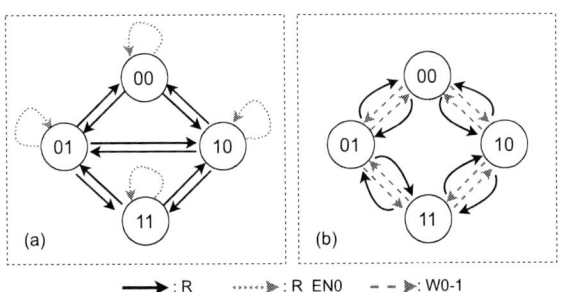

Fig. 4: Address orders of TF detection patterns for each ATPG run, where in (a) DC = 60%, (b) DC = 100%.

C. Comparison Results

To compare functional and structural approaches for testing the address decoder, the MATS++ algorithm is chosen as a reference due to its ability to detect all AFs while maintaining a low complexity [10]. The March elements in MATS++ were translated to ATPG-compliant patterns for fault simulations in our Verilog memory description, corresponding to write and read operations in the specified address order. Four fault simulations for SAFs (W0R0 and W1R1 operations in ascending and descending orders) yielded a 92% defect coverage for both intra- and inter-cell defects. For TFs, two simulations using transition patterns ($\Uparrow (r0w1); \Downarrow (r1w0r0)$) resulted in a 51% defect coverage. The undetected defects are analyzed and we have observed that the defects are related to the falling TF and SAFs in the WLEN, as well as falling TFs in each WL.

Table IV shows ATPG results for both functional (MATS++) and structural (CA methodology) approaches. The MATS++ algorithm's complexity is 6x16 for the 4x4 SRAM case study architecture. The number of patterns necessary to detect all defects using the CA approach is less than 96. This is because the ATPG considers only CLM0 when targeting the generation of patterns for row decoder defects. We have observed that the 52 stimuli necessary to detect TFs include the 12 necessary static stimuli for the detection of SAFs. For validation purposes, a 3-bit decoder in an 8x4 SRAM design has also been tested using both methodologies. The results are shown in the two last lines of Table IV. Similar results have been obtained for the case of the 3-bit decoder as well, proving the feasibility and efficiency of our CA test methodology.

	ATPG		MATS++	
	SAF+SCA	TF+DCA	SAF+SCA	TF+DCA
2-bit Dec.	100%	100%	92%	51%
# of patt.	12	52	6x16 = 96	
3-bit Dec.	100%	100%	93.44%	49.2%
# of patt.	26	100	6x64 = 384	

TABLE IV: Proposed CA approach vs. MATS++

V. Conclusion

This study presents a modified CA methodology for structurally testing of address decoders in SRAM architecture. By using a custom CA model to map UD defects causing 2-cell dynamic faults onto the memory array, we achieved a complete defect coverage for both inter- and intra-cell defects in address decoders. Comparing this with the MATS++ algorithm shows significant improvement using CA methodology with minimal test patterns. Future work will implement this methodology on the entire SRAM memory architecture, generating CA models for all memory blocks to compare with more complex March-like algorithms targeting static and dynamic defects.

References

[1] IRDS. "International roadmap for devices and systems." in *https://irds.ieee.org/editions/2020*, 2020.

[2] S. Borkar *et al.*, "Microarchitecture and design challenges for gigascale integration," in *MICRO*, vol. 37, pp. 3–3, 2004.

[3] A. J. Van de Goor and Z. Al-Ars, "Functional memory faults: a formal notation and a taxonomy," in *Proceedings 18th IEEE VLSI test symposium*, pp. 281–289, IEEE, 2000.

[4] P. Girard *et al.*, "A survey of test and reliability solutions for magnetic random access memories," *Proceedings of the IEEE*, vol. 109, no. 2, pp. 149–169, 2021.

[5] F. Hapke *et al.*, "Cell-aware test," *IEEE Transactions on Computer-Aided Design of Integrated Circuits and Systems*, pp. 1396–1409, 2014.

[6] X. Xhafa *et al.*, "On using cell-aware methodology for sram bit cell testing," in *2023 IEEE European Test Symposium (ETS)*, pp. 1–4, 2023.

[7] S. Hamdioui, *Testing static random access memories: defects, fault models and test patterns*, vol. 26. Springer Science, 2004.

[8] A. Bosio *et al.*, *Advanced test methods for SRAMs: effective solutions for dynamic fault detection in nanoscaled technologies*. Springer Science & Business Media, 2009.

[9] F. Hapke *et al.*, "Cell-aware analysis for small-delay effects and production test results from different fault models," in *International Test Conference*, pp. 1–8, 2011.

[10] A. J. Van de Goor, *Testing semiconductor memories: theory and practice*. John Wiley & Sons, Inc., 1991.

Special Session: Security and RAS in the Computing Continuum

Martí Alonso[1], David Andreu[1], Ramon Canal[1], Stefano Di Carlo[2], Odysseas Chatzopoulos[3], Cristiano Chenet[2], Juanjo Costa[1], Andreu Girones[1], Dimitris Gizopoulos[3], George Papadimitriou[3], Enric Morancho[1], Beatriz Otero[1], Alessandro Savino[2]

[1]Universitat Politècnica de Catalunya, Barcelona, Spain
[2]Politecnico di Torino, Torino, Italy
[3]University of Athens, Athens, Greece
Contact email: ramon.canal@upc.edu

Abstract—**Security and RAS are two non-functional requirements under focus for current systems developed for the computing continuum. Due to the increased number of interconnected computer systems across the continuum, security becomes especially pervasive at all levels, from the smallest edge device to the high-performance cloud at the other end. Similarly, RAS (Reliability, Availability, and Serviceability) ensures the robustness of a system towards hardware defects. Namely, making them reliable, with high availability and design for easy service.**

In this paper and as a result of the Vitamin-V EU project, the authors detail the comprehensive approach to malware and hardware attack detection; as well as, the RAS features envisioned for future systems across the computing continuum.

Index Terms—**RISC-V, Security, Malware, Hardware Attack, Computing Continuum, Simulation**

I. INTRODUCTION

RISC-V is a revolutionary open-source instruction set architecture (ISA) designed to offer simplicity, modularity, and extensibility [1]. This exciting development brings many benefits over proprietary processor architectures, including the potential for customization and lower licensing costs [2].

Despite these advantages and the fact that RISC-V applications have started to see their birth in the embedded domain [3], several challenges still need to be addressed before RISC-V can be widely adopted beyond conventional ones like performance or standardization. Among these challenges, security and RAS are of utmost importance. As RISC-V processors become more widely adopted, there is an increasing potential for security attacks. Ensuring the security of RISC-V-based applications will be an important challenge that needs to be addressed as technology develops.

Another key challenge is RAS. Enterprise and cloud data centers are increasingly integrating complex System-on-Chip (SoC) architectures to meet the demands of modern computing workloads. However, the widespread deployment of these devices raises the risk of undetected faults, which can lead

Funded by the European Union under the Horizon Europe Programme. Project name: Vitamin-V. Project number: 101093062. Views and opinions expressed are, however, those of the authors only and do not necessarily reflect those of the European Union or the HaDEA. Neither the European Union nor the granting authority can be held responsible for them. This paper is also funded by HFRI with title REDESIGN and Project Number 16973 and project SERICS (PE00000014) under the MUR National Recovery and Resilience Plan funded by the European Union - NextGenerationEU.

to critical issues like system crashes or silent data corruptions (SDCs). These faults, stemming from manufacturing defects and in-field reliability concerns, present significant challenges for data centers. While cosmic ray-induced soft errors have been extensively studied [4]–[11], modern data centers must also consider other potential fault sources, such as manufacturing defects and thermal variations [12], which may lead to SDCs during normal operations [13]–[17].

This paper presents AI-based malware and hardware attack detection. With special emphasis on the reproducible and cross-platform methodology; as well as RAS.

The paper is organized as follows: Section II describes the step-by-step methodology used for AI-based security and Section III provides it for RAS. Then, Section IV presents the performance and detection capabilities of the AI-based security and the RAS analysis. Finally, Section V summarizes the main contributions of the paper.

II. METHODOLOGY FOR AI-BASED SECURITY

This section describes the methodology for the AI-based malware and hardware attack detection. We propose two different sources of information to detect the malign behaviour: hardware performance monitors (HPM) and instruction opcodes. While hardware performance monitors are only available during the execution of the program (thus, dynamically), opcodes are both available statically and dynamically. Consequently, in this section we describe the process of using each of these sources of information to process through ML algorithms to detect the attacks.

A. Hardware Performance Monitoring

Modern processors include hardware performance monitoring units to track the CPU performance, necessary due to increased processor complexity, such as hierarchical cache subsystems, non-uniform memory, and out-of-order execution. Software that adapts to resource utilization benefits from better performance and efficiency. The HPM includes registers and counters for microarchitectural events accessible by the Operating System (OS) using libraries like Linux perf [18] and PAPI [19].

HPMs track events such as retired instructions, branch predictions, cache hits/misses, and clock ticks, but only a limited number of hardware counters can be active at a time due to design and cost constraints [20]–[22]. Developers access counters via performance monitoring instructions, reading, and writing counter values.

B. Static Analysis of applications

Evaluating software trustworthiness early is essential. We developed a machine learning (ML) tool that analyzes static executable content to determine if it is benign or malicious. Inspired by previous work on detecting software bugs and security threats using static analysis [23], [24], we incorporated deep learning (DL) techniques to identify complex patterns. In cases where the dataset is insufficient or a zero-day attack is involved, transfer learning (TL) can be used to enhance detection accuracy by leveraging pre-trained models [25].

Following Haddadpajouh et al. [24] approach, we breakdown the process in four steps: dataset creation, feature extraction, model training, and deployment. To train the model, we gather a balanced dataset of benign and malicious programs, including hardware attacks like Spectre [26], Meltdown [27], viruses, and malware. Malicious programs will be sourced from platforms like VirusTotal [28], VirusShare [29], or SourceFinder [30]. Benign programs will come from regular Linux applications in the Debian repository.

Programs are converted to feature vectors by disassembling ELF binaries to extract operation codes (OpCode) sequences. These sequences are analyzed to generate feature vectors to train the DL model. The model developed and benchmarked in Section IV uses AMD64. Yet, the same methodology can be applied to other ISAs.

C. Dynamic Analysis using Hardware Performance Counters

Anomaly detection using hardware monitoring and AI involves dynamic analysis of microarchitectural events via ML algorithms to identify abnormal behavior. This approach, first introduced by Demme et al. in 2013 [31], has been applied to malware detection [32] and hard and soft errors [33], [34] but not on RISC-V.

Programs exhibit phase behaviors [35], [36], allowing for anomaly detection through patterns in hardware performance counters. The proposed anomaly detection framework includes three main components: (i) a CPU with HPM, (ii) data collection, and (iii) anomaly detection via ML classifiers. Challenges arise due to the cost and complexity of monitoring speculative execution events, especially in resource-constrained devices, where balancing hardware events and detection accuracy is critical. To overcome limited counter availability, some methods run applications multiple times to capture more events [31], [37], [38], but this impacts runtime applicability.

Data collection involves selecting events, extracting features, and reducing dimensions [32]. Feature extraction captures HPCs into vector space, while dimensionality reduction minimizes redundant data that can decrease detection accuracy. When empirical event selection is not feasible, techniques like Principal Component Analysis (PCA), Fisher Score, and Information Gain are used to identify relevant features [39]. The

final block, anomaly detection, is carried out by ML classifiers. These classifiers can vary by type (multi-class or one-class), learning method (supervised, unsupervised, semi-supervised), or underlying algorithm (Neural Networks, Decision Trees, etc.). Multi-class classifiers assume multiple normal classes, while one-class classifiers identify anomalies based on a single normal class boundary [40]. Given the challenges of labeling data, unsupervised learning is often preferred [40]. To improve accuracy, advanced methods like Ensemble Learning [41], Boosting [42], and multi-stage classifiers [43] are used.

III. METHODOLOGY FOR RAS

A. Reliability, Availability, and Serviceability in Large-Scale SoC Deployment

Detecting and managing faults that result in SDCs is particularly challenging due to the specific conditions required for them to manifest [44]–[46]. These conditions can include particular machine instruction sequences, variations in operating voltages [47], temperature fluctuations, and platform-level behaviors like interrupts. This complexity results in low repeatability in SDC testing, necessitating extended testing periods to uncover potential issues. Developing effective testing methodologies to identify and address SDCs is therefore crucial for maintaining reliability, availability, and serviceability (RAS) in data centers. These strategies may involve repeated execution of specific code sequences to trigger SDCs or the use of pseudo-random instruction sequences to increase variability and expose latent faults during testing [48].

B. Sources of Faults and Impact on Reliability

SoCs can experience faults from various sources, including radiation, electrical marginalities, and silicon defects that may not be detectable during manufacturing. These issues may manifest in the field, impacting system reliability. The effects of such faults vary depending on where they occur within the SoC circuitry. For example, faults in error detection and correction-equipped circuits, such as caches, can be corrected at the hardware level, preventing system disruption [49]. However, SDCs where data errors propagate without triggering an interruption can have unpredictable consequences depending on the application [50]. A minor data error in a graphical operation might be negligible, but the same issue in a financial transaction could have serious implications.

Managing SDCs at scale is particularly crucial when deploying millions of processing cores, as even a single defect can cause significant operational disruptions. For instance, a modest-sized data center with 100,000 SoCs could experience at least one SDC per month at a 10 failures in time (FIT) rate [13] (where 1 FIT corresponds to one failure per billion hours of operation) [46], [48]. This occurrence rate underscores the need for rigorous reliability, availability, and serviceability (RAS) strategies to mitigate the effects of SDCs in large-scale computing environments. Hyperscale data centers, with millions of deployed SoCs, face a more acute risk, making reliability assessments and fault management an ongoing priority.

979-8-3503-6689-1/24 $31.00 © 2024 IEEE

C. RAS-Oriented Approach

Ensuring the reliability, availability, and serviceability of data center infrastructure requires more than simply detecting data corruption and defective chips. It demands a comprehensive strategy that encompasses both hardware and software design. Reliability can be enhanced by identifying and isolating defective components through regular testing, while availability ensures systems continue to function smoothly even in the presence of faults. Serviceability emphasizes quick fault diagnosis and repair to minimize downtime and provide useful information for system debug and defect root causing.

One effective strategy for early-stage mitigation is simulation-based testing. Researchers can use architectural and microarchitectural models to simulate SoC behavior under various conditions, identifying potential faults before physical chips are manufactured. These models allow prediction of FIT rates early in the design process, enabling adjustments when they are less costly. However, early-stage models only approximate final hardware behavior, as they often lack complete design details.

In the late stages of development, more precise measurements can be made using gate-level models and actual silicon. However, these methods are more resource-intensive and time-consuming, limiting their applicability to scenarios where extremely high accuracy is required, such as in critical systems or hyperscale environments. By combining early predictive models with late-stage testing, data centers can implement a robust RAS framework to proactively address SDC risks.

IV. RESULTS

A. AI-based security

The preliminary results on detecting malware and hardware attacks using HPCs are presented in this section, leveraging both supervised and unsupervised learning classifiers.

1) Supervised learning: With supervised classifiers, the detection of a specific type of hardware attack, i.e., side-channel attacks, is analyzed during runtime by monitoring HPCs.

Dataset creation: We selected 7 hardware-based attacks (i.e., Meltdown [27], Spectre [26](V1, V2 and V4), ZombieLoad [51], Fallout [52] and Crosstalk [53]) and 7 benign programs (i.e., Matrix multiplier, Debian stress tool [54], MiBench Bitcount [55], STREAM benchmark [56], bzip2 [57] and FFmpeg [58]) to construct the dataset.

We executed these applications and recorded multiple HPCs during binary execution using the *perf* tool. To streamline operations, we restricted execution to a single CPU core using the `taskset` Linux tool, ensuring that the collected HPC data remains unaffected by workload distribution across cores.

Some hardware attacks, like those in the Spectre family, exploit wrong speculative execution, triggered when the branch predictor mispredicts the outcome of a branch instruction. Therefore we selected both `branch-instructions` and `branch-misses` generic *perf* events, which provide the ratio between total branches and those where the predictor

TABLE I
SAMPLES DISTRIBUTION FOR THE 3 SCENARIOS: BALANCED, ONLY MALIGN, AND ONLY BENIGN.

Dataset	Train		Test	
	Malign	Benign	Malign	Benign
Balanced	11200	11200	2800	2800
Malign	11200	0	2800	14000
Benign	0	11200	14000	2800

missed. This selection is supported by previous work, such as Congmiago Li et al. [59].

Additionally, side-channel hardware-based attacks heavily stress the computer's cache memory. A high count of cache misses on the last-level cache (LLC) memory may indicate the presence of a FLUSH+RELOAD side-channel attack, known as the most effective and popular among hardware-based attacks. The first-level cache is also a common target in other attacks, as used by Stefano Carnà et al. [60]. Thus, we also selected the `LLC-load-misses` and `L1-dcache-load-misses` events.

Previous works have used sampling rates ranging from 1 ms per sample to 100 ms per sample. Congmiago Li et al. [59] even dynamically change the sample rate to prevent evasive malware. To generate a large number of samples for the machine learning model, the aim was to use the lowest possible sample rate. However, experimental results showed that rates below 10 ms caused anomalies in *perf*, such as missed samples. As a result, we chose a 1 ms sample rate. We extract 2,000 samples from each application, either by running the application for 2 seconds or repeating the execution until we get those samples.

Dataset preparation: The dataset for the 14 applications contains 28,000 samples, we created 3 different scenarios: Balanced, Only malign, and Only benign. Each scenario was split using 80% for training and 20% for testing (as shown in Table I).

Model evaluation: We used different models for training: Support Vector Machine (SVM) and One-Class SVM. The SVM model shows the best results for detecting side-channel attacks with a 99% accuracy, where only 2 of the attacks were misdetected (as observed in the confusion matrix of Figure 1 left).

Figure 1 (center and right) shows the scenarios where a single class dataset is used for training (either benign or malign). In these cases, we observed many false positives because the model encounters samples that were unseen during training and therefore misclassifies them as malware.

2) Unsupervised learning: To evaluate the effectiveness of unsupervised learning techniques, an experiment was designed to detect Stack Buffer Overflow (SBO) attacks. Four applications from the MiBench suite were selected—AES, RSA, SHA, and Dijkstra—and were subjected to SBO attacks. These attacks triggered the execution of a malicious function, with its size parametrized relative to the original code (i.e., the smaller the malicious function, the more stealthy its execution).

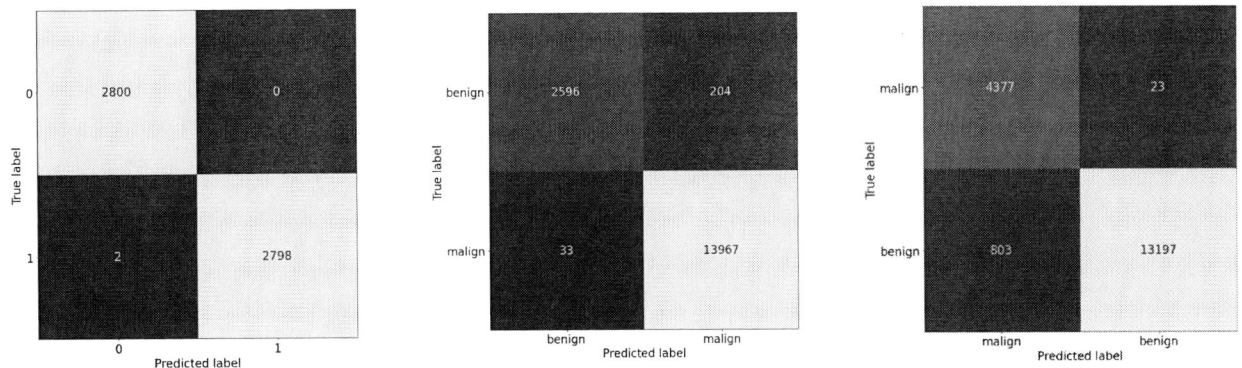

Fig. 1. Confusion matrices for detecting malware using a balanced data set (left), a benign dataset (center) and a malign dataset (right).

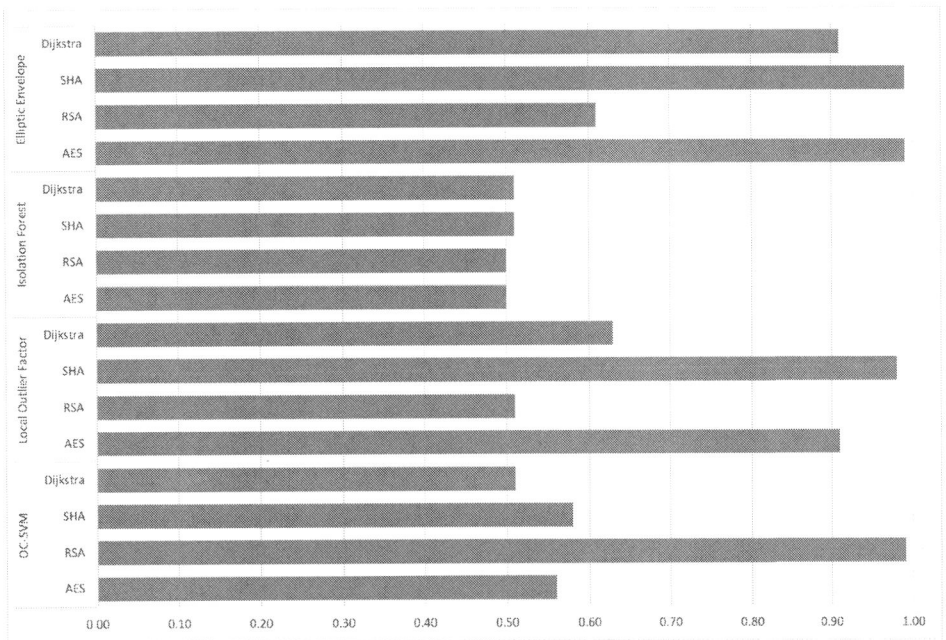

Fig. 2. Accuracy of unsupervised SBO detection for different benchmarks and classifiers. The malicious function runs a number of instructions lower than 1% of those of the original application.).

For each application, 20,000 executions were collected (10,000 for training and 10,000 for testing), with benchmark inputs randomly varied. The training dataset consisted solely of benign executions, while the testing dataset included 50% benign executions and 50% with attacks. Four anomaly detection models—OC-SVM, LOF, Isolation Forest, and Elliptic Envelope—were trained to assess their effectiveness on the dataset. Preliminary results, shown in Figure 2, indicate promising performance for the Elliptic Envelope classifier, except the RSA benchmark, which remains under investigation.

B. RAS

This subsection showcases the effects of *Permanent* faults on program execution. We focus on SDC outcomes, showing the probability that a fault in a specific hardware unit results

TABLE II
MAJOR SIMULATOR CONFIGURATIONS FOR EACH ISA.

Parameter	Value
ISA	RISC-V / Arm / x86
Pipeline	64-bit OoO (8-issue)
L1 Instruction Cache	32KB, 64B line, 128 sets, 4-way
L1 Data Cache	32KB, 64B line, 128 sets, 4-way
L2 Cache	1MB, 64B line, 2048 sets, 8-way
Physical Register File	128 Int; 128 FP
LQ/SQ/IQ/ROB entries	32/32/64/128

in an SDC. The basic parameters of the gem5 configuration we use in this paper can be seen in Table II.

1) SDCs due to Permanent Faults in L1 instruction and data caches: Fig. 3 illustrates the SDC probability outcomes for permanent faults in the L1 Instruction Cache across fifteen

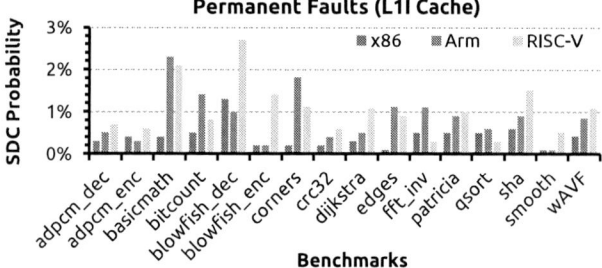

Fig. 3. SDC probability due to permanent faults in L1 instruction cache [46].

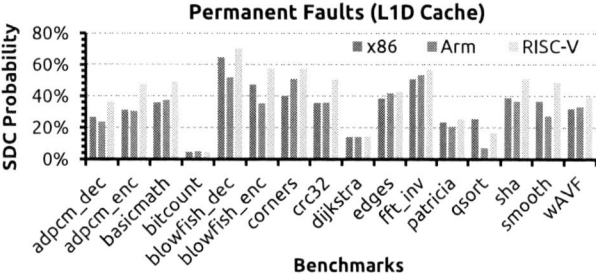

Fig. 4. SDC probability due to permanent faults in L1 data cache [46].

benchmarks of the MiBench [61] suite for the three ISAs (Arm, x86, RISC-V). As shown in Fig. 3, the SDC probability ranges from 0.1% to 2.3% for Arm ISA, 0.1% to 1.3% for x86, and 0.3% to 2.7% for RISC-V ISA. These results are expected because a workload running with a persistent fault in any level of cache memory that stores instructions is unlikely to survive to the end and produce a corrupted output. Faults in most fields of instruction will primarily impact the execution flow or the instruction operands, and thus, lead to a crash [10]. On average across all benchmarks, the x86 ISA demonstrates the lowest SDC probability among the ISAs studied in this paper, while the RISC-V ISA shows the highest SDC probability in most benchmarks.

Fig. 4 displays the SDC probability results for permanent faults in the L1 Data Cache across the same fifteen MiBench benchmarks for the three ISAs (Arm, x86, RISC-V). As shown in Fig. 4, the SDC probability varies from 5.1% to 53.3% for Arm ISA, 4.4% to 64.7% for x86, and 4.4% to 70.8% for RISC-V ISA. On average across all benchmarks, the RISC-V ISA exhibits the highest SDC probability among all ISAs studied in this paper. It is important to note that the L1 Data Cache is considered unprotected in our experiments, i.e., there is no ECC-related protection scheme. The actual SDC probability can be much lower in real systems due to these protection mechanisms.

Overall, for the microarchitecture and workloads analyzed, the RISC-V ISA demonstrates a significantly higher probability of SDCs due to permanent faults compared to the other ISAs, i.e., Arm and x86.

V. CONCLUSIONS

This paper provided an overview of AI-based security and RAS solutions in the computing continuum. The paper describes the comprehensive approach to malware and hardware attack detection; as well as, the RAS features envisioned for future systems across the computing continuum. AI-based detection is shown to be a highly effective way to detect malware either statically or dynamically and with several ML methods. The paper also provides an analysis of the vulnerability to SDCs of L1 data and instruction caches for the same core but different ISA. The results show the importance of RAS features in future systems as the vulnerability to SDCs increases in each technology.

REFERENCES

[1] A. Waterman, Y. Lee et al., "The RISC-V instruction set manual volume ii: Privileged architecture version 1.9," EECS Department, University of California, Berkeley, Tech. Rep., 2016.

[2] S. Greengard, "Will RISC-V revolutionize computing?" Communications of the ACM, vol. 63, no. 5, pp. 30–32, 2020.

[3] A. Dörflinger, M. Albers et al., "A comparative survey of open-source application-class RISC-V processor implementations," in Proceedings of the 18th ACM International Conference on Computing Frontiers, ser. CF '21. New York, NY, USA: Association for Computing Machinery, 2021, pp. 12–20.

[4] G. Papadimitriou and D. Gizopoulos, "Anatomy of on-chip memory hardware fault effects across the layers," IEEE Transactions on Emerging Topics in Computing, vol. 11, no. 2, pp. 420–431, 2023.

[5] G. Papadimitriou and D. Gizopoulos, "Avgi: Microarchitecture-driven, fast and accurate vulnerability assessment," in 2023 IEEE International Symposium on High-Performance Computer Architecture (HPCA), 2023, pp. 935–948.

[6] G. Papadimitriou and D. Gizopoulos, "Demystifying the system vulnerability stack: Transient fault effects across the layers," in 2021 ACM/IEEE 48th Annual International Symposium on Computer Architecture (ISCA), 2021, pp. 902–915.

[7] A. Chatzidimitriou, G. Papadimitriou et al., "Multi-bit upsets vulnerability analysis of modern microprocessors," in 2019 IEEE International Symposium on Workload Characterization (IISWC), 2019, pp. 119–130.

[8] P. Bodmann, G. Papadimitriou et al., "The Impact of SoC Integration and OS Deployment on the Reliability of Arm Processors," in 2021 IEEE International Symposium on Performance Analysis of Systems and Software (ISPASS), 2021, pp. 223–225.

[9] P. Bodmann, G. Papadimitriou et al., "Impact of cores integration and operating system on arm processors reliability: Micro-architectural fault-injection vs beam experiments," in 2020 20th European Conference on Radiation and Its Effects on Components and Systems (RADECS), 2020, pp. 1–4.

[10] G. Papadimitriou and D. Gizopoulos, "Silent data corruptions: Microarchitectural perspectives," IEEE Transactions on Computers, vol. 72, no. 11, pp. 3072–3085, 2023.

[11] P. R. Bodmann, G. Papadimitriou et al., "Soft error effects on arm microprocessors: Early estimations versus chip measurements," IEEE Transactions on Computers, vol. 71, no. 10, pp. 2358–2369, 2022.

[12] P. Koutsovasilis, C. D. Antonopoulos et al., "The impact of cpu voltage margins on power-constrained execution," IEEE Transactions on Sustainable Computing, vol. 7, no. 1, pp. 221–234, 2022.

[13] D. P. Lerner, B. Inkley et al., "Optimization of tests for managing silicon defects in data centers," in 2022 IEEE International Test Conference (ITC), 2022, pp. 578–582.

[14] H. D. Dixit, S. Pendharkar et al., "Silent Data Corruptions at Scale," 2021. [Online]. Available: https://arxiv.org/abs/2102.11245

[15] P. H. Hochschild, P. Turner et al., "Cores That Don't Count," in Proceedings of the Workshop on Hot Topics in Operating Systems, ser. HotOS '21. New York, NY, USA: Association for Computing Machinery, 2021, p. 9–16.

[16] S. Wang, G. Zhang *et al.*, "Understanding silent data corruptions in a large production cpu population," in *Proceedings of the 29th Symposium on Operating Systems Principles*, ser. SOSP '23. New York, NY, USA: Association for Computing Machinery, 2023, p. 216–230.

[17] D. Gizopoulos, "Sdcs: A b c," Sep 2024. [Online]. Available: https://www.sigarch.org/sdcs-a-b-c/

[18] J. M. Domingos, P. Tomas, and L. Sousa, "Supporting RISC-V performance counters through performance analysis tools for linux (perf)," in *5th Workshop on Computer Architecture Research with RISC-V (CARRV '21)*, 2021, pp. 935–948. [Online]. Available: https://arxiv.org/pdf/2112.11767.pdf

[19] S. Browne, J. Dongarra *et al.*, "A scalable cross-platform infrastructure for application performance tuning using hardware counters," in *SC '00: Proceedings of the 2000 ACM/IEEE Conference on Supercomputing*, 2000, pp. 42–42.

[20] B. Sprunt, "The basics of performance-monitoring hardware," *IEEE Micro*, vol. 22, no. 4, pp. 64–71, 2002.

[21] C. Malone, M. Zahran, and R. Karri, "Are hardware performance counters a cost effective way for integrity checking of programs," in *Proceedings of the Sixth ACM Workshop on Scalable Trusted Computing*, ser. STC '11. New York, NY, USA: Association for Computing Machinery, 2011, pp. 71–76.

[22] N. C. Doyle, E. Matthews *et al.*, "Performance impacts and limitations of hardware memory access trace collection," in *Design, Automation & Test in Europe Conference & Exhibition (DATE), 2017*, 2017, pp. 506–511.

[23] Y. Ding, W. Dai *et al.*, "Control flow-based opcode behavior analysis for malware detection," *Computers & Security*, vol. 44, pp. 65–74, 2014.

[24] H. HaddadPajouh, A. Dehghantanha *et al.*, "A deep recurrent neural network based approach for internet of things malware threat hunting," *Future Generation Computer Systems*, vol. 85, pp. 88–96, 2018.

[25] E. Rodríguez, P. Valls *et al.*, "Transfer-learning-based intrusion detection framework in iot networks," *Sensors*, vol. 22, no. 15, p. 5621, 2022.

[26] P. Kocher, J. Horn *et al.*, "Spectre attacks: Exploiting speculative execution," in *40th IEEE Symposium on Security and Privacy (S&P'19)*, 2019.

[27] M. Lipp, M. Schwarz *et al.*, "Meltdown: Reading kernel memory from user space," in *27th USENIX Security Symposium (USENIX Security 18)*, 2018.

[28] "VirusTotal," https://www.virustotal.com/.

[29] "VirusShare," https://www.virusshare.com/.

[30] M. O. F. Rokon, R. Islam *et al.*, "{SourceFinder}: Finding malware {Source-Code} from publicly available repositories in {GitHub}," in *23rd International Symposium on Research in Attacks, Intrusions and Defenses (RAID 2020)*, 2020, pp. 149–163.

[31] J. Demme, M. Maycock *et al.*, "On the feasibility of online malware detection with performance counters," *SIGARCH Comput. Archit. News*, vol. 41, no. 3, pp. 559–570, jun 2013.

[32] C. P. Chenet, A. Savino, and S. Di Carlo, "A survey on hardware-based malware detection approaches," *IEEE Access*, vol. 12, pp. 54 115–54 128, 2024.

[33] S. Dutto, A. Savino, and S. Di Carlo, "Exploring deep learning for in-field fault detection in microprocessors," in *2021 Design, Automation & Test in Europe Conference & Exhibition (DATE)*, 2021, pp. 1456–1459.

[34] D. Kasap, A. Carpegna *et al.*, "Micro-architectural features as soft-error induced fault executions markers in embedded safety-critical systems: a preliminary study," 2023.

[35] T. Sherwood, E. Perelman *et al.*, "Discovering and exploiting program phases," *IEEE Micro*, vol. 23, no. 6, pp. 84–93, 2003.

[36] C. Isci, G. Contreras, and M. Martonosi, "Live, runtime phase monitoring and prediction on real systems with application to dynamic power management," in *2006 39th Annual IEEE/ACM International Symposium on Microarchitecture (MICRO'06)*, 2006, pp. 359–370.

[37] B. Singh, D. Evtyushkin *et al.*, "On the detection of kernel-level rootkits using hardware performance counters," in *Proceedings of the 2017 ACM on Asia Conference on Computer and Communications Security*, ser. ASIA CCS '17. New York, NY, USA: Association for Computing Machinery, 2017, pp. 483–493.

[38] H. Sayadi, N. Patel *et al.*, "Machine learning-based approaches for energy-efficiency prediction and scheduling in composite cores architectures," in *2017 IEEE International Conference on Computer Design (ICCD)*, 2017, pp. 129–136.

[39] H. Peng, F. Long, and C. Ding, "Feature selection based on mutual information criteria of max-dependency, max-relevance, and min-redundancy," *IEEE Transactions on Pattern Analysis and Machine Intelligence*, vol. 27, no. 8, pp. 1226–1238, 2005.

[40] V. Chandola, A. Banerjee, and V. Kumar, "Anomaly detection: A survey," *ACM Comput. Surv.*, vol. 41, no. 3, jul 2009.

[41] H. Sayadi, N. Patel *et al.*, "Ensemble learning for effective run-time hardware-based malware detection: A comprehensive analysis and classification," in *2018 55th ACM/ESDA/IEEE Design Automation Conference (DAC)*, 2018, pp. 1–6.

[42] Y. Freund and R. E. Schapire, "A decision-theoretic generalization of on-line learning and an application to boosting," *Journal of Computer and System Sciences*, vol. 55, no. 1, pp. 119–139, 1997.

[43] H. Sayadi, H. M. Makrani *et al.*, "2smart: A two-stage machine learning-based approach for run-time specialized hardware-assisted malware detection," in *2019 Design, Automation & Test in Europe Conference & Exhibition (DATE)*, 2019, pp. 728–733.

[44] D. Gizopoulos, G. Papadimitriou, and O. Chatzopoulos, "Estimating the failures and silent errors rates of cpus across isas and microarchitectures," in *2023 IEEE International Test Conference (ITC)*, 2023, pp. 377–382.

[45] T. Macieira, S. Gurumurthy *et al.*, "Silent data corruptions in computing: Understand and quantify," in *2024 IEEE 30th International Symposium on On-Line Testing and Robust System Design (IOLTS)*, 2024, pp. 1–7.

[46] D. Gizopoulos, G. Papadimitriou *et al.*, "Silent data corruptions in computing systems: Early predictions and large-scale measurements," in *2024 IEEE European Test Symposium (ETS)*, 2024, pp. 1–10.

[47] G. Papadimitriou, M. Kaliorakis *et al.*, "A system-level voltage/frequency scaling characterization framework for multicore cpus," in *IEEE Silicon Errors in Logic – System Effects (SELSE 2017)*, 2017.

[48] N. Karystinos, O. Chatzopoulos *et al.*, "Harpocrates: Breaking the silence of cpu faults through hardware-in-the-loop program generation," in *2024 ACM/IEEE 51st Annual International Symposium on Computer Architecture (ISCA)*, 2024, pp. 516–531.

[49] D. Agiakatsikas, G. Papadimitriou *et al.*, "Impact of voltage scaling on soft errors susceptibility of multicore server cpus," in *Proceedings of the 56th Annual IEEE/ACM International Symposium on Microarchitecture*, ser. MICRO '23. New York, NY, USA: Association for Computing Machinery, 2023, p. 957–971.

[50] G. Papadimitriou, D. Gizopoulos *et al.*, "Silent data corruptions: The stealthy saboteurs of digital integrity," in *2023 IEEE 29th International Symposium on On-Line Testing and Robust System Design (IOLTS)*, 2023, pp. 1–7.

[51] M. Schwarz, M. Lipp *et al.*, "ZombieLoad: Cross-privilege-boundary data sampling," in *CCS*, 2019.

[52] C. Canella, D. Genkin *et al.*, "Fallout: Leaking data on meltdown-resistant cpus," in *Proceedings of the ACM SIGSAC Conference on Computer and Communications Security (CCS)*. ACM, 2019.

[53] H. Ragab, A. Milburn *et al.*, "CrossTalk: Speculative Data Leaks Across Cores Are Real," in *S&P*, May 2021, intel Bounty Reward. [Online]. Available: https://download.vusec.net/papers/crosstalk_sp21.pdf

[54] R. O. S. Projects, "stress," https://github.com/resurrecting-open-source-projects/stress.

[55] U. of Michigan, "Mibench version 1.0," https://vhosts.eecs.umich.edu/mibench/, 2002, accessed: 14-05-2024.

[56] J. D. McCalpin, "Stream: Sustainable memory bandwidth in high performance computers," https://www.cs.virginia.edu/stream/, accessed: 28-05-2024.

[57] "bzip2," https://sourceware.org/bzip2/, accessed: 28-05-2024.

[58] "Ffmpeg," https://ffmpeg.org/, accessed: 28-05-2024.

[59] C. Li and J.-L. Gaudiot, "Detecting spectre attacks using hardware performance counters," *IEEE Transactions on Computers*, vol. 71, no. 6, pp. 1320–1331, 2022.

[60] S. Carnà, S. Ferracci *et al.*, "Fight hardware with hardware: Systemwide detection and mitigation of side-channel attacks using performance counters," *Digital Threats*, vol. 4, no. 1, 3 2023.

[61] M. R. Guthaus, J. S. Ringenberg *et al.*, "Mibench: A free, commercially representative embedded benchmark suite," in *Proceedings of the fourth annual IEEE international workshop on workload characterization. WWC-4 (Cat. No. 01EX538)*. IEEE, 2001, pp. 3–14.

979-8-3503-6689-1/24 $31.00 © 2024 IEEE

RAPPER: **R**obust and **APP**roximate **ER**ror Tolerating Communication

Somayeh Sadeghi-Kohan
Paderborn University
33098 Paderborn, Germany
somayeh.sadeghi@upb.de

Sybille Hellebrand
Paderborn University
33098 Paderborn, Germany
sybille.hellebrand@upb.de

Hans-Joachim Wunderlich
Consultant
71083 Herrenberg, Germany
wu@hjwunderlich.de

Abstract— **Effective error-resilient communication is crucial, especially in autonomous systems. However, some applications can accept a certain level of approximation within a defined range. This opens up opportunities for optimizing system objectives, like minimizing error rates, extending operational lifetimes or reducing power consumption. In this study, we introduce an innovative approach called RAPPER (Robust and Approximate Error Tolerating Communication) which combines approximation and error correcting codes for optimizing the transmitted data. The underlying communication scheme exploits an extension of differential encoding to enhance communication performance. Simulation results show an increased interconnect mission time and performance, while simultaneously achieving significant reductions in error rates and power consumption.**

Keywords— *Approximate communication, Error-resilient communication, Interconnect lifetime, Electromigration, Power consumption, SECDED code, Data compression*

I. INTRODUCTION

This paper presents a new approach for efficient and robust data streaming combining differential encoding with optimized approximation and dual protection by a single error correcting and double error detecting (SECDED) code. The new communication scheme addresses both the high performance and the strict quality requirements for the streaming of sensor data in many applications, for example in the automotive domain [1]. It exploits the fact that small deviations in the transmitted data stream are often acceptable as long as a given approximation threshold A_{th} is not violated.

The quality requirements are typically specified in terms of power consumption, error rate and achievable lifetime. Power consumption depends on the switching activity and also on crosstalk between wires. The error rate is directly related to the frequency of crosstalk delays and glitches, and the lifetime can be shortened by crosstalk-induced electromigration [2] - [4].

On the other hand, approximation within the acceptable bounds provides a means to tune the communication. While early work on approximate communication has specifically targeted performance or power consumption [5] - [7], it has been shown in [8] that focusing on a single objective may lead to degradations with respect to other objectives. Therefore, the communication schemes in [9] and [10] have combined data compression with an optimized approximation scheme to reduce crosstalk effects

and switching activity. Although the schemes in [9] and [10] considerably reduce the error rates, they cannot completely avoid crosstalk effects. Therefore, an additional protection by error correcting codes is required to ensure error-resilient communication. However, if error correcting codes are applied to approximated data in a straightforward way, crosstalk effects on the check bits may have an adverse impact on the approximation objectives.

In this work, an extension of differential data encoding is used for performance improvement [11], and the protection of data is already considered during the approximation process. To increase the optimization potential, two variants of a SECDED code are employed. The details of this approach called RAPPER (**R**obust and **APP**roximate **ER**ror Tolerating Communication) are explained in the following sections. The necessary background on crosstalk and electromigration, bus encoding, reliability modeling and data compression is briefly summarized in Section II. Subsequently, in Section III, the core concepts of dual SECDED and dual streaming are introduced. Experimental results achieved with RAPPER are reported in Section IV before the paper concludes with Section V.

II. BACKGROUND

A. Crosstalk and electromigration

Coupling capacitances between interconnect lines can lead to crosstalk effects, if specific pairs of input vectors, so-called crosstalk patterns, are applied to a *victim line* and its adjacent *aggressor lines* [4]. Possible crosstalk effects and the respective crosstalk patterns are summarized in Table 1, where the symbols ↓ and ↑ represent falling and rising transitions, respectively, while 0 and 1 denote constant values over two clock cycles.

Table 1. Crosstalk effects and patterns.

Crosstalk effect	Delayed transition	Glitch	Speedy transition	Over- or undershoot
Crosstalk patterns	↑↓↑, ↓↑↓	↑0↑, ↓1↓	↑↑↑, ↓↓↓	↑1↑, ↓0↓

Delay and glitch patterns can cause transient errors and impact the error rate observed on the bus. Depending on the error handling scheme, this may lead to an increased number of retransmissions and impact the performance. While speedy and shoot patterns do not cause functional errors, they lead to an increased switching activity and a higher current density on the interconnect lines, which is even more pronounced than for delay and glitch patterns. The current density is defined as the average current divided by the cross-section of the interconnect.

The increased current density can also aggravate the material degradation due to electromigration. Black's formula quantifies this problem by estimating the median time to failure t_{50}, where half of the interconnects fail [3][12][13].

B. Bus encoding and protection

Many research activities have addressed the problem of power consumption and crosstalk in interconnects so far. A common strategy in most of the published approaches is to encode the information sent over the bus, such that the switching activity is reduced for low power applications [14] - [17], and crosstalk patterns as defined in Table 1 are avoided whenever possible [18] - [22]. The Gray code is a prominent example for low power bus encoding [14][16], and also some crosstalk avoidance codes are based on the Gray code [21]. To protect the data against remaining crosstalk faults, error correcting codes can be used [23]. In this work SECDED codes are employed [24].

C. Reliability and mission time of interconnects

In the following, the reliability $R(t)$ of a component or system is defined as the probability of survival from time 0 to t. Assuming a constant failure rate λ, the reliability is given by $R(t) =$ exp(-λt), and the Mean Time to Failure (MTTF) can be computed as MTTF $= 1/\lambda$ [25]. For a given threshold R_{th}, the mission time $TM(R_{th})$ denotes the time when the reliability drops to this threshold, i. e.

$$TM(R_{th}) = \max\{t \geq 0 \mid R(t) \geq R_{th}\}. \tag{1}$$

The median time to failure t_{50} obtained from Black's formula provides the failure rate for a single interconnect line as

$$\lambda = \ln(2)/t_{50}, \tag{2}$$

and from this the reliability $R(t)$, the mean time to failure, as well as the mission time for single interconnects can be computed.

For a bus with n interconnect lines, the reliability and the mission time depend on the used fault tolerance scheme. If no protection is used, then all bus lines must survive to ensure the survival of the bus. Assuming that each bus line has its specific failure rate λ_i and reliability $R_i(t)$, the reliability of an n-bit bus is obtained as

$$R_{Bus}(t) = \prod_{i=1}^{n} R_i(t) = \exp\left(-t \sum_{i=1}^{n} \lambda_i\right). \tag{3}$$

Consequently, the mission time of the bus is obtained as

$$TM_{Bus}(R_{th}) = -\frac{\ln(R_{th})}{\sum_{i=1}^{n} \lambda_i}. \tag{4}$$

If the bus is protected by an error correction scheme, then one or more broken bus lines may be tolerable. In the case of a SECDED scheme, the permanent failure of one bus line can be compensated by the code and an additional transient error on the bus can still be detected. However, this will lead to a degradation of the bus performance because a transient error will then always require a retransmission. Therefore, we do not consider a life-

time extension by error correcting codes in this paper and compute the mission time using formula (4).

D. Increasing performance by differential encoding

The performance of bus communication can be increased by data compression. In the literature various schemes can be found including run-length coding, base-delta compression and differential encoding [11][26]. This work relies on differential encoding, which is therefore briefly explained in the following. A sequence of numbers $N = (n_0, n_1, n_2, \ldots)$ is encoded as a stream of successive differential values $(n_0, \Delta_1, \Delta_2, \ldots)$ defined by $\Delta_i = n_i - n_{i-1}$ as long as the differential value Δ_i is within a predefined range $[\Delta_{min}, \Delta_{max}]$ determined by the available bit width. The first value n_0 is also referred to as the master of the differential sequence. If a differential value Δ_k is no longer in the range $[\Delta_{min}, \Delta_{max}]$, then a new differential sequence $(n_k, \Delta_{k+1}, \Delta_{k+2}, \ldots)$ with the new master n_k is started.

III. RAPPER

The presented RAPPER scheme relies on differential encoding for performance improvement and assumes that approximation is possible within a given range specified by an approximation threshold A_{th}. More precisely, a sequence of numbers $N = (n_0, n_1, n_2, \ldots)$ can be approximated by $\alpha(N) = (\alpha(n_0), \alpha(n_1), \alpha(n_2), \ldots)$ as long as $n_i - A_{th} \leq \alpha(n_i) \leq n_i + A_{th}$ for all $i \geq 0$. The differential encoding is applied to the approximated sequence, i. e. $\Delta_i = \alpha(n_i) - \alpha(n_{i-1})$. The approximation constraints can also be expressed as constraints for Δ_i by

$$n_i - \alpha(n_{i-1}) - A_{th} \leq \Delta_i \leq n_i - \alpha(n_{i-1}) + A_{th}. \tag{5}$$

The approximation must also ensure that Δ_i is within the interval $[\Delta_{min}, \Delta_{max}]$ as explained above, i. e.

$$\Delta_{min} \leq \Delta_i \leq \Delta_{max}. \tag{6}$$

Overall, equations (5) and (6) define the *approximation space* $A(\Delta_i)$ from which the differential Δ_i can be selected. If the approximation space is empty, then n_i is selected as the master of a new differential sequence. Following the work in [16] and [21], the differentials Δ_i are transmitted in the Gray code representation.

As shown in previous work, approximation can be exploited to optimize the transmitted data with respect to power, performance, error rate and mission time [9][10]. Nevertheless, despite reduced error rates, the data must still be protected against remaining crosstalk errors. While in [9][10] error detection and correction has not been considered during the optimization step, this paper presents an integrated approach for optimizing the complete data words including the check bits. Since the check bits are uniquely determined by the selected error correcting code, approximation is only possible for the information bits of a data word. To overcome this limitation, the optimization potential is increased by maintaining two streams of differentials and by employing a dual SECDED scheme as explained in the following.

A. Dual SECDED

Combining approximation with SECDED protection means that for each differential $\Delta \in A(\Delta_i)$, a string of check bits $p(\Delta)$ is determined and appended to Δ, which results in a string $\Delta\_p(\Delta)$.

As explained above, the check bits in $p(\Delta)$ are uniquely determined by Δ, and optimization with respect to crosstalk effects is not possible. To increase the optimization potential, also the string $\Delta \_ \overline{p(\Delta)}$ is added for all $\Delta \in A(\Delta_i)$, where $\overline{p(\Delta)}$ denotes the bitwise complement of $p(\Delta)$. The obtained approximation space $A_{dp}(\Delta_i)$ with dual SECDED protection is formally defined as

$$A_{dp}(\Delta_i) = \{\Delta \_ \mathrm{p}(\Delta) \mid \Delta \in A(\Delta_i)\} \cup \{\Delta \_ \overline{p(\Delta)} \mid \Delta \in A(\Delta_i)\}. \ (7)$$

As illustrated in Figure 1, this doubles the size of the approximation space and helps to avoid contradictory effects on the information and check bits.

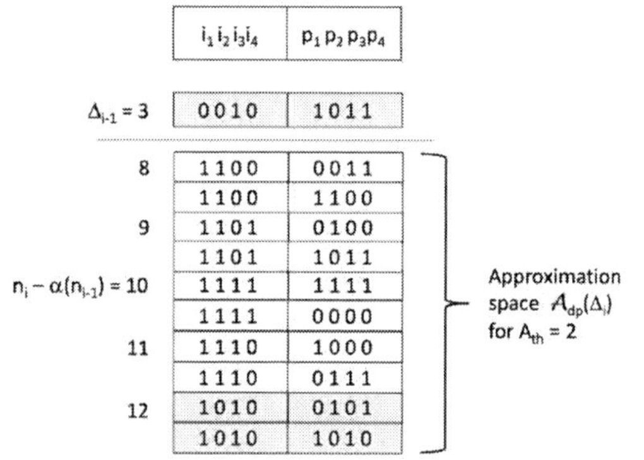

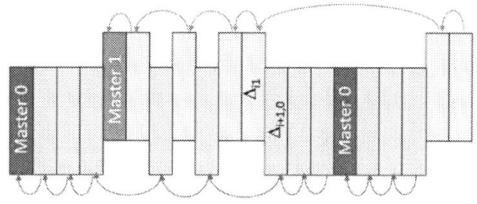

Figure 1. Approximation with dual SECDED.

Here, 4 data bits (i_1, i_2, i_3, i_4) in the Gray code representation are protected by a SECDED Hamming code with check bits $p_1 = i_1 \oplus i_2 \oplus i_3$, $p_2 = i_1 \oplus i_2 \oplus i_4$, $p_3 = i_1 \oplus i_3 \oplus i_4$, and the overall parity check $p_4 = i_1 \oplus i_2 \oplus i_3 \oplus i_4 \oplus p_1 \oplus p_2 \oplus p_3$, where $\oplus$ denotes an EXOR operation. Assume that differential encoding and approximation with $A_{th} = 2$ is used, and the first part of the data sequence $\alpha(N) = (\ldots, \alpha(n_{i-2}) = 150, \alpha(n_{i-1}) = 153)$ has already been transmitted. If the next value is $n_i = 163$, then Δ_i can be selected from the approximation space $A(\Delta_i) = \{8, 9, 10, 11, 12\}$.

Adding the check information $p(\Delta)$ and $\overline{p(\Delta)}$ to each value in $A(\Delta_i)$ provides the extended approximation space $A_{dp}(\Delta_i)$ shown in Figure 1. Selecting $\Delta_i = 12$ with inverted check bits minimizes crosstalk effects both on the information and on the check bits.

Figure 2. Dual stream differential encoding.

B. Dual streaming

The approximation space is further enlarged by maintaining two interleaved streams S_0 and S_1 of differential values as shown in Figure 2. More formally, the sequence $N = (n_0, n_1, n_2, \ldots)$ is

approximated by $\alpha(N) = (\alpha(n_0), \alpha(n_1), \alpha(n_2), \ldots)$, and each differential value Δ_i is associated with a stream S_j, which is expressed by Δ_{ij} for $j \in \{0, 1\}$.

During approximation and differential encoding, for each element n_i both streams of differentials are considered, and the approximation space $A_{ds}(\Delta_i)$ for dual streaming is obtained as $A_{ds}(\Delta_i) = A(\Delta_{i0}) \cup A(\Delta_{i1})$. $A(\Delta_{ij})$ is described by

$$n_i - \alpha(n_{\mathrm{pred}(i,j)}) - A_{th} \leq \Delta_{ij} \leq n_i - \alpha(n_{\mathrm{pred}(i,j)}) + A_{th}, \text{ and}$$

$$\Delta_{min} \leq \Delta_{ij} \leq \Delta_{max}, \quad (8)$$

where $\mathrm{pred}(i, j)$ denotes the index of the predecessor of n_i in the stream S_j. To minimize crosstalk effects, the differentials must be selected from $A_{ds}(\Delta_i)$, such that the number of crosstalk patterns between two successive differentials transmitted over the bus are minimized. As illustrated in Figure 2 for Δ_{i1} and $\Delta_{i+1,0}$ successive differentials on the bus do not necessarily belong to the same stream.

If the approximation space is empty, then one of the streams must be continued with n_i as a new master. The stream S_j is selected based on the activity and the overlap of the streams. If at least one of the streams has been inactive during at least τ steps for a predefined inactivity threshold τ, then the stream with longer inactivity is selected. This ensures that a frequently used stream can be continued in the case of noisy data. If both streams have been used during the last τ steps, then the stream S_j with minimum difference between $n_{\mathrm{pred}(i,j)}$ and n_i is selected. This ensures that the overlap between the two streams is kept as small as possible.

C. Data representation

If differential encoding is combined with dual streaming and dual SECDED protection, the control signals specified in Table 2 are required.

Table 2. Control signals for RAPPER.

Control signal	Meaning of value	
	0	1
MD	Master	Differential
SC	Original SECDED Code	Inverted SECDED Code
DS	Differential Stream S_0	Differential Stream S_1

In the following, the data format is explained in more detail for image data in RGB format, which will be used for the evaluation of the method. The color of a pixel in RGB format is represented by the contributions of the basic colors red, green and blue. If we assume an 8-bit code for these components, then a pixel is described by concatenating these codes to a 24-bit word. Accordingly, if 4-bit differentials Δ_R, Δ_G, Δ_B are acceptable for the individual components, then the 12-bit word $\Delta = (\Delta_R, \Delta_G, \Delta_B)$ represents the differential value Δ. The approximation threshold A_{th} applies to the individual 4-bit differentials Δ_R, Δ_G, Δ_B in this case.

Both master and differential values are combined with a SECDED code and the necessary control signals. For the 12-bit differentials, 5 check bits are needed for single error correction, and an overall parity check ensures the detection of double errors. So overall, the SECDED code requires 6 check bits.

979-8-3503-6689-1/24 $31.00 © 2024 IEEE

Combining information, check and control bits results in a 21-bit word as sketched in Figure 3.

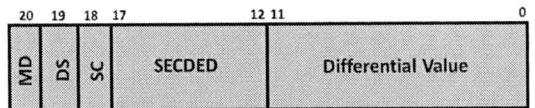

Figure 3. Data format for differentials.

For a master with 24 information bits, the SECDED code also has 6 check bits and is used in original format only. Together with the control bits MD and DS, 32 bits must be transmitted. If the master is distributed to two 21-bit words, then the remaining 10 bits can be used for shielding as shown in Figure 4.

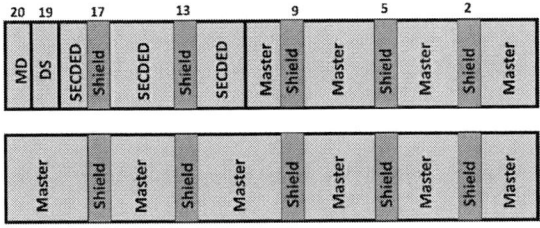

Figure 4. Data format for masters.

If more shielding is required, then three 21-bit words can be used for a master.

D. Approximation algorithm

In the RAPPER approach, the main objective of approximation is to minimize crosstalk effects. Therefore, the approximation goal is to minimize the number of crosstalk patterns between successive differentials transmitted over the bus. In case of several solutions, the switching activity is further minimized. An exact solution of this optimization problem is highly complex and only possible with information about the complete sequence $N = (n_0, n_1, n_2, \ldots)$. Therefore, a simple incremental heuristic is employed as outlined by the pseudo-code in Algorithm 1.

When the transmission starts, the first element n_0 is defined as the first master of S_0. In addition, both the sender and the receiver store the bitwise complement of n_0 as the first (virtual) master of S_1. Then in each iteration step, the complete approximation space for dual SECDED and dual streaming is explored by exhaustive search, and the differential Δ_{ij*} with the minimum number of crosstalk patterns between Δ_{i-1} and Δ_{ij} is selected.

The size of the approximation space depends on the approximation threshold A_{th}, as the information bits of Δ_{ij} can be selected from the interval $[n_i - \alpha(n_{\mathrm{pred}(i,j)}) - A_{th}, n_i - \alpha(n_{\mathrm{pred}(i,j)}) + A_{th}]$ of length $2A_{th} + 1$. With dual streaming and dual SECDED, the search space is bounded by $4(2A_{th} + 1)$.

IV. EXPERIMENTAL RESULTS

To evaluate the proposed methods, the interconnect parameters for a 32 nm technology shown in Table 3 are used.

To mimic the sensor data, serial traces produced by 2 representative benchmark RGB images B0 and B1 from the Kodak database are used [27]. The 21-bit interconnect structure described

in the previous chapter is assumed, where the available bit width for the differentials Δ_R, Δ_G, Δ_B is set to 4.

```
Algorithm Optimized_Approximation
Input:      N = (n₀,n₁,…),Aₜₕ, Δₘᵢₙ, Δₘₐₓ,
            Generator matrix for SECDED code
Output:     α(N), differential streams S₀,S₁
α(n₀) = n₀
S₀ = (α(n₀))
S₁ = (α(n₀) after bitwise inversion)
While (a new number nᵢ is available)
    Build 𝒜_dp(Δᵢ₀) ∪ 𝒜_dp(Δᵢ₁)   //Eq.(7)&(8)
    If 𝒜_dp(Δᵢ₀) ∪ 𝒜_dp(Δᵢ₁) is empty
        use nᵢ as a new master of Stream Sⱼ
        // select Sⱼ as described in III.B
    Else
        For j ∈ {0,1}
            Find Δᵢⱼ, such that (Δᵢ₋₁, Δᵢⱼ)
            has a minimum number of crosstalk
            patterns          //exhaustive search
        Select Δᵢⱼ* ∈ {Δᵢ₀, Δᵢ₁} with best
            properties and add it to stream Sⱼ*
End-While
```

Algorithm 1. Heuristic for optimized approximation.

Table 3. Interconnect parameters.

Parameter	Value or Range
Height (space between wire and reference plane)	0.09 μm
Width	84.4 nm
Height	151 nm
Space (nominal)	84.4 nm
Length	2000 μm (100 μm distance between buffers)
Conductor resistivity for copper	1.7e-8 Ωm
Dielectric constant	1.36
Supply voltage	0.9 V
n (material parameter, see [12])	1.1

In all experiments, the dual streaming in RAPPER is applied with an inactivity threshold $\tau = 5$. The results are compared to the original data stream protected with a SECDED code, with and without shielding mechanism (EXACT, EXACT_Shield). Furthermore, the results are also compared to an exact version of RAPPER with $A_{th} = 0$ (DIFFERENTIAL).

A. Error rates

To analyze the effects of RAPPER on functional safety, the total set of transition errors (TE) has been determined and three different types of errors have been identified in TE. *Correctable errors* (CE) are detected and corrected with the SECDED code. *Detectable errors* (DE) are detected but not corrected. These errors may result in a repetition of the transferred data and degrade the interconnect performance. *Undetectable errors* (UDE) are neither detected nor corrected and degrade the functional safety of the interconnect. Table 4 and Table 5 show the error detection profiles for the benchmarks B1 and B2. The first column of each table shows the applied scheme, the second column presents the total number of transition errors, and the remaining three columns list the sizes of CE, DE, and UDE relative to the total number of errors.

Table 4. Error detection profile for B0.

	# TE	# CE/#TE	#DE/#TE	#UDE/#TE
EXACT	354978	0.207	0.561	0.232
EXACT_Shield	169026	0.733	0.253	0.014
DIFFERENTIAL	47346	0.730	0.259	0.012
RAPPER				
$A_{th} = 1$	13026	0.605	0.365	0.030
$A_{th} = 2$	10452	0.560	0.410	0.030
$A_{th} = 3$	9360	0.600	0.375	0.025
$A_{th} = 4$	8892	0.614	0.360	0.026
$A_{th} = 5$	8424	0.639	0.343	0.019
$A_{th} = 6$	7332	0.681	0.309	0.011
$A_{th} = 7$	6162	0.759	0.228	0.013
$A_{th} = 8$	5616	0.764	0.222	0.014

Table 5. Error detection profile for B1.

	# TE	# CE/#TE	#DE/#TE	#UDE/#TE
EXACT	348816	0.393	0.389	0.218
EXACT_Shield	236340	0.721	0.232	0.046
DIFFERENTIAL	131898	0.512	0.433	0.055
RAPPER				
$A_{th} = 1$	93678	0.440	0.495	0.064
$A_{th} = 2$	86424	0.449	0.486	0.064
$A_{th} = 3$	78390	0.479	0.456	0.066
$A_{th} = 4$	69888	0.500	0.448	0.052
$A_{th} = 5$	66066	0.504	0.445	0.051
$A_{th} = 6$	63726	0.499	0.452	0.049
$A_{th} = 7$	60528	0.505	0.456	0.039
$A_{th} = 8$	55678	0.474	0.490	0.036

First of all, it can be observed that RAPPER substantially reduces the total number of errors. Already for $A_{th} = 1$, the total number of errors is reduced down to 27.5 % compared to the basic differential scheme without approximation and down to 3.6 % compared to the original scheme for B0. For B1 a reduction down to 71.0 % and 26.8 % is obtained. For the largest approximation threshold $A_{th} = 8$ these values are further improved to 11.8 % and 1.5 % for B0, as well as 42.2 % and 15.9 % for B1.

Furthermore, the detection profile is improved compared to the original scheme. The portion of correctable errors is increased while the share of undetectable errors is reduced down to a few percent. The profile shows similar characteristics as in EXACT_Shield and DIFFERENTIAL, however, the total number of uncorrectable errors is considerably smaller in RAPPER, because of the reduction of the total number of errors.

B. Mission time

For the analysis of the mission time formula (4) has been used with the reliability threshold $R_{th} = 0.99999$. Table 6 and Table 7 show the achieved mission times in years for B0 and B1 at varying temperatures.

The reported results show that, in line with Black's formula, the mission times decrease for higher temperatures. However, by applying the RAPPER scheme, the mission times are signifi-

cantly improved versus the EXACT scheme and even versus the EXACT_Shield scheme. Already for $A_{th} = 1$, the mission time of the interconnect at 140°C increases from 1.99 years to 3.13 years for B1. For the largest approximation threshold $A_{th} = 8$ the improvement in mission time reaches 485 % of the original mission time for B0 and 253 % for B1.

Table 6. Mission times for B0 at varying temperatures.

	40°C	80°C	120°C	140°C
EXACT	9084.78	208.08	10.28	2.84
EXACT_Shield	11210.89	256.77	12.68	3.51
DIFFERENTIAL	20398.49	467.20	23.08	6.38
RAPPER				
$A_{th} = 1$	21503.81	492.52	24.33	6.73
$A_{th} = 2$	25921.60	593.71	29.32	8.11
$A_{th} = 3$	26304.55	602.48	29.76	8.23
$A_{th} = 4$	26987.89	618.13	30.53	8.44
$A_{th} = 5$	31382.30	718.78	35.50	9.82
$A_{th} = 6$	32223.05	738.03	36.45	10.08
$A_{th} = 7$	41513.67	950.82	10.28	12.98
$A_{th} = 8$	44113.77	1010.38	12.68	13.80

Table 7. Mission times for B1 at varying temperatures.

	40°C	80°C	120°C	140°C
EXACT	6354.20	145.54	7.19	1.99
EXACT_Shield	9561.70	219.00	10.82	2.99
DIFFERENTIAL	9574.52	219.29	10.83	2.99
RAPPER				
$A_{th} = 1$	10015.32	229.39	11.33	3.13
$A_{th} = 2$	10282.02	235.50	11.63	3.22
$A_{th} = 3$	10286.26	235.60	11.64	3.22
$A_{th} = 4$	11901.61	272.59	13.46	3.72
$A_{th} = 5$	13522.61	309.72	15.30	4.23
$A_{th} = 6$	13902.76	318.43	15.73	4.35
$A_{th} = 7$	14926.34	341.87	16.89	4.67
$A_{th} = 8$	16103.74	368.84	18.22	5.04

C. Power consumption and performance

RAPPER also aims at reducing power consumption and supporting a high performance, as evidenced by a reduced number of transitions and clock cycles, respectively. As shown in Table 8, compared to the exact scheme without shielding, RAPPER reduces the number of transitions down to 12.9 % (B0) and 32.6 % (B1) for $A_{th} = 1$. A further reduction down to 6.2 % (B0) and 18.5 % (B1) is obtained for $A_{th} = 8$.

In Table 9, the number of clock cycles required for communication are reported. The numbers include the repetition of erroneous words which are detected but cannot be corrected (DE). The results clearly show that the integration of dual streaming and differential encoding in RAPPER provides the best performance. By increasing the approximation threshold A_{th}, the number of transmitted words is further reduced. Our findings reveal that, the required clock cycles are significantly reduced to 41.2 % (B0) and 56.2 % (B1) for $A_{th} = 1$. For $A_{th} = 8$, a further reduction down to 40.6 % (B0) and 50.5 % (B1) is achieved.

V. CONCLUSIONS

The RAPPER approach outlined in this paper integrates differential encoding with dual streaming and exploits the approximation potential of the application to substantially enhance the error resilience and lifespan of interconnect systems. Furthermore, it considerably reduces power consumption and improves the overall interconnect performance.

Table 8. Number of transitions.

	B0		B1	
	Total # transitions	Normalized to EXACT	Total # transitions	Normalized to EXACT
EXACT	8334300	1.000	8174946	1.000
EXACT_Shield	6147024	0.738	5725044	0.700
DIFFERENTIAL	1237548	0.148	3082482	0.377
RAPPER				
$A_{th} = 1$	1064076	0.128	2662998	0.326
$A_{th} = 2$	1044888	0.125	2410746	0.295
$A_{th} = 3$	944580	0.113	2163174	0.265
$A_{th} = 4$	769704	0.092	1962636	0.240
$A_{th} = 5$	738270	0.089	1869348	0.229
$A_{th} = 6$	590928	0.071	1702428	0.208
$A_{th} = 7$	550914	0.066	1645254	0.201
$A_{th} = 8$	513162	0.062	1513122	0.185

Table 9. Required number of clock cycles.

	B0		B1	
	Total # clock cycles	Normalized to EXACT	Total # clock cycles	Normalized to EXACT
EXACT	985722	1.000	921996	1.000
EXACT_Shield	829254	0.841	841344	0.913
DIFFERENTIAL	415506	0.422	536484	0.582
RAPPER				
$A_{th} = 1$	405756	0.412	518388	0.562
$A_{th} = 2$	405132	0.411	507390	0.550
$A_{th} = 3$	402480	0.408	495300	0.537
$A_{th} = 4$	401700	0.408	483678	0.525
$A_{th} = 5$	399906	0.406	478842	0.519
$A_{th} = 6$	398034	0.404	473226	0.513
$A_{th} = 7$	399438	0.405	469092	0.509
$A_{th} = 8$	399828	0.406	465816	0.505

ACKNOWLEDGMENT

This work has been partially supported by the Deutsche Forschungsgemeinschaft (DFG) under grant WU 245/22-1 (ACCROSS).

REFERENCES

[1] Y. Li and J. Ibanez-Guzman, "Lidar for Autonomous Driving: The Principles, Challenges, and Trends for Automotive Lidar and Perception Systems," *IEEE Signal Processing Magazine*, Vol. 37, No. 4, pp. 50-61, July 2020.

[2] S. Sadeghi-Kohan and S. Hellebrand, "Dynamic Multi-Frequency Test Method for Hidden Interconnect Defects," 38th IEEE VLSI Test Symp. (VTS'20), pp. 1-6, 2020.

[3] J. Lienig and M. Thiele Fundamentals of Electromigration-Aware Integrated Circuit Design. Springer Int. Publishing AG, 2018.

[4] M. Cuviello, et al., "Fault modeling and simulation for crosstalk in system-on-chip interconnects," IEEE/ACM Int. Conf. on Computer-Aided Design (ICCAD'99), pp 297-303, 1999.

[5] Y. Kim, et al., "AXSERBUS: A quality-configurable approximate serial bus for energy-efficient sensing," 2017 IEEE/ACM Int. Symp. on Low Power Electronics and Design (ISLPED'17), pp. 1-6, 2017.

[6] D. J. Pagliari, E. Macii and M. Poncino, "Serial T0: Approximate bus encoding for energy-efficient transmission of sensor signals," 53nd ACM/EDAC/IEEE Design Automation Conf. (DAC'16), pp. 1-6, 2016.

[7] J. R. Stevens, A. Ranjan and A. Raghunathan, "AxBA: An approximate bus architecture framework," IEEE/ACM Int. Conf. on Computer-Aided Design (ICCAD'18), San Diego, CA, USA, pp. 1-8, 2018.

[8] A. Badran, et al., "Approximate Communication: Balancing Performance, Power, Reliability, and Safety," IEEE European Test Symp. (ETS'23), Venice, Italy, pp. 1-6, May 2023.

[9] S. Sadeghi-Kohan, S. Hellebrand, H.-J. Wunderlich, "Low Power Streaming of Sensor Data Using Gray Code-Based Approximate Communication," Workshop on Approximate Computing at DSN'23, Porto, Portugal, pp. pp. 203-206, June 2023.

[10] S. Sadeghi-Kohan, et al., "Optimizing the Streaming of Sensor Data with Approximate Communication," 32nd IEEE Asian Test Symposium (ATS'23), Beijing, China, pp. 1-6, 2023.

[11] K. Sayood, "Introduction to Data Compression," 5th Edition, Morgan Kaufmann Publishers (Elsevier), Cambridge, MA, USA, 2018.

[12] J. R. Black, "Electromigration – A brief survey and some recent results," IEEE Trans. Electron. Devices, Vol. 16, No. 4, pp. 338-347, 1969.

[13] P. Livshits and S. Sofer, "Aggravated electromigration of copper interconnection lines in ULSI devices due to crosstalk noise," IEEE Trans. on Device and Materials Reliability Vol.12, No.2, pp. 341-346, 2012.

[14] F. Gray, "Pulse code communications," U.S. Patent 2 632 058, March 17, 1953.

[15] C. L. Su, C. Y. Tsui, and A. M. Despain, "Low power architecture design and compilation technique for high-performance processors," COMPCON'94, San Francisco, CA, USA, pp. 209-214, Feb. 1994.

[16] M. R. Stan and W. P. Burleson, "Bus-invert coding for low-power I/O," IEEE Trans. on VLSI Systems, Vol. 3, No. 1, pp. 49-58, Mar. 1995.

[17] L. Benini, et al., "Asymptotic zero-transition activity encoding for address busses in low-power microprocessor-based systems," Great Lakes Symp. on VLSI, Urbana-Champaign, IL, USA, pp. 77-82, 1997.

[18] C. Duan, A. Tirumala and S. P. Khatri, "Analysis and avoidance of crosstalk in on-chip bus," 9th Symp. High Performance Interconnects (HOTI), pp. 133-138, 2001.

[19] B. Victor and K. Keutzer, "Bus encoding to prevent crosstalk delay," IEEE/ACM Int. Conf. on Computer Aided Design (ICCAD'01), San Jose, CA, USA, pp. 57-63, 2001.

[20] Z. Shirmohammadi and S. G. Miremadi, "Crosstalk avoidance coding for reliable data transmission of network on chips," Int. Symp. on System on Chip (SoC'13), Tampere, Finland, pp. 1-4, 2013.

[21] Z. Shirmohammadi and S. G. Miremadi, "Using binary-reflected gray coding for crosstalk mitigation of network on chip," 17th CSI Int. Symp. on Computer Architecture & Digital Systems (CADS'13), pp. 81-86, 2013.

[22] A. Najafi, et al., "Integer-Value Encoding for Approximate On-Chip Communication," IEEE Access, Vol. 7, pp. 179220-179234, 2019.

[23] W. W. Peterson, E. J. Weldon, Jr., "Error-Correcting Codes," Second Edition; Cambridge: MIT Press, 1972.

[24] R. W. Hamming, "Error detecting and error correcting codes," Bell System Technical Journal, vol. 29, no. 2, pp. 147-160, 1950.

[25] I. Koren and C. Mani Krishna, "Fault-tolerant systems," Morgan Kaufmann, 2020.

[26] G. Pekhimenko, et al., "Base-delta-immediate compression: Practical data compression for on-chip caches," 21st Int. Conf. on Parallel Architectures and Compilation Techniques (PACT'12), Minneapolis, MN, USA, pp. 377-388, 2012.

[27] https://r0k.us/graphics/kodak/, accessed on May 3, 2024.

Optimizing Waveform Accurate Fault Attacks Using Formal Methods

Devanshi Upadhyaya and Ilia Polian
University of Stuttgart, Stuttgart, Germany
devanshi.upadhyaya@informatik.uni-stuttgart.de,
ilia.polian@informatik.uni-stuttgart.de

Abstract—State-of-the-art fault attacks demand either a large number of low-precision fault injections (statistical attacks) or very few injections using sophisticated equipment (algebraic attacks) to break modern cryptosystems. For example, a popular attack breaks AES-128 with one injection, but the fault effects must be restricted to one 4-bit nibble of its state. This paper combines the advantages of the two by optimizing the probability of achieving the desired failing state bit patterns, and thus the attack's success rate, during conventional, low-cost clock manipulation. The problem bears similarities with small-delay fault test generation, and we extend formal (Boolean satisfiability, or SAT) models that were initially developed for waveform-accurate SDF ATPG procedures. A fundamental distinction of our analysis is the presence of fixed-yet-unknown secret bits, which influence the failing state bit patterns. For this reason, we use a model-counting (#SAT) approach to estimate the success probability as an average across secret bit combinations.

Index Terms—cryptographic circuits, SBox, AES, delay faults, clock manipulation, fault injection, ATPG, Boolean satisfiability, model counting

I. INTRODUCTION

The cybersecurity landscape is constantly evolving as new threats and vulnerabilities emerge with alarming frequency. In order to protect sensitive information and prevent unauthorized access, it is essential to establish secure and reliable hardware implementations of cryptographic systems. One of the most significant threats to these systems comes in the form of fault attacks, which manipulate hardware to cause functional errors and compromise the integrity of the system. These attacks have grown increasingly sophisticated over time, with advanced laser equipment capable of precisely injecting single faults that undermine even the most complex and widely used encryption standards, such as AES-128.

In the past, various attack scenarios that directly target the hardware implementation of an algorithm have been proposed and demonstrated. These attacks establish and utilize side channels over which protected data can be written or read. Techniques used to access information include analyzing power consumption, electromagnetic emissions, or the timing of operations such as cache accesses. Physical faults are also injected into the hardware to manipulate the data being processed, and the effects of the modified calculation are observed. Fault injection attacks are particularly alarming as they exploit hardware vulnerabilities to circumvent conventional security measures. Successful fault-based attacks have been reported for various systems, including RSA, AES, and LED. Despite the effectiveness of sophisticated fault injection methods, their deployment often necessitates specialized and costly equipment alongside a high degree of technical expertise.

In the context of fault injection attacks, clock manipulation is a notably cost-effective method compared to invasive techniques requiring elaborate setups, such as laser or electromagnetic interference equipment. Clock manipulation exploits the sensitivity of digital circuits to timing variations, intentionally introducing faults by altering the clock signal. This method utilizes the inherent vulnerabilities of cryptographic algorithms to timing disparities, allowing attackers to induce faults without the need for direct physical access to the hardware.

Our research introduces a novel vulnerability index—a quantifiable measure derived using formal models—that assesses the security robustness of cryptographic algorithms against fault attacks. This index is computed using a sophisticated combination of waveform-accurate fault modeling and formal methods, including Boolean satisfiability (SAT) and model counting (#SAT), to offer a detailed and quantifiable assessment of potential vulnerabilities, similar to [1]. The vulnerability index measures how different delays caused by clock manipulation impact the cryptographic algorithm's output. This metric helps us better understand how susceptible the algorithm is to faults and allows us to optimize fault injection parameters such that the fault effects can be easily analyzed for differential analysis.

We evaluate the vulnerability of a cryptographic circuit to fault attacks as shown in Figure 1. The proposed methodology involves calculating the likelihood of a modeled fault being detected at a primary output at a predefined observation time, referred to as t_{obs}. In order to formalize this concept, a Boolean satisfiability (SAT) instance for every modeled fault is generated. Each feasible solution of this instance corresponds to a system state and an input sequence, resulting in the fault being observable at the output. As such, the faulty output could later be used for differential analysis. The next step is to use a SoA model counting algorithm to count the number of such solutions for each modeled fault. By dividing the result by the total number of states and inputs, the probability of a successful attack can be determined, assuming that the states and inputs are random and equally probable.

The remainder of this paper is structured as follows: Section II gives an overview of the relevant background. Then, Section

979-8-3503-6689-1/24 $31.00 © 2024 IEEE

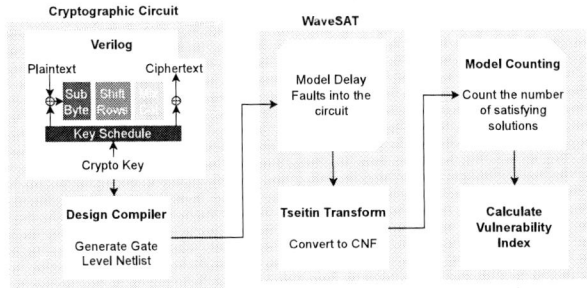

Fig. 1. Overview of attack methodology.

III presents the methodology in detail. Section IV shows our experimental results, and Section V concludes the paper.

II. BACKGROUND

One of the most prevalent types of attacks on cryptographic primitives is fault-injection attacks. These attacks use fault injections to produce faulty cipher executions, which can then be used for differential cryptanalysis. These types of attacks are quite strong, as they can achieve complete key recovery even for the most advanced and commonly used ciphers like AES, LED, and PRESENT [2]–[4] with only a single or very few fault injections. In this work, we employ waveform accurate state-of-the-art automatic test pattern generation (ATPG) techniques to analyze the effects of delay faults on cryptographic hardware. We generate SAT instances modeling different delay faults using WaveSAT [5] and calculate the number of satisfiable solutions for these instances using a model counter. The rest of this section describes the required background and toolset used in this work.

A. Fault Injection Attacks

Fault-injection attacks can be used to weaken cryptographic primitives by physically interfering with their operations. The outputs of the circuit can be used to observe faulty results from cryptographic operations. By comparing the faulty and fault-free results, one can determine secret information like the plaintext or secret cryptographic key. Physical fault injections into the circuit and mathematical analysis of the recorded data are two components of fault attacks. The physical fault injections can be carried out in several ways, from glitching and underpowering to overheating [6], optical, and electromagnetic fault injections [7], [8]. These attacks are generally active and invasive. For example, it is necessary to decapsulate the chip in order to shoot accurately at a specific location with a laser.

Power and clock glitches are the most frequent types of glitches that can occur in cryptographic circuits. The first involves changing the voltage that the targeted device supplies at a specific moment in the encryption process. This can cause some instructions to be skipped or some values to be stored incorrectly. The necessary equipment is typically inexpensive and widely accessible to cause these kinds of glitches. However, power manipulations typically result in imprecise

flaws. They are nevertheless frequently utilized since it is still possible to achieve fault injection with a reasonable level of precision. Using an external clock or altering the internal clock by adding a clock glitch are examples of other methods of injecting faults. Clock glitches are essentially the same as power glitches in that they can lead to incorrect data access or misreading of instructions. They typically lack precision and are also inexpensive. One such clock glitch example is the fault injector proposed in [9]. The cryptographic key bits or a smaller key space that allows for brute-force search can then be obtained by statistical or mathematical analysis of the flawed encryption caused by these fault injections.

B. Clock Manipulation

Due to their low cost and lack of technical expertise, clock glitches are typically regarded as the simplest fault injection technique. With low-end field programmable gate array (FPGA) boards, for instance, clock glitches can be easily produced [10]. Recently, devices for fault injections using clock manipulations that cost as little as $130 USD have been proposed. One such device is the multifault evaluation platform called TRAITOR [11]. It is common for devices like smart cards to be attacked when the adversary has direct control over the clock generator, which is necessary for clock glitches. To maintain synchronization across all of their computations, the majority of computing devices typically rely on an internal or external clock. The wrong instruction may be executed, or data may be tainted in the computation, which results from a change in the clock signal. Erroneous clock signals can be supplied to devices that need an external clock generator, e.g. a signal with fewer pulses compared to the standard signal [12]. However, devices that have internal clock generators are immune to clock glitching techniques.

Remote attacks can also be effectively executed through clock glitching. Vulnerabilities arise from a relatively new class of fault attacks that follow the development of effective energy management. Due to the hardware complexity of the devices, software execution issues, cost, and time-to-market constraints, energy management designers seldom take security into account [13]. [14] developed CLKSCREW by taking advantage of Dynamic Voltage and Frequency Scaling (DVFS), which allows an attacker to manipulate the frequency and voltage of a Nexus 6 phone, thereby forcing the processor to operate beyond recommended limits. The possibility of a one-byte random fault was confirmed experimentally. Even without physical access to the target device, CLKSCREW attack can be accomplished through software control of the energy management hardware regulators in those devices.

C. Formal Methods in Cryptographic Fault Analysis

Proving that a Boolean formula is satisfiable (SAT) is considered an essential and challenging task in computer science. It is a fundamental problem since many real-world problems can be converted into SAT problems. By solving the SAT problem, we can also obtain a solution to the corresponding original problem.

Over the years, several SAT algorithms have been developed, including glucose [15], lingeling [16], and cryptominisat [17]. These algorithms are fundamentally based on the classical Davis-Putnam-Logemann-Loveland method (DPLL) [18], [19], which was introduced in the early 1960s. However, they have been equipped with performance enhancements such as lazy clause evaluation, conflict-driven learning, efficient decision heuristics, and preprocessing. The significant progress made in practical SAT algorithms has opened up the possibility of solving related and often more complicated logical problems. One such problem is model counting (#SAT), which involves counting the number of different satisfying assignments that exist for a given CNF formula. It can be applied to various problems, including inference in Bayesian networks, combinatorial problems, probabilistic planning, and diagnosis.

Many of the latest #SAT solvers utilize modified versions of the DPLL algorithm. The process of solving a #SAT problem is more complex than traditional SAT solving due to the fact that the solver cannot stop its search after finding a single solution. The primary modification that enhances the efficiency of DPLL for #SAT involves dynamically breaking down the given CNF formula into separate components, solving these components independently, and then storing the number of solutions for each component. This allows each component to represent the remaining portion of a formula that still needs to be solved after a certain number of variables have been assigned. From a high-level perspective, both component caching and classical conflict-driven learning prevent a #SAT solver from exploring parts of the overall search space multiple times. Conflict clauses eliminate portions of the search space that have already been proven unsatisfiable, while component caching calculates the number of solutions for satisfiable portions of the search space that have already been examined. The #SAT solver utilized in this study is ApproxMC [17], [20], a state-of-the-art model counting algorithm.

For a given Boolean formula ϕ, the problem of model counting, also referred to as #SAT, as stated earlier is to compute the number of solutions of ϕ. However, exact model counting can be computationally infeasible for large instances due to its complexity, so ApproxMC provides an efficient approximation. ApproxMC uses hashing-based techniques to partition the solution space into smaller, more manageable segments. This is achieved using universal hash functions, which are designed to distribute solutions evenly across various partitions. Each partition is then explored using a SAT solver. The SAT solver checks whether there exists a solution within the current partition. If the partition contains too many solutions, it is further subdivided. For each partition that the SAT solver determines to be small enough (i.e., contains fewer solutions than a predefined threshold), ApproxMC estimates the number of solutions within that partition. The total number of solutions across all partitions is then scaled up based on the size of the partitions and the hashing parameters used, giving an estimate of the total number of satisfying assignments. To ensure accuracy and confidence in the results, ApproxMC repeats the above process multiple times and typically uses median or mean values of the estimates from several iterations. This repetition helps mitigate the effects of any particularly poor partitions and provides a statistically robust estimate.

D. Waveform Accurate ATPG

Small delay faults (SDF), which are common in nanoscale technologies, arise from defect mechanisms like resistive opens, resistive shorts, and marginal parameter shifts of transistors in logic gates. Researchers have dedicated significant effort to understanding SDF modeling and simulation, as well as developing automatic test pattern generation (ATPG) methods to detect these faults [21]. Achieving high coverage of SDFs is crucial for ensuring the quality of integrated circuits. An SDF involves a particular logic gate g within the circuit, with a size s and an affected transition (rising or falling). If this transition occurs at g's output, it experiences a delay of $(\delta + s)$, where δ is the nominal delay of gate g. When a test pair $(v1, v2)$ is applied, it typically triggers multiple transitions at different circuit lines, including their outputs. The term "waveform on l" in WaveSAT refers to the complete set of transitions on a line l, along with their corresponding times, under the test pair $(v1, v2)$ [5]. An SDF is detected by the test pair $(v1, v2)$ if the waveforms on the output o of the circuit under $(v1, v2)$ differ at the observation time t_{obs} between the faulty and fault-free circuit configurations.

The detection of a size s small delay fault at a gate g requires the presence of a sensitized path from an input i of the circuit to an output o, passing through the gate g, and having an accumulated fault-free delay of more than $t_{obs} - s$. In simple terms, a sensitized path consists of gates with their off-path inputs set to values that allow transitions from their on-path inputs to propagate to the output of the gate. Therefore, if a transition initiated at i can reach gate g, the single stuck-at fault will result in differences in the waveforms between the faulty and the fault-free circuits. These differences could then propagate through the rest of the path from g to o, potentially becoming observable.

In the research presented in this paper, we utilize WaveSAT, which accurately captures timing by integrating specific delays for each gate with discrete resolution. This approach enhances fault detection precision in comparison to traditional methods by considering the relationships between potential waveforms on various circuit lines. WaveSAT generates a consistent set of waveforms on all circuit lines necessary for the detection of a given SDF. The test pair is finally obtained from the input waveforms, with the restriction that they can only switch once at time 0. If no test pair is found, this serves as a formal proof of the untestability of the given fault based on the model assumptions. The waveform-based method is similar to unrolling the circuit but employs additional optimizations to produce a more compact and precise circuit representation.

The WaveSAT algorithm for small delay fault ATPG involves several key steps in accurately modeling timing, analyzing waveforms, and detecting faults in circuits. The timing model of the circuit is established with discrete time points and gate delays expressed in integer time units. All inputs are

979-8-3503-6689-1/24 $31.00 © 2024 IEEE

assumed to switch simultaneously at time 0 and stabilize their logical values under specific input conditions before time 0. Two Boolean variables are defined for each gate output in the circuit: Initialization value (I) and Stabilization value (S). The logical and temporal relationships of the circuit, along with timing details and temporal behaviors, are then encoded into a Boolean satisfiability instance. The waveforms and their propagation on each relevant line of the circuit are examined to capture the behavior of the signals accurately. A SAT instance representing the circuit's behavior in the presence of a small delay fault is then generated. Finally, a SAT solver is employed to analyze the SAT instance and ascertain the existence of a pair of assignments to the primary inputs of the circuit that leads to a faulty value at an observable output at a specific observation time. The SAT solver either returns a solution containing the test pair or proves unsatisfiability.

III. WAVEFORM ACCURATE FAULT ATTACKS

This section outlines the methodology used to enhance the precision and efficacy of fault injection attacks on cryptographic algorithms using waveform-accurate modeling combined with formal methods. Our approach integrates clock manipulation techniques with sophisticated Boolean satisfiability tools to quantify and optimize the vulnerability of cryptographic systems to fault attacks.

A. Security Analysis vs. Delay Fault Test Generation

Injecting a fault into a cryptographic circuit using clock manipulation has similarities, but also two important differences compared to delay fault testing. In both applications, the test engineer or the attacker is aiming at imposing values at the circuit's outputs that deviate from the ones during regular, disturbance-free operation. When testing delay faults, the deviations are due to manufacturing defects that modify the timing of the circuit's elements, while during an attack, the clock manipulation is intentionally performed by an attacker. In both cases, the values applied to the circuit's inputs must trigger transitions that propagate along one or several paths with a negative slack when the delay modifications are taken into account.

The first difference is the more stringent requirement on the effect of an attack. During testing for manufacturing defects, a defect is considered detected no matter whether it manifests itself on one or on multiple outputs of the circuit, as the automatic test equipment stores the reference test response and reports as detect any deviations from it. In contrast, a fault attack is followed by mathematical post-processing, which makes assumptions about the fault effect. For example, the popular attack on AES-128 [2] requires a fault within precisely one byte of the cipher's state, whereas any further bit-flips in the remaining 15 bytes would invalidate the attack. Applying delay fault ATPG for a fault attack on byte i would necessitate a restriction that the fault effect is propagated to (a subset of) outputs $\{8i, \ldots, 8i+7\}$, which is easy to achieve. However, the requirement that the *other* outputs outside this interval stay unaffected is hard to fulfill by conventional, structure-oriented

ATPG algorithms that often ignore switching events outside the considered location or path.

The second difference is the fact that delay fault ATPG assumes controllable, or at least known, circuit inputs, whereas the behavior of a cryptographic circuit subject to a fault attack depends on the internal state that is unknown, because it in turn depends on the cryptographic secret key. Cryptographic functions have diffusion and confusion properties, such that the state of, e.g., the AES cipher is totally unpredictable after a few rounds for an attacker who does not know the key. Therefore, it is in general not possible to find a test pair that produces an exploitable fault with certainty, as the propagation will depend on side-inputs determined by unknown secret key bits.

The formal model used in this work addresses all the above-mentioned challenges. WaveSAT has variables that represent values assumed by lines (including circuit outputs) during different time instances in both fault-free and fault-affected circuits, thus achieving waveform-accuracy. A fault effect on output o when the clock period has been reduced from T_{nom} to the glitch width T_{attack} can be expressed by $W_o \oplus S_o$, where S_o is a variable describing the value to which o stabilizes in the fault-free case and W_o is the variable describing the value which the faulty circuit assumes at the observation time $t_{obs} = T_{attack}$. To demand a detectable fault on byte i but not other bytes, as described above, we can add to the formula the condition

$$\bigvee_{j \in \{8i,\ldots,8i+7\}} (W_o \oplus S_o) \land \bigwedge_{j \notin \{8i,\ldots,8i+7\}} (W_o \equiv S_o)$$

(appropriately converted into the conjunctive normal form required by the SAT solver). Solving a formula with this addition for a regular combinational circuit would have resulted in a test pair that produces the desired failure pattern on the outputs when observed at time point T_{attack}, or prove that no such test pair exists. However, in security analysis, we are not interested in a test pair but instead in the *likelihood* of producing an exploitable fault by a random input sequence, assuming that the secret key bits are uniformly distributed. This leads to the definition of the *Variability Index* (VI).

B. Vulnerability Index (VI) Calculation

A key component of our methodology is the introduction of a vulnerability index designed to quantify the susceptibility of cryptographic algorithms to fault attacks. This index is calculated using the results from a model counting (#SAT) algorithm, applied to waveform-accurate faulty models of the cryptographic circuit under study. The vulnerability index measures the likelihood or the probability that a fault injection can successfully disrupt the encryption process and result in an observable faulty output.

We calculate the VI for the circuit under analysis, referred to as C, for each considered glitch width. The higher the value of VI, the better the glitch width from an attacker's perspective. We define VI as follows: For the circuit C,

WaveSAT produces a Boolean formula, ϕ, including the conditions for an exploitable fault corresponding to each injected delay, T_{attack}. The total number of satisfying assignments to this formula, obtained using the model counter, can be represented as $\#SAT(\phi(C, T_{attack}))$. Note that we consider delay faults due to manipulations of the clock, but it would be possible to incorporate the localized effects of, e.g., electromagnetic fault injection. If k is the total number of inputs to the circuit, the probability of input assignments that successfully inject a fault is calculated as

$$VI(T_{attack}) = \frac{\#SAT(\phi(C, T_{attack}))}{2^{2k}}.$$

Here, the denominator corresponds to the number of possible test pairs and secret bit combinations.

IV. EXPERIMENTAL RESULTS

This study systematically evaluated the vulnerability of AES cryptographic circuits to fault injection attacks, specifically through clock manipulation. By applying the methodology outlined in Figure 1, we generated and analyzed SAT instances modeling potential delay faults in order to quantify their impact. Our initial experiments focused on manipulating the clock signal to induce delay faults within the AES circuit. The faults were modeled across the AES encryption process, specifically the SubBytes layer and a whole round of AES. In the first case, we assume the SubBytes layer is an isolated layer separated by a register. The use of WaveSAT enabled the precise modeling of delay faults and their potential impact on the cryptographic output.

The generation of SAT instances for each modeled fault revealed a wide variance in the circuit's vulnerability across different stages of the AES process. Notably, faults induced in the SubBytes stage were more likely to result in observable changes at the output compared to those injected in the entire round, suggesting a higher susceptibility of the SubBytes operation to fault injection attacks.

A. Comparative Analysis: SubBytes vs Round

In order to show the applicability of our methodology, we injected different delays of varying widths to a hardware implementation of AES. The fault location is chosen to match that of a known 128-bit AES fault attack [2]: a fault is injected into 8 data bits (out of the total 128 data bits) during the SubBytes operation in the eighth encryption round, which is the target fault location. All other outputs must be fault-free.

We perform two sets of experiments. First, we focus only on the SubBytes layer. The operation consists of 16 parallel SBoxes, each handling 8 bits of data. Accordingly, the fault must be injected during one of the SBox operations. It is assumed that the control logic is not on the critical path. Subsequently, only the combinational SubBytes circuit is analyzed. Since the cryptographic key is unknown, v_1 is independent of the plaintext input from an attacker's perspective. Additionally, all possible values for v_1 can occur.

TABLE I
VULNERABILITY INDEX (VI) BY T_{ATTACK} FOR SUBBYTES AND AN AES ROUND

T_{attack}(ps)	VI	
	SubBytes	Round
10	0	0
20	0.083	0
30	0.261	0.002
40	0.467	0.016
50	0.318	0.073
60	0.294	0.112
70	0.075	0.095
80	0.028	0.081
90	0.003	0.029
100	0	0.005
110	0	0
120	0	0

The following input assignment v_2 to SubBytes depends on v_1 and the cryptographic key. The dependency on the key allows all possible values for v_2. Therefore, it is assumed that all (v_1, v_2) combinations occur in the circuit and are uniformly distributed. The Synopsys Design Compiler is used to synthesize the hardware implementation from the Verilog source and is mapped to a NanGate open cell library. The resulting circuit has a critical path of 350ps, which is chosen as the nominal clock period for the fault-free circuit. Next, the modified WaveSAT algorithm is used to generate a Boolean formula for the circuit with multiple injected glitches of lengths varying from 10 ps up to 340 ps. We use ApproxMC to count the number of satisfying solutions for each CNF generated in the previous step and calculate the VI.

For our second set of experiments, we apply the same methodology to a round of AES. The circuit then consists of a one-cycle implementation of the different AES operations: SubBytes, ShiftRows, MixColumns, and AddRoundKey. Once again, we analyze a purely combinational circuit. The resulting circuit has a critical path of 470ps, which is chosen as the nominal clock period for the fault-free circuit. We consider glitch widths from 10ps up to 460ps and calculate the vulnerability index for each.

Table I shows the results for both considered circuits. For the SubBytes circuit, the highest VI of 46.7% is observed at $T_{attack} = 40$ps while for the AES Round circuit, the highest VI of 11.2% is observed at $T_{attack} = 60$ps. We notice that the probability of a successful fault injection for both circuits is negligible above delay values larger than $100ps$. A probability of successful fault injection of 11.2% for a round of AES implies a sufficient opportunity for a successful key retrieval using a fault attack that requires a single fault injection.

B. Summary

A comparative analysis of the vulnerability index between a round of AES and a SubBytes layer highlighted the differential impact of delay faults. The vulnerability index, as shown in Figure 2, was notably higher for faults in the SubBytes layer compared to the whole round. This finding highlights the importance of targeting security measures on the most susceptible components of the encryption process. The findings from our experiments show the critical nature of delay faults as a vector for fault injection attacks on cryptographic circuits. The vulnerability index defined through this work offers a valuable metric for assessing and comparing the susceptibility of cryptographic circuits to such attacks.

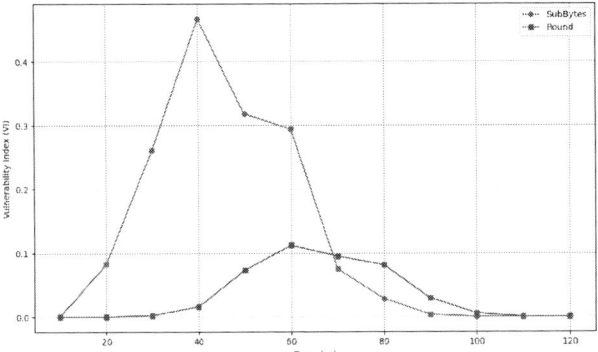

Fig. 2. Vulnerability Index vs T_{attack}

V. Conclusion and Future Work

In this work, we have presented a comprehensive methodology for evaluating the vulnerability of cryptographic circuits, focusing on AES, to fault injection attacks facilitated via clock manipulation. Our findings show that even minor manipulations in clock signals, resulting in small delays, can lead to significant vulnerabilities, allowing for fault injection attacks that compromise the integrity and confidentiality of encrypted data. By utilizing a combination of WaveSAT for modeling delay faults and ApproxMC for analyzing the satisfiability of these faults, we have quantitatively assessed the susceptibility of cryptographic circuits to such attacks.

Our work has two practical implications. First, it emphasizes the need for cryptographic hardware designers to consider and mitigate the risks associated with delay faults. Second, it offers a systematic approach for security analysts to evaluate and strengthen cryptographic systems against fault injection attacks. However, our research has its limitations. The scope of our experiments was confined to a single round of AES and a specific type of fault injection method. Future work could extend this research in several directions. Firstly, exploring the vulnerability of other cryptographic algorithms and hardware implementations would improve our understanding of the broader implications of delay faults. Secondly, investigating faults other than delay faults could uncover additional vulnerabilities. Lastly, developing more sophisticated metrics and analysis frameworks would further enhance our ability to assess and improve the security of cryptographic circuits.

VI. Acknowledgment

The authors would like to thank Matthias Sauer, Maël Gay, and the Computer Architecture Group at the University of Freiburg for their useful discussions and providing access to their tools.

References

[1] L. Feiten, M. Sauer, T. Schubert, A. Czutro, E. Böhl, I. Polian, and B. Becker, "sat-based vulnerability analysis of security components — a case study," in *2012 IEEE International Symposium on Defect and Fault Tolerance in VLSI and Nanotechnology Systems (DFT)*, 2012, pp. 49–54.

[2] M. Tunstall, D. Mukhopadhyay, and S. Ali, "Differential fault analysis of the advanced encryption standard using a single fault," in *IFIP international workshop on information security theory and practices*, 2011, pp. 224–233.

[3] P. Jovanovic, M. Kreuzer, and I. Polian, "A fault attack on the led block cipher," in *Constructive Side-Channel Analysis and Secure Design: Third International Workshop, COSADE 2012, Darmstadt, Germany, May 3-4, 2012. Proceedings 3.* Springer, 2012, pp. 120–134.

[4] S.-H. Park, K.-T. Jeong, Y.-S. Lee, J.-C. Sung, and S.-H. Hong, "Improved differential fault analysis on block cipher present-80/128," *Journal of the Korea Institute of Information Security & Cryptology*, vol. 22, no. 1, pp. 33–41, 2012.

[5] M. Sauer, A. Czutro, I. Polian, and B. Becker, "Small-delay-fault atpg with waveform accuracy," in *Proceedings of the International Conference on Computer-Aided Design*, 2012, pp. 30–36.

[6] H. Bar-El, H. Choukri, D. Naccache, M. Tunstall, and C. Whelan, "The sorcerer's apprentice guide to fault attacks," *Proceedings of the IEEE*, vol. 94, no. 2, pp. 370–382, 2006.

[7] J. G. Van Woudenberg, M. F. Witteman, and F. Menarini, "Practical optical fault injection on secure microcontrollers," in *2011 Workshop on Fault Diagnosis and Tolerance in Cryptography*, 2011, pp. 91–99.

[8] A. Dehbaoui, J.-M. Dutertre, B. Robisson, and A. Tria, "Electromagnetic transient faults injection on a hardware and a software implementations of AES," in *2012 Workshop on Fault Diagnosis and Tolerance in Cryptography*, 2012, pp. 7–15.

[9] M. Matsubayashi, A. Satoh, and J. Ishii, "Clock glitch generator on sakura-g for fault injection attack against a cryptographic circuit," in *2016 IEEE 5th Global Conference on Consumer Electronics*, 2016, pp. 1–4.

[10] S. Endo, T. Sugawara, N. Homma, T. Aoki, and A. Satoh, "An on-chip glitchy-clock generator for testing fault injection attacks," *Journal of Cryptographic Engineering*, vol. 1, pp. 265–270, 2011.

[11] L. Claudepierre, P.-Y. Péneau, D. Hardy, and E. Rohou, "Traitor: a low-cost evaluation platform for multifault injection," in *Proceedings of the 2021 International Symposium on Advanced Security on Software and Systems*, 2021, pp. 51–56.

[12] D. Karaklajić, J.-M. Schmidt, and I. Verbauwhede, "Hardware designer's guide to fault attacks," *IEEE Transactions on Very Large Scale Integration (VLSI) Systems*, vol. 21, no. 12, pp. 2295–2306, 2013.

[13] S. Pinto and N. Santos, "Demystifying arm trustzone: A comprehensive survey," *ACM computing surveys (CSUR)*, vol. 51, no. 6, pp. 1–36, 2019.

[14] A. Tang, S. Sethumadhavan, and S. Stolfo, "CLKSCREW: Exposing the perils of Security-Oblivious energy management," in *26th USENIX Security Symposium (USENIX Security 17)*, Aug. 2017, pp. 1057–1074.

[15] G. Audemard and L. Simon, "On the glucose sat solver," *International Journal on Artificial Intelligence Tools*, vol. 27, no. 01, p. 1840001, 2018.

[16] A. Biere, "Lingeling and friends at the sat competition 2011," 2011.

[17] M. Soos, K. Nohl, and C. Castelluccia, "Extending SAT solvers to cryptographic problems," in *International Conference on Theory and Applications of Satisfiability Testing*, 2009, pp. 244–257.

[18] M. Davis and H. Putnam, "A computing procedure for quantification theory," *Journal of the ACM (JACM)*, vol. 7, no. 3, pp. 201–215, 1960.

[19] M. Davis, G. Logemann, and D. Loveland, "A machine program for theorem-proving," *Communications of the ACM*, vol. 5, no. 7, pp. 394–397, 1962.

[20] M. Soos, S. Gocht, and K. S. Meel, "Tinted, detached, and lazy cnf-xor solving and its applications to counting and sampling," in *International Conference on Computer Aided Verification*. Springer, 2020, pp. 463–484.

[21] M. Sauer, B. Becker, and I. Polian, "Phaeton: A sat-based framework for timing-aware path sensitization," *IEEE Transactions on Computers*, vol. 65, no. 6, pp. 1869–1881, 2016.

Malware Detection on Linux Using Runtime Opcode Tracing

Martí Alonso, Juan José Costa, Enric Morancho
Department of Computer Architecture
Universitat Politècnica de Catalunya - BarcelonaTech (UPC)
Barcelona, Spain
{malonso, jcosta, enricm}@ac.upc.edu

Abstract—The fast-paced evolution of cyberattacks to digital infrastructures requires new protection mechanisms to counterattack them. Malware attacks, a type of cyberattacks ranging from viruses and worms to ransomware and spyware, have been traditionally detected using signature-based methods. But with new versions of malware, this approach is not good enough, and new machine learning tools look promising. In this paper we evaluate five machine learning models (SVM, KNN, Naive Bayes, Decision Tree and Random Forest) to detect Linux malware on applications during its runtime by tracking their executed instructions (opcodes). We show the methodology, the initial dataset preparation, the infrastructure used to obtain the traces of executed instructions, and the evaluation of the results for the different models used. The obtained results show that Random Forest classifier gets the best results, with 90% accuracy or higher.

Index Terms—Security, Linux Malware Detection, Dynamic Analysis, Opcodes, Machine Learning

I. INTRODUCTION

With the increasing use of digital systems, the rise of digital transformation in many businesses and the digital interconnection of every aspect of our lives, doubts arise about the security of these systems. To be able to safeguard sensitive information and preserve the integrity of digital infrastructure, it is of crucial importance to protect against the fast-paced evolving cyberattacks.

One of these cyberattacks are malware attacks. Malware is a type of software specifically built to perform malicious activity on any type of computer-based system. It can take many forms, ranging from viruses and worms to ransomware and spyware. The consequences of a malware infection in a computer system can be catastrophic, such as sensitive data breaches, privacy compromise, financial losses, or full exposure of a digital system. Although first malicious actors just wanted to prove their capabilities, nowadays, cyberattacks are a form of organized crime, and look for financial gain. Given such incentive, malware techniques don't stop to expand, adapt and innovate, therefore requiring proactive and adaptive cybersecurity measures to stay ahead of emerging threats.

Traditionally, anti-malware software has used signature-based detection methods. Such methods consist of matching known patterns of malicious code, URLs, metadata, etc., against a database. This method works great for constant and repetitive malware. However, when it comes to new or altered versions of malware, such as polymorphic or metamorphic, this approach doesn't perform good enough. Another approach must be employed, and one that has proven to be effective is the use of machine learning models [1]–[3].

Malware detection using machine learning models is an established technique that has demonstrated to be effective. The classification models and the feature vectors used for such can vary between proposals, however one that stands out is the use of runtime traces of opcodes, the specific operations that are executed in a computer by any program.

This strategy has worked well for the x86 architecture. However, new computer architectures have emerged recently, one of such is RISC-V [4], which is not yet fully established but is gaining momentum quickly, especially in the cloud space. Although some researchers have started to investigate this approach on the RISC-V ISA [5], there is still work to do to validate these techniques with a general malware dataset. But the main problem is that nowadays this dataset is missing.

Our goal is to validate malware detection techniques using machine learning models and runtime traces of opcodes for the x86 architecture. Hopefully we will be able to apply this knowledge to RISC-V architectures in future work.

II. BACKGROUND

Malware detection has been a field of study for a long time. Especially in the last decades, machine learning has been used to create classification models with the hope that they will effectively detect new malware. These studies can be divided into two groups: static analysis and dynamic analysis.

Static analysis consists of examining the characteristics of a program without executing it. Santos et al. [1] presented a method to represent malware based on the opcode sequence and construct a vector representation of each executable. Through experimentation with various classifiers such as decision trees, SVM, KNN and Bayesian network, they achieved a 95% accuracy. However, the dataset for this study consisted of x86 executables for the Windows operating system. Sharma and Sahay [2] proposed the use of number of occurrences of opcodes as a feature, and trained 13 different classifiers. Their top result, a Random Forest model, obtained a 97% accuracy.

Dynamic analysis consists of executing a program and capture the executed trace. Carlin et al. [3] carried out research on dynamic analysis of runtime traces of opcodes. Their study found that, with this type of analysis, a classifier can distinguish malware from benignware with a 99% accuracy using a sequence of only 32k opcodes. Its main limitation is that this result has only been validated on Windows malware. Koranek et al. [6] experimented with a LSTM classification model and dynamic runtime program traces of Return Oriented Programming (ROP) exploitations on the RISC-V ISA. Using opcode and operand sequences, they obtained a 97% accuracy.

Malware specific to Linux is not so common but it exists. Shahzad et al. [7] presents a framework to analyze and detect at run-time if a Linux application is benign or malware by just looking at kernel structures information. Cozzi et al. [8] does not use machine learning but presents a comprehensive premier on linux malwares using a hybrid analysis combining both, static and dynamic analysis. It obtains the dataset from VirusTotal [9] and describes and categorizes the different available malwares. Carrillo et al. [10] also presents a hybrid analysis. For the static part analyzes the opcode sequences, calculates n-grams, and detects similarities between those n-grams. For the dynamic part, it uses machine learning to process a mix of the number of system calls used, ioctls, created processes and process renaming.

This work tries to extend the work from Carlin et al. [3] using the dynamic analysis of runtime traces generated by Linux Malware applications.

III. PROPOSAL

Although malware detection using machine learning approaches have already been validated for established architectures like x86, there is still work to do to prove the effectiveness of this technique for the RISC-V architecture.

It won't be long enough until we start to see malware specifically crafted for the RISC-V architecture, which poses a threat to the security of those systems. As with any other digital system, security is one of the most important requirements for any computer architecture in order to be globally adopted.

Our approach is to analyze the opcode sequence that a program executes during its runtime. The hypothesis is that it is possible to identify patterns that can be distinguished as anomalies related to malicious activity. Therefore our goal is to collect and transform opcode sequences into appropriate features vector to train classification models that can detect malware activity during a program execution. The main goals are:

- **Develop the execution infrastructure.** Custom infrastructure is needed to capture the dynamic runtime opcode traces. Since it will execute malign software, this infrastructure must use a safe and isolated environment to be protected of any malicious threats the executed software might pose. Additionally, the platform has to be computer architecture independent, so executables from both x86 and RISC-V architectures may be tested. To save time and human errors this infrastructure must also be able to automatically process all the executables and capture their opcode traces.

- **Create a malware detection framework.** We will build a malware detection framework focused on the x86 architecture that uses a classifier model at its core. This classification model must be able to distinguish between a malware sample and a benign sample. Any machine learning model heavily depends on the dataset used to train itself, therefore a meaningful dataset containing samples from both malware and benign executable files has to be generated. Overall, the model's performance has to be evaluated taking into account that previous studies have successfully achieved great performances.

IV. METHODOLOGY AND IMPLEMENTATION

The methodology followed during this work is shown in Figure 1, where after an initial *Data Collection* step, there are two phases, 1) a Training phase with a *Data Preprocesing* and a *Model Training* steps to train the machine learning model, and 2) a Testing phase where the model is evaluated.

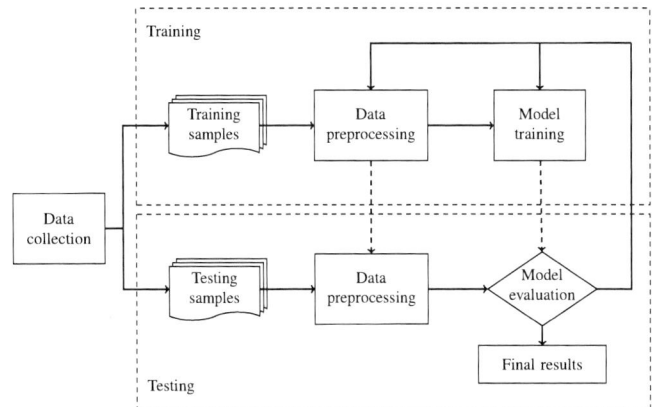

Fig. 1: Methodology process overview.

A. Data collection

The Data Collection is responsible to generate the dataset. In the first place, we select the set of Benign/Malware applications to use and, in the second place, we generate a set of samples for each application.

Select Benign/Malware applications: In this step, we gather benign and malware applications. Proper choice of these applications is crucial to ensure a representative data set.

For this work, 20 benign programs have been selected: *as, b2sum, base32, base64, basicmath* [11], *bitcount* [11], *bzip2* [12], *cat, cksum, cmp, colrm, cut, dpkg, factor, ffmpeg* [13], *gcc, matrixmult, STREAM* [14], 'stress -c' [15] and 'stress -m' [15]. These programs include basic user applications or system utilities from a basic Ubuntu installation, and some specific benchmarks that stress CPU and memory.

For the malware files, a package of malware from VirusShare [16] containing ELF malware binaries from 2019 to 2020 was used to select an equal amount of malware files as the benign files to keep the dataset well-balanced. The 20 executables that ran for the most time were selected to be included in the dataset. The list includes *CoinMiners, Backdoor/mirai, Backdoor/Tsunami, Backdoor/Dofloo,* and some *Trojans* [17].

Generate application samples: We need to execute the selected applications, and generate their runtime opcode traces (samples). Due to the dynamic nature of the analysis, we must execute malicious programs that will potentially infect the host machine. In order to isolate the host machine, we execute all applications (benign and malicious) in a virtual machine (VM). This VM has 2 vCores, 4 GB of RAM, a bootable 50GB disk drive with an Ubuntu LTS 22.04.2 installation. The VM does not have any network access, nor access to any other host device. Therefore, in order to share information between the host and the VM, we use a secondary virtual disk image that it is not used by both at the same time.

First, at the host, we prepare a disk image with information about the application to run and its parameters. Second, we start the VM, which automatically mounts the disk image locally (the secondary disk) and runs the selected application with its parameters. Third, the VM writes the generated samples (the trace of the executed instructions) in the mounted disk. Finally, at the host, after the application execution ends, we destroy the VM and we gather the samples from the disk image. The diagram in Figure 2 shows an overview of the whole data collection process.

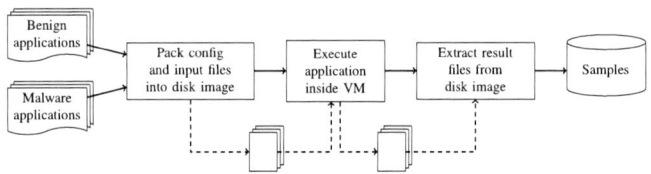

Fig. 2: Data collection process.

Config/Input files: The first step to generate samples for an application is to create a disk image that contains all the necessary information to execute the application. This information includes the path to the application executable, the arguments required for a correct execution, any necessary input files and other information such as whether it should be run with admin privileges or whether a timeout is needed.

Application execution: In the VM, a script is run right after start up that prepares the VM for executing the application: mounts the secondary disk, reads the config files and copies the input files to the right location. After that, the application is executed within a QEMU [18] user-space instance.

This QEMU instance is used as a Just-In-Time compiler to instrument and run the application. Each instruction to be executed is disassembled—using the Capstone framework [19]—to obtain the mnemonic of the opcode. Right before each instruction execution, a callback function is inserted with a reference to the instruction mnemonic. Within this function, the mnemonic is saved sequentially into a data stream. When this data stream reaches a threshold on the number of executed instructions (currently set to 10 millions), the data stream is saved as a sample in the secondary disk.

Result extraction: The final step in the process obtains the samples generated in the previous step from the secondary disk and stores them in a database with the corresponding label of the application type (benign or malware).

TABLE I: Parameters used to configure each classifier model.

Model	Parameters used
SVM	kernel: 'rbf', C: 1.0, gamma: 'scale'
KNN	n_neighbors: 5, weights: 'uniform', distance: 'euclidean'
Gaussian NB	-
Multinomial NB	alpha: 1.0
Decision Tree	criterion: 'gini', splitter: 'best', max_depth: None
Random Forest	criterion: 'gini', n_estimators: 100, max_depth: None
One Class SVM	kernel: 'rbf', gamma: 'scale', nu: 0.5

B. Data preprocessing

Generated samples must be converted to an appropriate format for the various classification models to be trained on. This conversion follows a two-step process:

1) Generate a histogram for each trace sample. This results in a fixed size vector representing the occurrences of each opcode for each sample.

2) Normalize the histogram—so each component is the percentage of the total number of executed instructions for that sample—to obtain a vector of the different opcode densities within a sample.

The resulting vectors are used as feature vectors to train the machine learning models.

C. Machine learning models

We are going to use five different classifier models to detect if an application is malware: k-Nearest Neighbor (KNN) [20], Support Vector Machine (SVM) [21], Naive Bayes (NB) [22], Decision Tree [20], and Random Forest [23]. We will also use a One-Class (OC) SVM—a SVM variant used for anomaly or outlier detection—which is trained with data from a single class. The rationale for OC SVM is that if we just have two classes to detect, then it makes sense to think that if an application does not belong to a class it will belong to the other. So we will compare the usefulness of training the model with a dataset containing: 1) malign and benign software, 2) only malign and 3) only benign.

Table I summarizes the parameters used for each model.

D. Model evaluation

To evaluate the performance of each model, we use a *confusion matrix*, a matrix to visualize the classification results. It shows the following four properties: *True Positive* (*TP*) and *True Negative* (*TN*), meaning the number of malware and benignware correctly classified; and *False Positive* (*FP*) and *False Negative* (*FN*), meaning the number of malware and benignware incorrectly classified.

Using these properties, we define the following metrics:

1) Accuracy: Ratio of correctly classified predictions over the total number of evaluated instances.

$$Accuracy = \frac{TP + TN}{TP + TN + FP + FN} \qquad (1)$$

2) Precision: Ratio of correctly classified items from the total of predicted items.

$$Precision = \frac{TP}{TP + FP} \qquad (2)$$

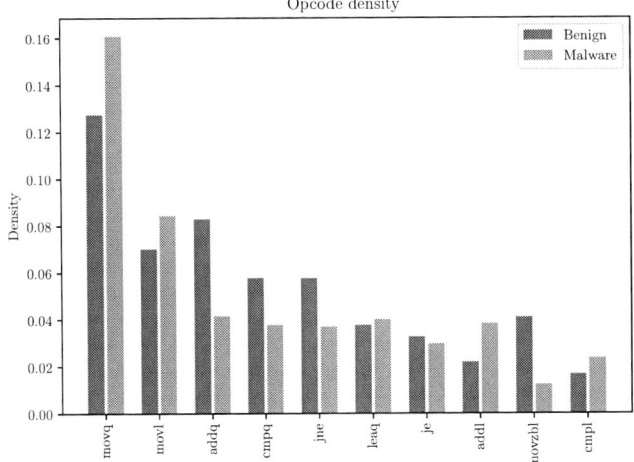

Fig. 3: Most frequent mnemonics in the final dataset.

3) Recall: Ratio of correctly classified items from the total of correct items.

$$Recall = \frac{TP}{TP + FN} \quad (3)$$

4) F1-Score: Harmonic mean of *Precision* and *Recall*.

$$F1\text{-}Score = \frac{2 * (Precision * Recall)}{Precision + Recall} \quad (4)$$

We calculate these metrics for all the models so we can compare the effectiveness for each one.

V. EVALUATION

This section details the experiments conducted in this study, based on the theoretical approach described earlier. In this scenario, we analyze 20 benign and 20 malware applications. Additionally, traces are constructed using the instruction mnemonics of the disassembled instructions. Due to the random nature of the applications, the number of samples per application may be quite diverse. This is problematic as it may cause an unbalanced number of events for each class, resulting in a very poor performance. So to avoid this problem, we have decided to fix the number of samples per application to use (1,000). To begin with, our collected dataset consists of 40,000 samples, each containing 10 million instructions. Figure 3 displays the 10 most frequent mnemonics in the dataset.

The classifier models require two datasets, one to train the model and the other to test its effectiveness. We want to evaluate two scenarios: 1) *Known* attacks, where we want to detect a previously seen attack; and 2) *Zero-day* attacks, with no previous knowledge about them. So a dataset is created for each scenario and Table II summarizes their characteristics.

For the *Known* dataset, we split our dataset into two separated sets, with a samples distribution of 80% and a 20% respectively (randomly selected from the 40 applications).

To simulate zero-day attacks, we remove all the samples from some applications from the training set, so the model will

TABLE II: Distribution of samples per category in each dataset

Dataset Name	Applications		Samples		
	Benign	Malign	Training	Testing	Total
Known	20	20	32,000	8,000	40,000
Zeroday	16	16	32,000	–	32,000
	4	4	–	8,000	8,000

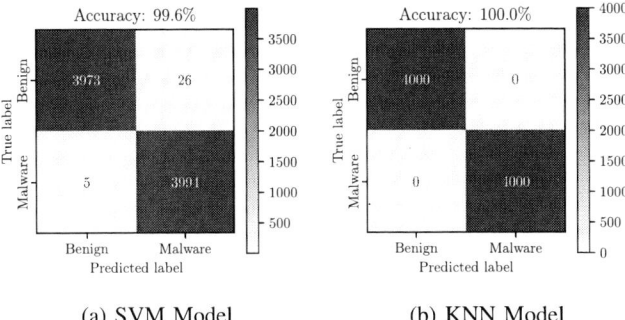

(a) SVM Model   (b) KNN Model

Fig. 4: Confusion matrices for SVM and KNN models trained with *Known* dataset (all applications).

be unaware of them. We want to keep the same distribution between datasets and, in particular, this means that the training dataset will contain samples from 32 applications (16 benign and 16 malware) and the testing dataset will contain samples from the remaining 8 applications.

A. SVM and KNN

Training SVM and KNN models with the *Known* dataset that includes samples from all applications leads to excellent results, as shown in Figure 4: about 99.6% and 100% accuracy respectively.

However, to simulate zero-day attacks, we have trained these models using samples from a subset of the applications. As the subset of applications used for training affects the final performance, we have evaluated 40 random splits and averaged the final results. These results, shown in Figure 5, are less impressive: the average accuracy is about 80%, with a standard deviation about 11%.

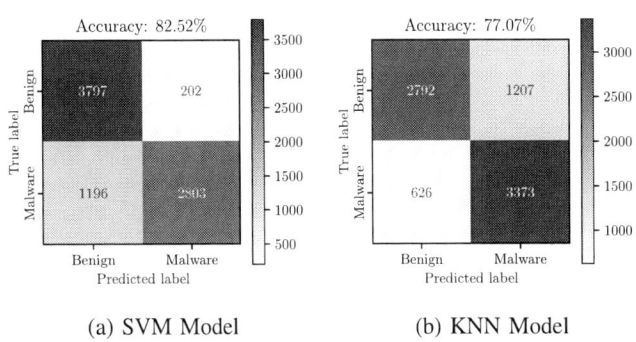

(a) SVM Model   (b) KNN Model

Fig. 5: Confusion matrices for SVM and KNN models trained with *Zero-day* dataset (a subset of applications).

979-8-3503-6689-1/24 $31.00 © 2024 IEEE

TABLE III: Summary table of all basic models metrics trained with *Zero-day* dataset (a subset of applications).

Model	Accuracy (%)	Precision (%)	Recall (%)	F1-Score (%)
SVM	82.52	93.43	70.09	78.57
KNN	77.07	75.89	84.34	78.91
Gaussian NB	72.06	66.86	93.62	77.52
Multinomial NB	68.83	65.57	86.46	73.72
Decision Tree	77.73	78.06	83.21	78.79
Random Forest	**91,19**	94,35	88,15	**90,46**
OC SVM (B)	53.88	52.40	**98.17**	68.21
OC SVM (M)	62.70	**96.66**	25.72	38.16

B. Naive Bayes

The results for both Gaussian and Multinomial Naive Bayes models are summarized in Table III. On one hand, it can be clearly seen that the results are not very good, which drastically drops the precision metric to 66,86% and 65,57% respectively (due to specially the false positive rate). On the other hand, false negative rate is relatively decent, specially for the Gaussian model, which has a recall rate of 93,62%. This is really important in this context because we want the model to detect malware even if it means that sometimes, good programs will be flagged as malware. These numbers, however, are a bit too extreme to be justifiable.

C. Decision Tree and Random Forest

Other model types that have also been tested are a Decision Tree model and a Random Forest model. The results for such models are summarized in Table III. While the accuracy of 77,73% for the Decision Tree model is not great, the Random Forest model shows superior results with 91,19%, the best one so far (within the zero-day attack scenario). Although the recall rate at 88,15% might not be optimal, the overall performance looks promising.

D. One-Class SVM

Finally, the last type of machine learning models are the One-Class SVM. These models are only trained with one class of samples, in this context, either only with benign samples (B) or with malware samples (M). The results for these models are summarized in Table III.

When trained with only benign samples (B), the model can easily classify malware as outlier since the model has not seen this data during training, however, benign samples that have not been seen either during training are also classified as malware. Although the results are terrible, the conclusions we can obtain from such are really interesting. The data from each program, either benign or malware, is distinct enough from each other that a model is not able to relate it to the same class of programs that has been trained with. From this, we can say that, a lot more data from a lot more programs would be needed for the model to be able to relate different programs with the same class.

While the accuracy of the model trained with only benign data is only of 53,39%, its recall metric is at an outstanding 98,17%. This, however, is completely useless in practical

terms, because a malware detector with such a model would flag practically all programs as malware.

Although the model trained with only malware samples (M) shows a bit better results, the practical conclusions are the same.

E. Final remarks

The models perform exceptionally well when detecting applications that have representation in the training phase. However, when not all applications are used for training, the results vary significantly depending on the split of applications. This is especially critical in the scenario of zero-day attacks: the ability to detect such an attack will depend on the similarity to the applications used to train the model.

A summary of the performance of all models using the *Zero-day* dataset is shown in Table III. We observe that Random Forest model behaves best in both *Accuracy* and *F1-Score* metrics. However, the models with higher *Recall* or *Precision* metric values are significantly less accurate than Random Forest model. Finally, One-Class models are practically useless because they tend to classify most applications as malware (OC SVM B) or benign (OC SVM M).

VI. CONCLUSION AND FUTURE WORK

In this work we have evaluated whether machine learning models are able to detect Linux malware using run-time sequences of opcodes. First, we have generated a dataset containing benign and malign Linux applications. Second, we have executed these applications in a virtualized environment, extracting their run-time sequence of executed opcodes. Third, we have used these sequences to train five machine learning models (SVM, KNN, Naive Bayes, Decision Tree and Random Forest) and evaluate their outcomes to categorize an unknown application into benign or malign class.

The results are promising, getting accuracies higher than 90% using Random Forest models. But, these results must have into account that the model is as good as the dataset used to train it. In particular, in the zero-day attacks scenario where an unknown application must be identified, the result will depend highly on the application similarity with other applications used to train the model.

Our future work will be focused on three issues: First, we want to explore if tuning the hyperparameters of the models may achieve significant improvements on their performance; although our preliminary tests show that the performance

979-8-3503-6689-1/24 $31.00 © 2024 IEEE

improvements are minimal, we should perform a deeper exploration. Second, we are interested in applying these models to detect RISC-V malware. To train the models we must collect RISC-V malware applications and generate their samples of executed opcodes. Finally, we are keen on the feasibility of a hardware implementation of these models, which one is most cost-effective? This path could open the way to implement an automatic malware detection mechanism in a RISC-V platform that may run in the background.

ACKNOWLEDGMENT

Funded by the European Union. Views and opinions expressed are, however, those of the authors only and do not necessarily reflect those of the European Union or the HaDEA. Neither the European Union nor the granting authority can be held responsible for them. Project number: 101093062 This work is also partially supported by the Spanish Ministry of Science and Innovation under contract PID2021-124463OB-IOO and by the Generalitat de Catalunya under grant 2021SGR00326.

REFERENCES

[1] I. Santos, F. Brezo, X. Ugarte-Pedrero, and P. G. Bringas, "Opcode sequences as representation of executables for data-mining-based unknown malware detection," en, *Information Sciences*, vol. 231, pp. 64–82, May 2013. DOI: 10.1016/j.ins.2011.08.020. [Online]. Available: https://linkinghub.elsevier.com/retrieve/pii/S0020025511004336 (visited on 02/24/2024).

[2] A. Sharma and S. K. Sahay, "An Effective Approach for Classification of Advanced Malware with High Accuracy," *International Journal of Security and Its Applications*, vol. 10, no. 4, pp. 249–266, Apr. 2016. DOI: 10.14257/ijsia.2016.10.4.24. [Online]. Available: http://article.nadiapub.com/IJSIA/vol10_no4/24.pdf (visited on 02/24/2024).

[3] D. Carlin, P. O'Kane, and S. Sezer, "Dynamic Analysis of Malware Using Run-Time Opcodes," in *Data Analytics and Decision Support for Cybersecurity: Trends, Methodologies and Applications*, Cham: Springer International Publishing, 2017, pp. 99–125. DOI: 10.1007/978-3-319-59439-2_4.

[4] A. S. Waterman, "Design of the RISC-V Instruction Set Architecture," en, Ph.D. dissertation, UC Berkeley, 2016. [Online]. Available: https://escholarship.org/uc/item/7zj0b3m7 (visited on 02/24/2024).

[5] T. Lu, "A Survey on RISC-V Security: Hardware and Architecture," *CoRR*, 2021. [Online]. Available: https://arxiv.org/abs/2107.04175.

[6] D. F. Koranek, S. R. Graham, B. J. Borghetti, and W. C. Henry, "Identification of Return-Oriented Programming Attacks Using RISC-V Instruction Trace Data," *IEEE Access*, vol. 10, pp. 45 347–45 364, 2022. DOI: 10.1109/ACCESS.2022.3170479. [Online]. Available: https://ieeexplore.ieee.org/document/9762913/ (visited on 02/24/2024).

[7] F. Shahzad, S. Bhatti, M. Shahzad, and M. Farooq, "In-Execution Malware Detection Using Task Structures of Linux Processes," in *2011 IEEE International Conference on Communications (ICC)*, 2011, pp. 1–6. DOI: 10.1109/icc.2011.5963012.

[8] E. Cozzi, M. Graziano, Y. Fratantonio, and D. Balzarotti, "Understanding Linux Malware," in *2018 IEEE Symposium on Security and Privacy (SP)*, 2018, pp. 161–175. DOI: 10.1109/SP.2018.00054.

[9] *VirusTotal.com*. [Online]. Available: https://virustotal.com/ (visited on 05/27/2024).

[10] J. Carrillo-Mondéjar, J. L. Martínez, and G. Suarez-Tangil, "Characterizing Linux-based Malware: Findings and Recent Trends," *Future Generation Computer Systems*, vol. 110, pp. 267–281, 2020. [Online]. Available: https://suarez-tangil.networks.imdea.org/papers/2020fgcs-iot.pdf.

[11] U. of Michigan, *Mibench version 1.0*, https://vhosts.eecs.umich.edu/mibench/, Accessed: 14-05-2024, 2002.

[12] *Bzip2*, https://sourceware.org/bzip2/, Accessed: 28-05-2024.

[13] *Ffmpeg*, https://ffmpeg.org/, Accessed: 28-05-2024.

[14] J. D. McCalpin, *Stream: Sustainable memory bandwidth in high performance computers*, https://www.cs.virginia.edu/stream/, Accessed: 28-05-2024.

[15] R. O. S. Projects, *Stress*, https://github.com/resurrecting-open-source-projects/stress.

[16] *VirusShare.com*. [Online]. Available: https://virusshare.com/ (visited on 03/03/2024).

[17] M. Alonso Garcia, "Malware detection using opcodes and machine learning," M.S. thesis, Facultat d'Informàtica de Barcelona, 2024. [Online]. Available: http://hdl.handle.net/2117/411787.

[18] *QEMU*. [Online]. Available: https://wiki.qemu.org/Main_Page (visited on 02/24/2024).

[19] *Capstone Engine*. [Online]. Available: https://github.com/capstone-engine/capstone (visited on 05/29/2024).

[20] T. Pn, M. Steinbach, and V. Kumar, *Introduction to data mining*. Addison-Wesley, 2005. [Online]. Available: https://thuvienso.hoasen.edu.vn/bitstream/handle/123456789/12544/Contents.pdf?sequence=1.

[21] V. N. Vapnik, "The support vector method," in *Artificial Neural Networks — ICANN'97*, W. Gerstner, A. Germond, M. Hasler, and J.-D. Nicoud, Eds., Berlin, Heidelberg: Springer Berlin Heidelberg, 1997, pp. 261–271.

[22] C. D. Manning, P. Raghavan, and H. Schütze, "Introduction to information retrieval," 2008.

[23] T. K. Ho, "Random decision forests," in *Proceedings of 3rd international conference on document analysis and recognition*, IEEE, vol. 1, 1995, pp. 278–282.

Special Session: Impact of Compiler Optimizations on the Reliability of a RISC-V-based Core

Pegdwende Romaric Nikiema, Marcello Traiola, Angeliki Kritikakou

Univ Rennes, CNRS, Inria, IRISA - UMR 6074, F-35000 and Institut Universitaire de France (IUF)
{pegdwende.nikiema, marcello.traiola, angeliki.kritikakou}@inria.fr

Abstract—The RISC-V Instruction Set Architecture (ISA) has gained popularity among systems designers thanks to its open-source nature. Its high flexibility has allowed it to be preferred in various domains and used to target multiple use cases, from embedded systems as co-processor to high-performance computers. Embedded systems, in general, and safety-critical ones, in particular, have strict requirements in terms of reliability and availability. The hardware is becoming less robust with the adoption of smaller technology nodes. The smaller transistor size, low operating voltage, and high switching frequency make transistors susceptible to Single-Event Upsets (SEU) faults, which can propagate to the application output and possibly cause catastrophic consequences. During the software design phase of the system, compilation optimizations can be made to improve the performance. Compilers have various flags that modify the source code to produce the binary. Although these flags can be crucial in assuring good performance, they can significantly impact the resilience to SEU. This work provides comprehensive insights into the impact of compiler optimizations on the reliability of safety-critical embedded systems. Specifically, a probabilistic fault injection campaign is conducted on various benchmarks running on a RISC-V core to evaluate the effect of several optimizations on reliability. The results are classified into functional and timing errors, offering a detailed understanding of the implications of these optimizations on reliability.

Index Terms—Reliability, Compiler optimizations, fault injection

I. INTRODUCTION

Modern electronic systems have adopted newer technologies that aim to improve efficiency and performance. Nowadays, transistors use a small node. Moreover, combined with their high switching frequencies and lower operating voltages [1], they become more powerful for almost equal power compared to a bigger counterpart. This scale-down has consequences on reliability. For example, the reduced critical charge makes the transistor more vulnerable to single-event upsets (SEU) caused by radiation [2]. In addition, the aging and wear-off of transistors also contribute to transient or permanent faults, which cause functional and timing errors on the system [3]. The radiation threat used to be a matter for space applications where the level of radiation is higher. Nowadays, even at sea level, we experience radiation issues, making regular consumer electronics vulnerable [4]. Different studies exist on vulnerability analysis on various architectures and systems from application level to hardware level, as well as techniques to reduce the impact of the SEU on the systems.

Nowadays, the RISC-V instruction Set Architecture (ISA) has gained popularity and is used in various application domains for various computation demands due to its open source, versatility, and extendability. To perform system deployment, the application is compiled for the specific processor ISA. During compilation, optimizations are applied to modify the application binary in order to improve performance and reduce resource utilization. Optimizations may imply instructions re-ordering, skipping redundant operations, etc. On one hand, such optimizations can negatively affect the system's vulnerability. On the other hand, it's worth noting that these optimizations can even reduce the exposure to fault due to their execution time reduction. The unpredictability of the impact of compiler optimizations in terms of reliability makes their application, without prior study, dangerous. Existing studies on showing the impact of compiler optimizations either focus on more fine-grained flags and try to find a good flag combination or target different processor architecture and functionalities such as ARM, Out of Order (OoO) processors, GPU [5]–[9]. Moreover, the fault injection campaigns conducted have a lower coverage due to the uniqueness of their input values.

In this paper, we exploit the impact of compiler optimization levels on RISC-V-based processors on reliability, considering both functional and timing effects. The impact of different optimizations on the system is obtained through a vulnerability analysis on a RISC-V-based processor core [10] using a Cycle Accurate Bit Accurate (CABA) simulator [11]. More precisely, we perform a probabilistic architectural-level single-fault injection on the whole processor. To have a high confident level of coverage, we consider different input values for the program, which are randomly selected. We perform experiments with various benchmarks to obtain the vulnerability metrics, to characterize the fault probability, fault propagation and the criticality of several compiler optimizations. The obtained results show that trends exist in the execution speed and the exposure to faults due to different optimizations. Such analysis is helpful during system deployment, driving the selection of an optimization level depending on the application's needs.

The remainder of this paper is structured as follows: Section II discusses compiler optimizations and how they affect reliability, along with the related work on the topic. Section III presents the proposed methodology and fault injection model in depth. Section IV presents the results, and section V concludes this study.

979-8-3503-6689-1/24 $31.00 © 2024 IEEE

II. BACKGROUND AND RELATED WORK

A. Compiler optimizations

Compiler optimization is a process of improving the compiled code. These improvements are performed by the compiler on demand using compilation flags. Such optimizations are helpful as they take advantage of the hardware depending on the needs, such as increasing execution speed, reducing binary size and memory usage, power consumption, etc. These optimizations are numerous. We can cite some examples, for instance, 1) loop optimizations, which apply techniques, such as unrolling, which increase the instructions in the loop body with the goal of reducing the loop condition tests, 2) loop tiling, which re-orders the iterations to improve cache efficiency, 3) function inlining, which reduces the function calls overhead by copying over the function instructions where the function has been called, 4) dead code elimination, which can reduce the binary size, 5) register renaming to deal with data hazards and dependencies, etc [12].

These optimizations have mainly explored regarding their impact on the binary code and its execution regarding timing. However, few works have been conducted to understand and quantify the impact of compiler optimizations on system reliability.

B. Related Work

The majority of existing works study the impact of compiler optimizations on reliability either target different processor architectures and functionalities, such as ARM, OoO processors etc, or focus on approaches that explore more fine-grained flags in order to find a more reliable flag combination. The main reliability threat discussed is single event upsets (SEU), which modify the control flow or some memory-related data and lead to errors.

Regarding the first category, approaches estimate the architectural vulnerability factor (AVF) of specific x86 processor structures such as Load/Store Queue and the Reorder buffer using zesto [13] simulator, and show the impact of optimization on these processor structures [6]. The AVF of the structures was estimated using equations. A similar study has been conducted for ARM based OoO processors [5], through an AVF analysis through microarchitectural-level fault injection considering the effects of optimization on eight structures of the processor, such as the Reorder-buffer, Load/Store buffers, processor caches, etc. The vulnerability factor of register files of an ARM cortex-A9 core is studied in [7] considering various optimizations and correlating the register file usage with reliability. The fault injection is carried on the user-accessible registers using interrupts and a heavy-ion radiation. The impact of optimizations on a High-Performance Computing (HPC) AMD Opteron application and the trade-off between performance and reliability are analyzed in [14]. Fault injection is conducted at the software level, showing that more optimization yields poor reliability. Early reliability analysis through fault injection with -O2 optimization level is performed in [15]. Last, a study on compiler optimizations for GPU and vulnerability assessment through beam experiments has been conducted [16].

Regarding the second category, studies are tailored into finding suitable flags for reliability using LLVM (Low-Level Virtual Machine) in [8], [9]. In [8], meta-heuristic methods are applied to optimize the reliability. In [9], a machine learning-based algorithm is used to derive the best set of flags in the context of real-time where the wcet is evaluated. Though the achieved results are better than the regular -Ox levels flags, it's worth noting that we are targeting only the regular optimizations due to increasing complexity with the required amount of fault injections.

This work belongs to the first category. Compared to existing approaches, it presents a vulnerability analysis through intensive microarchitectural-level fault injection on a RISC-V processor to characterize the criticality of optimizations and the fault probability. Furthermore, it studies the application execution profile, showing the impact of compiler optimization on the reliability.

III. PROPOSED METHODOLOGY

Figure 1 describes the overview of the reliability analysis methodology.

The inputs to the methodology are the different compilation flags that lead to different versions of the benchmarks, the benchmarks, and their inputs. The compiler optimizations are based on the processor ISA. Typical compiler optimizations under study consist of a set of individual optimizations applied in a specific order: -O0, -O1, -O2, -O3, -Ofast, -O, -Og, -Os, -Oz. Note that the level of optimizations is nested from one flag to another. For example, the -O2 flag enhances optimizations from -O1 with other optimization flags, -O3 enhances -O2 flag etc. We generate several input values for each selected benchmark to be analyzed for higher statistical confidence in the obtained results.

We consider, as a baseline, the execution of the benchmark binary without any optimization. For each compilation flag we run a set of experiments considering fault-free and faulty executions with different benchmark inputs. Regarding the faulty execution, fault injections are performed during execution using bit flip as the fault model. The injection is done on the processor registers, such as the pipeline registers and the register file. The process of generating the fault injection is as follows: we randomly select an area of the processor with respect to its size, as bigger elements are more likely to be targeted by radiation than smaller ones. Then, we randomly select a bit for a given area and then apply a logical xor with a random bit mask to flip it.

The comparison of these results provides insight into the processor's reliability and how it is affected by the compiler optimization flag. The reliability metrics are categorized into functional and timing errors. These errors include Silent Data Corruption (SDC), where the corrupted data is only detected at the end of a run, and Detectable Unrecoverable Errors (DUE), such as Hangs and Crashes. More precisely:

- *Execution Cycles Mismatch (ECM):* The execution cycles of the application are different than those of the golden reference.
- *Hang:* The execution time of the application has exceeded a threshold, and thus, it is assumed that it has entered

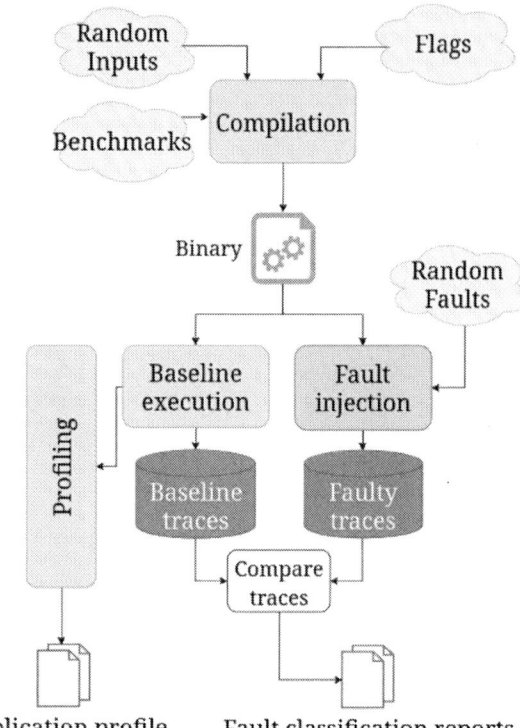

Fig. 1: Methodology

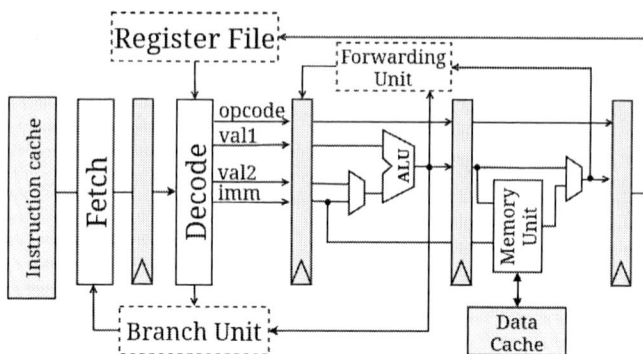

Fig. 2: RISC-V core with 5-stage pipeline [10].

IV. EXPERIMENTAL RESULTS

Our case study is a RISC-V processor and the `RISC-V` GCC compiler. The Device Under Test (DUT) is Comet [10], an open-source 32-bit RISC-V processor, which supports the RV32I base ISA[1]. It's written in High-Level Synthesis (HLS), which offers a unique high-level synthesis and simulation. A C++ model is used to design the processor. The model generates the hardware target design through High-Level Synthesis and a Cycle-Accurate Bit-Accurate (CABA) simulator. The processor consists of a standard 5-stage pipeline, including a forwarding mechanism and a register file with 32 registers in the write-back stage, as illustrated in Figure 2.

The studied benchmarks are taken from TacleBench [17]: `Bitonic, Binary search, Bsort, CountNeg-ative, Factorial, InsertSort, Matmul, QuickSort`. For each benchmark, we generate several input values, i.e., 650 different inputs based on [18] for higher statistical confidence in the obtained results. For each of the benchmarks, with one optimization flag, 385 faults are injected per binary, resulting in a margin error e=5% and a confidence level of 95% [19]. We repeat this for the 650 different inputs for higher statistical confidence, totaling `250'250` intensive faults injection per binary. In total, we inject `2,002,000` faults per binary. We first present the benchmarks' profiling regarding the different computations occurring during execution, followed by the criticality and fault probability of the obtained results.

A. Benchmarks profile

During profiling, we used Hardware Performance Counters to obtain the execution clock cycles and the instruction counts during a binary execution. Table III depicts the average clock cycles required to execute each benchmark, considering 650 different inputs. Tables I and II show the computation profile of the benchmarks, i.e., data related to the `instruction count` for each type of instructions for each benchmark. Each instruction count is divided each time by the corresponding execution clock cycles in order to obtain the computation profile. Note that the DUT doesn't support multi-cycle stages, which are used to implement multiplication and division instructions,

[1]https://gitlab.inria.fr/srokicki/Comet/-/tree/master

---

an infinite loop. A cycle counter is used to stop the current execution when the counted cycles exceed a given threshold.

- *Crash:* The execution of the application has terminated unexpectedly, and an exception has been thrown (out-of-bound memory access, misaligned PC, hardware trap, etc.)
- *Application Output Mismatch (AOM):* The application output is different than the golden reference.

In order to characterize the *criticality* of a given optimization on the system, we compute the aforementioned metrics considering the average impact of the compiler optimization on a given program. The bigger a metric, the more vulnerable an application is under a given compiler optimizations on the average case. In this scenario, the execution time duration is not taken into account. Furthermore, to characterize the *fault probability*, i.e., the probability of a fault impacting the system when it executes the optimized code, the execution time is taken into account. In this scenario, the longer the execution, the more the application is exposed to faults. Note that when optimizations reduce the execution time, the fault probability decreases as the application finishes earlier. As a result, resilience may be increased, depending on the optimization flags and the reliability metrics. Last, we perform application profiling in order to obtain the number of execution clock cycles and the number of different instructions.

979-8-3503-6689-1/24 $31.00 © 2024 IEEE

Instructions	LUI	LD	ST	OP_SLL	OP_ADD	OP_AND	OPI_SLLI	OPI_ANDI	OPI_ADDI	AUIPC
Bitonic	1.53 (-2.6)	20.6 (-20.7)	16.43 (+2.8)	0.02 (0.0)	8.12 (+0.9)	0.02 (0.0)	3.95 (+0.3)	0.07 (-0.9)	25.37 (+6.4)	0.1 (+0.1)
Binary Search	2.66 (-1.0)	19.05 (-10.9)	14.41 (+1.0)	0.27 (0.0)	3.86 (+0.6)	0.27 (0.0)	2.46 (+0.4)	1.6 (+0.9)	26.0 (+1.9)	1.87 (+0.8)
Bubble Sort	0.02 (0.0)	25.87 (-30.5)	17.26 (+11.5)	0.0 (0.0)	0.01 (-7.6)	0.0 (0.0)	0.01 (-8.6)	0.01 (0.0)	22.0 (+10.2)	0.01 (0.0)
Count Negative	0.27 (-7.7)	21.31 (+2.5)	1.98 (-3.8)	0.02 (0.0)	15.39 (-2.9)	0.02 (0.0)	0.15 (-15.5)	0.15 (+0.1)	32.91 (+7.0)	0.17 (+0.1)
Factorial	1.15 (+0.2)	13.75 (-6.0)	10.59 (-1.7)	0.13 (0.0)	5.15 (+1.9)	0.13 (0.0)	4.32 (+1.7)	4.32 (+1.7)	26.21 (-5.8)	0.89 (+0.4)
Insert Sort	1.21 (+0.7)	22.4 (-21.2)	20.9 (+8.8)	0.11 (+0.1)	3.29 (-4.6)	0.11 (+0.1)	3.07 (-4.8)	0.66 (+0.4)	24.64 (+4.7)	0.77 (+0.5)
Matrix mult.	0.06 (-0.6)	1.14 (-5.5)	1.02 (-0.6)	0.01 (0.0)	11.45 (+0.1)	0.01 (0.0)	11.15 (-0.2)	11.15 (+1.5)	13.25 (-0.8)	0.03 (0.0)
Quick Sort	1.71 (-5.3)	21.18 (-17.8)	12.82 (+0.5)	0.13 (+0.1)	7.19 (+1.7)	0.13 (+0.1)	8.06 (+2.6)	0.75 (+0.5)	25.1 (+4.7)	0.88 (+0.5)

TABLE I: Benchmark's computation profile on -O3 compared to -O0

Instructions	OPI_SRLI	OPI_SRAI	BR_BLT	BR_BNE	BR_BEQ	BR_BLTU	BR_BGE	BR_BGEU	JALR	JAL
Bitonic	2.04 (+1.2)	1.66 (+0.8)	0.43 (-1.1)	6.68 (+5.7)	2.08 (+2.0)	0.08 (+0.1)	3.59 (+2.7)	0.02 (0.0)	1.78 (-0.0)	1.83 (-0.6)
Binary Search	N/A	1.66 (+0.3)	0.53 (+0.1)	2.66 (-0.3)	6.19 (+2.0)	1.33 (0.0)	2.04 (+0.6)	0.53 (+0.3)	5.62 (+1.9)	4.31 (+0.3)
Bubble Sort	N/A	0.01 (0.0)	0.0 (-1.9)	17.35 (+17.3)	8.62 (+8.6)	0.01 (0.0)	8.59 (+4.7)	0.0 (0.0)	0.18 (+0.2)	0.02 (-2.0)
Count Negative	N/A	0.07 (+0.1)	0.05 (-2.6)	10.35 (+10.3)	5.36 (+5.2)	0.15 (+0.1)	10.03 (+7.2)	0.05 (0.0)	6.42 (+1.0)	3.88 (-1.6)
Factorial	5.15 (+1.8)	0.38 (+0.1)	0.38 (+0.2)	8.7 (+3.8)	6.68 (+2.4)	0.64 (0.0)	0.0 (-0.5)	0.25 (+0.1)	6.42 (+1.0)	3.88 (-1.6)
Insert Sort	N/A	0.33 (+0.2)	0.33 (+0.3)	2.74 (+2.2)	2.41 (+1.7)	1.1 (-0.7)	2.52 (+1.2)	5.56 (+5.0)	3.72 (+2.8)	3.06 (+1.9)
Matrix mult.	16.49 (+2.2)	0.02 (0.0)	0.01 (0.0)	11.2 (+1.5)	11.23 (+1.5)	0.05 (0.0)	0.0 (-0.5)	0.01 (0.0)	11.24 (+1.5)	0.44 (-0.0)
Quick Sort	N/A	0.38 (+0.2)	0.81 (-0.7)	1.26 (+0.8)	2.51 (+1.6)	0.63 (+0.4)	8.78 (+6.7)	0.25 (+0.2)	3.17 (+1.4)	3.0 (+1.2)

TABLE II: Benchmark's computation profile on -O3 compared to -O0 (cont.)

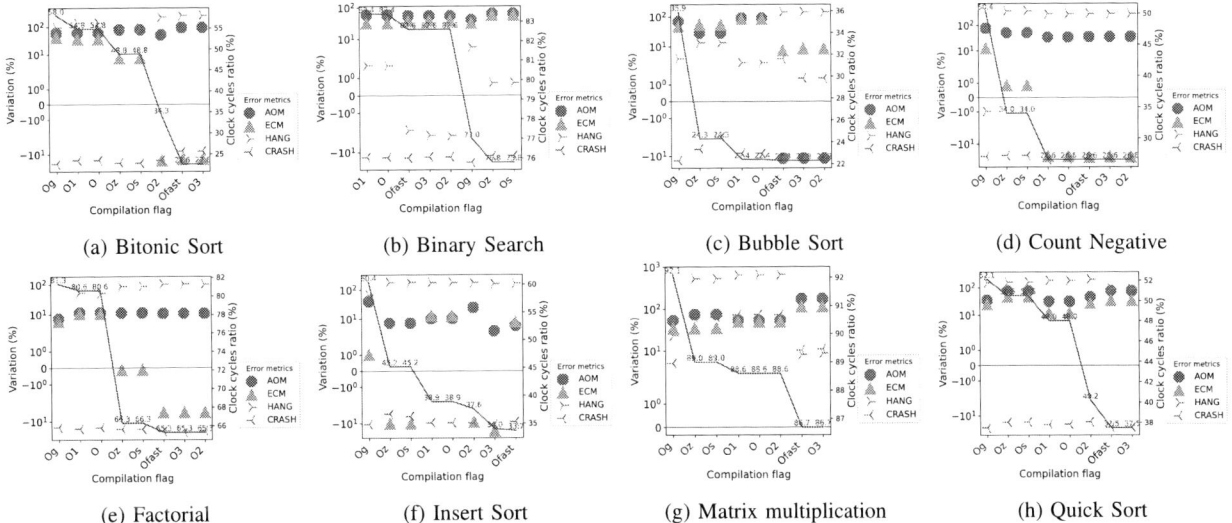

Fig. 3: Vulnerability results: Fault criticality

Benchmark	Average Clock Cycles
Bitonic	26,669
Binary Search	455
Bubble Sort	264,048
Count Negative	15,379
Factorial	1,969
Insert Sort	2,686
Matrix mult.	20,573
Quick Sort	2,134

TABLE III: Fault-free average execution cycles (-O0)

etc. Considering an implementation with a multiplication (mult) unit, the mult instruction will be split into smaller sub-stages, and the pipeline stalled – fetch and decode stages – until the multi-cycle operation instruction reaches the memory stage. Therefore, the benchmark profiling will be modified in order to take into account the stall caused by the multi-cycle additional cycles.. In order to determine the reliability of the system under different executions and different compiler optimizations, Tables I and II show the increase and decrease in terms of number of different types of instructions. More precisely, the percentage indicates how much is decreased or increased each of the instructions executed with a given optimization, e.g., -O3, compared to -O0 version.

B. Reliability analysis

The average impact of a compiler optimization on the functional and timing behavior of the program is shown in Fig. 3 (fault criticality) and Fig. 4 (fault probability). The x-axis corresponds to the compiler flag. The right-side y-axis is the execution time ratio (%) between the benchmark, compiled with an optimization flag, and its baseline -O0. For instance, a value of 100 for a compiler flag means that the optimized version has the same execution time as the baseline, and a value of 50 means that the optimized version finished in half of the time compared to the baseline -O0. The left-side y-axis is on a logarithmic scale and shows the effect that a flag has on a reliability metric. It shows the variation in the number

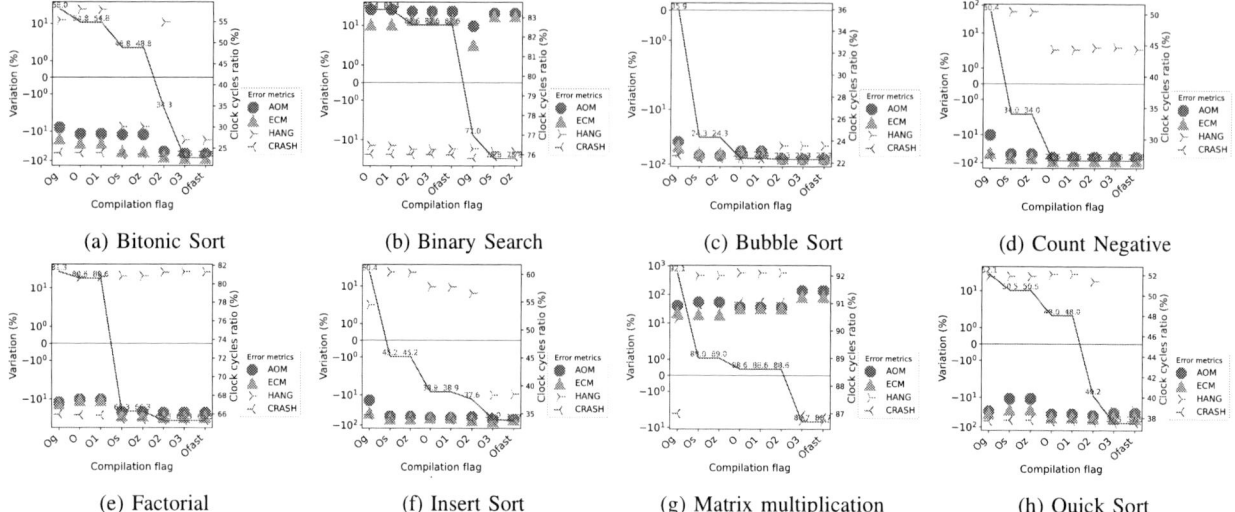

(a) Bitonic Sort　　(b) Binary Search　　(c) Bubble Sort　　(d) Count Negative

(e) Factorial　　(f) Insert Sort　　(g) Matrix multiplication　　(h) Quick Sort

Fig. 4: Vulnerability results: Fault probability

of faults observed between the benchmark, compiled with an optimization flag, and its baseline -O0.

1) Instructions and error correlations: Combining the computation profile and the vulnerability metrics information allows us to observe several correlations between the benchmark instructions and the observed errors.

The number of Crashes of a benchmark correlates with the number of load instructions (LD) during its execution. Overall, we observe that the more loads there are, the more crashes are reduced. This is expected since the more frequent loads occur, the more frequently the content of registers is overwritten with new data. For instance, with Matmul, the -O3 flag reduces the number of loads, which leads to an increase in the number of crashes (Fig. 3g).

We observe a correlation between the Branching instructions (BR) and the number of hang occurrences. The more branch instructions exist, the more susceptible the benchmark is to Hangs. The benchmarks, such as Bitonic 3a and count negative 3d, have an increased number of branch instructions when compiled with the -O3, and result in a higher number of Hangs. On the contrary, Bsearch 3b and Matmul 3g, benchmarks, have a lower number of branch instructions on average and show a lower number of Hangs.

For the AOM, we notice that the more the LUI instructions are reduced, the more the AOM. AOM is also correlated to the number of arithmetic operations (OP) that occurred during the execution. On average, Matmul has the highest number of operations, which exposes it to wrong computations and, thus, to AOM. The less operation-intensive benchmarks, such as Bubble Sort and Insert Sort, result in a lower AOM counts under -O3.

Last, but not least, the ECM is correlated to the branching and operation instructions. When a fault hits a branch instruction, either conditional or unconditional, aside from functional error, it also produces timing errors. For example, a loop index may be altered, leading to skipping or re-execution of iterations. This behavior can also appear when we alter the

condition test in conditional jumps. For instance, JALR for subroutines shows increased ECM for Bsearch, Matmul, etc.

2) Impact of compiler optimizations: The key findings from Fig. 3 and Fig. 4 indicate that applying compiler optimizations to the program consistently impacts the reliability.

In the overall trend regarding fault criticality (Fig. 3), we observe a significant increase in AOMs for most of the benchmarks and optimization flags. However, we observe the opposite trend for the Crash, with an overall average decrease, except the Matmul benchmark. The Hangs and the ECM vary depending on the benchmark, but the overall observed trend is an increase.

These observations reveal a trade-off between performance and vulnerability, enabled by the different compilation flags. While some flags have significantly reduced execution time, specific functional errors have worsened. For instance, on one side, with the Matmul benchmark, all selected flags increase the percentage of faults, considering the criticality. Conversely, most flags have increased the tolerance to Crashes and decreased the ECMs, especially for Insertsort, Cntnegativebenchmarks. This suggests that no universal flag can optimize for performance while maintaining robustness to faults. This is particularly evident in the Matmul benchmark, where the most optimized version in terms of execution cycles remains high at 86%, and the optimizations have not been achieved to reduce the observed errors.

Looking at a more fine-grained comparison, i.e., the impact between different optimization flags, we observe that specific errors are reduced depending on the flag. For instance, on the Count Negative benchmark, the less optimized version in terms of clock cycles (-Og) gives better results for hangs compared to more optimized versions, such as -O2. It's worth highlighting that, regarding the execution time reduction, -O3 is not always the best option.

These results suggest that the more the application is optimized, the more likely it is to produce wrong results, hangs, and timing errors.

To get the intrinsic reliability of an application, we have to analyze its exposure to faults. To achieve that, we present the fault probability, which takes into account the execution time of the application for the different optimization flags (Fig. 4). When applying optimization flags, the result usually yields a faster execution of the benchmark regarding clock cycles. Considering the time required to execute the benchmarks, the number of faults is reduced, on average, when the benchmark runs faster. When faults occur after the benchmark finishes, then they have no impact on the execution of the benchmark.

From the obtained results, we observe that `Bsearch` and `Matmul` give the least improvements in terms of functional and timing errors for the different optimization flags, as only a part of Hangs and Crashes is reduced with the different flags. Note that, this is also a consequence of the nature of these benchmarks, which are less optimized, and thus, more likely to have faults.

V. CONCLUSION

This work presented a study on compiler optimizations' impact and safety-critical systems' reliability. A vulnerability analysis was done by running probabilistic fault injections on the processor registers file and the pipeline registers. In order to have statistically sound results, we apply techniques in order to take into account the worst-case inputs in the fault injections by using different inputs. Our experiment, run on a RISC-V processor core with no hardware multiplier and disabled caches, shows that the optimization can be beneficial for some use cases. If, on one side, the goal of the design is to reduce the probability of faults, then using optimization that reduces the execution time is beneficial as it will reduce the exposure to fault. This can increase the Mean Work To Fault, for example. On the other hand, to reduce the fault criticality, a prior study is recommended as the impact changes with respect to the application computation profile. Some flags that optimize the application less show better resilience than others. In safety-critical real-time systems, where the system tasks have been scheduled with the unoptimized version, applying optimizations and maintaining the same schedule can lead to better reliability. It would be interesting to further assess the impact on the wcet estimation for real-time systems. In addition, a mean to estimate the reliability of a given benchmark by only studying it's application profile could be beneficial for system development.

ACKNOWLEDGMENTS

This work has been funded by the French National Research Agency (ANR) through the FASY research project (ANR-21-CE25-0008).

Experiments presented in this paper were carried out using the Grid'5000 testbed, supported by a scientific interest group hosted by Inria and including CNRS, RENATER and several Universities as well as other organizations (see https://www.grid5000.fr).

REFERENCES

[1] A. Dixit et al., "The impact of new technology on soft error rates," in *Int. Reliability Physics Symp.*, Apr. 2011, pp. 5B.4.1–5B.4.7.

[2] S. Rehman et al., *Reliable Software for Unreliable Hardware: A Cross Layer Perspective.* Springer Publishing, 2016.

[3] A. Kritikakou et al., "Functional and timing implications of transient faults in critical systems," in *IEEE Int. Symp. On-Line Testing and Robust System Design (IOLTS)*, 2022, pp. 1–10.

[4] F. Catthoor et al., "Will chips of the future learn how to feel pain and cure themselves?" *IEEE Design & Test*, vol. 34, no. 5, pp. 80–87, 2017.

[5] G. Papadimitriou et al., "Characterizing soft error vulnerability of cpus across compiler optimizations and microarchitectures," in *2021 IEEE International Symposium on Workload Characterization (IISWC)*, 2021, pp. 113–124.

[6] M. Demertzi et al., "Analyzing the effects of compiler optimizations on application reliability," in *2011 IEEE International Symposium on Workload Characterization (IISWC)*, 2011, pp. 184–193.

[7] F. M. Lins et al., "Register file criticality and compiler optimization effects on embedded microprocessor reliability," *IEEE Transactions on Nuclear Science*, vol. 64, no. 8, pp. 2179–2187, 2017.

[8] N. Narayanamurthy et al., "Finding resilience-friendly compiler optimizations using meta-heuristic search techniques," in *2016 12th European Dependable Computing Conference (EDCC)*, 2016, pp. 1–12.

[9] M. Dardaillon et al., "Reconciling compiler optimizations and wcet estimation using iterative compilation," in *2019 IEEE Real-Time Systems Symposium (RTSS)*. Los Alamitos, CA, USA: IEEE Computer Society, dec 2019, pp. 133–145. [Online]. Available: https://doi.ieeecomputersociety.org/10.1109/RTSS46320.2019.00022

[10] S. Rokicki et al., "What You Simulate Is What You Synthesize: Designing a Processor Core from C++ Specifications," in *IEEE/ACM Int. Conf. on Computer-Aided Design (ICCAD)*. IEEE, Nov. 2019.

[11] P. R. Nikiema et al., "Impact of transient faults on timing behavior and mitigation with near-zero wcet overhead," in *ECRTS 2023 - 35th Euromicro Conference on Real-Time Systems*, Vienna, Austria, July 2023.

[12] B. Gough et al., "An introduction to gcc : for the gnu compilers gcc and g++," 2005. [Online]. Available: https://api.semanticscholar.org/CorpusID:116776160

[13] G. H. Loh et al., "Zesto: A cycle-level simulator for highly detailed microarchitecture exploration," in *2009 IEEE International Symposium on Performance Analysis of Systems and Software*, 2009, pp. 53–64.

[14] R. A. Ashraf et al., "Exploring the effect of compiler optimizations on the reliability of hpc applications," in *2017 IEEE International Parallel and Distributed Processing Symposium Workshops (IPDPSW)*, 2017, pp. 1274–1283.

[15] N. Lodéa et al., "Early soft error reliability analysis on risc-v," *IEEE Latin America Transactions*, vol. 20, no. 9, pp. 2139–2145, 2022.

[16] F. F. D. Santos et al., "Assessing the impact of compiler optimizations on gpus reliability," *ACM Trans. Archit. Code Optim.*, vol. 21, no. 2, feb 2024. [Online]. Available: https://doi.org/10.1145/3638249

[17] H. Falk et al., "Taclebench: a benchmark collection to support worst-case execution time research," 01 2016.

[18] L. Cucu-Grosjean et al., "Measurement-based probabilistic timing analysis for multi-path programs," in *Euromicro Conference on Real-Time Systems (ECRTS)*, 2012, pp. 91–101.

[19] R. Leveugle et al., "Statistical fault injection: Quantified error and confidence," in *2009 Design, Automation & Test in Europe Conference & Exhibition*, 2009, pp. 502–506.

An experimental comparison of RISC-V processors: performance, power, area and security - Special Session Paper-

Elia Lazzeri[1], Bruno Endres Forlin[2], Gianluca Furano[3], Marco Ottavi[2], Luca Cassano[1]

[1]*Dipartimento di Elettronica, Informazione e Bioingegneria, Politecnico di Milano, Italy*
[2]*Faculty of Electrical Engineering, Mathematics, and Computer Science, University of Twente, The Netherlands*
[3]*European Space Research and Technology Centre, Europeran Space Agency, The Netherlands*
[1]elia.lazzeri@mail.polimi.it, luca.cassano@polimi.it
[2]{b.endresforlin, m.ottavi}@utwente.nl
[3]gianluca.furano@esa.int

Abstract—The RISC-V instruction set architecture (ISA) has garnered significant interest from both industry and academia because of its open source nature. Recent years have witnessed a surge in published implementations of the RISC-V ISA, with various companies actively pursuing its adoption in their products. However, a comprehensive analysis comparing these emerging RISC-V processors with each other and against established embedded computing platforms has not yet been published. This paper presents an experimental evaluation of three high-performance RISC-V processors: BOOM, NOEL-V, and CVA6 (formerly Ariane). The investigation is conducted on a Xilinx Kintex-7 Field-Programmable Gate Array (FPGA) platform. We perform a detailed analysis of critical performance metrics, covering area footprint, power consumption, and performance efficiency. Additionally, we assess the security posture of these cores against transient execution attacks, a prominent contemporary security threat. Finally, a comparative evaluation is undertaken between the RISC-V processors and two conventional application-level ARM processors to elucidate technological discrepancies and application suitability. This analysis aims to provide valuable insights into the current state and potential of RISC-V processors within the embedded computing domain.

Index Terms—Hardware Security, Microprocessors, Performance Benchmarking, RISC-V, Transient Execution Attacks

I. INTRODUCTION AND RELATED WORK

Processors implementing the RISC-V (Reduced Instruction Set Computer V) Instruction Set Architecture (ISA) are gaining increasing interest from both academy and industry [1], [2]. Several RISC-V implementations have emerged in recent years, such as BOOM from the Berkeley University [3], NOEL-V from Frontgrade Gaisler [4] and CVA6 (formerly Ariane) from ETH Zürich and the University of Bologna [5].

The RISC-V ecosystem offers a comprehensive software toolchain that includes compilers, simulators, and emulators, which simplifies and accelerates the design and implementation of deployable systems [6]. The free and open nature of several RISC-V implementations allows the customization and extension of the ISA and of the micro-architecture, enabling the design of ad-hoc cores and platforms. Moreover, several

RISC-V ISA extensions have been proposed to accelerate various functionalities, including signal processing, deep learning operators, cryptography and control-flow checking [7].

The drawback of such flexibility is that evaluating a single RISC-V core and then comparing it with others implementing the same ISA may be very difficult. Indeed, every microarchitecture may contain several tunable parameters, e.g., cache size, availability of out-of-order execution, depth of the pipeline; moreover, every core may implement different sets of extensions of the ISA and different versions of the RISC-V specifications, and, even worse, it may deviate from the specifications themselves [8]. In addition, open-source cores often come with a single example project for a single FPGA platform, necessitating tedious, difficult, and error-prone porting to adapt to other target platforms. Many platforms come with their own dedicated software toolchains, particularly compilers, and partially supported tools may not work out-of-the-box with different platforms. It may therefore occur that code that runs on one processor hangs on another, making comparisons even more difficult. This makes it challenging for users to identify the right processor for their specific requirements and application scenarios [9].

Several comparisons among existing RISC-V cores have already been published in recent years. Initial analyses focused on the feasibility of implementing RISC-V cores on FPGA devices and estimating their resource utilization [10], [11]. Other works addressed the problem of comparing different RISC-V implementations from the energy consumption and performance points of view, such as [12], [13]. However, a comparative analysis of RISC-V cores against processors implementing different ISAs is still lacking. Such an analysis is crucial for understanding the maturity of the RISC-V technology and its readiness for real-world adoption. Although the security of RISC-V processors has recently been analysed under several points of view, e.g., data confidentiality and integrity [14] and *traditional* side-channel attacks in [15], the feasibility of *Transient Executions Attacks* on such processors has not yet been evaluated [16], [17], [18].

979-8-3503-6689-1/24 $31.00 © 2024 IEEE

In this paper, we present an experimental comparison of three well-known RISC-V processors both from academy and industry, namely BOOM, NOEL-V and CVA6, which are meant for the high-performance computing market. The three processors have been synthesized and implemented on a Xilinx Kintex-7 FPGA to measure power consumption and resource utilization. Moreover, the performance of the considered cores have been measured by exploiting benchmarks for both embedded and high-performance computing, i.e., CoreMark, CoreMark-PRO and Dhrystone. Furthermore, the vulnerability to Transient Execution Attacks (TEAs) of the three considered microprocessors has been analysed by attacking through prime+probe (which is the basic attack on which all TEAs rely) and demonstrating that cache access times can be effectively measured. Finally, the performance obtained by the considered cores have been compared with those achieved by an ARM Cortex-A53 and an ARM Cortex-A72 in order to estimate the performance gap between the available high-end RISC-V processors and a state-of-the-art commercial microprocessor. We believe that our analysis may help in understanding how mature RISC-V processors are w.r.t. other platforms in terms of performance and security and how ready they are for adoption in the market.

The remainder of this paper is organized as follows: Section II introduces the RISC-V ISA and the Transient Execution Attacks; Section III presents the architectural details of the considered cores; Section IV presents the experimental setup that has been used for the analysis whose results are presented in Section V; Section VI concludes the paper.

II. BACKGROUND

A. The RISC-V ISA

RISC-V is an open-source ISA originating from UC Berkeley in 2010. Unlike proprietary ISAs, RISC-V offers a royalty-free and modular design. It provides a base integer instruction set with optional extensions for specific requirements like floating-point operations or vector processing. This flexibility enables customization for applications ranging from embedded systems to high-performance computing. The open development model and growing ecosystem driven by academia, industry, and open-source communities position RISC-V as a potential future dominant architecture, particularly well-suited for next-generation applications in machine learning, Internet of Things (IoT), and data centers.

B. Transient Execution Attacks

Transient Execution Attacks (TEAs), e.g., Spectre and Meltdown, rely on the fact that, due to features such as speculative execution, out-of-order execution and branch prediction, unauthorized computations may be carried out by the microprocessor before it determines that those computations were unauthorized, i.e., the *transient executions* [19]. Once the microprocessor detects the transient execution, it rolls back its architectural state in order to make the transient execution ideally invisible to the user(s). Unfortunately, most of the existing microprocessors do not roll back the microarchitectural state. Therefore, transient executions leave their traces in the microarchitecture, e.g., in the cache. Since the microarchitectural state is shared among the threads, once the transient execution completes, the attacker can retrieve data belonging to other threads by decoding the traces left in the microarchitectural state by the transient execution itself. For example, the prime+probe attack (which is the basis of almost all TEAs) relies on measuring the access time to several cache locations and distinguishing between long accesses (cache miss) and short ones (cache hit).

What makes TEAs particularly dangerous is that the attacker does not need either physical access to the attacked system or expensive tools/equipment; the only requirement to deploy a TEA is to be allowed to load and run programs in the system without even requiring any privilege level.

III. THE CONSIDERED RISC-V PROCESSORS

A. BOOM

The Berkeley Out-of-Order Machine (BOOM) [3] is a high-performance, out-of-order, superscalar RISC-V microprocessor with support for several ISA extensions including rv64imafdczicsr, Zifencei, and Zihpm. We adopted the Medium BOOM configuration that features a 2-wide pipeline with a fetch width of 4, a decode width of 2 and an issue width of 4. The 16KByte L1 instruction cache (L1-ICache) has an 8-byte cache line, a block size of 64 bytes, and is 4-way set-associative with 64 sets. The branch predictor boasts a memory size of 14 KB, enhancing its accuracy. The execution units support memory operations, arithmetic logic, multiplication, and floating-point operations, with specialized pipelines for each. The floating-point pipeline features 3 read ports, 2 write ports, and a cost-efficient design. The implemented Reorder Buffer (ROB) has 64 entries and 32 rows, facilitating out-of-order execution. Additional core parameters include 80 integer and 64 floating-point physical registers, accommodating complex computations. The integer register file has 6 read ports, 3 write ports, and advanced bypassing capabilities for efficient data flow. The translation look-aside buffer (TLB) features 8 ways for data and 32 ways for instructions. The 16KByte L1 data cache features an 8-byte cache line and is 4-way set-associative with 64 sets and 2 MSHRs.

B. NOEL-V

The NOEL-V core [4], developed by Frontgrade Gaisler, is a synthesisable VHDL model of a RISC-V core designed for space applications. We adopted the GP64-SC configuration that implements the rv64imafdbch ISA, providing comprehensive support for integer, multiplication, atomic, floating-point, double-precision, and bit manipulation operations. Its dual-issue pipeline is designed for high efficiency and performance in demanding applications. This particular configuration features a single core with an L1 cache of 16KByte for both instruction and data caches. The core integrates the nanoFPUnv floating-point unit, hardware performance monitoring interface and advanced power management.

979-8-3503-6689-1/24 $31.00 © 2024 IEEE

C. CVA6

CVA6 [5] (formerly known as Ariane) is a high-performance, 64-bit RISC-V core developed by the PULP platform implementing the rv64imafc, Zicsr, and Zifencei ISA extensions. It is an application-class processor capable of running Linux, making it suitable for various high-performance applications. Although primarily in-order, its six-stage pipeline features out-of-order write-back within the execution stage. The core's frontend parameters include a fetch width of 2, a decode width of 2, and an issue width of 2, an issue window size of 16, and a load/store unit size of 16/16. Additionally, the core has 128 integer physical registers and 64 floating-point physical registers, accommodating complex computational tasks. It supports a maximum branch count of 8, further enhancing its execution efficiency. The branch predictor features 32 BTB entries, 8 RAS entries, and 128 BHT entries. The CVA6 processor has various execution units, including memory, ALU, multiplication, division, and FPU units. The floating-point pipeline features 3 read ports and 2 write ports, with bypassable units to enhance data flow efficiency. The reorder buffer (ROB) has a machine width of 2, 64 entries, 32 rows, and 2 FPU FFlag ports, supporting out-of-order execution. Other core parameters include a fetch width of 2, a decode width of 2, an issue width of 2, a ROB size of 64, an issue window size of 16, and a load/store unit size of 16/16. The integer register file includes 6 read ports, 3 write ports, and bypassable units. The translation lookaside buffer (TLB) is configured with 4 ways each for data and instruction caches. The 32KByte L1 data cache is configured with 8 ways and 256 sets with 16 cache line bytes. The L1 instruction cache is configured as the data cache.

IV. THE ADOPTED EXPERIMENTAL SETUP

Each RISC-V processor has been implemented using a specific workflow. For BOOM Medium, we used the repository available at vivado-risc-v to generate the bitstream. This repository provides a comprehensive setup for integrating the BOOM core into Vivado, allowing us to streamline our board's synthesis and bitstream generation process. The GRLIB IP library from Gaisler was employed for the NOEL-V core to generate the VHDL code, followed by Vivado for the board-specific integration and bitstream generation. Regarding CVA6, we relied on its official repository, which offered documentation and the necessary files for implementation.

All processors have been implemented in their single-core version onto a Digilent Genesis 2 board equipped with a Xilinx Kintex-7 FPGA (Xilinx part number XC7K325T-2FFG900C), counting 326,080 logic cells and 16,020 memory blocks (such available resources did not allow us to consider a BOOM version larger than Medium BOOM). In this setup BOOM and CVA6 worked at 50MHz while NOEL-V worked at 100MHz.

Regarding the operating systems and software support, for BOOM the software environment was built including the RISC-V Open Source Supervisor Binary Interface (OpenSBI), U-Boot bootloader, Linux kernel, and a Debian root filesystem configuration as the operating system. For NOEL-V we

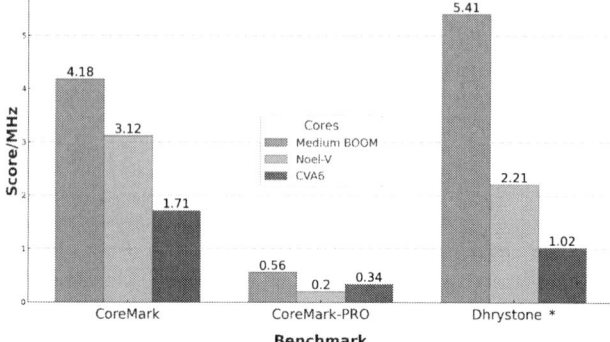

Figure 1: Summary of normalized performance (* Dhrystone's values shall be multiplied by 1,000).

generated a Buildroot image using the officially supported generator from Gaisler. Finally, for CVA6 we utilized the cva6-sdk official repository to generate the Buildroot image. Each core's implementation of the software environment was carefully chosen to maximize performance and compatibility for our experimental evaluations.

Concerning the performance evaluation, we employed three benchmark suites, namely CoreMark, which is most suited for embedded computing, and CoreMark-PRO and Dhrystone, that are more specific for high-performance computing.

Finally, for the power consumption measurements, we employed a laboratory bench power supply to power the board and run all the benchmarks. We first measured the board's consumption without any bitstream for a baseline power usage. After flashing the bitstream for each core, we ran the benchmarks and recorded the total power consumption. The power measurement for each core was determined by calculating the delta between the baseline power consumption and the total power consumption observed during the benchmark runs.

V. EXPERIMENTAL RESULTS

We present the performance of the considered cores, the power consumption and the resource occupation. Moreover, we report about the execution of a prime+probe attack on the considered cores, and we finally compare the benchmark score of the three RISC-V with those of two ARM processors.

A. Performance

As a first step of our comparative analysis, we measured the performance achieved by the three considered processors when running the CoreMark, CoreMark-PRO and Dhrystone benchmark suites. For each processor and benchmark, Figure 1 reports the normalized score, that is, the score achieved divided by the working frequency of the processor. It appears clearly that BOOM is always the best performing processor, with scores from 35% up to 180% higher than NOEL-V and from 64% up to 430% higher than CVA6. Such dramatic performance gap between BOOM and the other two cores can for sure be explained by taking into account the much

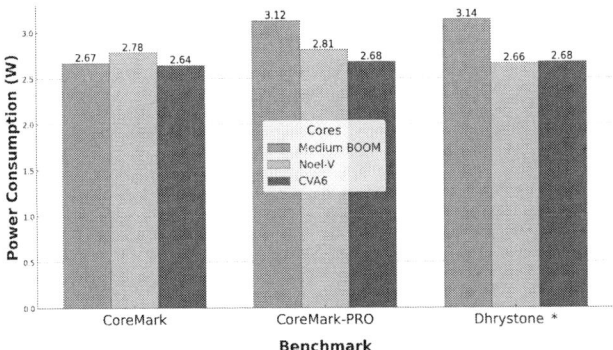

Figure 2: Summary of power consumption (* Dhrystone's values shall be multiplied by 1,000).

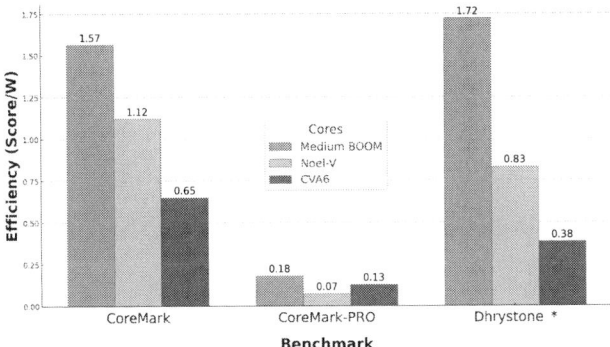

Figure 3: Summary of power efficiency, i.e., the benchmark score divided by the power consumption (* Dhrystone's values shall be multiplied by 1,000).

higher complexity (not only limited to availability of out-of-order execution) of BOOM w.r.t. both NOEL-V and CVA6. On the other hand, NOEL-V, which is an industrial project ready for the market, more than doubles the score of CVA6 for CoreMark and Dhrystone, but achieves worse performance when dealing with CoreMark-PRO.

B. Power Consumption

Figure 2 reports the power consumption values measured for the three considered cores for each of the three benchmark suites running on top of Linux following the methodology described in the previous section. The measured values are comparable to each other, with a slight increase for BOOM running CoreMark-PRO and Dhrystone that exceeds 3W in both cases. In particular, BOOM consumes 18% more than NOEL-V and CVA6 when running Dhrystone.

Interesting considerations can be drawn when considering an energy efficiency metric calculated as the normalized benchmark score divided by the power consumption. Figure 3 reports these values: when considering such an efficiency metric, we may see that, again, BOOM is the best among the three considered cores. In fact, although it consumes slightly more power, the much higher performance provided by BOOM

Table I: Summary of area usage

Core	LUTs	FFs	BRAMs	DSPs
BOOM	169,503	91,068	153	36
NOEL-V	65,004	27,402	35.5	18
CVA6	72,854	46,749	50	27

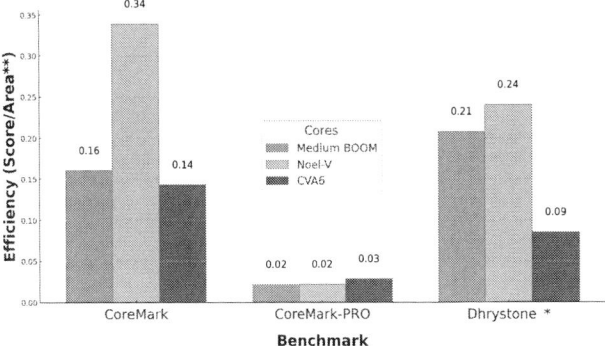

Figure 4: Summary of area efficiency, i.e., the benchmark score divided by the sum of all the occupied FPGA resources (* Dhrystone values shall be multiplied by 1,000).

provides a power efficiency from 40% up to 157% higher than NOEL-V and from 38% up to 353% higher than CVA6.

C. Area Occupation

As an estimation of the area occupation of the considered RISC-V cores, we measured the resource occupation within the adopted Xilinx Kintex-7 FPGA. Table I reports the amount of LUTs, FFs, BRAMs and DSPs occupied by the micro-processors. The first (more obvious) consideration that can be drawn is that, given its much higher complexity, BOOM is the largest core, using more than double LUTs and FFs and more than three times BRAMs w.r.t. to both the other cores. Nevertheless, it is worth mentioning that such high resource occupation is not only due to the complexity of BOOM but also to the fact that it is written in the Chisel high-level description language that is not able to achieve strong area optimization. What is also surprising is that the resource occupations of NOEL-V and CVA6 are comparable, although the former is an industrial, optimized product directly developed in VHDL, while the latter is still an academic project written in a mix of Verilog and System-Verilog.

To correlate the occupied area and achieved performance, Figure 4 shows the *area efficiency* for each processor and benchmark, calculated as the normalized benchmark score divided by the number of thousands of LUTs, FFs, BRAMs and DSPs occupied by the processor. If we look at this area efficiency metric, we may observe that NOEL-V is the core that better exploits the resources used with an area efficiency about three times higher than those of BOOM and CVA6 when dealing with CoreMark. In other words, we may say that the huge resource occupation gap between BOOM and NOEL-V does not lead to a correlated performance increase.

Table II: Comparison of normalized performance between RISC-V and ARM processors. (* Dhrystone saturated the counter registers in both ARM processors)

Core	CoreMark	CoreMark-PRO	Dhrystone
BOOM	4.18	0.56	5,407.34
NOEL-V	3.12	0.20	2,214.83
CVA6	1.71	0,34	1,023.54
Cortex-A53	6.53	1.41	11,111.11*
Cortex-A72	14.01	3.13	15,873.01*

D. Security

To asses if the considered RISC-V processors are prone to transient execution attacks or not, we implemented a prime+probe attack [20] (which represents the basic behavior of most TEAs) executable on all three RISC-V cores. The attack relies on several steps. First, we primed the cache by accessing specific memory locations to populate the cache with known data. Next, the victim program, executed instructions that accessed multiple memory locations influenced by a secret key, leading to the eviction of a number of cache lines. After the victim's execution, the attack program probed the cache again by measuring the access times to every memory location.

For the sake of space, we only report the graphs for CVA6, but the very same considerations could be drawn also for BOOM and NEOL-V. Figure 5 presents the access times measured for accessing every cache line over a single run of the attack. Certain cache lines showed significantly higher access times, indicating a cache misses. These cache misses unveiled the memory access patterns of the victim function. By analyzing the average access times on multiple runs to reduce the noise, we deduced the secret key based on the cache lines with the highest evictions, corresponding to the memory locations accessed by the victim as shown in Figure 6. This experiment effectively demonstrates how a prime+probe attack can extract sensitive information from a target RISC-V system by exploiting the cache behaviour. Thereby we confirm the viability of timing attacks (with particular emphasis on transient execution attacks) across different RISC-V implementations.

E. Comparison with ARM Processors

Finally, we were interested in understanding how ready RISC-V processors are for the real-world market. To do so, we compared the benchmark scores (normalized against the working frequency) achieved by the three considered processors with those of two conventional application-level ARM processors, namely, a Cortex-A53 (featured by a Raspberry 3 board) working at 600MHz and a Cortex-A72 (featured by a Raspberry 4 board) working at 700MHz. We point out that this comparison can be done although the RISC-V processors are implemented onto FPGAs while the ARM processors are ASICs and although they all work at different frequencies because, once normalized over the frequency, the benchmark scores only depend on the architectural details of the cores.

Table II reports the normalized (w.r.t. the working frequency) benchmark scores for the five analysed processors.

When looking at CoreMark (which is a benchmark mainly meant for embedded computing), RISC-V processors have comparable results w.r.t. ARM ones; on the other hand, if we consider CoreMark-PRO (which is a benchmark suite meant for high-performance computing) and Dhrystone, the score achieved by both ARM processors is at least one order magnitude higher then the score of RISC-V processors. It is worth mentioning that for both ARM processors Dhrystone saturated the counter registers, therefore, the scores reported in the table (and marked with *) represent a lower bound score.

To quantify the performance disparity between the RISC-V and ARM processors, our experiments revealed that all the RISC-V cores required execution time of the order of tens of hours for the CoreMark-PRO benchmark. In stark contrast, the execution time for this benchmark on the two ARM processors were measured in minutes. This significant difference further highlights the current performance gap between these architectures.

VI. CONCLUDING CONSIDERATIONS

We have presented a comparative experimental analysis in terms of performance and security against transient execution attacks of three high-end RISC-V processors meant for the high-performance computing market, namely, BOOM, NOEL-V and CVA6. Moreover, the performance of the considered RISC-V processors have been compared with the ones of two conventional application-class ARM processors. Based on the results presented, we may highlight the following points:

- Among the analysed RISC-V processors, BOOM achieved the highest performance on all the considered benchmarks (although for the limited resources available on the employed FPGA we considered Medium BOOM);
- Such higher performance makes BOOM the most power efficient core, although the absolute power consumption of BOOM is slightly higher than the other two cores;
- BOOM is far larger than NOEL-V and CVA6, but such dramatic area/resource occupation is not supported by a correlated performance gap; this allows us to argue that there is huge room for optimizing BOOM;
- None of the analysed RISC-V processors reported performance scores even comparable with those of the two considered ARM processors except when running the simpler CoreMark benchmark suite;
- All the considered RISC-V processors demonstrated to be prone to the prime+probe attack, thus opening the door to most of the existing Transient Execution Attacks.

In conclusion, our findings suggest that significant advancements are necessary for RISC-V processors to become fully competitive with established embedded computing platforms in terms of raw performance and resource efficiency. However, the open-source nature of RISC-V offers a unique advantage for niche markets like space and safety-critical systems. In these domains, complete control over the Instruction Set Architecture (ISA) is paramount, and the flexibility of RISC-V outweighs concerns about peak performance. Continued development efforts focused on performance optimization and

979-8-3503-6689-1/24 $31.00 © 2024 IEEE

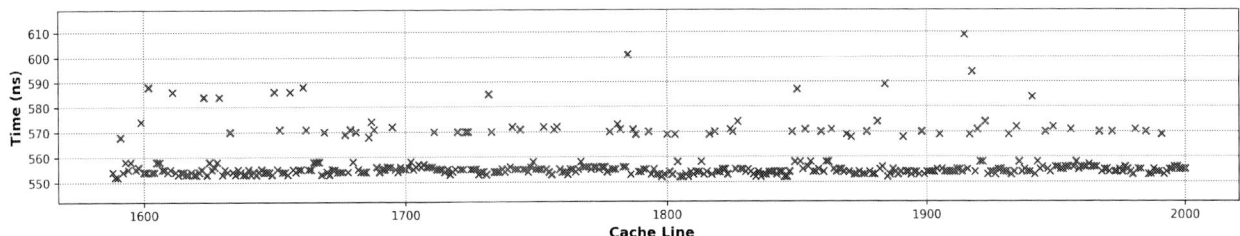

Figure 5: Time required for accessing each cache line during the probe phase on CVA6

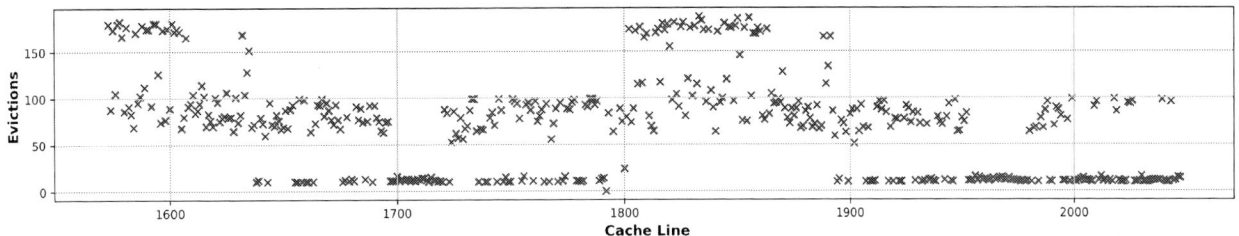

Figure 6: Cumulative evictions over multiple runs of the prime+probe attack on CVA6

resource reduction will be crucial for broader adoption of RISC-V processors in the general market, while the current strengths make them well-suited for specialized applications requiring high control and customization.

REFERENCES

[1] Semico Research & Consulting Group, "RISC-V Market Analysis: The New Kid on the Block," https://semico.com/content/risc-v-market-analysis-new-kid-block, 2019, accessed: 2024-06-04.

[2] G. Furano, S. Di Mascio, A. Menicucci, and C. Monteleone, "A european roadmap to leverage risc-v in space applications," in *2022 IEEE Aerospace Conference (AERO)*. IEEE, 2022, pp. 1–7.

[3] C. Celio, D. A. Patterson, and K. Asanovic, "The berkeley out-of-order machine (boom): An industry-competitive, synthesizable, parameterized risc-v processor," *EECS Department, University of California, Berkeley, Tech. Rep. UCB/EECS-2015-167*, 2015.

[4] J. Andersson, "Development of a NOEL-V RISC-V SoC Targeting Space Applications," in *2020 50th Annual IEEE/IFIP International Conference on Dependable Systems and Networks Workshops (DSN-W)*, 2020, pp. 66–67.

[5] The Integrated Systems Laboratory (IIS) of ETH Zürich and Energy-efficient Embedded Systems (EEES) group of the University of Bologna, "The parallel ultra low power (pulp) platform," https://github.com/openhwgroup/cva6/tree/master/docs, accessed: 2024-06-04.

[6] B. W. Mezger, D. A. Santos, L. Dilillo, C. A. Zeferino, and D. R. Melo, "A Survey of the RISC-V Architecture Software Support," *IEEE Access*, vol. 10, pp. 51 394–51 411, 2022.

[7] E. Cui, T. Li, and Q. Wei, "RISC-V Instruction Set Architecture Extensions: A Survey," *IEEE Access*, vol. 11, pp. 24 696–24 711, 2023.

[8] D. Richmond, M. Barrow, and R. Kastner, "Everyone's a Critic: A Tool for Exploring RISC-V Projects," in *2018 28th International Conference on Field Programmable Logic and Applications (FPL)*, 2018, pp. 260–2604.

[9] C. Heinz, Y. Lavan, J. Hofmann, and A. Koch, "A Catalog and In-Hardware Evaluation of Open-Source Drop-In Compatible RISC-V Softcore Processors," in *2019 International Conference on ReConFigurable Computing and FPGAs (ReConFig)*, 2019, pp. 1–8.

[10] E. Matthews, Z. Aguila, and L. Shannon, "Evaluating the Performance Efficiency of a Soft-Processor, Variable-Length, Parallel-Execution-Unit Architecture for FPGAs Using the RISC-V ISA," in *2018 IEEE 26th Annual International Symposium on Field-Programmable Custom Computing Machines (FCCM)*, 2018, pp. 1–8.

[11] R. Höller, D. Haselberger, D. Ballek, P. Rössler, M. Krapfenbauer, and M. Linauer, "Open-Source RISC-V Processor IP Cores for FPGAs — Overview and Evaluation," in *2019 8th Mediterranean Conference on Embedded Computing (MECO)*, 2019, pp. 1–6.

[12] P. Davide Schiavone, F. Conti, D. Rossi, M. Gautschi, A. Pullini, E. Flamand, and L. Benini, "Slow and steady wins the race? A comparison of ultra-low-power RISC-V cores for Internet-of-Things applications," in *2017 27th International Symposium on Power and Timing Modeling, Optimization and Simulation (PATMOS)*, 2017, pp. 1–8.

[13] A. Dörflinger, M. Albers, B. Kleinbeck, Y. Guan, H. Michalik, R. Klink, C. Blochwitz, A. Nechi, and M. Berekovic, "A comparative survey of open-source application-class RISC-V processor implementations," in *Proceedings of the 18th ACM International Conference on Computing Frontiers*, 2021, p. 12–20.

[14] E. B. Annink, G. Rauwerda, E. Hakkennes, A. Menicucci, S. Di Mascio, G. Furano, and M. Ottavi, "Preventing soft errors and hardware trojans in risc-v cores," in *2022 IEEE International Symposium on Defect and Fault Tolerance in VLSI and Nanotechnology Systems (DFT)*. IEEE, 2022, pp. 1–6.

[15] J. Anders, P. Andreu, B. Becker, S. Becker, R. Cantoro, N. I. Deligiannis, N. Elhamawy, T. Faller, C. Hernandez, N. Mentens, M. N. Rizi, I. Polian, A. Sajadi, M. Sauer, D. Schwachhofer, M. S. Reorda, T. Stefanov, I. Tuzov, S. Wagner, and N. Zidarič, "A Survey of Recent Developments in Testability, Safety and Security of RISC-V Processors," in *2023 IEEE European Test Symposium (ETS)*, 2023, pp. 1–10.

[16] W. Xiong and J. Szefer, "Survey of Transient Execution Attacks and Their Mitigations," *ACM Comput. Surv.*, vol. 54, no. 3, may 2021.

[17] L. Cassano, S. D. Mascio, A. Palumbo, A. Menicucci, G. Furano, G. Bianchi, and M. Ottavi, "Is RISC-V ready for Space? A Security Perspective," in *2022 IEEE International Symposium on Defect and Fault Tolerance in VLSI and Nanotechnology Systems (DFT)*, 2022, pp. 1–6.

[18] P. R. Nikiema, A. Palumbo, A. Aasma, L. Cassano, A. Kritikakou, A. Kulmala, J. Lukkarila, M. Ottavi, R. Psiakis, and M. Traiola, "Towards Dependable RISC-V Cores for Edge Computing Devices," in *2023 IEEE 29th International Symposium on On-Line Testing and Robust System Design (IOLTS)*, 2023, pp. 1–7.

[19] W. Xiong and J. Szefer, "Survey of transient execution attacks and their mitigations," *ACM Computing Surveys (CSUR)*, vol. 54, no. 3, pp. 1–36, 2021.

[20] V. Martinoli, E. Tourneur, Y. Teglia, and R. Leveugle, "Ccalk:(when) cva6 cache associativity leaks the key," *Journal of Low Power Electronics and Applications*, vol. 13, no. 1, p. 1, 2022.

Special Session: Reliability and Performance Evaluation of a RISC-V Vector Extension Unit for Vector Multiplication

Carolina Imianosky*, Douglas A. Santos*, Douglas R. Melo[†], Felipe Viel[†‡], Luigi Dilillo*

*IES, University of Montpellier, CNRS, Montpellier, France
[†]LEDS, University of Vale do Itajaí, Itajaí, Brazil
[‡]SpaceLab, Federal University of Santa Catarina, Florianópolis, Brazil
{carolina.imianosky, douglas.santos}@umontpellier.fr, {drm, viel}@univali.br, luigi.dilillo@umontpellier.fr

Abstract—The RISC-V Vector Extension (RVVE) enhances computational efficiency by exploring data-level parallelism, which can benefit Artificial Intelligence (AI) applications. The Zve32x subset is tailored for embedded systems, including those in space, in which AI applications are very useful to process data onboard. However, environments like space challenge the reliable functioning of electronic systems due to radiation exposure and extreme temperatures. Thus, fault tolerance techniques are crucial to mitigate potential failures. Therefore, we extended a previous implementation of a subset of the RVVE to the HARV-SoC, a fault-tolerant RISC-V-based system-on-chip, to add support to multiplication operations, which are essential in AI applications. We also added support to more configuration instructions that allow increasing the number of elements that are processed with a single instruction, further accelerating the processing. We evaluated the impact of the vector instructions over the scalar ones using C and Assembly applications to evaluate the impact of vector intrinsic functions. We estimated the reliability of the core through simulations based on fault injection campaigns with both baseline and hardened HARV-SoC. The results show that the vector unit offers a performance acceleration of up to 28.69 times with manual code optimization and can enhance reliability. Furthermore, this work assesses how different configurations of the vector parameters affect the performance and reliability.

Index Terms—RISC-V, System-on-Chip, Vector Instructions, Fault Tolerance, Space Systems

I. INTRODUCTION

The RISC-V is an open standard Instruction Set Architecture (ISA) that is simple and modular, enabling processor implementations to be designed to meet specific needs. The RISC-V Vector Extension (RVVE) is one of the modules available, and it enables systems to execute a single instruction on multiple data elements simultaneously, exploring Data-Level Parallelism (DLP). The ratified version 1.0 of the RVVE is specified in [1].

The results presented in this paper have been obtained in the framework of the EU project RADNEXT, receiving funding from the European Union's Horizon 2020 research and innovation programme, Grant Agreement no. 101008126, the Region d'Occitanie and the École Doctorale I2S from the University of Montpellier (contract no. 00137932/22009671), the Foundation for Support of Research and Innovation, Santa Catarina (FAPESC-2021TR001907), the Brazilian National Council for Scientific and Technological Development (CNPq - processes 408641/2023-1 and 350794/2023-5), and Project HARV (project PE24PR01) in the framework of the action "Accelerateur d'innovation" from the University of Montpellier.

Since RVVE is the largest RISC-V extension [1], its full implementation can be costly in terms of area and power, which is not ideal for embedded systems. To address this, the RISC-V committee introduced the Zve* extensions, five subsets of RVVE. These limit vector operations to 32-bit and 64-bit integers, fixed-point, and single- and double-precision floating-point. Notably, Zve32x restricts the element width to 32 bits and supports only integer data. This allows faster image and video processing, like AI applications [2], in embedded systems where low latency and power efficiency are crucial, such as space applications.

AI can be used in many space applications, such as remote sensing for image classification and target detection [3]. Additionally, AI can be used to process data onboard nano- and pico-satellites instead of sending them to the ground station, which reduces the demand for downlink communication and increases capabilities and dependability [4].

However, the harshness of the space environment imposes challenges on the design of dependable systems. Exposure to radiation, extreme temperatures, and several other conditions can affect the operation of electronics, leading to critical failures. Thus, applying fault tolerance techniques in these systems is crucial to meet reliability constraints [5].

HARV (HArdened RISC-V) [6] is a low-cost fault-tolerant RISC-V processor. It is hardened against SEUs (Single-Event Upsets) and SETs (Single-Event Transients) using Triple Modular Redundancy (TMR) and Error-Correcting Code (ECC) and has been characterized in different irradiation facilities [7], proving suitable for reliable applications.

In our previous work [8], we presented a Vector Extension Unit (VEU) for the HARV processor, supporting a limited subset of the vector instructions: sequential memory access, basic vector configurations, and integer arithmetic operation (*addition, subtraction, and, and xor*). However, it lacked support for the multiplication operation, which is crucial for AI algorithms [9], [10].

This work is based on the Zve32x and presents the improvement of the VEU proposed in [8]. We added support to vector multiplication operations and additional required vector configuration instructions. We also added support for the LMUL

979-8-3503-6689-1/24 $31.00 © 2024 IEEE

parameter, which groups the vector registers, allowing for fewer but longer vectors. This approach allows for a more cost-effective hardware implementation while supporting vectors of the same length as those in [8].

We extended the validation by comparing benchmarks using scalar and vector instructions in HARV with both baseline and hardened configurations. Additionally, this work presents a reliability characterization of the VEU in the HARV System-on-Chip (HARV-SoC), performed through fault injection simulations. We assessed different configurations of the VEU parameters to analyze how they impact the performance and reliability of applications that execute vector multiplications. This assessment provides a valuable analysis of the impact of using vector extensions for reliable applications.

The remainder of this paper is organized as follows. Section II discusses the related work. Next, Section III presents additional concepts on the RISC-V ISA and its vector extension, and Section IV describes our VEU design. Section V presents the experimental methodology employed in this study, while Section VI discusses the experimental results. Finally, the final remarks are shown in Section VII.

II. RELATED WORK

Some works in the literature have implemented the RVVE for embedded processors. For example, [11] describes a minimalist implementation of the RVVE aimed at embedded devices. Another example is the [12], which presents the design and implementation of a pluggable vector unit. This research extends the open-source RISC-V processor CAV6 to support the RVVE, specifically targeting the Zve64x extension.

In the context of reliable and space applications, [13] details the implementation of the RVVE on the HPP64 NOEL-V platform. This work aims to enhance the performance and capabilities of space processors while considering the specific constraints and requirements of satellite data systems. Additionally, the authors of [14] implemented fault tolerance techniques in the Vitruvius+ architecture, a RISC-V vector coprocessor for High-Performance Computing (HPC), to address the growing complexity and error vulnerability in modern HPC.

While [14] presents fault injection simulation results, it is not focused on embedded systems. Similarly, [13] is designed for space applications, utilizing a hardened processor core, but does not provide a reliability evaluation. In contrast, the works by [11] and [12] focus on embedded systems but do not address reliability or fault tolerance. Among the metrics analyzed in each implementation, operating frequency, performance, and hardware overhead are consistently evaluated. Besides that, all the mentioned works support multiplication operations, which is essential for AI applications [10], [15].

Given this context, we improved the implementation from [8], supporting multiplication operations and targeting the Zve32x extension. We assessed the performance gain and hardware overhead of the implementation. Additionally, we performed fault injection simulations to conduct a comprehensive reliability analysis.

III. RISC-V VECTOR EXTENSION

Vector architectures originated from SIMD processing, which partitions large registers into multiple elements and operates them in parallel. In SIMD instructions, the operation and data width are defined by the operation code, i.e., increasing the length of the vectors also implies an increase in the instruction set [16]. On the other hand, in vector architectures, the implementation defines the data width. Thus, it reduces the size of the instruction set and makes the hardware design more flexible for data parallelization without affecting algorithm development.

The RVVE specifies the ELEN and VLEN parameters. The ELEN parameter represents the maximum size in bits of a vector element. The VLEN indicates the number of bits in each of the 32 vector registers in the Vector Register File (VRF) specified by the RVVE. Also, seven Control and Status Registers (CSRs) are added to support the RVVE [17], and their values are modified by the Configuration-Setting instructions. The instruction *setvl* needs to be the first vector instruction to be executed since it sets three mandatory parameters: Vector Lenght (VL), vector Lenght Multiplier (LMUL), and the Selected Element Width (SEW).

The VL denotes the number of elements a specific instruction will operate on, and it can be a value between 1 and VLMAX (Equation 1). The LMUL is a parameter that can change the granularity of the register file since it allows multiple vector registers to be grouped and operated on as one vector. The SEW represents the width of the elements and acts as a divider, breaking down the vector register into multiple elements of the specified width [1].

$$\text{VLMAX} = \frac{\text{LMUL} \times \text{VLEN}}{\text{SEW}} \qquad (1)$$

IV. HARV VECTOR EXTENSION UNIT

This work proposes an improved architecture implementation introduced in [8] but based on the Zve32x. Thus, this VEU sets the following parameters: (i) ELEN is set to 32 bits, i.e., supports element widths of 8, 16, and 32 bits; and (ii) VLEN is set to 32 bits, i.e., the registers in the VRF are 32-bit wide.

In this work, we extended the supported subset of vector instructions. Thus, the supported instructions include (i) all vector configuration instructions, (ii) unit-stride load and store instructions, (iii) integer arithmetic operations (*addition, subtraction, multiplication, AND, XOR,* and *OR*). We also provided support for setting the LMUL parameter, allowing it to have up to 256-bit vectors (when LMUL = 8) even with the VLEN set to 32. With this, it is possible to decrease the overhead of logic resources while allowing the same maximum vector length as the old implementation.

The Fig. 1 shows an overview of the VEU. For this implementation, we separated the vector control from the main control unit of the core. Thus, the VEU comprises the control unit, Vector Register File (VRF), elements counter, execution unit, and interface unit.

The elements counter holds the number of vector elements that were already processed. Signals from the vector control unit specify the value and when the counter should be incremented. The Interface Unit is responsible for directing the

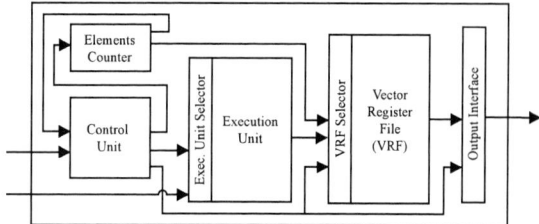

Fig. 1. Vector Extension Unit

outputs of the VEU. The outputs are a signal indicating if all the necessary elements were processed (v_done) and two arrays of data that serve for memory access.

The main control unit of the core integrates the instruction decoder. The instruction bits 6 to 0 are the input to a combinational logic component that determines the instruction format, according to the RISC-V ISA specification [16]. Afterward, an output of the control unit that signals if it is a vector instruction v_inst is either enabled (1) or disabled (0). In the case of a vector instruction, it waits for the signal v_done to proceed with the execution.

The vector control unit has as inputs the instruction to be executed, the values from the vector CSRs, and the v_inst, that is used to initiate the decode and execution of the vector instructions. To apply the same logic as the HARV microarchitecture [6], the vector ALUs are controlled by the vector control unit to simplify the implementation.

The execution unit, represented in Fig. 2, comprises one 32-, one 16-, and two 8-bit ALUs. These ALUs are simplified and exclusive to arithmetic vector instructions, operating only on the vector elements. With VLEN being 32 bits, it can perform either one 32-bit, two 16-bit, or four 8-bit operations simultaneously. Therefore, the level of parallelization of operations depends on the SEW value. The inputs and outputs are mapped to their respective ALUs according to the SEW value. The elements that are being operated on wider ALUs are zero-extended.

The VRF contains 32 VLEN-bit registers exclusive to the VEU and can perform two readings and one writing simultaneously. The data outputs are defined to have the size of VLEN so that all the vector elements are read at once. When LMUL > 1, adjacent vector registers are merged to create register groups, i.e., a register file with fewer but longer vector registers.

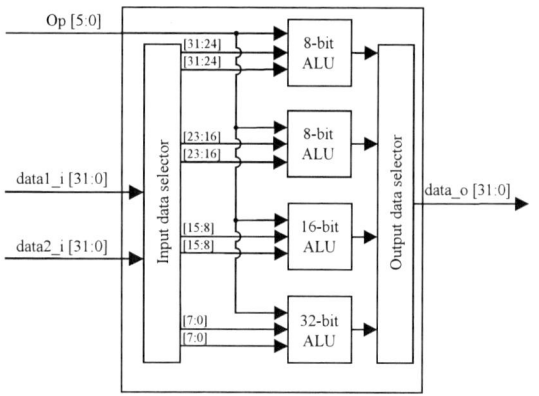

Fig. 2. Vector Execution Unit

V. EXPERIMENTAL METHODOLOGY

In this section, we describe the experimental methodology applied to this work. The following subsections detail the development and synthesis tools, the benchmark algorithms, and the performance and reliability evaluation.

A. Development and Synthesis Tools

The VEU was implemented using VHDL (VHSIC Hardware Description Language) since HARV-SoC was also implemented using this HDL. We used TCL (Tool Command Language) and Python scripts with the ModelSim SE-64 2019.01 software for the fault injection simulations and reliability evaluation.

All the algorithms were implemented in C or Assembly language. For the vector instructions in C, we used the RISC-V vector C intrinsics [18]. We compiled the applications using the GNU Compiler Collection (GCC) for RISC-V [19].

The synthesis data were collected using the Xilinx Vivado 2020.2 Design Suite tool and the Zynq ZC7020 FPGA SoC device. Thus, we analyzed the number of Look-up Tables (LUTs), Flip-Flops (FFs), and digital signal processing (DSP) blocks, as well as the maximum operating frequency (F_{max}), and the dynamic power dissipation (P_{dyn}) for the HARV-SoC.

B. Benchmark algorithms

We implemented an algorithm that reads elements of two vectors, multiplies them, and stores the result in memory. The number and size of the elements varied depending on the scenario that the algorithm was being executed. The benchmark algorithm was implemented in both scalar and vector versions using C language. Additionally, we implemented the scalar and vector versions in assembly language versions to maximize performance through low-level optimizations.

C. Performance evaluation

For the performance evaluation, we executed the benchmark algorithms to evaluate the performance variation using the baseline HARV (RV32I) and the HARV with VEU (RV32IZve32x), comparing both the C and assembly versions of the benchmark algorithm. This comparison allowed us to assess the impact of C and intrinsic functions on performance across different configurations. In all scenarios, the number of elements in each vector was set to 256. In the baseline HARV scenario, we varied the size of the elements between 8, 16, and 32 bits. For the executions with the VEU, we varied the LMUL parameter between 1, 2, 4, and 8 and the SEW between 8, 16, and 32 bits, with the VL parameter always being set to VLMAX.

To measure processing latency, we read the cycle counter register (mcycle) at both the start and end of the execution of the algorithm in each scenario. The performance variation was calculated relative to the scalar C version of the benchmark algorithm. This allowed us to evaluate the impact of vector instructions and Assembly-level optimizations on performance.

D. Reliability evaluation

We ran 1024 experiments for each of the four scenarios: baseline HARV-SoC, baseline HARV-SoC with the VEU, hardened HARV-SoC, and hardened HARV-SoC with the VEU.

979-8-3503-6689-1/24 $31.00 © 2024 IEEE

This way, we could assess the impact of the VEU in terms of fault tolerance for each SoC version.

We used the C language version of the benchmark algorithm for the reliability evaluation since it is more commonly used in practical applications. The number of elements in each vector was set to 16384. In the executions with the VEU, we set the LMUL parameter to 8 and the VL to VLMAX, aiming to exploit maximum parallelism. We varied the SEW between the best and worst parallelization levels, i.e., 8 and 32 bits, to explore how the different SEWs impact reliability. Thus, we also operated on 8- and 32-bit elements in the scalar scenarios.

At the end of each execution, the results stored in the memory are verified to check if they match the expected multiplication result. If not, a message is printed to inform that the execution experienced an error. Each execution creates a log file with all the UART output and the simulation time.

The simulations start by executing a golden run, i.e., executing the application without fault injections. This run provides reference parameters for comparison with the subsequent simulations, such as execution time, UART output, and all the FFs of the implementation. These indicators are also used as parameters for the fault injections. When the golden run is finished, we run the fault injection simulations.

For each simulation, we run a script that randomly selects locations and times to inject bit flips based on neutron-characterized flip-flop FIT_{NYC} (Failure In Time for a billion hours in the New York City's neutron flux at sea level) of 248 and a neutron flux of $5 \times 10^{11} n/cm^2/s$. This value of flux was set after a set of preliminary simulation runs in order to ensure statistically meaningful data. At first, the script fetches all register signals and sets the simulation time to zero. Then, it increments the simulation time by a random value and calculates the neutron fluence from the previous simulation time until the incremented time. The script uses the cross-section and calculates the FF error rate for a given fluence. After this, it iterates through all FFs and randomly decides whether it should inject an error based on the error rate. Finally, it calculates the random injection time between the previous simulation and the incremented time and stores this in a file with the corresponding FF signal. This flow runs in a loop, incrementing the simulation time until it reaches the set maximum simulation time.

We evaluated the reliability of the SoC by analyzing the internal registers of the processor and the outputs. The executions were classified into four categories: (i) correct, when the application was executed correctly; (ii) benchmark error, when the application is executed but has incorrect results; (iii) hang, when the application does not end the execution nor produces any result; (iv) uncorrectable, when it detected an uncorrectable error; and (v) exception, when an unscheduled event disrupts the program execution [20]. The considered exceptions were instruction address misaligned, instruction access fault, illegal instructions, load access fault, and store access fault.

To estimate the trade-off between performance and reliability, we analyzed the metrics Mean Work to Failure (MWTF) and the event cross-section. We considered as an event all the cases in which the execution was not classified as correct. Based on

[21], we calculated the MWTF by Equation 2, where a unit of work is considered one execution of the algorithm, thus the amount of work is 1024. The cross-section (XS) is a calculation of the error rate of the device based on the fluence expressed in cm^2/device.

$$MWTF = \frac{\text{amount of work}}{\text{number of errors encountered}} \qquad (2)$$

We used a Python script to analyze the files containing the outputs of each execution and determine the number of executions for each category, the MWTF, the event XS, the average number of injected faults, and the execution time.

VI. RESULTS

The following subsections introduce the results of the VEU implementation. First, we present the synthesis results, followed by an analysis of the performance achieved compared to the baseline processor. Finally, we show the results of the fault injection simulations.

A. Synthesis

Table I shows the resource usage for the HARV-SoC and the HARV-SoC with the VEU in both baseline and hardened scenarios. The table presents the number of FFs, LUTs, DSPs, maximum operating frequency, and dynamic power dissipation for each implementation.

TABLE I
SYNTHESIS RESULTS.

HARV-SoC	LUTS	FFs	DSP	F_{max} (MHz)	P_{dyn} (mW)
Baseline	3748	3583	4	65.40	138
Baseline with VEU	5399	4170	9	64.40	140
Hardened	5802	4381	12	41.91	164
Hardened with VEU	7555	4968	17	40.77	175

In the baseline scenario, adding the VEU increases the number of LUTs by 44%. The number of FFs increases by 16%, mainly because of the VRF, and the number of DSPs increases due to the vector execution unit. Using fault tolerance techniques increases the number of logical resources the processor uses. The number of DSPs in the hardened scenario increased 3 times compared to the baseline because of the TMR implemented in the ALU. Adding the VEU to the hardened system increases LUT usage by 30% and FFs by 13%.

The maximum operating frequency of the processor decreases as the critical path increases. In the baseline implementation, the SoC can operate at a maximum of 65.40 MHz, and it drops 35% when applying the hardening techniques. Adding the VEU slightly decreases F_{max} by 1.5% in both baseline and hardened scenarios. The addition of the VEU also results in a moderate increase in P_{dyn} by 1.4% and 6.7% in the baseline and hardened scenarios, respectively.

B. Performance

The plot in Fig. 3 presents the variation in the performance achieved by different configurations of the benchmark algorithm relative to the scalar C version. We varied the LMUL

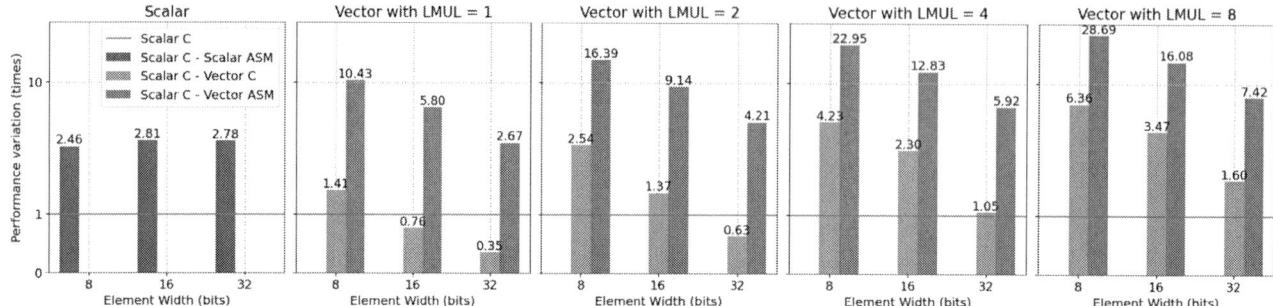

Fig. 3. Performance variation for different versions of the benchmark algorithm in comparison to the scalar C.

parameter between 1, 2, 4, and 8 and the SEW between 8, 16, and 32. The VL parameter was set to VLMAX in all scenarios. In cases where the performance factor is less than 1, the performance is worse than the scalar C.

The scalar Assembly implementation shows an acceleration between 2.46 and 2.81 times faster than scalar C across different configurations due to the manual optimization in Assembly, simplifying code execution. The vector Assembly achieves the highest speedup among all configurations (10.43 to 28.69 times faster than scalar C) since it benefits from the parallelization and optimized assembly coding. The LMUL parameter significantly influences the performance since it affects the number of elements that a single instruction will operate. Also, smaller SEW values explore greater parallelism, resulting in higher performance.

The scenario of vector C, using intrinsic functions, shows performance variations ranging from 0.35 to 6.36 times compared to scalar C. The cases where the VEU performed worse than the scalar C occur because, unlike in the vector Assembly, where we execute the configuration instructions (setting the VL, LMUL, and SEW) only once at the beginning of the execution, the intrinsic functions run it every time before each operation, introducing repetitive overhead. Thus, vector C can reduce the performance compared to scalar C when the parallelization provided by vector operations is not enough to offset the repeated configuration overhead.

C. Fault Injection Simulation

We evaluated the output of the UART of each execution by using the Python script and the five categories of results mentioned in Section V-D. The experimental results are presented in Table II, which compares the results for different configurations of the HARV-SoC and HARV-SoC with the VEU. Thus, it presents the number of executions of each category achieved by each approach for the 1024 executions. Also, it presents the average number of fault injections, the total simulated fluence, the event cross-section, and the MWTF for each scenario.

The event cross-section gives an indication of the susceptibility of devices to radiation-induced faults since lower cross-section values mean lower vulnerability to faults. The results show an increase in the event cross-section values in the scenarios with the VEU since it increases the sensitive area. However, in the cases for the execution with the VEU and SEW = 8, the Mean Work to Failure (MTWF) was higher. This is

because the executions with higher LMUL and smaller SEW values apply more parallelism and, therefore, have a shorter execution time. The difference in the execution time directly impacts the number of fault injections since the longer the execution lasts, the more faults are likely to be injected.

In the cases of SEW = 32, the amount of provided parallelism does not compensate for the increase in the sensitive area, resulting in fewer correct executions. In the hardened scenario, the difference in correct executions with and without the VEU is larger due to the application of hardening techniques, especially in the ALU and the register file, that are not applied for the vector ALUs and the VRF. Thus, even though fewer faults were injected, the MWTF is smaller.

Most of the benchmark errors in the scenarios with the VEU were caused by fault injections in the VRF. Unlike the main register file of the core, the VRF does not apply fault tolerance techniques due to the hardware overhead that would be required. Thus, the difference in the number of benchmark error executions between the vector and scalar executions is greater in the hardened scenario than in the baseline.

D. Discussion

Although adding the VEU to the HARV-SoC introduces hardware, frequency, and power consumption overhead, it can significantly increase performance and reliability. The VEU can significantly accelerate computations, as for the vector Assembly application, where it achieved an execution up to 28.69 times faster than scalar C.

However, achieving these advantages depends on selecting the appropriate parameters, such as LMUL and SEW. Incorrect parameter choices can lead to degraded performance and reliability rather than improvement. For instance, in the hardened scenario with the VEU, using 8-bit elements increased the number of correct executions by 2.3%, but using 32-bit elements decreased it by 7.9% compared to the hardened scenario without the VEU.

Nevertheless, the VEU enhances reliability by reducing execution times and improving the MWTF metrics under the right conditions, i.e., with smaller elements. Thus, the VEU is a promising solution for applications that use multiplication operations and dependability is critical, particularly for 8-bit elements. This enhancement is particularly advantageous in image processing, where an 8-bit element represents a byte, especially in AI applications such as image classification.

979-8-3503-6689-1/24 $31.00 © 2024 IEEE

TABLE II
FAULT INJECTION EXECUTIONS

Execution classification	Baseline				Hardened			
	Scalar 8b	Vector 8b	Scalar 32b	Vector 32b	Scalar 8b	Vector 8b	Scalar 32b	Vector 32b
Correct	722	849	699	694	913	934	911	839
Benchmark error	32	27	54	101	15	31	15	84
Hang	7	2	5	1	1	0	0	0
Uncorrectable	198	109	196	145	89	49	92	60
Exception	65	37	70	83	6	10	6	41
Reliability analysis								
Average # of injections	18.70	11.11	18.73	19.36	21.98	13.63	21.52	19.44
Simulated fluence [n/cm^2]	$1.84 \cdot 10^{14}$	$8.92 \cdot 10^{13}$	$1.78 \cdot 10^{14}$	$1.49 \cdot 10^{14}$	$1.74 \cdot 10^{14}$	$9.20 \cdot 10^{13}$	$1.72 \cdot 10^{14}$	$1.31 \cdot 10^{14}$
Event XS [cm^2/device]	$1.61 \cdot 10^{-12}$	$1.94 \cdot 10^{-12}$	$1.80 \cdot 10^{-12}$	$2.22 \cdot 10^{-12}$	$6.31 \cdot 10^{-13}$	$9.79 \cdot 10^{-13}$	$6.59 \cdot 10^{-13}$	$1.41 \cdot 10^{-12}$
MWTF	3.47	5.92	3.20	3.11	9.31	11.38	9.06	5.54

VII. CONCLUSION

In this work, we have enhanced the VEU proposed in [8] by introducing support for vector multiplication and additional configuration instructions based on the Zve32x subset of the RVVE. The main goal was to accelerate the processing in applications requiring multiplication operations and assess its reliability. Also, we explored how different configurations of the VEU parameters can affect the reliability and performance.

We evaluated the impact of vector instructions over scalar ones to measure the processing time among different configurations of the VEU parameters. We also analyzed the impact of using applications in C and Assembly languages, in which we performed low-level optimizations. The results showed that the VEU proved very efficient, accelerating the execution up to 28.69 times when using the Assembly version of the benchmark. For the C version, the maximum acceleration was 6.36. However, we noticed that for VEU configurations where less parallelism is being applied (i.e., smaller LMULs with bigger SEWs), the use of the C version of the benchmark was not worth it since the parallelization provided by the VEU does not compensate for the repeated instruction overhead introduced by the compiler. The impact of this overhead is also present in the reliability evaluation, where for a SEW of 8 bits, the VEU enhances the fault tolerance, and for 32 bits, it worsens it.

In future work, we intend to conduct a comprehensive analysis of the impact on the reliability of all VEU parameters previously assessed for performance. This analysis will enable us to evaluate the benefits of utilizing C abstraction versus low-level optimization techniques to ensure reliability. Also, we aim to implement all the Zve32x extension instructions.

REFERENCES

[1] RISC-V Organization, "RISC-V "V" Vector Extension," https://github.com/riscv/riscv-v-spec, 2021, (Accessed on 05/22/2024).

[2] S. D. Mascio, A. Menicucci, E. Gill, G. Furano, and C. Monteleone, "Leveraging the openness and modularity of RISC-V in space," *Journal of Aerospace Information Systems*, vol. 16, no. 11, pp. 454–472, 2019.

[3] G. Furano, A. Tavoularis, and M. Rovatti, "Ai in space: applications examples and challenges," in *2020 IEEE International Symposium on Defect and Fault Tolerance in VLSI and Nanotechnology Systems (DFT)*. IEEE, 2020, pp. 1–6.

[4] D. Cappellone, S. Di Mascio, G. Furano, A. Menicucci, and M. Ottavi, "On-board satellite telemetry forecasting with rnn on risc-v based multicore processor," in *2020 IEEE International Symposium on Defect and Fault Tolerance in VLSI and Nanotechnology Systems (DFT)*. IEEE, 2020, pp. 1–6.

[5] M. Yang, G. Hua, Y. Feng, and J. Gong, *Fault-tolerance techniques for spacecraft control computers*. Singapore: John Wiley & Sons, 2017.

[6] D. A. Santos, L. M. Luza, C. A. Zeferino, L. Dilillo, and D. R. Melo, "A low-cost fault-tolerant RISC-V processor for space systems," in *2020 15th Design Technology of Integrated Systems in Nanoscale Era (DTIS)*, Marrakesh, 2020, pp. 1–5.

[7] D. A. Santos, A. M. P. Mattos, D. R. Melo, and L. Dilillo, "Enhancing fault awareness and reliability of a fault-tolerant RISC-V system-on-chip," *Electronics*, vol. 12, no. 12, 2023.

[8] C. Imianosky, D. A. Santos, D. R. Melo, F. Viel, and L. Dilillo, "Implementation and reliability evaluation of a risc-v vector extension unit," in *2023 IEEE International Symposium on Defect and Fault Tolerance in VLSI and Nanotechnology Systems (DFT)*. IEEE, 2023, pp. 1–6.

[9] I. Goodfellow, Y. Bengio, and A. Courville, *Deep learning*. MIT press, 2016.

[10] S. D. Mascio, A. Menicucci, E. Gill, G. Furano, and C. Monteleone, "On-board decision making in space with deep neural networks and RISC-V vector processors," *Journal of Aerospace Information Systems*, vol. 18, no. 8, pp. 553–570, 2021.

[11] M. Johns and T. J. Kazmierski, "A minimal RISC-V vector processor for embedded systems," in *2020 Forum for Specification and Design Languages (FDL)*. Kiel: IEEE, 2020, pp. 1–4.

[12] V. Maisto and A. Cilardo, "A pluggable vector unit for risc-v vector extension," in *2022 Design, Automation & Test in Europe Conference & Exhibition (DATE)*. IEEE, 2022, pp. 1143–1148.

[13] S. Di Mascio, A. Menicucci, E. Gill, and C. Monteleone, "Extending the noel-v platform with a risc-v vector processor for space applications," *Journal of Aerospace Information Systems*, pp. 1–10, 2023.

[14] M. Barbirotta, F. Minervini, C. R. Morales, A. Cristal, O. Unsal, and M. Olivieri, "Enhancing fault tolerance in high-performance computing: A real hardware case study on a risc-v vector processing unit," *Authorea Preprints*, 2024.

[15] J.-G. Dumas, P. Lafourcade, J.-B. Orfila, and M. Puys, "Private Multiparty Matrix Multiplication and Trust Computations," in *Proceedings of the 13th International Joint Conference on e-Business and Telecommunications (ICETE 2016) - Volume 4: SECRYPT*. SciTePress, 2016, pp. 61–72.

[16] D. Patterson and A. Waterman, *The RISC-V Reader: an open architecture Atlas*. San Francisco: Strawberry Canyon, 2017.

[17] A. Waterman, Y. Lee, D. Patterson, and K. Asanovic, "The RISC-V instruction set manual," *Volume I: User-Level ISA', version*, vol. 2, 2014.

[18] RISC-V Organization, "Risc-v vector intrinsic document," https://github.com/riscv-non-isa/rvv-intrinsic-doc, 2024, (Accessed on 06/27/2024).

[19] ——, "Gnu toolchain for risc-v," https://github.com/riscv-collab/riscv-gnu-toolchain, 2024, (Accessed on 05/28/2024).

[20] J. L. Hennessy and D. A. Patterson, "Computer organization and design RISC-V edition: The hardware software interface," 2017.

[21] G. A. Reis, J. Chang, N. Vachharajani, S. S. Mukherjee, R. Rangan, and D. I. August, "Design and evaluation of hybrid fault-detection systems," in *32nd International Symposium on Computer Architecture (ISCA'05)*. IEEE, 2005, pp. 148–159.

979-8-3503-6689-1/24 $31.00 © 2024 IEEE

Special Session: Software-Based Self-Test Generation for RISC-V

– Stuck-At Generation, Functional Cell-Aware Untestability, and FPGA Demonstration –

Tobias Faller*, Nikolaos I. Deligiannis[†], Riccardo Cantoro [†], Matteo Sonza Reorda[†], Bernd Becker*

[†] Politecnico di Torino, Department of Control and Computer Engineering (DAUIN) — Turin, Italy
* University of Freiburg, Department of Computer Science — Freiburg, Germany

Abstract—Software-Based Self-Tests (SBST) allow at-speed, native online-testing of processors by running software programs on the processor core, requiring no Design for Testability (DfT) infrastructure. Traditionally, the generation of such SBST programs requires time-consuming manual labor in combination with in-depth knowledge of the processor's architecture to target hard-to-test faults. In contrast, encoding the SBST generation task as a Bounded Model Checking (BMC) problem allows using sophisticated, state-of-the-art BMC solvers to automatically generate an SBST for a given fault model and, moreover, prove the untestability of certain faults. During generation, a constraint specification Validity Checker Module (VCM) ensures correct, functional SBST behavior. The operating conditions in the field and the processor and system-specific constraints are included in this specification, too, in a highly configurable form. In this paper, we present an intermediate VCM interface to generalize the functional constraint specification on RISC-V processors and apply it to two in-field test-related areas. First, we perform SBST generation for detecting permanent hardware faults (stuck-at) in the ALU and the register file of multiple RISC-V processors. Second, we perform functional untestability analysis for the cell-aware fault model on the same processors. Finally, we synthesize the largest processor core on an FPGA with user-controlled fault injection to demonstrate the subsequent detection via the generated stuck-at SBST to a non-technical audience.

Index Terms—Software-Based Self-Test, Functional ATPG, Automatic SBST, Microprocessor Test, RISC-V

I. INTRODUCTION

Software-Based Self-Test (SBST) [1] programs enable at-speed, native testing of processors in the field without requiring any Design for Test (DfT) infrastructure. Many IP core and semiconductor companies provide SBSTs as Self Test Libraries (STLs) for use in safety-critical applications. Developing SBSTs is often time-consuming and costly, requiring skilled developers to manually construct test programs based on detailed knowledge of the processor's micro-architecture. This involves reasoning about the micro-architecture's behavior under fault conditions, which is complex and inefficient. Additionally, SBST programs must be embeddable into their target environments, making the creation process complex and necessitating repetition for each new design.

The introduction of the license-free RISC-V [2] instruction set architecture (ISA) facilitates the creation of a vast amount of new processor cores featuring different micro-architectures, base instruction sets, and extensions. With the adoption of high-level synthesis, IP core providers supply whole RISC-V processor families with architectures and ISA extensions depending on the targeted use case. This poses new challenges for SBST creation as development cycles shorten and a new adaptive, automated approach for SBST development is required.

Manually constrained automatic test pattern generation (ATPG) using Bounded Model Checking (BMC) [3] has been shown to allow semi-automatic generation of SBSTs for processors [4]–[7]. Extending that, [8] introduced an abstraction of the applied constraints by introducing the so-called *Validity Checker Module* (VCM). The VCM is a circuit written in a Hardware Description Language (HDL). It allows to specify constraints in a high-level language and is used during SBST generation to apply constraints to the processor, simplifying the development of constraints for complex SBST scenarios.

However, specifying SBST constraints for whole processor families requires an even higher abstraction level. In [9] we generalize the processor-specific signals to a well-defined, intermediate interface that sits between a set of generic SBST constraint and the targeted processor at hand. The relation between the processor signals and behavior and the interface is specified via an HDL specification that has to be manually adapted once. The generic SBST constraints are specified in a processor-agnostic way and are re-usable between different processor cores and families.

To show the effectiveness of the approach, we use our highly configurable SBST constraint set to (i) generate stuck-at SBST programs for multiple processor cores as shown in [9] and subsequently (ii) identify untestable cell-aware faults originally shown in [10] via the same constraint set on the same processor cores. Consequently, we tackle two challenging problems in the context of in-field testing, while keeping the effort for adaptation low. That is, we not only address these two problems for one processor, but we also address them for various RISC-V processors.

Subsequently, we present an FPGA-based SBST demonstration board that allows for intuitive user-controlled stuck-at fault injection. The firmware running on an industrial processor core visualizes the SBST result to the user by displaying a checksum on an LCD screen and multiple status LEDs. Thus, detection of the selected stuck-at fault(s) can be easily understood and verified by the user.

This work was supported in part by the German Federal Ministry of Education and Research (BMBF) within the project Scale4Edge under contract no. 16ME0132.

979-8-3503-6689-1/24 $31.00 © 2024 IEEE

The rest of the paper is organized as follows: Section II introduces the VCM concept, its application in SBST generation and the cell-aware fault model. Section III presents a VCM architecture that is then used for the generation of stuck-at SBSTs in Section III-B and identification of untestable cell-aware faults in Section III-C. Following that, we present the FPGA-based demonstration board in Section III-D. In Section IV we present the results for two example processor families, and lastly, we draw some conclusions.

II. BACKGROUND

A. Validity Checker Module (VCM)

The Validity Checker Module (VCM) [8] contains functional test constraints formalized as a circuit via a Hardware Description Language (HDL). The constraints are applied to the processor at hand during test generation and enforce a functional behavior. The VCM circuit is synthesized into a gate-level representation, as done with the processor at hand, and subsequently encoded together as a single miter circuit into the bounded model checking (BMC) problem. As shown in Figure 1, the VCM observes the miter circuit via its inputs. The VCM's inputs are connected to the processor's internal and external signals. This allows for simultaneously observing the processor's state and environment under fault-free and faulty conditions. The VCM behaves like a normal sequential circuit that evaluates the processors's state and behavior according to the encoded test constraints. However, the VCM's so-called validity outputs are treated as an indicator for the satisfaction of each test constraint in the form of a Boolean value. The output of 1 indicates that the corresponding constraint is held. When solving the BMC problem, all VCM validity outputs are constrained to have an output value of 1. This enforces the test constraints and with that a valid behavior and state of the processor.

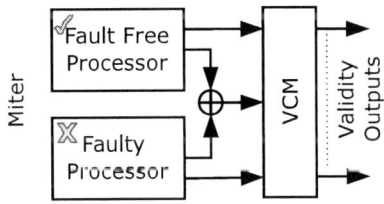

Figure 1. VCM observes miter circuit (left) and validates constraints (right)

The original concept of the VCM as Boolean constraint specification presented in [8] has been significantly extended. For example special, binary-encoded VCM inputs that do not correspond to miter circuit signals are used to pass application-defined binary encoded configuration parameters into the VCM. These parameters enable and disable the constraints applied by the VCM to the processor and allow for reconfigurability without the need for multiple gate-level variants originating from the same VCM HDL code. During the BMC process, these inputs are constrained according to the application-defined configuration.

B. Cell-Aware Fault Model

In recent years, statistics have reported an increasing number of failing devices returned to semiconductor suppliers, although their test flows have been sufficient in terms of achieved test coverage (thus complying with the thresholds mandated by the standards). The root cause of the failing devices was, in some cases, proved to be latent defects related to cell-internal faults. These defects are not sufficiently covered by the state-of-the-art fault models (e.g., stuck-at and transition delay faults) because in these models, the common assumption is that a fault can only be present on the connections between cells (i.e., the ports of the technology library cells). As a solution to this problem, the cell-aware test (CAT) was proposed [11].

Although currently used as an end-of-manufacturing test, cell-aware testing (CAT) can also benefit in-field testing. Most dominant fault models assume no faults occur within the internal parts of cells, leaving latent defects a potential threat in safety-critical systems. CAT addresses these defects and could be included as a complementary fault model in future safety standards. In advanced semiconductor manufacturing, where silicon aging is a concern, CAT provides a valuable solution. However, if CAT is used for in-field testing, identifying functionally untestable faults (i.e., those faults that never produce failures in the operational scenario) remains unresolved.

For test generation the cell-aware fault model specifies a set of faults per cell of the technology library in the form of defect matrices. Each fault can have one or multiple test alternatives that excite the fault. Table I contains a simplified defect matrix representing a defect in an AND gate with two inputs A and B, and one output O. This fault model only contains the input combinations for which the cell's outputs differ from the fault-free outputs. In this example, the cell has two different entries, called *test alternatives* in the following, with input combinations, called *fault conditions*, for which the fault effect of the respective combination is made visible to the gate's output port.

Table I
EXAMPLE DEFECT MATRIX FOR AN AND GATE

Alternative No.	A	B	O
1	1	0	1
2	1	1	0

III. APPROACH

In this section, we first focus on the constraint formulation mechanism i.e., the Validity Checker Module (VCM) that forms the basis for our functional test generation. We detail how it interacts with the RISC-V processor during test generation. Following, we elaborate on how we formulate the two problems at hand: the stuck-at SBST program generation and the cell-aware untestability analysis on top of the functional constraints in our BMC-based flow.

The test generation is implemented in our Automated Test Pattern Generation (ATPG) framework named FreiTest which is derived from the PHAETON [8] framework. However, even though the core concept of the VCM originates from PHAETON, the functionality of the VCM has been greatly extended, while the framework was fully rewritten and re-designed for RISC-V SBST generation. This includes the circuit import, the VCM handling, the fault simulation, data export and visualization, as well as Conjunctive Normal Form (CNF) generation and the whole ATPG process itself.

A. VCM Architecture and Constraint Specification

The concept of the VCM allows for convenient specification of test constraints. However, to support a fast adaptation to new processors cores and families a well-defined interface in the form of an abstraction layer between the processor's signals and the test constraints is introduced [9]. This abstraction layer is shown in Figure 2 where the processor is depicted on the left-hand side. Although only the RISC-V processor is shown here, the left-hand side represents the whole miter circuit with a fault-free and faulty version of the processor. The miter signals of the processors are related to the well-defined interface via the comparatively small mapping layer (middle). The interface is used to formalize generic constraint specifications on the right.

The generic constraint set defines the valid processor state and behaviour including the set of RISC-V instructions the processor is allowed to execute. Therefore, a generated RISC-V ISA decoding module is embedded into the VCM constraints. This ISA decoding module's source code is automatically generated from multiple sources, including the formal specification used by the MINRES DBT-RISE-RISC-V instruction set simulator [12], the official RISC-V ISA opcodes specifications [13] and additional specifications in JSON format to support custom instructions. The sources are selected according to the instruction set of the processor core at hand.

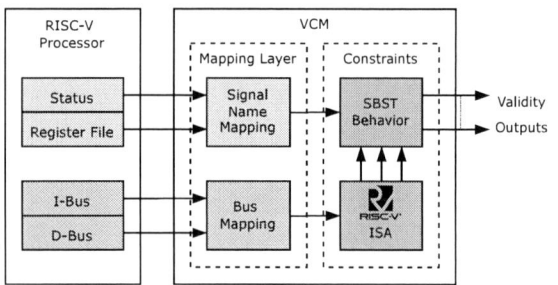

Figure 2. Interaction of processor(s) (left) and VCM (right) with mapping layer in between (middle)

The interface between the mapping layer and the generic constraint set provides the signals necessary to observe the processor's state and behavior. In total, the following necessary signals have been identified for SBST development and are provided via the well-defined interface:

The relation of processor signals to this well-defined interface is achieved by either (I) forming them from logical

- Processor control (reset, halt, run)
- Processor state (resetting, halted, running)
- Pipeline state (bubble, flush, halted)
- Program counter and register file
- Instruction and data bus transactions

combinations of processor signals or (II) using a translation module. For example, the processor control signals are typically exposed via the top-level processor inputs and are directly related via method (I). However, the instruction and data bus interfaces are typically too complex to allow direct mapping due to the required bus timing according to the protocol used. Therefore, they are related using an embedded bus mapping module that translates between the processor's bus protocol and the interface's signals. For each supported bus protocol, such a module has to be developed only once and can be reused thereafter. So far, the Advanced High-performance Bus (AHB), Open Bus Interface (OBI), PicoRV32, and Dark-RISCV protocols have been implemented.

B. Stuck-At SBST Generation

Due to the intricate complexity of the SBST generation process, we partition the approach into eight distinct steps. 1) The processor and VCM gate-level description are imported. 2) A stuck-at fault list is created based on the processor gate-level description. 3) A reset sequence is generated using BMC to initialize the processor into a well-defined state. 4) For each fault, multiple BMC-based checks are performed to generate a functional testability verdict. 5) For each potentially testable fault, an instruction sequence is generated using BMC. 6) Redundant instruction sequences are eliminated. 7) The instruction sequences are concatenated to a full SBST program. 8) The SBST program is evaluated on the full fault list, and statistics are derived.

Three of these steps (the reset sequence generation, testability check, and instruction sequence generation) each make use of a single VCM using the previously described architecture containing configurable generic SBST constraints. These three steps each put the VCM into a distinct mode that is unique to the step while keeping the VCM implementation constant. Hence, this configurable, generic VCM is a core component for the SBST generation process. In the following, we will elaborate on the necessary constraints used by each of the modes:

3) Reset sequence generation: During this step, the processor, starting from a fully undefined initial state, is forced to reset via its control inputs. Subsequently, the processor is forced to run and finally fetch a NOP instruction on the instruction bus. At this point the program counter and register file is forced to be fully initialized, concluding the reset sequence. To summarize, the following constraints are defined inside the VCM:

4) Functional testability verdict: This step solves multiple (Boolean satisfiability) SAT and BMC problems in increasing complexity to identify untestable stuck-at faults in a functional scenario. If one of the SAT or BMC problems is proven to be

---

979-8-3503-6689-1/24 $31.00 © 2024 IEEE

- Processor control: initially reset, then running
- Instruction bus: always, only fetch NOP instructions
- Data bus: always, no transactions
- Program counter: finally, fully initialized
- Register file: finally, all registers initialized

unsatisfiable by the solver, the fault is marked as functionally untestable. Note that, for solving the BMC problems, an unbounded model checker with Craig interpolation is used. Table II shows the four configurations the VCM applied during the four distinct identification steps: Constraints marked with a star are enforced via FreiTest directly. (a) Combinational testability check (equivalent to full-scan ATPG). The VCM applies no constraints. (b) Combinational testability check with VCM enforcing a RISC-V instruction executing processor. (c) Sequential testability check with VCM enforcing a RISC-V instruction executing processor and a defined initial state. (d) Sequential testability check with VCM enforcing a fault propagation to an observation point (data bus, program counter and register file).

Table II
CONSTRAINTS FOR TESTABILITY CHECK

VCM Constraints	a	b	c	d
Processor control: running	-	X	X	X
Data bus: only RISC-V instructions	-	X	X	X
Data bus: fault propagation	-	-	-	X
Program counter: fault propagation	-	-	-	X
Register file: fault propagation	-	-	-	X
Framework constraints				
Initial state *	-	-	X	X
Fault activation and propagation *	X	X	X	X

ᵃ SAT full-scan   ᵇ SAT partial SBST
ᶜ BMC partial SBST   ᵈ BMC SBST

The summarize, the VCM applies the following core set of constraints during the funcional testability verdict generation:

- Processor control: always, running
- Instruction bus: always, only fetch valid instructions
- (Data bus: finally, fault propagation)
- (Program counter: finally, fault propagation)
- (Register file: finally, fault propagation)

5) Instruction sequence generation: For each stuck-at fault, an instruction sequence must be generated to activate the fault effect and propagate it to an observable point. As an observable point, the register `x1` of the processor's register bank is used. This register accumulates all fault effects, which, by the end of the SBST execution, form a signature. This signature must be compared with the expected golden signature by the firmware to verify that the system is not faulty.

This implies that since all sequences are concatenated one after the other, a mechanism abstracting from the current system state must be applied. Furthermore, to prohibit fault masking by cancellations occurring in the `x1` register, which

serves as the fault effect accumulator, a scrambling sequence must take place after the completion of each generated sequence. Also, the contents of the said register must never be changed by, e.g., being considered as a destination register of a generated instruction within a sequence, as that would cancel out all fault effects that have been previously accumulated.

To summarize, the following constraints are defined inside the VCM:

- Processor control: always, running without interrupts
- Program counter: always, increases linearly
- Register file: always, registers initialized before read
- Data bus: always, neither read nor written
- Instruction bus: always, only supported RISC-V instructions, no control flow instructions, scramble sequence applied to register x1, only XOR and XORI instructions using register x1

C. Functional Cell-Aware Untestability Analysis

Based on the previously descibed VCM architecture we devise an identification of functionally untestable cell-aware faults in RISC-V processors. Note that, our approach directly uses the VCM architecture to specify the functional test constraints. With that, contrary to [10], the constraint set can be adapted to different processors with minimal effort. The used functional constraints applied here are shown below. Additionally, a well-defined initial state is directly imposed for the BMC problem by the framework.

- Processor control: always, running
- Instruction bus: always, only fetch valid instructions
- Data bus: finally, fault propagation

Each cell-aware fault is encoded into a BMC problem according to the defect matrix. If the solver deems the problem unsatisfiable the fault is declared untestable. Note that, for solving the BMC problems, an unbounded model checker with Craig interpolation is used.

$$1: \qquad A \wedge \neg B \rightarrow O$$
$$2: \qquad A \wedge B \rightarrow \neg O$$
$$\neg(A \wedge \neg B) \wedge \neg(A \wedge B) \rightarrow \underbrace{(A \wedge B \leftrightarrow O)}_{\text{fault-free behavior}}$$

Figure 3. Encoding of faulty AND gate for example model

In the case of the example cell-aware fault from Table I, both test alternatives are encoded according to the formula shown in Figure 3, where the first two lines encode the faulty behavior according to the defect matrix; the third line encodes the behavior of the gate when none of the test alternatives is applied to the gate's inputs and is equal to the fault-free gate behavior. An exemplary activation of test alternative 1 in the middle timeframe is shown in Figure 4, while the first and last timeframes show a fault-free behavior.

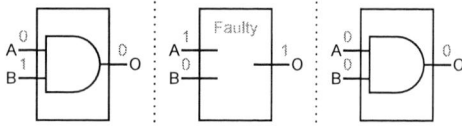

Figure 4. Fault injection on AND gate for example model

D. FPGA-based Stuck-At SBST Demonstration

To showcase the concept of an SBST to a general audience, we create an FPGA-based demonstration board (Altera DE2-115 Development board with a Cyclone IV FPGA), that allows users to manually inject stuck-at faults into a RISC-V processor core. Figure 5 shows the exemplary system architecture chosen. A firmware running on the processor core signals its activity via LEDs and an LCD with marquee text shown on the left. This allows the user to verify that the firmware is running correctly. The SBST embedded into the firmware produces a checksum that is shown by the firmware on an 7-segment display, such that the user can identify deviating checksums under fault influence. The fault injection mechanism for randomly selected stuck-at faults is manually added to the gate-level description of the processor core and linked to 18 user-accessible switches on the demonstration board.

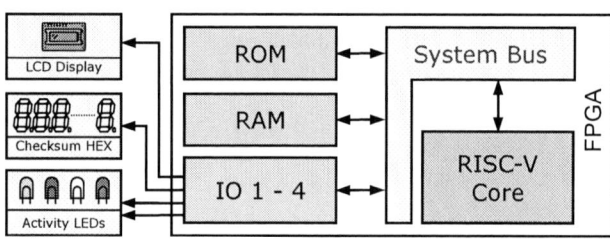

Figure 5. FPGA architecture

IV. RESULTS

The SBST generation and cell-aware untestability identification was evaluated for two processor families with four configurations in total. The DarkRISCV (3-stage scalar pipeline) and a proprietary, industrial core (5-stage scalar pipeline) were chosen to show the effectiveness but also the limitations of the presented approaches and are synthesized using the Nangate 45nm PDK [14]. The BMC depth was set to 15 timeframes for the SBST geneartion, and 50 for the cell-aware untestability identification. Both use a timeout of 5 minutes per fault. All experiments were conducted using a dual AMD EPYC 7343 system (2x16 cores, 2x32 threads) with 2 TB of RAM and a computationally equivalent AMD Threadripper 3970X system (32 cores, 64 threads) with 256 GB of RAM.

Based on the data of the Nangate 45 nm PDK, we used a commercially available tool to generate a cell-aware fault model. Sequential cells and sequential test alternatives were removed to allow for a simplified encoding in our ATPG tool, as described in the previous chapter.

Table III contains the resulting fault statistics and generation times for (I) the stuck-at SBST generation and (II) the identification of functionally untestable cell-aware faults. Note, that the stuck-at fault list and the cell-aware fault list are not directly comparable as the cell-aware fault model was stripped of test alternatives with sequential behavior.

The SBST results earlier presented in [9] show a fault coverage of 75.85 % to 82.42 % for the DarkRISCV processor variants with 22,810 to 35,645 stuck-at faults. The number of timeouts is comparatively low from 2 to 7, while the test generation time is 16.78 h to 54.18 h. This shows that the SBST generation approach is applicable to small-sized processor cores with acceptable computation time.

The industrial processor core provides more detailed statistics broken down into the pipeline stages and essential modules. There are 151,531 stuck-at faults, where many of them fall into the miscellaneous category containing proprietary units and faults modeled on module interconnects. The rest of the faults are distributed over instruction fetch, instruction decode, and execute stages containing many architecture and control-and-status registers (CSRs). Most of the timeouts of 10,260 and 6,242 for the execute and instruction fetch stage are assumed to be caused by forbidding CSR accesses, and interrupt and exception handling. With the generation of the SBST allowing only mostly arithmetic instructions, the register file with 93.40 % fault coverage and the ALU with 99.70 % fault coverage is the primary target for the generated SBST type. The two modules show 3 timeouts each, and a generation time of 34.19 h and 5.36 h, respectively. This shows the strength of our approach for targeting the register file and ALU, while for other modules, the SBST generation should be adjusted.

The functional cell-aware untestability experiments on the DarkRISCV processor show 3.89 % to 8.97 % of functionally untestable faults and no timeouts. The runtime varies from 4.82 h to 15.22 h (0.28 s to 0.58 s per fault), depending on the processor size. This is significantly lower than the 2.64 s to 5.47 s per fault for generating the stuck-at SBST. This shows that the cell-aware untestability check is computationally simpler as no timeouts are present either. A reason for the decreased complexity can be a simple fault propagation via the memory bus, while at the same time the behavior-limiting constraints for the stuck-at SBST generation are not present. With that, easy-to-generate solutions are allowed.

Analyzing the results for the industrial core shows the inverse effect on the runtime. On average, it increased significantly to 3.91 s per fault versus 2.29 s for the stuck-at SBST generation. This is reflected in 36.17 % of timeouts versus 13.99 % of timeouts for the stuck-at SBST generation. Interestingly, the number of timeouts shows a different distribution compared to the stuck-at SBST generation. A possible explanation is the fault list containing only combinational faults, where each fault can contain many test alternatives depending on the cell type. This contrasts to the stuck-at model where only one port of each gate is affacted at a time. Hence, the generation complexity can heavily depend on the

979-8-3503-6689-1/24 $31.00 © 2024 IEEE

Table III
FREITEST RESULTS FOR RISC-V PROCESSOR CORES

Processor and ISA		Stuck-At SBST				Cell-Aware Functional Untestability			
		Faults	Coverage	Timeouts	Runtime	Faults	Untestable	Timeouts	Runtime
DarkRISCV	**RV32E**	22,810	75.85 %	2	16.78 h	57,452	6.55 %	0	4.82 h
	RV32I	34,039	82.42 %	3	46.80 h	91,236	3.89 %	0	7.02 h
	RV32I_Zicsr	35,645	79.18 %	7	54.18 h	94,612	8.97 %	0	15.22 h
Proprietary	IF Stage	11,455	6.54 %	6,242	15.59 h	30,823	6.43 %	21,898	62.60 h
RV32I	ID Stage	21,541	40.29 %	1,417	4.08 h	10,056	1.35 %	2,605	7.89 h
_Xunknown	└─Register File	17,704	93.40 %	3	34.19 h	47,630	0.25 %	26,677	71.68 h
	EX Stage	25,534	11.66 %	10,260	29.82 h	49,301	5.46 %	6,747	27.84 h
	└─ALU	7,872	99.70 %	3	5.36 h	19,574	0.30 %	163	3.56 h
	MEM Stage	2,353	12.88 %	985	2.45 h	5,492	7.21 %	2,512	7.64 h
	WB Stage	515	25.63 %	261	0.84 h	1,510	5.50 %	66	0.43 h
	Miscellaneous	64,557	4.27 %	2,029	4.06 h	6,261	23.75 %	1,047	3.62 h
	Sum	151,531	45.40 %	21,200	96.39 h	170,647	4.08 %	61,715	185.26 h

cell types present in the respective units. Additionally, the advanced pipeline of the core could affect the runtime too, but reasoning about the influence on the essential constraints, including the fault propagation, is difficult due to the intricate interactions of all pipeline components.

However, the majority of timeouts concern the register file and the instruction fetch stage, which show the highest runtime of the processor units too. The timeouts for the instruction fetch stage can be attributed to the handling of interrupts and exceptions, showing significant complexity as these might affect the behavior of the whole processor pipeline. The timeouts in the register file can be attributed to synchronous reset logic that is mostly untestable due to the essential functional constraints of requiring the processor to be running. Hence, it can be assumed, that the timeouts represent untestable faults where additional computation time is required to show their untestability. Note, that faults at flip-flops are only present in the stuck-at SBST generation.

Summarizing the results for the functionally untestable cell-aware faults, the DarkRISCV core shows a differing computation complexity versus the industrial core. It can be assumed that the architecture has a non-negligible influence on the computation requirement. In the case of the DarkRISCV processor the runtime per fault is lower compared to the stuck-at SBST, while for the industrial core the runtime is higher. This can be explained by the advanced architecture in combination with differing cell types, and the complexity of proving certain cell-aware faults for the reset logic untestable.

V. CONCLUSION

SBSTs allow at-speed, native online testing of processors without requiring DfT infrastructure. Creating SBST programs is a complex process that is repeated for every new processor design. By introducing an abstraction layer for our presented SBST generation, we can adapt a generic constraint set to the processor at hand and the desired generation target. We presented results for (I) stuck-at SBST generation and (II) functionally untestable cell-aware fault identification based on our single, highly configurable SBST constraint set. The results were analyzed for fault coverages and runtimes and showed the potential of BMC-based SBST generation and

functionally untestable fault identification while also showing the limits of our constraint-based approach. The intersection of SBST generation and cell-aware testing shows a non-trivial step in the direction of cell-aware SBST generation. As future work, we focus on increasing coverages in less tested units and implementing test generation with dynamic cell-aware faults.

ACKNOWLEDGMENT

This work was supported in part by the German Federal Ministry of Education and Research (BMBF) within the project Scale4Edge under contract no. 16ME0132 and by the Italian ICSC National Research Centre for High Performance Computing, Big Data and Quantum Computing within the NextGenerationEU program.

REFERENCES

[1] M. Psarakis et al., "Microprocessor Software-Based Self-Testing," *IEEE Design & Test of Computers*, vol. 27, 2010.

[2] RISC-V Foundation, *The RISC-V Instruction Set Manual, Volume I: Unprivileged ISA, Document Version 20191213*, 2019.

[3] A. Bierre, *Handbook of Satisfiability*. IOS Press, 2009.

[4] Y. Zhang et al., "Automatic Test Program Generation for Out-of-Order Superscalar Processors," in *IEEE Asian Test Symposium (ATS)*, 2012.

[5] R. Cantoro et al., "Effective techniques for automatically improving the transition delay fault coverage of Self-Test Libraries," in *IEEE European Test Symposium (ETS)*, 2022.

[6] A. Ruospo et al., "On-line Testing for Autonomous Systems driven by RISC-V Processor Design Verification," in *IEEE International Symposium on Defect and Fault Tolerance in VLSI and Nanotechnology Systems (DFT)*, 2019.

[7] P. Bernardi et al., "Software-based self-test techniques of computational modules in dual issue embedded processors," in *European Test Symposium (ETS)*, 2015.

[8] A. Riefert et al., "A Flexible Framework for the Automatic Generation of SBST Programs," *IEEE Transactions on Very Large Scale Integration (VLSI) Systems*, vol. 24, 2016.

[9] T. Faller et al., "Constraint-Based Automatic SBST Generation for RISC-V Processor Families," in *IEEE European Test Symposium (ETS)*, 2023.

[10] N. I. Deligiannis et al., "Automatic Identification of Functionally Untestable Cell-Aware Faults in Microprocessors," in *IEEE Asian Test Symposium (ATS)*, 2023.

[11] F. Hapke et al., "Cell-Aware Test," *IEEE Transactions on Computer-Aided Design of Integrated Circuits and Systems*, vol. 33, 2014.

[12] E. Jentzsch, *Minres DBT-RISE-RISCV*, https://github.com/Minres/DBT-RISE-RISCV.

[13] RISC-V Foundation, *RISC-V Opcodes*, https://github.com/riscv/riscv-opcodes/, 2010.

[14] Silvaco, *Open-cell 45nm freepdk*, https://si2.org/open-cell-library/.

Special Session: A mixed simulation-, emulation-, and formal-based fault analysis methodology for RISC-V

Endri Kaja[*†], Nicolas Gerlin[*†], Ares Tahiraga[*], Jad Al Halabi[*†], Sebastian Prebeck[*],
Dominik Stoffel[†], Wolfgang Kunz[†], Wolfgang Ecker[*‡]

[*]Infineon Technologies AG, Germany
[†]Rheinland-Pfälzische Technische Universität Kaiserslautern-Landau, Germany
[‡]Technische Universität München, Germany

Abstract—As the semiconductor industry rapidly expands, there is a need for new development methods, especially in the domain of safety-critical designs. The ISO 26262 standard plays a crucial role in the automotive industry by requiring systems to operate safely and reduce the chance of critical failures. Advanced, automated approaches are essential to meet these challenges effectively. This paper presents an automated framework for safety verification using the principles of model-driven architecture, which aims to increase the efficiency, quality, and trustworthiness of these systems. It introduces the concept of creating models with different levels of granularity for various components of a design such as gate-level models for the safety-critical parts and Register-Transfer Level (RTL) models for the rest of the designs. The innovation includes adding fault injectors to these models to inject faults directly in the design. Through a holistic generation flow, various safety verification techniques have been explored, including methods based on simulation, emulation, and formal verification. The effectiveness of these methods has been demonstrated in many different industrial design applications.

Index Terms—Fault Simulation, Fault Emulation, Formal Verification, Model-Driven, RISC-V.

I. INTRODUCTION

Over the last years, there has been a rapid growth in technology which includes the development of AI accelerators, the evolution of RISC-V ecosystem, Smart Systems, Cyber-Physical Systems and other numerous innovations. This intricate development process necessitates the use of intelligent and automated digital design solutions alongside advanced methods for verification to guarantee their functionality. However, despite progress within the semiconductor industry, the verification phase remains a critical hurdle that constrains design efficiency and affects the time frame for introducing products to the market. The increasing complexity in integrated circuit (IC) and application-specific integrated circuit (ASIC) design and verification can also be attributed to the rise of extra layers of design imperatives. Beyond the basic functional specifications, these additional criteria encompass aspects such as *safety*. Following the integration of the ISO 26262 safety standard into the development workflow for safety-critical automotive systems, these considerations have become a mandatory aspect of the development lifecycle. The ISO 26262 standard mandates that such systems operate within established safety parameters by reducing the occurrence of dangerous malfunctions. It advises the use of *fault injection* techniques for the assessment and verification of these critical systems; nonetheless, these techniques tend to be cumbersome and susceptible to errors. To address these issues, there is a need for more progressive and automated strategies.

In this paper, we propose the automation of the safety verification process through the application of model-driven architecture principles, aiming to improve efficiency, quality, and reliability. We introduce a holistic approach for the analysis of digital faults. The methodology involves processing models at the gate level, extending them with fault injection mechanisms, establishing control systems for these injectors, creating a duplicate design for generating reference values, and integrating checkers. The advanced model preserves a register-transfer level (RTL) abstraction, meaning it operates with clock-related timing and utilizes bits or vectors of bits, allowing for simulation using any standard RTL simulator, whether open-source or proprietary. The RTL abstraction also ensures that the models are synthesizable, permitting the fault analysis to be carried out on field-programmable gate arrays (FPGAs). For FPGA deployments, only the end results of the analysis are required, formatted specifically for FPGA use. The integrity of the model extension process is validated through formal proofs, which confirm the equivalence between the original design and the modified version. These proofs verify both the correct operation of the fault injectors and the unaltered functionality of the design when faults are not active, ensuring the modified design behaves identically to the original in the absence of faults. This approach accommodates various applications, such as testing, or safety verification, by varying the configuration of the fault injection control mechanism.

The rest of the paper is organized as follows: Section II summarizes the most related works, and Section III describes the problem and challenges that we try to solve in this paper. Section IV provides a general overview of the framework and the requirements it fulfills. Section V discusses the framework in details, section VI provides results and Section VI concludes the paper.

II. RELATED WORK

Significant research work has focused on safety verification, especially highlighting the use of fault injection techniques and formal methods to verify safety-critical designs. This section offers an overview of related works.

Baraza et al. [1], [2] made considerable progress in the domain of VHDL-based fault injection techniques compared to previous works. The techniques they introduced were capable of injecting both permanent and transient fault models. This injection is facilitated using automated methods that incorporate mutants and saboteurs within the VFIT tool. Ferrareto et al. [3] developed an automated, non-intrusive fault injection framework that utilizes QEMU, an emulator compatible with a range of microprocessor architectures. This framework enables the induction of faults into the processor's operation by altering data structures within QEMU, thereby mimicking the effects of the faults. The authors in [4] present a method for fault injection within microarchitectural simulators. They enhanced

979-8-3503-6689-1/24 $31.00 © 2024 IEEE

the MARSS and Gem5 simulation platforms, introducing to each the ability to perform fault injection, named MaFIN and GeFIN, respectively. A key component they developed is the Fault Mask Generator, which has the capability to produce random fault masks for various fault categories, such as bit-flips, permanent, and intermittent faults. Furthermore, they established an Injection Campaign Controller tasked with handling these masks and orchestrating the injection processes by forwarding requests to the Injector Dispatcher. This module is specifically designed to interact with the simulation platforms for effective fault injection. Various studies, such as [5], have introduced techniques for fault modeling at the register-transfer level (RTL) that draw upon a gate-level depiction of the design. Espinosa et al. [6] delve into the relationship between RTL and Instruction Set Simulator (ISS) fault injections. Their research yields highly precise outcomes for permanent fault models through the examination of data from executed applications on a microcontroller and by estimating the likelihood of fault manifestations. Bagbaba et al. [7] present a technique for representing gate-level Single Event Transient (SET) faults by employing multiple Single Event Upset (SEU) faults at the RTL.

Fault injection methodologies that leverage emulation have been developed to improve the fault injection process, with a predominant focus on the use of Field Programmable Gate Arrays (FPGAs) for fault emulation. Lopez-Ongil et al. [8] developed an FPGA-based automated fault emulation framework to carry out SEU fault injection, coordinated by an external host computer. This system integrates a suite of components critical for its operation: an emulation controller, a module to control fault injection, a unit for classifying faults, an application module for testbenches, onboard RAM, and an interface module for interaction with the host computer. The fault induction mechanism operates by substituting standard flip-flops with specially designed mutants and saboteurs using diverse techniques. A significant experimental finding is the system's capability to process over one million faults per second. Entrena et al. [9] further refine this architecture by introducing a multilevel fault emulation technique. This approach employs two congruent models of the target design, one at the gate-level and the other at the RTL, allowing dynamic switching between models to facilitate fault emulation. The gate-level representation is particularly beneficial for fault injection activities, offering enhanced precision in the estimation of the susceptibility to errors. In a similar vein, Grinschgl et al. [10] propose a method for emulation-based fault injection that focuses on the automation of placing fault injectors.

A notable limitation of fault simulation and emulation techniques is their limited fault coverage, which is constrained to only those faults that have been explicitly defined and modeled within the system. Seshia et al. [11] have presented a technique for soft error resilience that employs comprehensive fault injection. This method acts as an addition to existing functional verification methods. Sauer et al. [12], [13] have developed a SAT (Satisfiability)-based methodology for Automatic Test Pattern Generation (ATPG) that is designed to generate comprehensive testing for the identification of delay faults. In [14], [15], the authors integrate fault simulation, ATPG, and formal verification methods to improve the dependability of fault analysis procedures.

III. MOTIVATION, CHALLENGES AND REQUIREMENTS

The advancement of technology leads to more intricate designs, which inherently complicates the process of ensuring system reliability. Meeting high safety standards is one important aspect of these designs, particularly for safety-critical systems. The adherence to these standards adds another layer of challenge to the reliability assurance process. Wilson Reseach Group [16] estimates that in 2022, only 24% of the projects achieved success on their first silicon attempt. A direct consequence of the low first silicon success rate is the need for additional tapeouts. The majority of projects, i.e., 76% [16], required more than one tapeout, leading to increased production costs due to the additional need for wafers and masks. A small but significant portion of design respins is due to safety-related issues. These flaws account for between 5-11% of total respins. Therefore, the development and implementation of thorough safety verification techniques are vital. Such techniques help in the early detection and correction of safety-related errors, preventing them from progressing to later stages of development. As recommended by ISO 26262 standard, fault injection is used to test safety-critical designs. However, this process is intricate and consumes considerable time, which can delay the overall development process. The combination of design complexity, stringent safety standards, and the rigorous nature of safety verification processes forms several challenges as following:

1) **Informal specifications**: Safety requirements and specifications are often documented in natural language or through informal means, which can give rise to misunderstandings, lack of clarity, and discrepancies among various engineering teams involved in the product development lifecycle, including concept engineers, design engineers, and verification engineers.

2) **Human errors**: System defects or oversights can have significant implications for safety. Since humans predominantly conduct the verification procedures, there is an intrinsic possibility for human mistakes. Such errors might lead to critical issues going undetected during the verification stage.

3) **Changing design requirements**: Safety-critical designs experience changes in their requirements, necessitating updates to the design itself. This, in turn, may require alterations to the current verification processes or the incorporation of new steps for verification and analysis.

4) **Compliance to safety standards**: Meeting the diverse and evolving requirements of safety regulations poses a significant challenge. Safety-critical designs, which include numerous elements, necessitate a thorough verification process to ensure they meet the established standards. Fulfilling all the specified criteria within these frameworks can be a demanding task.

5) **Verification requirements and efforts**: Ensuring the integrity of safety-critical designs demands thoroughness, which may be constrained by available resources and expertise levels. As a result, the approach to safety verification is often iterative, involving a variety of methods such as analytical models, simulations, testing, and validation. This multi-faceted process presents a considerable challenge in terms of the effort required.

6) **Limited accuracy**: The process of safety verification spans multiple layers of design abstraction. When evaluating safety-critical designs, one must scrutinize it from the broad scope of system requirements down to the more granular details like transistor or gate-level characteristics. Each tier of abstraction introduces its unique set of suppositions, simplications, and approximations, potentially impacting the precision of the verification process.

7) **Time and cost constraints**: A variety of safety verification techniques require a lengthy amount of time to

execute, because of the abstraction level of the design at which they operate, which can lead missing project deadlines. In addition, the need for specific licenses and specialized tools to perform these verification processes can pose a substantial challenge to adhering to the set budget.

The primary goal of this paper is to tackle the obstacles faced in the area of safety verification. To do so, we propose a model-driven framework that automatically performs safety verification based on different techniques. In this section, we have compiled a list of necessary requirements for a proper safety verification as following:

1) **Formalization**: Specifications and requirements for safety verification should be explicit and unambiguous. It is essential that these specifications are represented through formal models.

2) **Compliance with safety standards**: The techniques used for safety verification must comply with applicable safety standards, like ISO 26262. The standard mandates fault injection as a necessary procedure to evaluate the safety-critical levels, and it also requires the use of formal verification methods to verify safety-critical components.

3) **Automation**: Maximizing automation within the safety verification process is crucial to reduce the necessity for human input. By automating the workflow, the dependability, productivity, and overall impact of the safety verification process is enhanced, which, in turn, reduces the reliance on manual efforts.

4) **Reusability and adaptability**: The framework for safety verification should be constructed with reusability and flexibility in mind to accommodate evolving design specifications. A structured, automated verification flow is capable of efficiently managing modifications to safety-critical designs. It should include carrying out an assessment of the potential effects of changes and executing re-verification when necessary to maintain the integrity and dependability of the verification process.

5) **Extensibility and flexibility**: The design of the safety verification framework should prioritize extensibility and flexibility to seamlessly adjust to new requirements and design modifications. Such a design will facilitate the smooth integration of the verification process with current workflows and tools. Additionally, it will support the incorporation of diverse verification methods such as simulation, emulation, and formal verification, enhancing the trustworthiness and efficiency of the verification activities.

6) **Correctness**: The design of the automated safety verification framework should be correct by construction, crafted in such a manner that it prevents bugs or defects that might compromise the integrity of the verification process.

7) **Data collection and analysis**: The safety verification framework must provide straightforward methods for collecting and analyzing safety-relevant data throughout the verification phase and operational lifetime. Such protocols are critical to identify and categorize the impact of different faults on the design's functionality, a key step to guarantee an all-encompassing and efficient safety verification process.

8) **Structured documentation**: Documentation of safety verification specifications and results should be documented with clarity and precision, following an organized format to comply with safety standards. It is imperative to have structured documentation to establish traceability and transparency between the safety requirements and the verification tasks that correspond to them.

9) **Resource scalability**: The resources allocated for the safety verification process should be capable of scaling to respond to changing requirements, ensuring that essential hardware and software resources remain accessible. Additionally, the safety verification strategy must be tailored to align with the financial limitations set by the project's budget.

10) **Accuracy and reliability**: The safety verification method must deliver accuracy, reliability, and efficacy across different layers of system abstraction. Outcomes from the verification should be exact and trustworthy, enabling early detection and mitigation of risks within the development cycle, which contributes to cost savings.

11) **Performance**: The execution of safety verification should be fast. Performance is key to an efficient and successful safety verification process. Additionally, it is vital to carry out safety verification at the analysis stage. A verification workflow characterized by high performance meets safety challenges, supports design expandability, and adheres to time constraints, all of which contribute to improved safety results.

IV. OVERVIEW OF THE FRAMEWORK

In this paper, we provide a model-driven methodology, which facilitates the automated safety verification and analysis focusing on RISC-V through a spectrum of techniques such as simulation, emulation, and formal verification. It's imperative to mention that each method of safety verification put forth in this paper is in accordance with the fault injection criteria established in ISO 26262. Fig. 1 displays a general overview of our framework.

The framework introduces an innovative technique that employs a mixed granularity representation for the Design-under-Test (DUT) to optimize the fault injection process. Within this framework, chosen design modules that undergo fault injection are depicted at the fine-grained gate level, while the remaining parts of the design maintain their higher-level Register-Transfer Level (RTL) form, which is deemed appropriate and precise for analyzing fault propagation, similar to [17]. Traditionally, creating RTL models, which incorporate elements of gate-level granularity, is known to be a time-consuming and error-prone task. To address these difficulties, the framework known as MetaRTL [18], [19] has been incorporated into the fault injection flow. MetaRTL consists of three different layers. The first layer defines and encodes the specifications through a metamodel, while the subsequent layer, which is an instance of the metamodel, describes the design abstractly, independent of any particular target language. This formalized approach within the model layer satisfies *Requirement 1*.

As the next step, synthesis is performed on the generated RTL and this synthesized RTL is converted into a gate-level netlist that represents the design. This netlist is then transformed back into design models. The inputs and outputs for each gate within the netlist remain visible in the resulting RTL model. Utilizing the capabilities of the MetaRTL framework, RTL models with gate-level granularity are created. Fault injection at this mixed granularity within the RTL models provides sufficient accuracy concerning the types of faults being modeled. At the same time, this technique increases the performance of the fault simulation process, thus fulfilling the criteria laid out in *Requirements 10,11*. The process for creating RTL models that combine varying levels of detail is a fully automated process, offering flexibility to accom-

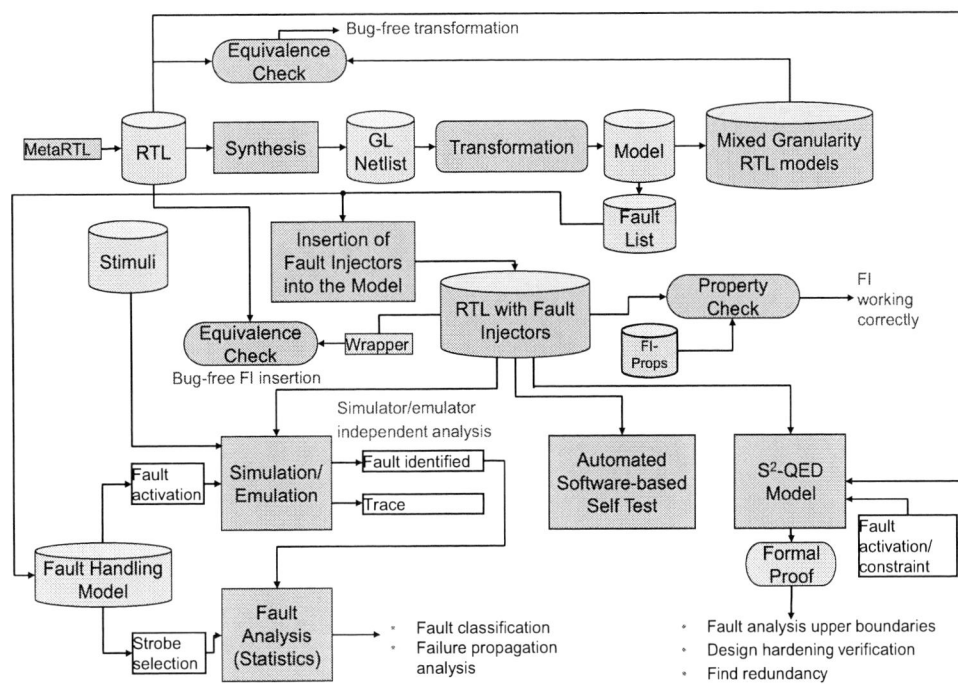

Fig. 1. Overview of the framework

modate various design configurations, which is in line with *Requirements 3, 4*. Once the RTL models of mixed granularity are created, a formal equivalence check between the original RTL model and its mixed granularity counterpart is conducted. This verification step is crucial to ensure the integrity of the transformation process, thereby guaranteeing the correctness as demanded by *Requirement 6*.

Next, the design model, which features a combination of different granularity levels, undergoes an automated transformation to integrate fault injectors. These injectors are capable of injecting various fault models into the design. The RTL generation framework is utilized again to produce mixed RTL models that now contain these fault injectors. Subsequently, a second round of equivalence checking is carried out, under the specific condition that the fault injectors are inactive, ensured by a wrapper. This step is important to validate that the fault injectors themselves have not introduced any errors during the model transformation process, thus addressing and fulfilling *Requirement 6* once more.

The RTL models, which now incorporate fault injectors, are designed to be compatible with different open-source and commercial simulation tools. This compatibility ensures that the framework meet the demand for resource scalability as outlined in *Requirement 9*. Additionally, the framework incorporates an innovative fault emulation framework that not only improves the efficiency of the fault injection process but also offers a system that is both scalable and adaptable. Such a system provides designers with the means to evaluate various aspects of system reliability early in the design phase, employing techniques rooted in both simulation and emulation. Consequently, this approach successfully fulfills the criteria specified in *Requirements 5, 11*.

The creation of the **Fault Handling Model** facilitates the automatic generation of a variety of testbenches. Each of these testbenches is custom-designed to execute different fault injection campaigns, which is in accordance with *Requirement 3*. Within this framework, specific faults can be selectively

activated based on the particular fault injection campaign being run. Additionally, users have the option to choose which design strobes (or signals) they wish to monitor closely during the analysis. These testbenches become a part of the simulation/emulation framework, following which a thorough analysis is conducted to assess the impact of the faults. The fault analysis phase also produces detailed documentation, which comprehensively captures the effects of the faults, thus fulfilling *Requirements 7, 8*.

Additionally, a formal-based methodology based in the S^2-QED model [20], is utilized to verify the hardening of safety-critical designs. The injected faults are constrained according the specifications of the safety mechanisms. This formalized strategy establishes upper-boundaries for fault analysis and identifies potential redundancies in the design, thereby revisiting *Requirement 5*. In addition, an automated Software-based Self Test (SBST) has been developed. This SBST utilizes formal techniques to derive deterministic testing patterns, which result in high fault coverage. This approach is further enhanced through integration with Program Flow Checking (PFC), ensuring that the fault detection rate corresponds with the various Automotive Safety Integrity Levels (ASILs) specified by the ISO 26262 standard. The entire procedure is automated and addresses a suite of requirements, specifically *Requirements 1, 2, 3, 4, 5, 10*.

The combination of fault injection methodologies with a range of safety verification strategies aligns with ISO 26262 guidelines, effectively satisfying *Requirement 2*.

V. FAULT ANALYSIS METHODOLOGY

Section IV gives a general overview of our fault analysis methodology which can be divided into two main parts. The first part consists of generating mixed granularity RTL models which enable fault injection directly on the design. The second part consists of utilizing these mixed granularity models and enabling automated fault simulation/emulation and formal verification of hardened designs. The integral component of the fault simulation and emulation is the fault handling metamodel as shown in Fig. 2. Fault injection campaigns require various features and parameters that are described by this metamodel including *Fault Controller*, *Fault List* , and *Fault Analyzer*, similar to [21].

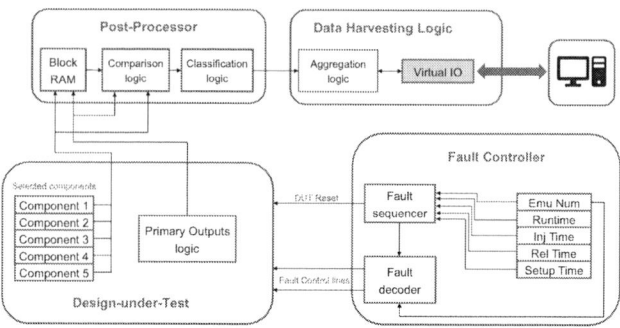

Fig. 3. Fault emulator architecture [22]

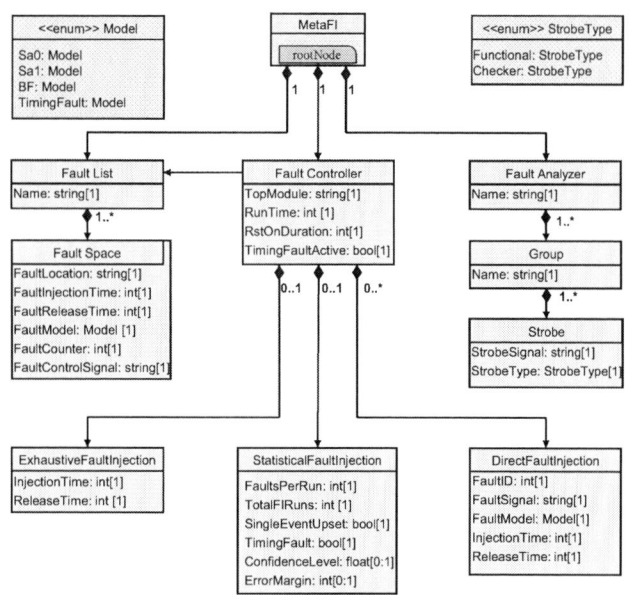

Fig. 2. Fault handling metamodel

The *Fault Controller* contains a set of classes and class attributes that define the type of the fault injection campaign, i.e., exhaustive fault injection, statistical fault injection, or direct fault injection. The *Fault List* class describes all the features of the injected faults, i.e., the fault space. The *Fault Analyzer* is responsible for collecting and analyzing the data obtained from fault simulation/emulation, i.e., classifying the faults according to their effect.

After the user fills these metamodel attributes accordingly, the framework generates automatically fault simulation and emulation environment. The automation is done via python and Tool Command Language (TCL) scripts. The fault simulation environment includes a generated SystemVerilog testbench that automatically injects faults on the mixed RTL models and any simulator can be utilized to perform the fault simulation. For fault emulation, extra hardware components are generated to perform fault injection and analysis as shown in Figure 3. The Fault Controller drives fault control lines, determines injection and the release of individual faults into the DUT with precise time points. The Post-Processor captures and stores the emulation traces, including primary outputs and selected internal registers. Furthermore, it classifies the faults into different classes such as failure, latent or silent faults. The Data Harvesting Logic receives fault classification values and transmits this data to the Host PC.

A general flow of the fault simulation/emulation is shown in Fig. 4. The flow starts with the instantiation of the design, i.e., DUT. The DUT for simulation and emulation environment is also automatically generated and it consists of a RISC-V processor. Next, a golden simulation/emulation is performed without injecting any faults and the results are stored in a log file. Afterwards, all faults are injected sequentially according to the data that the user specifies complying with the metamodel attributes. As the next step, simulation or emulation

is performed and the selected strobes (signals) are compared with the golden run strobes and injected faults are classified.

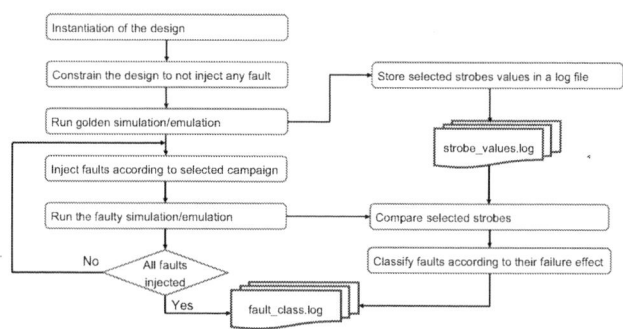

Fig. 4. Fault simulation/emulation flow

The formal-based approach incorporates the application of mixed granularity models and builds upon the S^2-QED methodology to verify the hardening of safety-critical designs [20]. This technique fundamentally relies on a comparative analysis of a pair of identical RISC-V processor cores. Within this framework, these cores are synchronized to fetch and execute an identical set of instructions. Despite both processors possessing the capability to inject faults, the methodology delineates one processor as the 'golden' reference unit, i.e., fault free. Concurrently, the second core is subjected to fault injection. The formal-based property then checks that both processors behave identically even after the faults are injected. In principle, as long as safety mechanisms work properly, the faults should be corrected thus the faulty design should behave as the non-faulty one.

VI. RESULTS AND OBSERVATIONS

In this section, we provide some experimental results and data regarding our framework. As DUT, we considered a CPU subsystem which contains a 5-stage pipelined RISC-V processor, as well as Instruction Memory, Data Memory, and peripherals such as UART, PIC, and different buses.

The fault emulation is performed utilizing an FPGA, therefore it is necessary to comply with area constraints. Table I displays the resource utilization of the emulation framework for various CPU components after inserting fault injectors into the RTL models. The data clearly illustrates a linear correlation between the fault control lines and the corresponding utilization of LUTs. Therefore resource utilization displays a linear pattern as the design size with fault injectors increases, and it does not experience exponential expansion.

979-8-3503-6689-1/24 $31.00 © 2024 IEEE   142

TABLE I
RESOURCE UTILIZATION OF DUT WITH FAULT INJECTORS

Selected component for fault injection	Fault control lines	Slice LUTs	Slice Registers	BRAM Primitives
HDU	63	9061	6791	14
Program Counter	109	9223	6790	14
WB stage	262	9615	6791	14
ALU result	613	10291	6796	14
Forwarding Unit	786	10519	6795	14
MEM stage	954	10708	6794	14
Instruction Decoder	1865	12485	6790	14
Event Counters	1919	12667	6795	14
Prefetcher	2125	13033	6790	14
ALU	2708	14672	6796	14
Exception Unit	5008	18766	6766	14
CSRs	6371	21791	6795	14
Register File	7679	24389	6793	14
IF stage	7952	24389	6793	14
ID stage	10948	29867	6792	14
EX stage	15826	43044	6793	14

We also performed fault simulation to analyze the fault propagation rates, i.e., the amount of fault that propagate to the primary outputs. To do that, we utilized statistical fault injection with 99.8%, 99% and 95% confidence rate. The faults were injected into three different components such as Arithmetic Logic Unit (ALU), Instruction Decoder and Branch Control Unit (BCU) and as benchmark we utilized Dhrystone. Table II displays the fault propagation rates for these components.

TABLE II
FAULT PROPAGATION RATE

Confidence	ALU	Decoder	BCU
95%	32.7%	45.8%	34.7%
99%	31.9	47.8%	30.8%
99.8%	32.9%	47.5%	29.9%

Further results can also be found on previous publication on which our complete simulation, emulation and formal-based fault analysis framework builds upon such as [17], [20]–[23].

VII. CONCLUSION

In this paper, we presented a holistic approach for fault analysis focusing on RISC-V processors. Our methodology utilizes various fault injection techniques such as simulation-based, emulation-based, and formal-based. The methodology is fully automated following model-driven principles. As a result, with very minimal manual efforts, various fault injection techniques can be utilized. Therefore, the user can easily select the fault injection technique considering factors such as cost, performance, resources etc..

VIII. ACKNOWLEDGEMENTS

Part of the work described herein is funded by the German Federal Ministry of Education and Research (BMBF) as part of the research project Scale4Edge (16ME0122K). Part of the work has also been performed in the project ISOLDE under grant agreement No 101112274.

REFERENCES

[1] J. Baraza, J. Gracia, D. Gil, and P. Gil, "Improvement of fault injection techniques based on vhdl code modification," in *Tenth IEEE International High-Level Design Validation and Test Workshop, 2005.*, pp. 19–26, 2005.

[2] J.-C. Baraza, J. Gracia, S. Blanc, D. Gil, and P.-J. Gil, "Enhancement of fault injection techniques based on the modification of vhdl code," *IEEE Transactions on Very Large Scale Integration (VLSI) Systems*, vol. 16, no. 6, pp. 693–706, 2008.

[3] D. Ferraretto and G. Pravadelli, "Efficient fault injection in qemu," in *2015 16th Latin-American Test Symposium (LATS)*, pp. 1–6, 2015.

[4] M. Kaliorakis, S. Tselonis, A. Chatzidimitriou, N. Foutris, and D. Gizopoulos, "Differential fault injection on microarchitectural simulators," in *2015 IEEE International Symposium on Workload Characterization*, pp. 172–182, 2015.

[5] P. Thaker, V. Agrawal, and M. Zaghloul, "A test evaluation technique for vlsi circuits using register-transfer level fault modeling," *IEEE Transactions on Computer-Aided Design of Integrated Circuits and Systems*, vol. 22, no. 8, pp. 1104–1113, 2003.

[6] J. Espinosa, C. Hernandez, J. Abella, D. de Andres, and J. C. Ruiz, "Analysis and rtl correlation of instruction set simulators for automotive microcontroller robustness verification," in *2015 52nd ACM/EDAC/IEEE Design Automation Conference (DAC)*, pp. 1–6, 2015.

[7] A. C. Bagbaba, M. Jenihhin, R. Ubar, and C. Sauer, "Representing gate-level set faults by multiple seu faults at rtl," in *2020 IEEE 26th International Symposium on On-Line Testing and Robust System Design (IOLTS)*, pp. 1–6, 2020.

[8] C. Lopez-Ongil, M. Garcia-Valderas, M. Portela-Garcia, and L. Entrena, "Autonomous fault emulation: A new fpga-based acceleration system for hardness evaluation," *IEEE Transactions on Nuclear Science*, vol. 54, no. 1, pp. 252–261, 2007.

[9] L. Entrena, M. Garcia-Valderas, R. Fernandez Cardenal, A. Lindoso, M. Portela-Garcia, and C. Ongil, "Soft error sensitivity evaluation of microprocessors by multilevel emulation-based fault injection," *IEEE Transactions on Computers*, vol. 61, pp. 313–322, 03 2012.

[10] J. Grinschgl, A. Krieg, C. Steger, R. Weiss, H. Bock, and J. Haid, "Automatic saboteur placement for emulation-based multi-bit fault injection," in *6th International Workshop on Reconfigurable Communication-Centric Systems-on-Chip (ReCoSoC)*, pp. 1–8, 2011.

[11] S. A. Seshia, W. Li, and S. Mitra, "Verification-guided soft error resilience," in *2007 Design, Automation & Test in Europe Conference & Exhibition*, pp. 1–6, 2007.

[12] M. Sauer, A. Czutro, T. Schubert, S. Hillebrecht, I. Polian, and B. Becker, "Sat-based analysis of sensitisable paths," in *14th IEEE International Symposium on Design and Diagnostics of Electronic Circuits and Systems*, pp. 93–98, 2011.

[13] M. Sauer, Y. M. Kim, J. Seomun, H.-O. Kim, K.-T. Do, J. Y. Choi, K. S. Kim, S. Mitra, and B. Becker, "Early-life-failure detection using sat-based atpg," in *2013 IEEE International Test Conference (ITC)*, pp. 1–10, 2013.

[14] F. Augusto da Silva, A. C. Bagbaba, S. Hamdioui, and C. Sauer, "Combining fault analysis technologies for iso26262 functional safety verification," in *2019 IEEE 28th Asian Test Symposium (ATS)*, pp. 129–1295, 2019.

[15] F. A. da Silva, A. Cagri Bagbaba, S. Hamdioui, and C. Sauer, "An automated formal-based approach for reducing undetected faults in iso 26262 hardware compliant designs," in *2021 IEEE International Test Conference (ITC)*, pp. 329–333, 2021.

[16] "Part 12: The 2022 wilson research group functional verification study." https://blogs.sw.siemens.com/verificationhorizons/2023/01/09/part-12-the-2020-wilson-research-group-functional-verification-study-2/. Accessed: 2023-09-13.

[17] E. Kaja, N. Gerlin, M. Vaddeboina, L. Rivas, S. Prebeck, Z. Han, K. Devarajegowda, and W. Ecker, "Towards fault simulation at mixed register-transfer/gate-level models," in *2021 IEEE International Symposium on Defect and Fault Tolerance in VLSI and Nanotechnology Systems (DFT)*, pp. 1–6, 2021.

[18] J. Schreiner, R. Findenigy, and W. Ecker, "Design centric modeling of digital hardware," in *2016 IEEE International High Level Design Validation and Test Workshop (HLDVT)*, pp. 46–52, 2016.

[19] W. Ecker and J. Schreiner, "Introducing model-of-things (mot) and model-of-design (mod) for simpler and more efficient hardware generators," in *2016 IFIP/IEEE International Conference on Very Large Scale Integration (VLSI-SoC)*, pp. 1–6, 2016.

[20] E. Kaja, N. Gerlin, B. Zhao, D. Sanchez Lopera, J. Al Halabi, K. Sher Azam, , S. Prebeck, D. Stoffel, W. Kunz, and W. Ecker, "An automated exhaustive fault analysis technique guided by processor formal verification methodss," in *25th International Symposium on Quality Electronic Design*, 2024.

[21] E. Kaja, N. Gerlin, M. Bora, K. Devarajegowda, D. Stoffel, W. Kunz, and W. Ecker, "Metafs: Model-driven fault simulation framework," in *2022 IEEE International Symposium on Defect and Fault Tolerance in VLSI and Nanotechnology Systems (DFT)*, pp. 1–4, 2022.

[22] E. Kaja, N. Gerlin, M. Bora, G. Rutsch, K. Devarajegowda, D. Stoffel, W. Kunz, and W. Ecker, "Fast and accurate model-driven fpga-based system-level fault emulation," in *2022 IFIP/IEEE 30th International Conference on Very Large Scale Integration (VLSI-SoC)*, pp. 1–6, 2022.

[23] E. Kaja, N. Gerlin, U. Yun, J. Al Halabi, S. Prebeck, D. Stoffel, W. Kunz, and W. Ecker, "A statistical and model-driven approach for comprehensive fault propagation analysis of risc-v variants," in *Proceedings of the Design and Verification Conference and Exhibition (DVCon)*, 2024.

Special Session: In-Field ML-Assisted Intermittent Fault Localization and Management in RISC-V SoCs

Hardi Selg, Konstantin Shibin, Anton Tsertov, Maksim Jenihhin, Peeter Ellervee, Jaan Raik

Department of Computer Systems, Tallinn University of Technology, Estonia
hardi.selg@taltech.ee

Abstract—This paper proposes a novel in-field ML-assisted fault localization and fault management approach for intermittent faults in RISC-V microprocessor cores operating in tandem in a dual-core lockstep setup. The approach is nonintrusive with respect to the RISC-V cores, as it relies on existing trace buffer data for fault localization. The localization is carried out by a lightweight pre-trained Neural Network (NN) optimized by a Neural Architecture Search (NAS) framework. It provides in-field localization of faults with a resolution of RISC-V processor modules. Furthermore, the paper proposes a Fault Management Architecture (FMA) to process the localized fault information and control the graceful degradation of the SoC, e.g. through fault-aware task scheduling. The proposed approach is validated on RISC-V Pulpissimo-based implementation with various workload programs. Experiments demonstrate a localization accuracy at 80-94% level.

Index Terms—fault localization, fault management, processor architecture, machine learning, neural architecture search.

I. INTRODUCTION

Dual-core lockstep (DCLS) [1], [2] is a redundancy technique for high-reliability computing used in safety-critical systems [3], [4], such as aerospace, automotive, and industrial control systems. In this configuration, two identical processor cores, a primary core and its identical copy, a shadow core, operate in parallel, executing the same set of instructions simultaneously. The key feature of DCLS is that both cores execute the same instructions and compare their results at every step to ensure their match. If a discrepancy is detected, it indicates that a fault has occurred in one of the cores, triggering a fault-handling mechanism. Further resilience can be added by delaying one core with respect to the other, adding temporal separation to prevent both cores from being affected by the same fault.

The DCLS approach provides a high degree of detection to faults. However, it will not guarantee fault resilience, in the sense that not allowing localizing and isolating them in the processor cores. At the same time, an increasing number of commercial processors are including on-chip neural accelerators [5]. Taking into account the above, in this paper, we investigate whether on-chip neural accelerators can be applied in making the dual lockstep based architectures resilient to

intermittent faults. The main contributions of the paper are as follows:

- A nonintrusive in-field ML-assisted fault localization approach for intermittent faults in RISC-V microprocessor cores operating in tandem in a dual-core lockstep setup;
- Neural Architecture Search (NAS) trained lightweight Multi-Layer Perceptron (MLP) optimized for fault localization by an on-chip neural accelerator with a resolution of RISC-V processor modules;
- An in-field Fault Management Architecture (FMA) that processes the fault information and can be applied for controlling the system by a graceful degradation approach, e.g. fault-aware task scheduling.

The proposed approach is validated on RISC-V Pulpissimo based implementation with different workload programs. Experiments demonstrated a localization accuracy close to 80-94% when predicted fault-location was within 3 to 5 top candidates with the highest probabilities.

The rest of the paper is structured as follows. Section II provides an overview of the related works. In Section III, the proposed methodology is introduced. Section IV explains in detail a practical case study based on a complex CPU design. Section V presents the experimental results and Section VI concludes the work.

II. RELATED WORKS

Several works which are addressing on-chip fault management and health/resource management do address health data collection but do not consider historical health data, i.e. only rely on immediately detected faults [6], [7]. Hierarchical agent framework as a monitoring layer for self-aware system is proposed in [8], but the authors do not detail the collection of health data required for consistent self-awareness.

There exist works on in-field machine-learning-based localization of faults and errors. [9] proposes an embedded neural accelerator for the localization of design bugs. In [10], in-field localization of hardware faults is addressed, and different neural networks for that purpose are evaluated. To the best of the authors' knowledge, this paper is the first work to address in-field intermittent fault localization and management for DCLS processors utilizing on-chip artificial intelligence (AI) accelerator cores.

979-8-3503-6689-1/24 $31.00 © 2024 IEEE

III. PROPOSED METHODOLOGY

It is assumed that the System-on-Chip has a working Lock-Step processor system, a Neural Accelerator (NA), and has the capability of signal trace buffers, as shown in Figure 1.

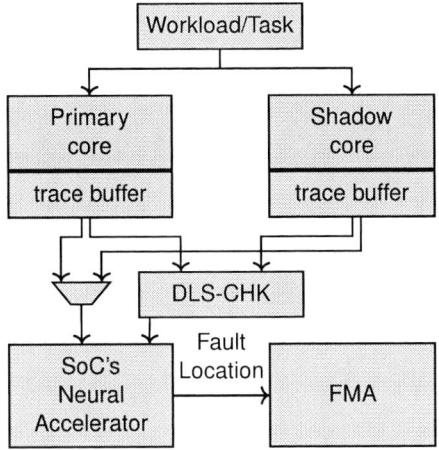

Fig. 1: Dual-lockstep architecture

This work aims to monitor non-intrusively a set of processing core signals via Lock-Step and keep a snapshot of them in the trace buffer. On fault detection, e.g., Lock-Step mismatch, which is detected by the Dual Lock-Step checker (DLS-CHK), the on-chip NA is reconfigured for the Fault Localization task, and stored traces in the trace buffer are used as the input for the NA. The System Health Monitor (SHM) collects NN localization statistics and compiles the result for the Fault Manager (FM). The latter, in return, gives feedback to the Scheduler (S) on which parts of the processor core are still usable, semi-trustworthy, or have been deemed unusable. The methodology is divided into three main parts: *Fault Trace Computation Platform (A)*, *Fault localization (B)*, and *Fault Management (C)*.

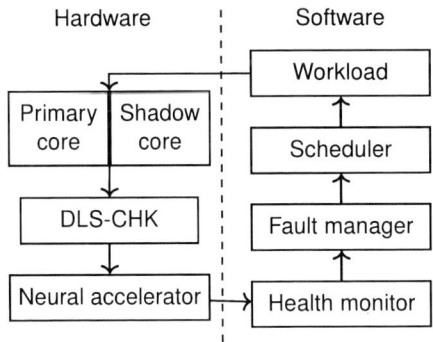

Fig. 2: Hardware-software architecture

A. Methodology overview

The methodology architecture consists of distinct hardware and software components used during the device's active operation, shown in Figure 2. The hardware portion handles fault detection and localization by utilizing the following hardware components: Lock-step processors, Lock-step checker, and Neural Accelerator (NA). It is performed during the device operation to localize the fault in the processor as accurately as possible because of the availability of in-situation information. Creating a full snapshot of a system during runtime and transmitting the information to a database would be too costly, both in energy and bandwidth. The software portion of the methodology statistically analyses the frequency of faults by utilizing the System Health Monitor (SHM) and Fault Manager to update the system's scheduler accordingly. The idea is to keep the device running as long as possible by *I* mapping out faulty regions of Lock-Step Processors, *II* keeping the device running in a degraded state by allowing only certain workloads to run.

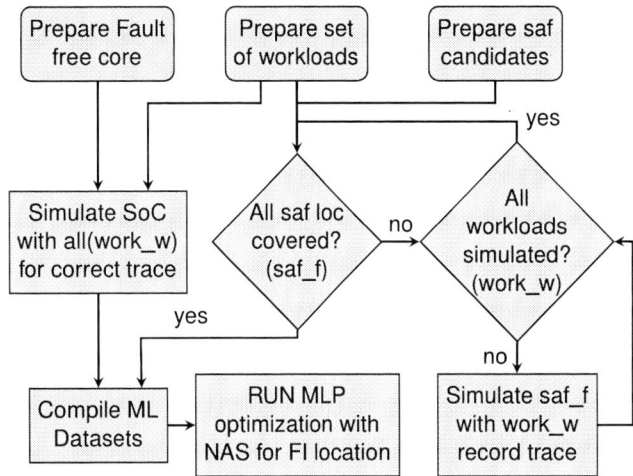

Fig. 3: Fault Trace Collection Procedure

B. Preparation and optimization of the Neural Network

To produce such a system, the target Lock-Step processors must be profiled by running various workloads to record signal activations, which is expected to be deterministic. The same profiling exercise must be repeated by a predefined fault injection campaign. Doing so makes it possible to distinguish wrong behavior from the correct one. This gives two pieces of important information: **the exact moment of fault detection** and **how the signal states differ from the correct one**. With the information about the exact moment of detection, it is possible to model the Lock-Step Checker activation without co-simulating both injected and injection-free processor cores, which consumes much more time. Secondly, the recorded differences between correct and wrong signal activations produce the required data for Neural Network (NN) training and optimization.

During the NN optimization, the following aspects are optimized: a set of signals to monitor, signal trace buffer sizes, MLP network architecture, and its size. MLP was chosen because it is the most lightweight NN type for classification. To automate this optimization step, the JÄNES framework is used [11]. The resulting most accurate NN is then possible to be implemented in the SoC for fault localization.

C. Fault localization

As mentioned, the main goal is to perform fault localization on the device, re-using the existing SoC's Neural Accelerator.

The SoC is assumed to have Lock-Step cores, an appropriate Checker, Trace buffers (TB) for signal monitoring, and a Fault Manager Architecture (FMA) for on-board fault analysis. Therefore, only a marginal hardware overhead is introduced to the SoC, including a localization procedure controller and some additional memory for Localizer MLP weights. The amount of extra memory used is highly dependent on the processor core and the number of signals that have been monitored; MLP's overall size is highly dependent on these features and is subject to optimization.

On the SoC, the localization procedure begins with Fault detection by the Lock-Step Checker. The procedure consists of the following steps:

1) Stop updating Trace Buffer contents
2) Reconfigure onboard NA for fault localization
3) Flush TB contents to the trained NN as input data for classification
4) Update FMA based on prediction.
5) Reconfigure the NA for the previous application
6) Invoke fault management
7) Update information for the task scheduler
8) Continue operation

D. Fault Management

The fault management architecture (FMA) [12] provides valuable insight into the nature of the faults, as well as the remaining "health" of the system.

For complex systems with redundant resources, the FMA approach can provide the possibility for graceful degradation, i.e., prolonging the lifetime of the system despite the failure of some of its parts. The data collected for FMA is stored, analyzed and summarized, in addition it becomes very useful in higher-level processes like the mapping of tasks to available resources to ensure graceful degradation. In the case of a SoC with multiple DCLS core clusters FMA could allow to map the software tasks to the processing cores while taking into consideration the localized faults.

FMA approach centrepiece is a special structure called Health Map (HM) that holds detailed information about functional resources, health information in terms of occurring faults and operating parameters and maps the relationships between these.

This health information can then be used at the SoC level for task scheduling (new workload preparation) as well as at the higher levels of system hierarchy for distributed hierarchical health management.

IV. CASE STUDY - RISC-V PULPISSIMO CORE

For the case study, a RISC-V processor implementation PULPissimo [13] was chosen to run the experiments. For the experiments, the processor was divided into sub-instances, resulting in 26 modules in total, which corresponded one-to-one to the elaborated HDL entities within the processor project.

All experiments were executed on a Linux workstation with the latest Python libraries. The specification of the workstation is as follows: CPU AMD Ryzen™ 9 7950X running on 4.5GHz under load, 128Gb of system memory, NVIDIA GPU RTX 3090 24Gb that was isolated for the ML application only.

A. Experimental overview

For this case study, the processor core PULPissimo CV32P40E was simulated with a predefined set of workloads ($w \in W$) with and without fault injections. Stuck-at-0 ($i_{sa0} \in I_{sa0}$) and Stuck-at-1 ($i_{sa1} \in I_{sa1}$) faults where placed at emulated HDL output signals one at a time for all targeted ($f \in F$) fault injection locations. The set of Fault Locations F included all possible output signals from all possible emulated processor core modules, resulting in a total of **2462** possible injection locations. Because each location was injected with both SA0 and SA1 cases, the total number of different injections was **4924**. A fault was injected to the target location after the bootloader, i.e., right before the workload start, to emulate a real use case as closely as possible.

After completing the fault injection campaign, the collected processor activities profiles, e.g., signal traces, both for correct and incorrect behaviors, were compiled into different ML data sets. It ensured that every *Evaluation set*, set to compare prediction performance with all other NN, contained exactly the same trace instances. The inference results were collected together and analyzed to decide which signal configuration to implement and train MLP in the target SoC.

B. Preparation for experiment

As mentioned previously, in total, 4924 fault injections were performed for all fault locations (F) for *SA0* and *SA1* fault models. For each of the fault injections ($fi \in FI, where FI = \{F_{sa0}, F_{sa1}\}$) the target processor core was simulated with a set of **38** different workloads W, resulting in ~ 187000 unique traces with fault injections, although not all FI produced incorrect behavior, due to masking or not using target processor module during a workload. Therefore, usable, e.g., traces with faulty behaviors amounted to ~ 100000 unique cases, of which had a distribution of 45%-55% of *SA0* and *SA1* fault detections, respectively.

JÄNES run (JR)	Correct (C) & Faulty (F) trace	Added "correct" output	Correct trace in eval data	32bit split to 4x8bit	batch size
JR#1		No	No	No	2048
JR#2	Concatenation of C and F trace	No	No	Yes	256
JR#3		No	No	Yes	256
JR#4		No	No	Yes	256
JR#5		No	No	Yes	2048
JR#6	Only F trace	No	No	Yes	2048
JR#7		Yes	No	No	2048
JR#8	1x C trace for every F	Yes	Yes	No	2048
JR#9		Yes	No	Yes	2048
JR#10		Yes	Yes	Yes	2048
JR#11		Yes	No	No	2048
JR#12	2x C trace for every F	Yes	No	Yes	2048
JR#13		Yes	Yes	Yes	2048
JR#14	3xC trace for every F	Yes	No	No	2048
JR#15		Yes	No	Yes	2048

TABLE I: Input data and JÄNES framework configurations

979-8-3503-6689-1/24 $31.00 © 2024 IEEE

The usable traces were compiled into 15 different data sets and JÄNES framework [11] configurations, e.g. **JÄNES runs** (JR) as shown in Table I. These configurations were used as input for the Neural Architecture Search framework "JÄNES," which generated and trained 200 unique MLP networks for each configuration. The most considerable difference is whether and how correct behavior is introduced to the data set:

- Concatenation - Faulty and Correct traces for single fault injection are contacted and given to NN as a single item.
- Only Faulty trace is used for training and evaluation. However, this assumes that there is another NN for deciding whether the Fault originated from the Primary or Shadow core.
- For every Faulty trace item, there are either 1, 2, or 3 times of unique Correct trace items. With this setup, the aim is to combine the Faulty core decider with the Fault localization NN.

An additional "correct" output neuron was added for selected data sets to differentiate between correct and faulty behavior. This also meant that these data sets included "correct" traces. Furthermore, separate data sets were generated in which correct traces were either added or not to the Evaluation data set, which assumed the existence of "correct" output.

In select **JR** configurations, primarily concerning the data and address buses, all 32-bit signals were split into four 8-bit signals. This was done to increase the impact of 32-bit signals compared to 1-bit signals. However, doing so considerably increased the input neuron count, directly affecting the training batch size. In most cases, the JÄNES framework was run with a batch size of **2048**; however, for configurations JR#2-4, in combination with concatenation and splitting 32-bit signals, the training batch size had to be lowered to **256** to fit all the active data into memory.

C. Fault Localization

For current experiments, 26 fault locations were chosen, corresponding to distinct processor core instances where multiple instantiations of the same module can exist. As mentioned, fault injection targets were all the output signal bits for these 26 instances. The trained and optimized NN is tasked to predict the faults' location online based on the contents of the trace buffers. Because the monitoring is done non-invasively, it was assumed that there would be some monitoring issues; therefore, during experiments, the five highest prediction confidences were recorded for statistical analysis. For example, suppose the NN constantly places the faulty module within the top 3 most probable locations. In that case, it can be assumed that this module is unsafe to be used in mission-critical applications.

D. Fault Management

1) Health Map: The collected fault data cannot be immediately used to perform health management. In most cases, the information about the occurred fault needs to be analysed in the context of the previously reported fault events. For example, single fault detection data is often not sufficient to classify the fault, and analysis of previous occurrences of the fault is essential.

For this purpose, an FMA maintains and utilises a fault occurrence database, called Health Map (HM), which holds information about functional resources, occurring faults and the relationship between those two. HM preserves the statistics of fault occurrences, which can be used for fault classification and better prediction of resource reliability. This analysis and classification is performed by the software part of the FMA.

Health Map describes the functional and diagnostic resources in a specific system, and the initial map must be prepared in advance based on the system's specification. This is done offline before deploying FMA to the target system. This description is compiled into a binary form for efficient handling by FM and storage in memory. The serialized HM binary data structure is loaded into the non-volatile memory of the target system.

2) Resource Map: Resource Map (RM) is a data structure prepared by FM at run-time that holds the information about the current health status of hardware resources. All modules defined in HM could be added to RM together with the information about the worst severity and persistence of faults detected and attributed to corresponding modules. This represents a quick summary of module's current health status and is used in calculating the mapping of software tasks to hardware resources.

In contrast to HM, Resource Map is available only during run-time. Each entry in RM corresponds to a module in HM and it is populated with data by scanning HM and searching for worst faults registered for a given module. During calculation, criticality is also taken into account to calculate the propagation of faults to higher levels of module hierarchy (from child to parent). After the initialization procedure RM may be updated whenever a new fault is detected.

3) Health-aware task scheduling : CPU cores or processing cores are part of the system's hardware resources of a special type because, from the operating system (OS) perspective, they run the software tasks. In multicore systems, several programs can be run in parallel, and each of the latter is assigned to a certain processing core ID. In the scope of FMA, the faults are detected and attributed to certain modules in HM. If the accumulated fault(s) rendered the affected processing core unusable for some or all tasks, it is important that the OS scheduler does not schedule tasks to these cores anymore. It is essential that FM could translate a particular processing core module in HM to core ID understood by OS and instruct the OS scheduler not to use certain cores for certain tasks. This is achieved by comparing the status of processing cores in RM to the tasks' requirements. The requirements can include sub-modules like FPU. Based on the comparison result, FM fills a core affinity mask for each task. Some OS schedulers (e.g. Linux) accept this information without modification of the scheduler code.

V. EXPERIMENTAL RESULTS

This section presents the experimental results of JÄNES. The aim is to identify the most performant set of signals to monitor, the sizes of trace buffers, and to train and optimize an NN. First, the results of different JÄNES runs and the corresponding configurations are discussed, giving the desired set of signals and buffer sizes. In the next section, the JÄNES run, with the most accurately trained network, is discussed in more detail.

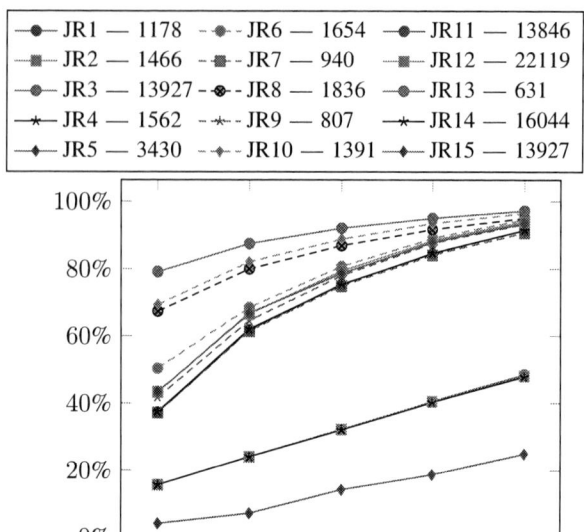

Fig. 4: Prediction accuracy comparison of most accurate MLPs from the 15 JÄNES runs

A. Match accuracy comparison of different data sets

For each JÄNES configuration, 200 NNs were trained by the JÄNES framework [11]. Figure 4 shows the inference results of the most accurate NNs for each of the 15 run configurations, where **five** highest probabilities and corresponding fault-location candidates were recorded. The data field **top***n* shows whether the actual fault location was within *n*-highest predicted locations, which gives us extra statistical information for FMA to ingest.

The reason behind the statistical data is that if the correct fault location is constantly predicted within the top 5 with the most confident results and rarely as top 1, it would still be possible to deter by the confidence frequency that a processor module might not be reliable anymore. From the results shown in Figure 4, it could be said that **JR#13** has the highest match accuracy over all other configurations; however, considering that JR#13 also includes correct trace behavior also in the evaluation set, same goes for JR#8 and JR#10, it was analyzed that the NNs, which had extra "correct-trace" output neuron and correct behavioral data also included in the inference, were highly biased to predict "correct-trace" condition which in return raised the classical match accuracy artificially. Therefore, **JR#6** is chosen, with the accuracy of **51.61%**, for a more thorough analysis because it does not include correct behavior in training or evaluation data, thus avoiding "correct-trace" bias. Additionally, this choice defines that **all monitored signals** must be observed, and **trace buffers** must be able to contain **21** clock cycles of data for each of the monitored

signals. These 21 clock cycles are partitioned as 19 cycles for trailing, one cycle for failure detection, and one extra cycle for leading trace.

NN#	top 1	top 2	top 3	top 4	top 5	Total Neurons
148	51.61%	69.71%	81.03%	89.58%	94.77%	1894
179	50.40%	69.41%	81.25%	89.45%	94.29%	2150
193	50.27%	68.42%	80.67%	89.05%	94.56%	1654
137	50.19%	68.44%	80.45%	89.22%	94.75%	2022
147	50.17%	68.94%	81.03%	89.90%	94.60%	1894
187	49.73%	68.09%	80.23%	88.58%	94.27%	2150
142	49.57%	67.89%	80.68%	88.86%	94.61%	1894

TABLE II: Seven most accurate NNs from JÄNES run JR#6

In Table II, experimental results of the seven most accurate NN of JÄNES run JR#6 can be seen, including the predicted fault-location was within 1 to 5 highest, e.g. top 1 to 5 probabilities. In addition, total neuron counts are shown, which indicate the NN architectures are relatively small, ranging from 1654 to 2150 total neurons, including the input and output neurons. The most accurate NN, named **NN#148**, reached the match accuracy of **51.61%**, with total neurons of 1894. Statistical accuracies for top 2 to top 5 were 69.71%, 81.03%, 89.58%, and 94.77%, respectively, and were almost the same for all other NN shown in the table. For practical fault management and analysis of intermittent faults, top 3 to top 5 fault location candidates are stored in the Health Map.

B. Analysis of JR#6 prediction confidences

In this section, the seven most accurate NNs out of 200, from JR#6, are compared based on the individual output neurons, e.g., fault locations prediction accuracies. Figure 5 shows the accuracies and corresponding fault injection distributions. Prediction match accuracies shown in the figure are normalized based on their expected occurrence frequency and sorted from the most inaccurate to the most accurate location neuron based on experimental results of **JR#6-NN#148**. Two potential fault locations have no data associated with them because they are not present in the evaluation set. The reasons here are two-fold: insufficient simulation data and, in some cases, fault injection effects were masked by other modules.

Figure 5 shows that injections from *instruction_obi_i* to *cs_registers_i* instances produce unique enough trace behaviors so that NN can relatively efficiently learn these patterns, even if the fault injection distribution might be low. Additionally, this figure shows that locating faults for select modules is challenging. However, this information can be used to target these select modules as candidates for custom fault monitors.

Analyzing the JR#6 seven most accurate NNs highlighted a pattern that might be a reason for less precise prediction results. The two modules, *core_i* and *FC_CORE_i*, had considerably lower prediction confidences in some cases. The hypothesis is that because one module is essentially a wrapper for the other module, the NN had difficulty distinguishing trace differences from each other. Therefore, in future work, it should be considered that it might be meaningful to group wrappers and corresponding modules together.

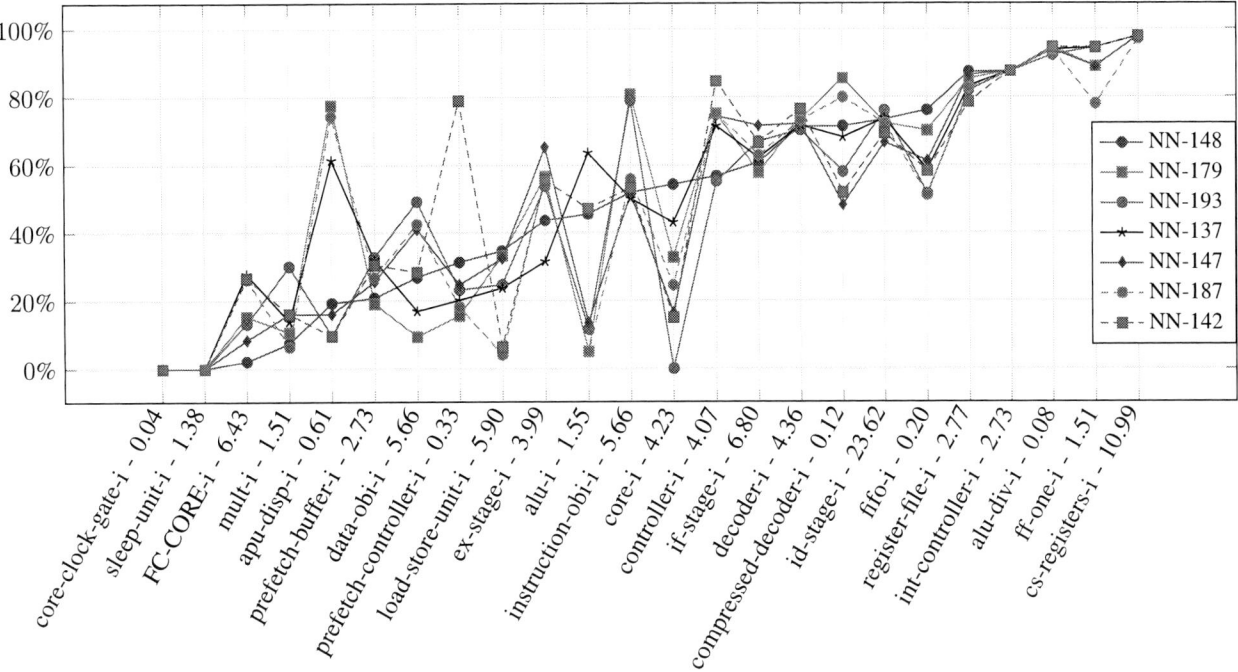

Fig. 5: 7 best performing NN comparison of JÄNES configuration JR#6

VI. CONCLUSION

In this paper, we propose a methodology for the immediate localization of intermittent faults detected in microprocessors in the field during operation. Localization is achieved by utilizing an on-SoC Neural Accelerator, which reports its results to a Fault Manager and updates the System Health Map. This methodology provides the designer with methods to select an appropriate set of signals to monitor noninvasively, fine-tune trace buffer sizes for optimal performance, and train and optimize a fault-localizing MLP network for predictions. However, it is assumed that the targeted SoC already has a Dual-Lockstep processor core, Dual-Lockstep checker, and Neural Accelerator available. The proposed approach was validated with a fault-injection campaign on a widely used RISC-V implementation, PULPissimo. The experiments demonstrate that an MLP-type network can predict fault locations by non-invasively monitoring processor signals. As a result, the best-performing MLP network could predict correct fault-injection locations with a match accuracy of 51.61%. Because this localization system is directly tied to the Fault Management Architecture, the statistical frequency of predictions should be considered. The experiments demonstrate that by considering **3** to **5** highest prediction confidences for statistical analysis brings the match accuracy to **81.03 - 94.77%**.

ACKNOWLEDGMENTS

This work was supported in part by the Estonian Research Council grant PUT PRG1467 "CRASHLESS", EU through the project TAICHIP #101160182 and by Estonian-French PARROT project "EnTrustED".

REFERENCES

[1] Z. Kalbarczyk, S. Bagchi, K. Whisnant, and R. K. Iyer, "Lockstep processor design for multithreaded error detection and recovery," *IEEE Transactions on Computers*, vol. 48, no. 6, pp. 603–618, 1999.

[2] Q. G. Qiu, H. Qi, W. Wang, and X. C. Lu, "Safety mechanisms of dual-core lock-step in infineon microcontroller," in *Proceedings of the 2006 IEEE International Symposium on Circuits and Systems*. IEEE, 2006, pp. 4–7.

[3] F. A. da Silva, A. Cagri Bagbaba, A. Ruospo, R. Mariani, G. Kanawati, E. Sanchez, M. S. Reorda, M. Jenihhin, S. Hamdioui, and C. Sauer, "Special session: Autosoc - a suite of open-source automotive soc benchmarks," in *2020 IEEE 38th VLSI Test Symposium (VTS)*, 2020, pp. 1–9.

[4] M. Jenihhin, M. S. Reorda, A. Balakrishnan, and D. Alexandrescu, "Challenges of reliability assessment and enhancement in autonomous systems," in *2019 IEEE International Symposium on Defect and Fault Tolerance in VLSI and Nanotechnology Systems (DFT)*, 2019, pp. 1–6.

[5] R. van Loon, "Scalable Processors with Built in AI Accelerators," April 18, 2023.

[6] T. D. T. Braak, S. T. Burgess, H. Hurskainen, H. G. Kerkhoff, B. Vermeulen, and X. Zhang, "On-line dependability enhancement of multiprocessor socs by resource management," *2010 International Symposium on System-on-Chip Proceedings, SoC 2010*, pp. 103–110, 2010.

[7] A. M. Ibrahim and H. G. Kerkhoff, "An on-chip ieee 1687 network controller for reliability and functional safety management of system-on-chips," *Proceedings - 2019 IEEE International Test Conference in Asia, ITC-Asia 2019*, pp. 109–114, 9 2019.

[8] L. Guang, E. Nigussie, P. Rantala, J. Isoaho, and H. Tenhunen, "Hierarchical agent monitoring design approach towards self-aware parallel systems-on-chip," *ACM Trans. Embedd. Comput. Syst*, vol. 9, 2010.

[9] H. Selg, M. Jenihhin, P. Ellervee, and J. Raik, "Ml-based online design error localization for risc-v implementations," in *2023 IEEE 29th International Symposium on On-Line Testing and Robust System Design (IOLTS)*, 2023, pp. 1–7.

[10] S. Dutto, A. Savino, and S. Di Carlo, "Exploring deep learning for in-field fault detection in microprocessors," in *2021 Design, Automation Test in Europe Conference Exhibition (DATE)*, 2021, pp. 1456–1459.

[11] H. Selg, M. Jenihhin, and P. Ellervee, "JÄnes: A nas framework for ml-based eda applications," in *2021 IEEE International Symposium on Defect and Fault Tolerance in VLSI and Nanotechnology Systems (DFT)*, 2021, pp. 1–6.

[12] K. Shibin, M. Jenihhin, A. Jutman, S. Devadze, and A. Tsertov, "On-chip sensors data collection and analysis for soc health management," in *2023 IEEE International Symposium on Defect and Fault Tolerance in VLSI and Nanotechnology Systems (DFT)*, 2023, pp. 1–6.

[13] P. D. Schiavone, D. Rossi, A. Pullini, A. Di Mauro, F. Conti, and L. Benini, "Quentin: an ultra-low-power PULPissimo SoC in 22nm FDX," *2018 IEEE SOI-3D-Subthreshold Microelectron. Technol. Unified Conf. S3S 2018*, jul 2018.

Exploring Total Ionizing Dose Radiation Effects Across Generations of NVIDIA Jetson Devices: A Comparative Analysis

Ivan Rodriguez-Ferrandez[*†‡], Maris Tali[‡], Leonidas Kosmidis[†*], Alessandra Costantino[‡], David Steenari[‡]

[*]Universitat Politècnica de Catalunya (UPC), Barcelona, Spain
[†]Barcelona Supercomputing Center (BSC), Barcelona, Spain
[‡]European Space Agency (ESA), Noordwijk, The Netherlands

Abstract—With the rise of autonomous systems, high performance commercial embedded systems have emerged as powerful and energy-efficient solutions, holding significant promise for small satellites and spacecraft, as well as for payload processing. However, the typical commercial devices are not inherently engineered for space operations, necessitating thorough evaluation. NVIDIA Jetson embedded multicore and GPU devices are among these promising systems. In this study, we assess the impact of radiation ageing, specifically through a Total Ionizing Dose (TID) campaign, on three models of the NVIDIA Jetson family: Jetson Orin NX, Jetson Orin Nano and Jetson Xavier NX. Through comprehensive evaluation, we aim to uncover insights into the resilience and suitability of these devices for space applications.

I. INTRODUCTION AND BACKGROUND

The space industry is witnessing a remarkable surge in demand for payload processing performance, driven by the anticipation of future missions. These upcoming endeavors necessitate what are termed *next-generation data handling systems*, projected to handle significantly larger data volumes compared to current missions. This surge is predominantly fueled by the adoption of advanced on-board processing methods, aimed at achieving higher resolutions in earth observation, increased precision and sampling frequencies, thereby further amplifying the volume of the acquired data. Given the inability of transmitting such vast amounts of data to the ground, onboard processing or compression before transmission emerge as the only viable solutions. Additionally, with the advent of Artificial Intelligence (AI), there is a growing trend towards extracting actionable information on-board with minimal latency [1][2][3].

The current rad-hard processor technologies deployed in space, such as the radiation-hardened LEON [4] or PowerPC [5] families of processors, fall short of meeting these evolving performance requirements. Consequently, Commercial Off-The-Shelf (COTS) devices, despite not being specifically designed for space applications, are increasingly recognized as enabling solutions [6][7][8][9].

Embedded Graphics Processing Units (GPUs) have emerged as promising candidates for high-performance processing in temperature- and battery-constrained devices, such as spacecraft [10]. Recently, the European Space Agency (ESA) initiated exploration into the potential utilization of Commercial Off-The-Shelf (COTS) devices in Class IV missions, targeting lower-cost endeavors like micro-satellites or constellations. To facilitate the integration of innovation into this mission class, the ESA COTS EEE Working Group is actively pursuing the standardization of COTS device usage. This initiative entails defining categories, establishing testing guidelines, and outlining interactions with space-qualified components, all aimed at ensuring that COTS systems do not compromise the integrity of other subsystems [9]. The outcome of these efforts is reflected in the updated version of the ECSS-Q-ST-60-13C standard [11]. Such devices have already been proposed for future systems [10][12][13], with some system demonstrations already underway [14][15].

In this paper we focus in the evaluation, comparison and analysis of the effects of Total Dose Ionization (TID) on multiple generation and models of the NVIDIA Jetson family of multicore and GPU devices.

II. EXPERIMENTAL SETUP

For this experiment we used three different devices from the NVIDIA Jetson family of devices. Table I shows the device selection and characteristics of such embedded GPUs.

	Jetson Xavier NX	Jetson Orin NX	Jetson Orin Nano
Technology	12nm	7nm	7nm
Cores	6 - core NVIDIA Carmel	8 - core Arm A78AE	6 - core Arm A78AE
Max Frequency	1.9 GHz	2 GHz	1.5 GHz
GPU SMs	6 SM	8 SM	8 SM
GPU Max Frequency	1.1 GHz	0.918 GHz	0.625 GHz
Main Memory	16 GB LPDDR4x	16 GB LPDDR5	8 GB LPDDR5
Main Memory Max Frequency	1.6 GHz	3.2 GHz	2.13 GHz
Internal Storage	16 GB eMMC	N/A	N/A
Max Power	20W	25W	15W

TABLE I: Comparison of the Devices Under Test (DUT)

For our testing, we utilized the European Space Agency's ESTEC Co-60 irradiation facility [16].

As previously mentioned, our experiment entails testing three distinct systems, with the aim of having two samples

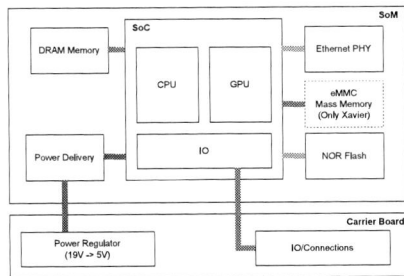

Fig. 1: NVIDIA Jetson System Architecture.

of each under different bias conditions. This results to a total of 12 devices placed within the irradiation area. The complete list of devices, along with their corresponding IDs and bias conditions, is provided in Table II. The unbiased samples are connected to the ground, while the biased are connected to the power supply and operate normally during the test.

ID	Device Name	Generation	Test Bias
1	NVIDIA Xavier NX	Xavier, 12 nm FinFet	BIAS
2	NVIDIA Xavier NX	Xavier, 12 nm FinFet	BIAS
3	NVIDIA Xavier NX	Xavier, 12 nm FinFet	UNBIAS
4	NVIDIA Xavier NX	Xavier, 12 nm FinFet	UNBIAS
5	NVIDIA Orin NX	Orin, 7 nm FinFet	BIAS
6	NVIDIA Orin NX	Orin, 7 nm FinFet	BIAS
7	NVIDIA Orin NX	Orin, 7 nm FinFet	UNBIAS
8	NVIDIA Orin NX	Orin, 7 nm FinFet	UNBIAS
9	NVIDIA Orin Nano	Orin, 7 nm FinFet	BIAS
10	NVIDIA Orin Nano	Orin, 7 nm FinFet	BIAS
11	NVIDIA Orin Nano	Orin, 7 nm FinFet	UNBIAS
12	NVIDIA Orin Nano	Orin, 7 nm FinFet	UNBIAS

TABLE II: Table of Devices used in the TID test

These devices are classified as System On Modules (SoM), indicating that the System on Chip (SoC), encompassing CPUs, GPU and other accelerators, is already integrated with the main memory and external peripherals such as power regulators or I/O. Consequently, a carrier board becomes necessary for the biased condition. To fulfil this, we selected the commercial off-the-shelf carrier board AVerMedia D131, due to the availability of suitable I/Os for the test of the board. Figure 1 illustrates the various blocks of these SoMs and their interaction with the carrier board.

Moreover, two custom in-house designed carrier boards with components selected for their TID testing heritage, were employed for two of the modules under test. For the unbiased set, a simple board was implemented to securely hold and ground all the pins of the SoM.

Figure 2 provides insight into the positioning and arrangement of the devices within the irradiation chamber. In order to minimise the exposure of the power regulators of the commercial carrier boards, which is their most sensitive part from radiation, a lead shield was put in place on the bottom row of the AVerMedia carrier boards, without shielding any part of the SoM devices, which are the actual devices of interest in our test.

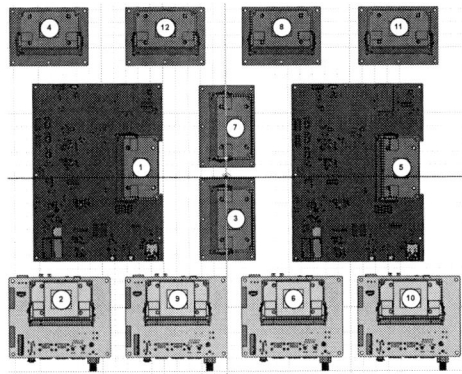

Fig. 2: TID Mechanical Fixation with ID numbers.

Before the irradiation test, one of the custom carrier boards, ID 1, suffered a software issue that did not allow to operate the Jetson Xavier over USB. Because of the short time before the test, the impacted carrier board was swapped with a spare AVerMedia board. Nothing more was changed in the setup.

A. *Setup and Measuring Methodology*

For this experiment, it was imperative to continuously monitor the biased devices during irradiation, concurrently with measuring the unbiased boards at each irradiation step. To accomplish this, a specific set of cables was required for the biased boards. These cables facilitated the provision of power, Ethernet for telemetry, debug port communication and mass memory for the operating system. To prevent the data storage devices from being affected by irradiation, a series of USB 3.0 cable extenders were installed to position the data storage devices outside of the chamber to host the operating system in discrete thumb drives. The remaining cables were sufficiently long to remain outside of the irradiation chamber.

Each biased board was linked to a distinct port on a DC power supply, while the Ethernet ports were consolidated into a network switch. The debug ports were connected to the DUT (Device Under Test) controller, which is a Raspberry Pi system responsible for collecting data from the bias devices, debug ports and power supply metrics. This data was then stored on the control PC. The setup for this specific test is illustrated in Figure 3.

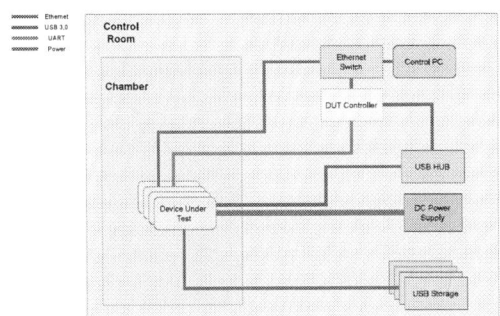

Fig. 3: TID test diagram.

979-8-3503-6689-1/24 $31.00 © 2024 IEEE

	Xavier NX			Orin NX			Orin Nano		
	15W	10W	SUB10W	15W	10W	SUB10W	15W	10W	SUB10W
Number Cores	2	2	2	4	4	2	4	4	2
Clock Frequency	1.9GHz	1.4GHz	0.345 GHz	1.65 GHz	1.19 GHz	0.729 GHz	1.65 GHz	1.19 GHz	0.729 GHz
Number of SM	6	4	4	8	8	8	8	4	4
GPU Frequency	1.1 GHz	0.8 GHz	0.114 GHz	0.612 GHz	0.612 GHz	0.612 GHz	0.612 GHz	0.408 GHz	0.408 GHz
Main Memory Frequency	1.6GHz	1.6GHz	0.200 GHz	3.2 GHz	2.133 GHz	2.133 GHz	2.133 GHz	2.133 GHz	2.133 GHz

TABLE III: Jetson power modes used during the testing. The 10W and 15W are standard power modes provided by NVIDIA, while the SUB10W is a custom power mode we created.

For the unbiased boards, we also used another AVerMedia board outside of the chamber with the same connections as the bias set for doing the measurements for each of the irradiation steps. In each of the measurements points, a matrix multiplication benchmark for multi-core and CUDA versions from the OBPMark Kernels / GPU4S Bench [17][18] was launched repeatedly to stress the CPUs and the GPU, in order to observe any anomalies. This benchmark is executed in three different power modes shown in Table III to detect possible anomalies in different operating conditions.

For all devices, we meticulously recorded data from all available internal sensors, including multi-point temperature sensors, system clock frequencies, operating system error reports, internal voltage and current measurements and more.

Throughout the entire irradiation campaign, this information was continuously extracted from the biased boards every second. During each irradiation step, both the bias and unbiased boards increased data acquisition to one sample every 100 milliseconds while executing tests on the CPU and GPU complex.

To ensure that the devices were not shielded from the gamma radiation, the heatsink of each device was removed, both for the biased and unbiased sets. To maintain a consistent temperature of around 65 degrees Celsius, the biased set was operated in the custom SUB10W power mode between measurement points.

For the radiation test, we aimed for a dose rate of 360 Rads per hour in silicon, in line with ESCC recommendations [19].

Regarding the measurement data points for the test, we based our selection on results from similar tests [20][21][22] to ensure an adequate number of data points for the unbiased set and representative doses relevant to the missions for which the devices will be utilized. These points were chosen at 1, 2, 8, 10, 16.5, 18.5, 25, 26.5, and 50 krads.

III. RESULTS AND ANALYSIS

During the irradiation, several DUTs failed at different stages. When this occurred, the affected devices were removed from the chamber. This was particularly evident at the 25 krad measurement point. At this stage, both biased devices of the Xavier NX (ID 1 and 2), Orin Nano 2 (ID 10), and Orin NX 2 (ID 6) failed and were subsequently removed from the irradiation facility. Additionally, Orin Nano 1 (ID 9) failed one hour before the 25 krad measurement. Upon inspection, it was discovered that the carrier board had failed, though the

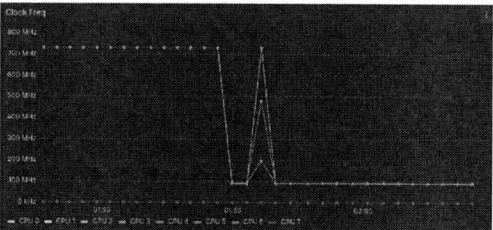

Fig. 4: Failure of the clock on a biased Orin NX board.

DUT continued to operate nominally. Consequently, the carrier board was replaced with a new one to proceed with the test.

At the conclusion of the test, at the 50 krad measurement point, none of the biased boards managed to withstand the radiation dose. In contrast, all the unbiased boards successfully booted into Linux and ran the benchmarks. However, it is noteworthy that all Orin family devices (Orin NX and Orin Nano) encountered an issue with clock frequency, only managing to operate at 200 MHz instead of the intended 1.65 GHz. The results for the Total Ionizing Dose (TID) and failure points are summarised in Table IV.

The carrier board played an important role in the failures modes, as shown in Table IV. Two of the DUTs failed due to voltage spikes, which originated from the commercial carrier board and propagated to the DUT, causing a failure. This is further supported by the fact the only two of the biased devices that passed the 25 krads limit were the Orin NX with ID 5 and Orin Nano with ID 9. The former had a custom carrier board with power regulators that already had been tested to more that 50 krads, and the later had the carrier board which was swapped out at 25 krads with a new one.

An interesting case was observed with DUT ID 6 which failed at 25 krad exactly. We stopped the irradiation to perform the measurements of this step, and the benchmark finished correctly. However, while we were testing other devices and DUT ID 6 was idle, it experienced a voltage spike without irradiation on the chamber. We do not know the exact cause of this failure. However, it seems to be related with the fact that in the Orin NX devices (both biased and unbiased), the failure of the main clock was observed before the complete failure of the device. Figure 4 shows this transition in the Orin NX, ID 5, before the failure of the DUT.

Based on the effect observed on the Orin unbiased boards, we did a deep analysis of the collected sensor data focused on the clock frequency. As the total ionising irradiation increased,

DUT ID	Device	Carrier board	State	TID Dose (Si)	Status (at 50krad)	Failure mode
1	Xavier NX	AVerMedia	BIASED	~24 krad	Failed	SoM stopped operating
2	Xavier NX	AVerMedia	BIASED	~24 krad	Failed	SoM stopped operating
3	Xavier NX	Ground plane	UNBIASED	50 krad	Nominal	-
4	Xavier NX	Ground plane	UNBIASED	50 krad	Nominal	-
5	Orin NX	Custom carrier board	BIASED	~40 krad	Failed	SoM stopped operating
6	Orin NX	AVerMedia	BIASED	25 krad	Failed	Voltage spike from carrier board damaged SoM
7	Orin NX	Ground plane	UNBIASED	50 krad	Degraded	Core frequency reduced to 200MHz
8	Orin NX	Ground plane	UNBIASED	50 krad	Degraded	Core frequency reduced to 200MHz
9	Orin Nano	AVerMedia	BIASED	~40 krad	Failed	SoM stopped operating
10	Orin Nano	AVerMedia	BIASED	~20 krad	Failed	Voltage spike from carrier board damaged SoM
11	Orin Nano	Ground plane	UNBIASED	50 krad	Degraded	Core frequency reduced to 200MHz
12	Orin Nano	Ground plane	UNBIASED	50 krad	Degraded	Core frequency reduced to 200MHz

TABLE IV: TID test results for Jetson SoM DUTs

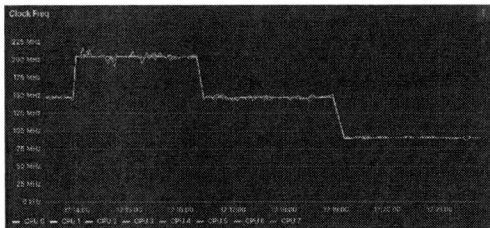

Fig. 6: Orin unbiased board under test after 50 krad.

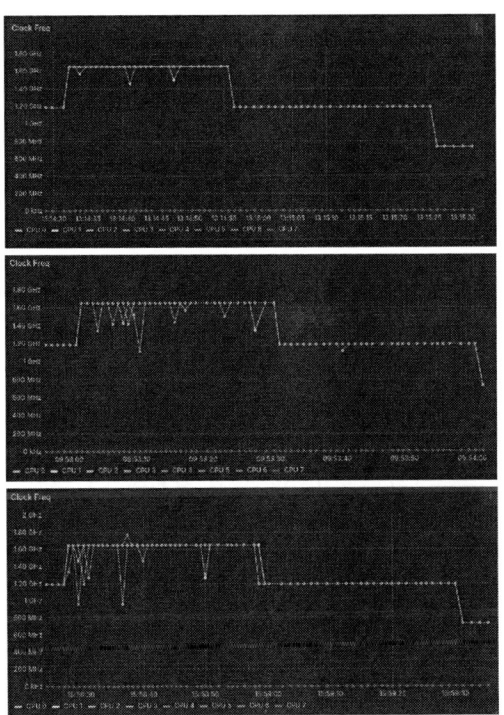

Fig. 5: Clock frequency measurements for the same device at 1 krad, 10 krad and 25 krad.

a higher clock frequency variability was observed, which was more pronounced in higher power modes. Figure 5 shows the same benchmark test at 1 krad, 10 krad and 25 krad. The effect is observed when the device is set in the 15W power mode. At 1 krad, the small variability is expected, but with a higher dose, this variability increases reaching the point that at 25 krads an overshoot of the frequency of the device is observed. Even though this seems concerning, the benchmark application was executed correctly.

Interestingly, this effect was only observed in the Orin NX and Orin Nano, but the Xavier NX did not suffer from clock fluctuations more than expected.

Finally we analysed the clock frequency crash on the unbiased Orin NX devices. Figure 6 shows the benchmark executed on one of the unbiased devices.

Interestingly we can observe the three distinct clock frequencies that seem to match the values of the previous test at different points, but the maximum frequency is now 200 MHz for the 15W power mode, 150 MHz for the 10W power mode, and 90 MHz for the SUB10W mode. In the Orin Nano there is a similar behaviour, but with frequencies of 150 MHz and 90 MHz respectively. Initially we thought that this was caused by an error in the hardware sensor that measures the clock frequency, but after checking the execution time of the benchmarks we confirmed that the issue was with the reduced frequency of the clock. In the case of Xavier devices we do not see this drop in frequency at 50 krad, In subsection III-A, we delve more into detail of the effects in execution time. Currently, the root cause of this effect is not clear due to the lack of transparency in the documentation of the DUT.

It is important to mention that other clock frequencies were not affected, e.g. the GPU clock frequency and main memory frequencies remained in nominal values and the benchmarks

ID	DUT	1 krad	18.5 krad	25 krad	50 krad
1	XANX1	2.19	2.06	No Data	No Data
2	XANX2	2.08	2.02	No Data	No Data
3	XANX3	2.64	3.60	2.71	2.34
4	XANX4	2.61	7.16	2.63	3.29
5	ORNX1	2.06	2.08	2.05	No Data
6	ORNX2	2.15	2.17	2.14	No Data
7	ORNX3	2.25	2.31	3.11	48.55
8	ORNX4	2.32	2.34	2.32	48.63
9	ORNA1	2.23	2.42	2.25	No Data
10	ORNA2	2.23	2.35	No Data	No Data
11	ORNA3	2.42	2.50	2.46	42.81
12	ORNA4	2.51	2.51	2.53	50.36

TABLE V: Execution time in seconds for Matrix Multiplication using OpenMP (multicore CPU execution) during the test points in 15W power mode.

ID	DUT	1 krad	18.5 krad	25 krad	50 krad
1	XANX1	0.35	0.40	No Data	No Data
2	XANX2	0.44	0.40	No Data	No Data
3	XANX3	0.40	0.39	0.40	0.46
4	XANX4	0.39	0.53	0.38	0.46
5	ORNX1	0.93	0.93	0.93	No Data
6	ORNX2	0.93	0.92	0.94	No Data
7	ORNX3	0.92	0.93	0.94	7.24
8	ORNX4	0.94	0.93	0.92	7.39
9	ORNA1	1.38	1.36	1.37	No Data
10	ORNA2	1.37	1.37	No Data	No Data
11	ORNA3	1.37	1.36	1.36	10.81
12	ORNA4	1.39	1.38	1.36	10.82

TABLE VI: Execution time in seconds of Matrix Multiplication with CUDA (GPU execution) during the test points in 15W power mode.

ID	DUT	1 krad	18.5 krad	25 krad	50 krad
1	XANX1	8.17	7.50	No Data	No Data
2	XANX2	7.18	7.79	No Data	No Data
3	XANX3	10.09	9.69	9.95	10.45
4	XANX4	9.95	9.82	9.39	9.48
5	ORNX1	4.82	4.61	4.61	No Data
6	ORNX2	4.69	4.81	4.82	No Data
7	ORNX3	5.50	5.19	5.49	27.18
8	ORNX4	5.16	5.61	5.53	24.81
9	ORNA1	4.64	4.80	4.61	No Data
10	ORNA2	4.61	4.93	No Data	No Data
11	ORNA3	5.13	5.54	5.38	41.56
12	ORNA4	5.29	5.44	5.35	61.34

TABLE VII: Execution time in seconds of Matrix Multiplication with OpenMP (multicore CPU execution) during the test points in SUB10W power mode.

ID	DUT	1 krad	18.5 krad	25 krad	50 krad
1	XANX1	0.45	0.45	No Data	No Data
2	XANX2	0.45	0.45	No Data	No Data
3	XANX3	0.46	0.46	0.47	0.42
4	XANX4	0.47	0.47	0.46	0.41
5	ORNX1	0.94	0.94	0.93	No Data
6	ORNX2	0.96	0.95	0.97	No Data
7	ORNX3	0.93	0.98	0.94	7.29
8	ORNX4	0.94	0.95	0.94	7.27
9	ORNA1	1.38	1.37	1.38	No Data
10	ORNA2	1.37	1.37	No Data	No Data
11	ORNA3	1.37	1.36	1.36	10.82
12	ORNA4	1.36	1.36	1.37	10.81

TABLE VIII: Execution time in seconds of Matrix Multiplication with CUDA (GPU Execution) during the test points in SUB10W power mode.

show that there is an increase in execution time with the GPU, but not at the same scale as on the CPU complex. This can be due to the slowdown of the CPU which issues the commands to the GPU, thus increasing the overall execution time.

A. Analysis of the TID effects on the benchmark execution

In this section, we analyse the effects of the TID on the execution time of benchmarks on the DUTs. In order to study the benchmark execution time for each power mode, we launched three executions of OBPMark kernels / GPU4S Bench [17][18] matrix multiplication, OpenMP for multi-core CPU execution and CUDA for GPU with data input size 2048. Both versions are the optimised code variants.

The size was chosen to allow sufficient execution time for the CPU and GPU. As the size of the calculation is still relatively small for the GPU, a bigger size was considered. However, we wanted to use the same size for CPU and GPU, and increasing it would increase the execution time too much.

Tables V and VII summarise the execution time results for the CPU, for the highest and lowest power mode tested respectively. A slight increase in the execution time on CPU for the unbiased board with respect to the biased set can be seen. This was due to the increase of data acquisition in the unbiased devices in order to have sufficient data points on the DUTs which are only connected and analysed during the measurement points between irradiation steps.

For the presented analysis we selected only four of the measurement points to reduce the amount of data to a manageable amount. These points are 1, 18.5, 25 and 50 krad. These were selected to represent the start of the test, at 1 krad, the last data point before any failure (18.5 krad), the point at which more of the biased boards failed (25 krad), and the last point (50 krad) of the test, with all unbiased boards.

First we analyse the effects on the CPU. There is minimal deviation from the average CPU execution time at 1, 18.5, and 25 krad, with the exception of ID 4 exhibiting an anomaly at 18.5 krad. However, this anomaly could be attributed to run-to-run variability since no other metric indicates any system changes. Additionally, the variability observed in the clock at these data points does not appear to significantly impact the execution time.

At the 50 krad point, a notable increase in execution time, up to 20 times, is observed due to the reduced clock frequency. In contrast, the Xavier remains unaffected by this change and maintains a consistent average execution time.

Shifting focus to the GPU effects, data for the 15W power mode are presented in Table VI, while data for the lowest power mode can be found in Table VIII.

We can observe a similar trend with the CPU execution time. For 1, 18.5 and 25 krad there is not much variability on the execution time, only an increase in execution time on the 50 krad for the Orin family. However, in this case the increase is only of $7\times$ increase instead of the $20\times$ of the CPU. This could be explained by the fact that the CPU is the one sending the commands to the GPU, so with a slowed down CPU, these commands take longer to been issued, causing an increase in the execution time of the GPU application even though the GPU clock frequency remains the same. Similar to the CPU, the Xavier does not suffer this slowdown.

IV. CONCLUSIONS

Our radiation analysis indicates that the three devices we studied exhibit a high Total Ionizing Dose (TID) performance, aligning well with typical mission profiles where high-performance Commercial Off-The-Shelf (COTS) devices would be used, e.g. few years in LEO orbits. More importantly, all tested devices demonstrate resilience, surviving exposure to at least 20 krad. Furthermore, this study elucidates various impacts on new generation nodes and discerns disparities between device generations. It also distinguishes itself by pioneering the comparison of radiation performance between biased and unbiased Jetson devices, offering valuable insights for future mission planning. Additionally, we delve into the influence of irradiation on execution time across different points. Moreover, our investigation emphasizes the significance of the carrier board, showcasing its pivotal role during testing. We advocate for future tests to adopt a tailored approach, utilizing custom carrier boards equipped with only essential components and selected with established radiation testing practices.

V. ACKNOWLEDGEMENTS

This work was supported by ESA through the 4000136514/21/NL/GLC/my co-funded PhD activity "Mixed Software/Hardware-based Fault-tolerance Techniques for Complex COTS System-on-Chip in Radiation Environments". Moreover, it was partially supported by the European Community's Horizon Europe programme under the METASAT project (grant agreement 101082622), by the Spanish Ministry of Economy and Competitiveness under grants PID2019-107255GB-C21 and IJC2020-045931-I (Spanish State Research Agency / http://dx.doi.org/10.13039/501100011033) and the HiPEAC Network of Excellence. The authors want to extend their gratitude to the people from ESA ESTEC TEC-QEC for their support during the execution of this test campaign.

REFERENCES

[1] G. Giuffrida, L. Fanucci, G. Meoni, M. Batič, L. Buckley, A. Dunne, C. Van Dijk, M. Esposito, J. Hefele, N. Vercruyssen *et al.*, "The Φ-Sat-1 Mission: The First On-board Deep Neural Network Demonstrator for Satellite Earth Observation," *IEEE Transactions on Geoscience and Remote Sensing*, 2021.

[2] F. Ouallouche, K. Labadi, Y. Mohia, M. Lazri, and S. Ameur, "Artificial Intelligence for Satellite Image Processing: Application to Rainfall Estimation," in *Intelligent Systems and Applications: Select Proceedings of ICISA 2022*. Springer, 2023, pp. 165–174.

[3] F. Ortiz, V. Monzon Baeza, L. M. Garces-Socarras, J. A. Vásquez-Peralvo, J. L. Gonzalez, G. Fontanesi, E. Lagunas, J. Querol, and S. Chatzinotas, "Onboard Processing in Satellite Communications Using AI Accelerators," *Aerospace*, vol. 10, no. 2, p. 101, 2023.

[4] J. Gaisler, "The SPARC History in Space," *Keynote at ESA Workshop on Avionics, Data, Control and Software Systems (ADCSS)*, 2017.

[5] R. W. Berger et al., "The RAD750 -A Radiation Hardened PowerPC Processor for High Performance Spaceborne Applications," in *IEEE Aerospace Conference Proceedings*, vol. 5, March 2001, pp. 2263–2272.

[6] D. Steenari, K. Forster, D. O'Callaghan, M. Tali, C. Hay, M. Cebecauer, M. Ireland, S. McBerren, and R. Camarero, "Survey of High-Performance Processors and FPGAs for On-Board Processing and Machine Learning Applications," in *Proceedings of the European Workshop on On-Board Data Processing (OBDP)*, 2021, pp. 14–17.

[7] M. Esposito, S. Conticello, M. Pastena, and B. C. Domínguez, "In-orbit Demonstration of Artificial Intelligence Applied to Hyperspectral and Thermal Sensing From Space," in *CubeSats and SmallSats for Remote Sensing III*, vol. 11131, 2019.

[8] J. Swope, F. Mirza, E. Dunkel, A. Candela, S. Chien, A. Holloway, D. Russell, J. Sauvageau, D. Sheldon, and M. Fernandez, "Benchmarking Space Mission Applications on the Snapdragon Processor Onboard the ISS," *Journal of Aerospace Information Systems*, vol. 20, no. 12, pp. 807–816, 2023.

[9] I. Rodriguez-Ferrandez, D. Steenari, M. Tali, L. Kosmidis, and F. Tonicello, "Case-Study for Integration of COTS SoC Devices in Reliable Space Systems for On-Board Processing," in *2023 European Data Handling & Data Processing Conference (EDHPC)*. IEEE, 2023.

[10] L. Kosmidis, I. Rodriguez, A. Jover-Alvarez, S. Alcaide, J. Lachaize, O. Notebaert, A. Certain, and D. Steenari, "GPU4S: Major Project Outcomes, Lessons Learnt and Way Forward," in *2021 Design, Automation & Test in Europe Conference & Exhibition (DATE)*. IEEE, 2021.

[11] European Cooperation for Space Standardization (ECSS), "ECSS-Q-ST-60-13C Rev.1 – Commercial Electrical, Electronic and Electromechanical (EEE) Components," European Space Agency, ECSS Standard ECSS-Q-ST-60-13C Rev.1, 2022, accessed: March 10, 2023. [Online]. Available: https://ecss.nl/standard/ecss-q-st-60-13c-rev-1-commercial-electrical-electronic-and-electromechanical-eee-components-12-may-2022/

[12] N. Destrycker, W. Benoot, J. Mattias, I. Rodriguez, and D. Steenari, "EDGX-1: A New Frontier in Onboard AI Computing with a Heterogeneous and Neuromorphic Design," in *2023 European Data Handling & Data Processing Conference (EDHPC)*. IEEE, 2023, pp. 1–6.

[13] F. C. Bruhn, N. Tsog, F. Kunkel, O. Flordal, and I. Troxel, "Enabling Radiation Tolerant Heterogeneous GPU-based Onboard Data Processing in Space," *CEAS Space Journal*, pp. 1–14, 2020.

[14] SpiralBlue. (2023, jul) SE-1 SPACE EDGE ONE. [Online]. Available: https://www.spiralblue.space/

[15] D. Posada, C. W. Hays, J. Jordan, D. Lopez, T. Yow, A. Malik, and T. Henderson, "EagleCam: a 1.5 U Low-Cost CubeSat Mission for a Novel Third-Person View of a Lunar Landing," in *2023 IEEE Aerospace Conference*. IEEE, 2023, pp. 1–12.

[16] ESA. (2024, may) ESA Co-60 Irradiation Facility. [Online]. Available: https://escies.org/webdocument/showArticle?id=251

[17] D. Steenari et al., "On-Board Processing Benchmarks," 2021, http://obpmark.github.io/.

[18] I. Rodriguez, L. Kosmidis, J. Lachaize, O. Notebaert, and D. Steenari, "GPU4S Bench: Design and Implementation of an Open GPU Benchmarking Suite for Space On-board Processing," Universitat Politècnica de Catalunya, Tech. Rep. UPC-DAC-RR-CAP-2019-1, https://www.ac.upc.edu/app/research-reports/public/html/research_center_index-CAP-2019,en.html.

[19] European Space Components Coordination (ESCC), "Total Dose Steady-state Irradiation Test Method," ESCC Basic Specification No. 22900, 2016. [Online]. Available: https://escies.org/escc-specs/published/22900.pdf

[20] W. S. Slater, N. P. Tiwari, T. M. Lovelly, and J. K. Mee, "Total Ionizing Dose Radiation Testing of NVIDIA Jetson Nano GPUs," in *2020 IEEE High Performance Extreme Computing Conference (HPEC)*, 2020.

[21] W. S. Slater, B. B. Rutherford, J. K. Mee, R. E. Pinson, M. Gruber, D. Sabogal, and I. A. Troxel, "Single Event Effects and Total Ionizing Dose Radiation Testing of NVIDIA Jetson Orin AGX System on Module," in *2023 IEEE Radiation Effects Data Workshop (REDW)(in conjunction with 2023 NSREC)*. IEEE, 2023, pp. 1–6.

[22] S. L. Katz, C. Heistand, and E. N. Miller, "NVIDIA Jetson TX2i TID and Proton SEE Testing: Results and a Comparison of Two Proton Beam Facilities," in *2023 IEEE Radiation Effects Data Workshop (REDW)(in conjunction with 2023 NSREC)*. IEEE, 2023, pp. 1–6.

An Enhanced Fault Injection Framework for FPGA-Based Soft-Cores

Tijmen T. Smit*, Bruno Endres Forlin*, Kuan-Hsun Chen*, Ioanna Souvatzoglou‡, Mihalis Psarakis‡, Marco Ottavi*†

*University of Twente, the Netherlands, †University of Rome Tor Vergata, Italy

‡University of Pireus, Greece

Email: {t.t.smit, b.endresforlin, k.h.chen, m.ottavi}@utwente.nl

Abstract—**Contemporary space system architectures necessitate rigorous validation to ensure robust performance post-deployment. Fault injection is a critical methodology that improves confidence in these systems by simulating errors under controlled conditions. Traditional fault injection approaches, such as simulation and emulation, often require costly resources or invasive alterations to the Device Under Test (DUT). The FREtZ tool addresses some of these challenges by facilitating non-invasive bit flip injections into user-bits and Configuration RAM (CRAM) bits via the FPGA's JTAG interface. However, its integration with soft-cores remains limited. This paper introduces a novel fault injection framework that extends the capabilities of the FREtZ tool. Our framework improves both the precision and the efficiency of fault injections in soft-core processors by enabling targeted fault injections at specific clock cycles or program counter locations. Hence, the injection space can be reduced to the DUT, enabling the execution of a thorough injection campaign. The proposed method not only refines the granularity of the fault injections but also streamlines the emulation process, thereby providing a more efficient and less intrusive approach to system testing. This advancement represents a significant step forward in emulation-based fault injection, particularly for complex space system architectures where reliability is paramount.**

Index Terms—**Fault Injection, Emulation, Soft-Cores, Reliability**

I. Introduction

The rapid expansion of the aerospace industry has exposed the limitations of traditional space-grade computing systems, which are burdened by outdated architectures, declining support, and an increasingly obsolete software ecosystem. Researchers are increasingly turning to innovative solutions, such as the RISC-V architecture, to address these challenges. Additionally, the diversity of future processing cores presents significant opportunities for advancement. To ensure the reliability of these new systems in space, comprehensive ground testing is essential. A crucial aspect of design and testing with hardware-in-the-loop is the ability to conduct fault injection campaigns quickly and efficiently.

Fault injection (FI) in complex RTL designs is challenging due to the vast injection space, encompassing all possible injection points and numerous time instances for fault introduction. Each combination of location and time can yield different effects, such as masking, Silent Data Corruption

(SDCs), or critical failures like Detectable Uncorrectable Errors (DUEs). Determining the injection points and developing the fault injection campaign is a complex and expansive area of research [3]. Equally important to deciding where to inject faults is determining how to inject them, as the various methods of fault injection can significantly impact the efficiency and coverage of testing.

FI techniques are categorized into hardware-based, simulation-based, and emulation-based approaches. Hardware-based FI, while the fastest and closest to real operating conditions, risks damaging the IC and requires expensive equipment [12]. Simulation-based FI allows for early reliability assessments without extra hardware, using a virtual prototype to introduce faults, but is time-consuming and depends on model quality [18]. Emulation-based FI offers faster testing since the hardware is executed at native speeds. However, non-invasive integration becomes problematic, requiring significant effort to minimize the probe effect on the system. Furthermore, some injection techniques are so invasive that full coverage would bloat the design beyond the capacity of any FPGA.

To avoid this bloating while still delivering the advantages of emulation speed, we utilize the FREtZ tool. The tool allows for injection in user-bits via the FPGA's JTAG interface. We extend the FREtZ tool to enable easy testing with soft-cores via a standard wrapper around the DUT. Only a few signals are connected, and the DUT itself remains intact, minimizing integration efforts. Therefore, **the main contributions of this paper are:**

- Demonstrate a means to perform automated campaign generation from a fault model.
- Develop a framework for dynamically injecting specific registers at certain program milestones.
- Achieve this without modifications to the DUT netlist.
- Describe a framework and methodology for low-effort integration of a softcore as the DUT.
- Experimentation with real hardware accelerator developed for space applications.

II. Background and Related Work

A. Fault Injection tools

Fault injection (FI) is a well-known method used to quantify the reliability and resilience of a system, the device under test

---

979-8-3503-6689-1/24 $31.00 © 2024 IEEE

(DUT), against soft errors, by assessing the system's ability to detect, locate, or mitigate fault occurrences [25], [3]. FI can be categorized in three main groups, hardware-based, simulation-based, and emulation-based fault injection approaches [12].

1) Hardware-based Fault Injection: Hardware-based FI is the fastest technique among all, providing results close to the actual operating environment of the circuit. It allows for injecting actual faults via pin-forcing, irradiation beams, or with a laser. However, it comes with a high risk of damage to the target IC. In addition, this method requires the use of expensive equipment, which may not be available to all users.

2) Simulation-based Fault Injection: Simulation-based FI enables early reliability assessments without need for a prototype of the design or additional hardware. This makes it cost-effective, but each experiment consumes a significant amount of time. In this setup, use is made of a virtual prototype to which fault are introduced. This is an executable software model of a hardware component that can be run on a host computer and mirrors the functionality of the hardware component. The quality of the model implementation can affect the results obtained, and real faults in the working circuit may be overlooked during experiments [18].

vRTLmod [13] is a simulation-based FI tool, which takes Verilator output and automatically adds fault injection capability at flip-flop level. Faults are injected by modifying variable values between evaluation cycles. Urban [24] extended vRTLmod to achieve campaign speed-up by leveraging runtime switching between different virtual prototype abstraction layers. This is a similar approach as presented by ETISS-ML [16], however utilizing SystemC/TLM. Another widely used open-source system simulator is Gem5, which has served as the foundation for the development of several FI tools, including GemFI [20] and gem5-FIM [8], which both extend gem5 with an injection API.

However, as highlighted by the authors in [9], gem5 exhibits a significant disparity in performance compared to hardware. This discrepancy is also evident in the findings of Elsabbagh et al. [11] regarding Verilator, with a 200-fold speed-up from circuit simulation to emulation.

3) Emulation-based Fault Injection: Emulation-based fault injection (FI) offers early reliability assessments by emulating and injecting faults into a hardware design on an FPGA. It is cost-effective, flexible in choosing fault locations, and does not require physical components or special facilities. This method allows for the study of faults at the architectural level, providing insight into hardening strategies and maintaining the probability relationship between soft errors and software errors [3]. Furthermore, observing circuit behaviour in a real-time environment helps to consider real-time interactions [2].

Various tools and environments exist for emulation-based FI, utilizing techniques such as logic modification, reprogramming after netlist modification, or bitstream modification [2]. Each technique has different trade-offs between campaign time and implementation area:

- **FI by Reconfiguration:** Involves recompiling a netlist and fixing selected injection locations to a static output,

limiting injections to stuck-at faults and resulting in long compilation times. However, as recompilation and reconfiguration are timely processes, state-of-the art injectors resort to other techniques.

- **Bitstream Modification:** Alters LUT contents, allowing for temporal injection freedom via read-modify-write operations. For example, FT-UNSHADES uses read-modify-write operations and partial reconfiguration, achieving an injection time of 10 μs[1]. Ullah et al. present an FI environment that injects stuck-at faults via Xilinx's ICAP interface, with bit inversion taking a minimum of 64 μs [23].

- **Instrumented Circuit Technique:** Enables dynamic fault injection by replacing flip-flops with saboteur circuitry, facilitating transient and stuck-at fault injections via externally coordinated signals [7]. Time synchronization between the DUT and the controller is achieved through fault-mask shift registers connected to a scan-path chain. NETFI applies saboteurs to both single registers and full register files, injecting faults without time penalties during real-time execution [15]. Lopez-Ongil et al. add a saboteur to each flip-flop along with replicated flip-flops to check fault effects at runtime, minimizing campaign times to around 20 μs per injection but with significant area overhead [14].

From the presented emulation techniques, we can see that emulating faults at the user-bits of the design (not on the CRAM) is costly, requiring invasive circuitry and modifications of the DUT. Although this is a possible strategy, it has adverse effects on the performance and usability of the techniques, and might require extensive intervention to adapt to new designs.

B. FREtZ fault injection emulation

Yet another emulation-based FI is the FREtZ framework. The license-free FREtZ framework is developed by Sari et al. [22]. Contrary to most emulation-based FI tools, which support injection of upsets in the Configuration RAM (CRAM), FREtZ also supports injections into user registers. It provides access to the FPGA configuration memory and circuit logic via the JTAG protocol. JTAG is advantageous because of its universality. The FREtZ framework provides a Python API to a Tcl interface handler communicating with a JTAG communication engine to read and write configuration memory and configuration registers. This JTAG communication engine is running on an external dedicated FPGA board, communicating to the host PC via a TCP connection.

Injecting faults in this manner, omits the need to either resynthesize and reprogram the DUT for every campaign or having to introduce new hardware in the DUT. The alternatives, as described in the previous section, pose a time-area tradeoff. The presented tool introduces only a small and fixed area overhead. Furthermore, no logical changes to the DUT are required, ensuring that the results obtained can be extrapolated to setups beyond the fault injection environment.

C. Fault model

When establishing a FI campaign, a fault model is employed to describe the specific faults that are introduced, along with their respective locations and timing [21]. It describes, the type of real-world error being simulated, either transient or persistent faults. Besides, as faults occur in components that make up systems, a campaign can target specific parts of the system. Therefore, temporal and spatial characteristics of the campaign are described, which allows the modelling of single-bit upsets (SBU), multiple-bit upsets (MBU), or single event upset (SEU). Common models include the bit-flip model for simulating SEUs, the set-and-reset model that sets bits to a fixed value, and the stuck-at model that permanently fixes a bit's value to simulate persistent faults.

D. AVF

Soft errors in a processor structure may not always impact the final program output, with the likelihood of a fault leading to a visible error defined as the structure's architectural vulnerability factor (AVF) [17]. This is used to calculate the total error rate of micro-architectural components by multiplying the AVF with the raw fault rate, aiding in assessing if the design aligns with error rate objectives. Estimating AVF involves identifying architecturally correct execution (ACE) bits, where soft errors in ACE bits affect the program output, while the remaining bits are un-ACE.

The AVF of a storage cell is the fraction of time an upset in that cell will cause a visible error in the final output of a program. For a full hardware structure, the AVF is the average AVF of all the bits in the structure, assuming the same raw upset rate in the structure.

$$\text{AVF} = \frac{\bar{r}}{B} = \frac{\sum_{i \in B} r_i}{BC} \qquad = \frac{1}{n} \sum_{i=1}^{n} f(X_i) \qquad (1)$$

The AVF is calculated with (1), where B are the bits in a hardware structure, r a bit's ACE cycles count, $\bar{r}$ the average number of ACE bits per cycle, and C the total execution cycles. And, with FI, the AVF of a system can be approximated with the right-hand side of (1) [10]. An evaluation function $f(X)$, returns 0 for a correct output, else 1. With n faults being evaluated in a campaign.

E. SPARROW hardware

SPARROW is a low-cost SIMD (single instruction, multiple data) accelerator for AI operations, and will serve as DUT in later experiments. As its developer, Bonet [4], observed, matrix multiplications serve as the foundation for the majority of machine learning operations, whose computation is based on the dot product. This resulted in SPARROW's two stages architecture, (see Figure 1). The first stage executes four arithmetic operations (e.g. *add, mul, max, shift* or binary operations) in parallel, whereafter these four results are combined in reduction operations (*sum, max, min* or *xor*) to yield a 32-bit outcome. A saturation option is included in both stages to mitigate possible overflow of 8-bit values. Additionally, the

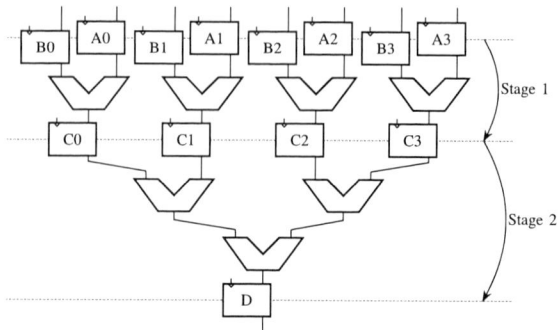

Figure 1: Simplified overview of the SPARROW module [5].

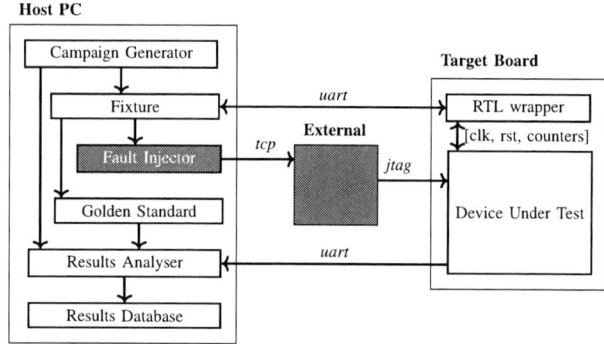

Figure 2: Tools components, hardware, and their relations, the grey parts are existing components from the FREtZ environment.

unit includes masking and swizzling of the input operands, but this is not explored in this text.

Furthermore, Bonet points out that the precision of arithmetic operations involved in machine learning is often limited to 8-bit integers. This accommodates the reuse of the integer register file, by dividing one 32-bit register into four lanes. The use of the integer pipeline results in no performance penalty in the rest of the operations of the base processor.

III. EMULATION-BASED INJECTION CAMPAIGNS

An FI environment is generally composed of the same elements [7], see Figure 2. A campaign generator creates a fault list from a fault model specified by a user. This list is provided to a fault injector and fixture. The latter controls the execution of a DUT, while the former injects the faults according to the fault list. A result analyser checks the execution and output of the DUT, and discriminates between a DUE, timeout or SDC. For this, a golden standard is required, which can be either a clean copy of the DUT, or a recorded execution run. Lastly, a results database saves the results, for direct or later analysis.

A. Campaign generator

The tool offers a customizable and user-friendly campaign generator in the Python programming language. A logic location file from a DUT implementation is supplied, which

979-8-3503-6689-1/24 $31.00 © 2024 IEEE   158

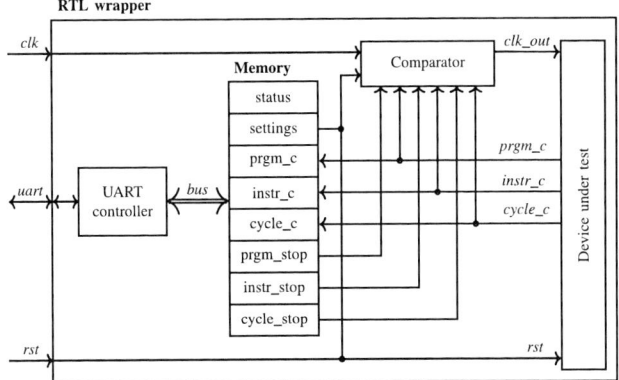

Figure 3: Simplified overview of the RTL wrapper included on the target FPGA interacting with the DUT.

can be exported from Xilinx Vivado. This file provides details of nodes in the design with bit positions, frame addresses, offsets, logic resources, and component names. The generator processes this file to create a dataset of injectable memory elements and latches, associating the latter with nets and hierarchy from the DUT design. Users can specify nets or components for inclusion in the fault model. Additionally, temporal injection points can be defined using the RTL wrapper, as discussed in Section III-C.

B. Results analyser and database

The results analyser categorizes campaign outcomes as correct, SDC, or timeouts by comparing the DUT's UART output with the golden standard. It measures campaign execution time, excluding fault injection time, marking runs exceeding a user-defined limit as timeouts. These results, along with information from the fault list, are stored in the results database.

C. RTL wrapper and fixture

The DUT can be any hardware, although the tool setup discussed focuses on injecting faults into (subsystems of) a soft core CPU. To facilitate FI at specific instructions, the DUT's execution must be halted by interrupting its clock signal. For accurate synchronization, a minor modification to the DUT is required, involving extracting machine counters from the CPU to the fixture. This modification, crucial for precise FI, does not alter the DUT's logic.

For a tight integration, the fixture includes a wrapper integrated on the same FPGA as the DUT, as shown in Figure 3. Communication between the fixture and the wrapper is facilitated through UART, enabling the configuration of a few registers. These registers' values are compared with the CPU's program counter, cycle counter, and instructions-retired counter to pause the clock. The UART connection also permits DUT reset. The incremental hardware overhead from these improvements is outlined in Table I.

Table I: Hardware utilization of extra introduced hardware for FI tool, as obtained from Vivado after implementation.

	LUT	FF	IO	BUFG
Utilization	208	88	16	2

Figure 4: The physical system setup. From left to right: host PC, target FPGA and external board.

D. Physical Setup

For the setup, see Figure 4, two Digilent Zedboards, containing a Zynq-7000 SoC are used. One as the target FPGA running the DUT and fixture wrapper, and one as the external FPGA for the fault injector. Two PmodUSBUART modules are utilized for connecting the target to a Windows 11 host PC, running Python 3.10. The external board is connected to the target via JTAG and to the host PC via an Ethernet connection.

IV. CASE STUDY

The presented methodology is evaluated through two experiments, which are presented and analysed in this section. Firstly, the AVF of the SPARROW unit is calculated and then verified with an injection campaign. Secondly, registers in the same hardware are identified with high and low ACE.

In these experiments, instructions utilizing SPARROW are selected from two benchmarks and supplied to the campaign generator. Bit-flips are introduced solely in SPARROW's registers during their usage, limiting the injection area to the DUT. This method enables a comprehensive injection campaign without the need to exhaustively inject in the entire design, as injections are targeted only at the necessary instructions.

A. Experimental setup

For all experiments, the same setup is used, revolving around the environment and hardware as presented in the previous section.

1) Target CPU and DUT: In the experiments, SPARROW is used as DUT, implemented in the NEORV32 CPU. The NEORV32 [19] processor is an open-source RISC-V compatible processor system. The CPU utilizes a 2-stage pipelined multi-cycle architecture. Besides, the system's customizability allows easy implementation of custom RISC-V instructions, as also used for implementing SPARROW. Similarly, other DUTs

979-8-3503-6689-1/24 $31.00 © 2024 IEEE

```
210: 37 25 15 0a    lui  a0, 41298
214: 13 05 15 a1    addi a0, a0, -1519
218: b7 15 0e 07    lui  a1, 28897
21c: 93 85 95 30    addi a1, a1, 777
220: 0b 54 b5 04    add_usum s0, a0, a1
```

Listing 1: Machine code for summing eight values using SPARROW. `add_sum` is a custom SPARROW instruction.

are straightforward to implement. Also, Böhmer et al. [6] has performed a characterization of the NEORV32 core.

From this CPU, the machine counters are extracted as described in subsection III-C. Additionally, the NEORV32 allows the instruction memory to be pre-initialized with a benchmark program, and load this on startup.

2) Benchmark: When choosing a benchmark, it is important to make heavy use of the specific DUT. Moreover, having temporal synchronization with the counter requires deterministic program execution with fixed inputs. In the experiments, two different programs are run on the DUT. In the first, the values of the eight SIMD registers from SPARROW are summed as shown in Listing 1.

A second benchmark aims to utilize SPARROW's machine learning capabilities, with a matrix multiplication and convolutional kernel operation. For the matrix multiplication, a series of dot product of two 4D vectors are calculated. All with one SPARROW instruction. Likewise, each convolution is calculated in three steps, every row of the kernel separately by doing 3D dot products.

The operation consists of $C = A^T B$ and $D = C * \omega$, where D is the output matrix. Low values for the input and kernel matrices are used to prevent the final matrix from overflowing the 8-bit integers. From the values of matrix D a 32-bit cyclic redundancy code with polynomial `0xF8C9140A` is calculated, which is printed to a UART output for verification.

B. Experiment results

This section will discuss the theoretical AVF of SPARROW, verify it with an experiment, and explore SPARROW's vulnerability during benchmark execution.

1) Theoretical AVF: With (1) the AVF of SPARROW is calculated. Figure 5 provides an overview of SPARROW's register, giving also the number of vulnerable bits per cycle. (2) shows the theoretical AVF. It can be observed that, because of the multi-stage design, with no registers shared between these stages, the expected AVF always yields 1/3.

$$\text{AVF}_{\text{sparrow}} = \frac{72 + 36 + 32}{140 \cdot 3} = \frac{1}{3} \quad (2)$$

2) AVF from fault injection: A fault injection campaign has been performed on SPARROW as DUT while running a program performing the calculation from Listing 1. From this compiled code, the target address is selected, `0x220`. Injections were done at 2, 3 and 4 clock cycles after the start of this instruction (clock offsets), to inject only during

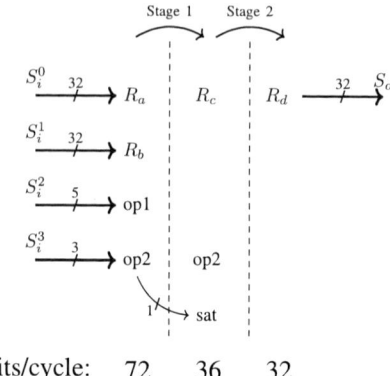

Bits/cycle: 72   36   32

Figure 5: Flow diagram of SPARROWs internal register. `sat` is a partial copy of the `op2` register.

the SPARROW stages. With these fault model parameters, a campaign was generated with the campaign generator.

This campaign resulted in 420 injections, of which 137 suffered SDC with no timeouts. This yields, an AVF of $\frac{137}{420} = 0.326$. Further analysis reveals that the last bit of the second stage operation is invulnerable. Injecting this bit alters the operation from unsigned to signed sum. However, due to the small input values, this change does not impact the final result significantly. Consequently, a slight deviation is observed between the discovered and calculated AVF.

3) Experiment statistics: The experiment described above was expanded to include clock offsets ranging from 0 to 7. It was conducted twice to allow for comparison of outputs. A total of 2752 injections were carried out during these two runs. Among these, the results of 5 injections did not align; they were deemed correct in one run and as SDC in the other.

The entire campaign involving these 2752 injections lasted 33 minutes, resulting in an average of 720 ms per injection. Of this, approximately 35 ms was dedicated to the execution of the benchmark.

4) AVF per register: A second FI campaign was run with the second benchmark program. Faults were injected on program counters where SPARROW instructions are executed, which includes 8 instructions, again with clock offsets 2 - 4. The rest of the fault model is the same as the previous experiment. This resulted in 106596 injections, of which 21179 were marked as incorrect and 113 as a timeout. This gives an AVF of $(21179+113)/106596 = 0.20$. This deviation from the theoretical AVF may be attributed to the influence of masking effects. Interestingly, however, is the appearance of time-out errors. This can be ascribed to an error-prone UART communication between the DUT and host PC.

V. DISCUSSION

The experiments demonstrated that the FI environment can efficiently set up an injection campaign with minimal effort, providing valuable insights into the AVF of a system. The initial experiment showed alignment with the calculated AVF, while the second experiment highlighted the masking effect's significant influence on the AVF. Moreover, the tool allows for targeted injections at precise clock offsets from specific instructions, enhancing its utility.

Injections take about 685 ms, which is longer than other emulation-based FI tools. However, the tool does not require changes to the DUT's netlist. Additionally, it supports a wide variety of fault models, unlike simulation-based tools, and does not necessitate lengthy simulations.

Besides, the design allows for easy testing of modular elements of the system without incurring penalties for executing the complete benchmark. Injections to the hardware accelerator were required only when in use, thereby limiting the injection space while maintaining functional fidelity.

VI. CONCLUSION

The presented tool enables researchers to generate FI campaigns effortlessly by programmatically defining a fault model, and providing significant flexibility for various DUT scenarios. In these fault models, it is possible to confine injections to specific locations and instructions, reducing the injection space to just include the DUT and therefore enabling thorough injection campaigns. These capabilities have been demonstrated by multiple experiments.

Furthermore, the framework and methodology facilitate seamless integration of DUTs in a soft core with minimal effort, requiring no alterations to the DUT's functionality and ensuring consistent operation beyond the FI environment. The target board needs to be programmed only once, leading to reduced campaign execution times compared to FI tools based on reprogramming, while leveraging the speed-up provided by circuit emulation.

REFERENCES

[1] M. Aguirre, D. Merodio, J. Tombs, F. Munoz, V. Baena, H. Guzman, J. Napoles, A. Torralba, A. Fernandez-Leon, and F. Tortosa-Lopez. Selective Protection Analysis Using a SEU Emulator: Testing Protocol and Case Study Over the Leon2 Processor. *IEEE Transactions on Nuclear Science*, 54(4):951–956, Aug. 2007.

[2] L. Antoni, R. Leveugle, and B. Feher. Using run-time reconfiguration for fault injection applications. In *IEEE Instrumentation and Measurement Technology Conference. Rediscovering Measurement in the Age of Informatics*, volume 3, pages 1773–1777, 2001.

[3] Y. B. Bekele, D. B. Limbrick, and J. C. Kelly. A Survey of QEMU-Based Fault Injection Tools & Techniques for Emulating Physical Faults. *IEEE Access*, 11:62662–62673, 2023.

[4] M. S. Bonet. Hardware-software co-design for low-cost AI processing in space processors. Master's thesis, Universitat Politècnica de Catalunya, Oct. 2021. Accepted: 2022-02-02T08:19:46Z.

[5] M. S. Bonet and L. Kosmidis. SPARROW: A Low-Cost Hardware/Software Co-designed SIMD Microarchitecture for AI Operations in Space Processors. In *2022 Design, Automation & Test in Europe Conference & Exhibition (DATE)*, pages 1139–1142, Mar. 2022. ISSN: 1558-1101.

[6] K. Böhmer, B. Forlin, C. Cazzaniga, P. Rech, G. Furano, N. Alachiotis, and M. Ottavi. Neutron Radiation Tests of the NEORV32 RISC-V SoC on Flash-Based FPGAs. In *2023 IEEE International Symposium on Defect and Fault Tolerance in VLSI and Nanotechnology Systems (DFT)*, pages 1–6, Juan-Les-Pins, France, Oct. 2023. IEEE.

[7] P. Civera, L. Macchiarulo, M. Rebaudengo, M. Reorda, and M. Violante. Exploiting circuit emulation for fast hardness evaluation. *IEEE Transactions on Nuclear Science*, 48(6):2210–2216, Dec. 2001.

[8] F. R. Da Rosa, R. Reis, and L. Ost. gem5-FIM: a flexible and scalable multicore soft error assessment framework to early reliability design space explorations. In *2018 IEEE 9th Latin American Symposium on Circuits & Systems (LASCAS)*, pages 1–4, Puerto Vallarta, Feb. 2018. IEEE.

[9] B. E. Forlin, P. C. Santos, A. E. Becker, M. A. Alves, and L. Carro. Sim 2 PIM: A complete simulation framework for Processing-in-Memory. *Journal of Systems Architecture*, 128:102528, July 2022.

[10] M. Ebrahimi, N. Sayed, M. Rashvand, and M. B. Tahoori. Fault injection acceleration by architectural importance sampling. In *2015 International Conference on Hardware/Software Codesign and System Synthesis (CODES+ISSS)*, pages 212–219, Amsterdam, Oct. 2015. IEEE.

[11] F. Elsabbagh, S. Sheikhha, V. A. Ying, Q. M. Nguyen, J. S. Emer, and D. Sanchez. Accelerating RTL Simulation with Hardware-Software Co-Design. In *56th Annual IEEE/ACM International Symposium on Microarchitecture*, pages 153–166, Toronto ON Canada, Oct. 2023. ACM.

[12] M. Eslami, B. Ghavami, M. Raji, and A. Mahani. A survey on fault injection methods of digital integrated circuits. *Integration*, 71:154–163, Mar. 2020.

[13] J. Geier and D. Mueller-Gritschneder. vRTLmod: An LLVM based Open-source Tool to Enable Fault Injection in Verilator RTL Simulations. In *Proceedings of the 20th ACM International Conference on Computing Frontiers*, pages 387–388, Bologna Italy, May 2023. ACM.

[14] C. Lopez-Ongil, M. Garcia-Valderas, M. Portela-Garcia, and L. Entrena. Autonomous Fault Emulation: A New FPGA-Based Acceleration System for Hardness Evaluation. *IEEE Transactions on Nuclear Science*, 54(1):252–261, Feb. 2007.

[15] W. Mansour and R. Velazco. An Automated SEU Fault-Injection Method and Tool for HDL-Based Designs. *IEEE Transactions on Nuclear Science*, 60(4):2728–2733, Aug. 2013.

[16] D. Mueller-Gritschneder, M. Dittrich, J. Weinzierl, E. Cheng, S. Mitra, and U. Schlichtmann. ETISS-ML: A multi-level instruction set simulator with RTL-level fault injection support for the evaluation of cross-layer resiliency techniques. In *2018 Design, Automation & Test in Europe Conference & Exhibition (DATE)*, pages 609–612, Dresden, Mar. 2018. IEEE.

[17] S. Mukherjee, C. Weaver, J. Emer, S. Reinhardt, and T. Austin. A systematic methodology to compute the architectural vulnerability factors for a high-performance microprocessor. In *22nd Digital Avionics Systems Conference*, pages 29–40, San Diego, CA, USA, 2003.

[18] R. Natella, D. Cotroneo, and H. S. Madeira. Assessing Dependability with Software Fault Injection: A Survey. *ACM Computing Surveys*, 48(3):1–55, Feb. 2016.

[19] S. Nolting and A. t. A. Contributors. The NEORV32 RISC-V Processor, Aug. 2023.

[20] K. Parasyris, G. Tziantzoulis, C. D. Antonopoulos, and N. Bellas. GemFI: A Fault Injection Tool for Studying the Behavior of Applications on Unreliable Substrates. In *2014 44th Annual IEEE/IFIP International Conference on Dependable Systems and Networks*, pages 622–629, Atlanta, GA, USA, June 2014. IEEE.

[21] H. M. Quinn, D. A. Black, W. H. Robinson, and S. P. Buchner. Fault Simulation and Emulation Tools to Augment Radiation-Hardness Assurance Testing. *IEEE Transactions on Nuclear Science*, 60(3):2119–2142, June 2013.

[22] A. Sari, M. Psarakis, and V. Vlagkoulis. An Open-source Framework for Xilinx FPGA Reliability Evaluation. In *Open Source Design Automation*, Florence, Mar. 2019.

[23] A. Ullah, P. Reviriego, and J. A. Maestro. An Efficient Methodology for On-Chip SEU Injection in Flip-Flops for Xilinx FPGAs. *IEEE Transactions on Nuclear Science*, 65(4):989–996, Apr. 2018.

[24] L. Urban. Development of a runtime switch behavior for a multi-level risc-v instruction set to register transfer fault injection simulator. Master's thesis, Technische Universität München, 2022.

[25] H. Ziade, R. A. Ayoubi, R. Velazco, and others. A survey on fault injection techniques. *Int. Arab J. Inf. Technol.*, 1(2):171–186, 2004.

Using High-Level Profiling Data to Early Assess the Robustness of Digital Systems

Luc Noizette[1,2], Florent Miller[2], Youri Helen[3], Régis Leveugle[1]

[1] Univ. Grenoble Alpes, CNRS, Grenoble INP*, TIMA, 38000 Grenoble, France
[2] Nucletudes, Les-Ulis, France
[3] French Ministry of Defense, Bruz, France

Abstract—Complex digital components such as high performance CPUs are increasingly used in embedded systems with major robustness requirements. Early robustness evaluations, and evaluations all along the development phases, are necessary when designing such systems to quickly converge towards the expected dependability level. These evaluations must take into account both the hardware platform and the application software, but at a time when the real application software is not yet available. First evaluations must therefore rely on expected characteristics of the target software and some knowledge about the robustness of similar programs. In this context, we propose a robustness evaluation methodology based on a set of metrics quantified using high-level software profiling and exploited to compute derating factors. We show that the early predictions are very close to results obtained from lengthy fault injections performed on the actual hardware platform, and still better when several metrics are combined with the usual consideration of data lifetime. We also discuss the impact of the compilation options on the quality of the evaluations.

Keywords—Digital Systems, Early Robustness Assessment, Software Profiling

I. INTRODUCTION

Commercial grade complex digital components (COTS) are growingly used in critical embedded systems in all sectors. Even aerospace manufacturers resort to such devices in their on-board control units due to the race for High-Performance Embedded Computing (HPEC) as illustrated by Vision Based Navigation (VBN) [1]. This tendency is supported by the scaling trend of technological nodes which allow ever more computing power in ever more integrated components. In practice, such a shift is reflected in space agencies roadmaps [2] where the integration of more COTS devices represents one of the highest challenges.

Embedded digital systems in the aerospace sector require a high level of robustness, i.e. a high capability to operate nominally or within user-acceptable degradation bounds in spite of faults due to harsh environments. Among the many constraints to which these systems are exposed, the radiation one is among the most impactful on the race for HPEC. In fact, through the interaction of particles with silicon, this harsh radiation environment can produce unwanted transient currents at transistor level, propagating in the component to generate at higher levels a wide range of fault types, from non-critical Silent Data Corruption (SDC), to worst cases like critical SDC or Single Event Functional Interrupt (SEFI). These phenomena have become more pronounced with the scaling trends [3] and the use of COTS components, resulting in an increased demand for robustness assessment.

Early robustness evaluation of such systems, and evaluations all along their development phases, are necessary to rapidly meet the specified dependability levels and achieve

the expected qualification without iterations after the costly standardized tests under beam. To be as accurate as possible, these early robustness evaluations must take into account both the static sensitivity of the hardware platform (i.e., the probability of fault occurrence due to disturbances) and the dynamic derating related to the possible propagation of a fault towards a user-noticeable failure when running the application software. Taking into account this derating is crucial, since the actual robustness can vary by several orders of magnitude for the same platform in the same radiation environment [4]. However, since the final program is rarely available during early assessment, evaluations must rely on expected characteristics of the target software and on available data about programs exhibiting structural and behavioral similarities e.g, benchmarks or previously developed software with similar purposes.

One way to improve these robustness estimations is to better understand the link between the robustness of digital systems and the programs they execute. A first step on this path, was to use the internal switching activity as a first metric of interest [5] but this approach was difficult to apply in the early stages of development. Following on from this work, the methodology presented here relies on a set of metrics quantified using high-level software profiling. The system robustness is then derived using derating factors computed from these metrics. Such metrics can be estimated at a very early stage of the software design and then incrementally refined all along the process thanks to very quick computations.

The organization of this paper is as follows. Section II positions the approach presented here with respect to related works in the state-of-the-art. Section III details the methodology proposed in this work. Section IV explains the flow used to obtain high-level profiling data. Section V describes the setup used to obtain the experimental results. Section VI and VII compare the robustness levels estimated by the methodology with those obtained from fault injections performed on the actual hardware platform respectively in the main memory and inside General Purpose Registers (GPR). The impact of the software optimizations on the quality of the evaluation is also discussed in these two last sections.

II. RELATED WORKS AND POSITIONING

From a global perspective, the methodology presented here can be split into two aspects: (1) Robustness evaluation, (2) High-level software profiling. For each of these domains, a summary of related works is presented.

The robustness evaluation of digital systems is generally performed at different stages of the design flow using fault injection methods. From virtual platforms such as QEMU [6] to microarchitectural simulations with e.g., GeFIN [7], down to electrical or physical simulators, all the main simulation

---

* Institute of Engineering Univ. Grenoble Alpes

979-8-3503-6689-1/24 $31.00 © 2024 IEEE

platforms have been enhanced to integrate such capabilities. When the digital system is an ASIC or a softcore, methods using e.g., saboteurs/mutants [8] or reconfigurations on a FPGA fabric [9] have been developed to inject faults at the flip flop level. In a complementary manner, injections can be driven by the software, for example at compile time with tools like LLFI [10] or at runtime leveraging exceptions [11].

As an alternative to fault injection, the use of derating factors to early assess the robustness of digital systems has been applied multiple times in the state of the art. For example, in their work, S. Houssany et al. [12] have developed a tool called Cache Analyzer that computes a robustness derating factor based on the analysis of cache memory transactions. More recently, T. S. Hsu et al. [13] have proposed to estimate the robustness of processors based on derating factors computed from the fraction of architecturally correct execution (ACE) bits. A similar approach where derating factors are computed from ACE bits estimated through fault injection has been used in [14] on Arm microprocessors.

High-level software profiling is also not new in the state of the art but is mostly used for benchmarks characterization. In their work, K. Hoste et al. [15] have characterized workloads of 118 programs from 6 benchmark suites using 8 different characteristics that are microarchitecture-independent. On their side, R. Panda et al. [16] have characterized the similarities inside SPEC benchmark suites with micro-architecture independent workload characteristics. In general, profiling techniques are used to select a representative subset of benchmarks based on clustering as presented in [17]. The other main aspect covered by software profiling concerns the generation of synthetic benchmarks [18] or clone morphing benchmarks [19]. On the robustness side of benchmark selection, H. Quinn et al. [20] have proposed a list of programs to use during robustness evaluation of digital components to cover a wide variety of software behaviors.

To the best of our knowledge, the few works that have tried to early assess the robustness of digital systems based on derating factors are based on fault injections or only focus on the Life Time criterion, evaluated for a given microarchitecture and memory hierarchy. None of them have tried to use a combination of profiling metrics obtained at high abstraction level, associated to a fit established for a set of data available from previous program characterizations. The methodology presented here aims to fill this gap in the state of the art. The proposed approach does not rely on any fault injections and leverages high level simulation platform features for profiling and behavioral metrics quantification. These simulations are not cycle accurate to be very quick and easily repeated during the software development to support design decisions or accurately follow the robustness estimation trend. Benchmark characterization can be used to identify a class of programs similar to the target application software. For these reference programs, the metrics of interest are obtained from profiling and the robustness data (corresponding to the probability of no functional impact after a fault occurs) can result from e.g., previous fault injection campaigns with these reference programs, experiments under beam, operating data records or any data set that provides information on the reliability of the application. The derating factor for the target application is then estimated assuming that all programs in this selected set have correlating relationships between profiling metrics and robustness.

III. METHODOLOGY

In this work, we take into account only two levels in the memory hierarchy, composed of the Main Memory (MM) and General Purpose Registers (GPR). We do not consider microarchitectural registers, nor microarchitectural optimization features but only the sequence of executed instructions during the program execution. This also means that the profiling metrics used to estimate robustness levels are only linked to memory behaviors with approximate timing characteristics i.e., clock cycles are replaced by instruction sequence lengths. This approximation allows estimations even when the final processor hardware is not yet selected, provided that the target Instruction Set Architecture (ISA) has been decided. Moreover, this imprecision is not a real limitation in the context of this work, since it has been shown in [21] that the application has a significantly greater impact on the robustness than the CPU microarchitecture.

In our approach, the global process is divided into four stages. First, a reference set of benchmarks representative of the domain of interest is selected. For example, the set of benchmarks defined in [20] may be used for radiation testing of microprocessors or the method in [17] may be leveraged.

This reference set is then characterized in terms of profiles and robustness. Profiles may be already available in a database, or generated using a tool similar to the one presented in Section IV. From these profiles, the metrics of interest can be computed. In this study, we considered four different metrics that will be presented in Section IV, but another set of metrics may be chosen. As previously mentioned, the robustness data for each reference program can come from different origins (beam/laser experiments, faults injection campaigns, bibliography, mission data, etc...) and be collected in a database. Robustness is characterized as the probability of no functional impact after a fault occurs. This probability should be known for each program in the reference set, for both MM and GPR. The probability in MM can be refined, considering separately the different memory zones (e.g., program instructions, application data or stack data) that may exhibit different behaviors and sensitivity to faults. Each zone (or the global MM) and the GPR are denoted hereafter as a section.

For each profiling metric considered in the analysis, a regression formula (noted Rf_s) is then determined for each section. The derating formula (noted Df_s) is obtained for each section as the average of all the regressions.

Finally, when the estimated robustness (Er) of a new program must be estimated, only its profiling metric values are required to compute the value of Df_s for each section. The total number N of possible faults with respect to the selected fault model is split into subsets of N_s faults, one per section, and Er is computed from the weighted sum of estimated no impact faults in each section, as summarized in Eq. 1.

$$Er_{program} = \frac{\sum_s Df_s(program_{profile}) \times N_s}{\sum_s N_s} \quad (1)$$

Eq. 1. Generic expression of the estimated robustness (Er) for a new program. Ns represents the number of possible faults in the section s, considering the selected fault model.

This approach will be illustrated in the next paragraph, showing also the benefits of taking several profiling metrics into account.

979-8-3503-6689-1/24 $31.00 © 2024 IEEE

IV. OBTAINING HIGH-LEVEL PROFILING DATA

The process used in our work to compute high-level profiling data is divided into two parts. First, a platform is used to simulate the execution of a program and obtain its execution trace i.e., the sequence of executed instructions (without more precise timing information). Then, a custom program uses these traces to compute high-level profiling metrics.

In this case study, as our work targets RISC-V processors, we based our profiling tool on Spike. This ISA simulator, commonly referred to as the golden reference of the RISC-V ISA, is often the first platform used by the software development team to validate an application before switching to cycle-accurate simulation. Since this platform is only used to obtain execution traces, other simulators or virtual platforms like QEMU could have been used as alternatives to Spike as presented in [22].

Our custom profiling tool takes as input the execution traces provided by Spike and records every access to any memory element of the hierarchy. In the MM part, a memory element corresponds to a byte while in the GPR part, a memory element corresponds to one register. For example, an instruction like "add rd, rs1, rs2" performs three GPR accesses (one write to the destination register and one read on each of the source registers) and one read access in MM at each of the four bytes at the address in the main memory containing this instruction. Once this tracing step is completed, the profiling tool computes the four memory-related metrics described in Table I and visually illustrated in Fig. 1.

Table I. Listing of the memory profiling metrics used in this work, computed for each memory element and then averaged per considered memory zone.

Label	Full name	Description
Nacc	Number of accesses	Number of accesses (Read or Write) to a memory point
Rrat	Read Ratio	Number of Read accesses divided by Nacc
MDbwA	Mean dist. Between Access	Mean distance (in number of instructions) between two consecutive accesses to a memory point
Lt	Life Time	Sum of time intervals (in number of instructions) between a Write and the last associated Read

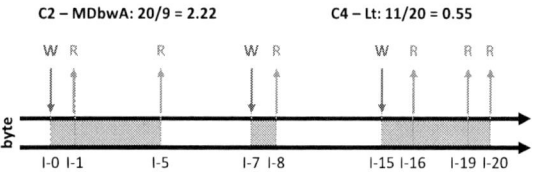

Fig. 1. Visual representation of the four profiling metrics used to describe accesses to each element of the memory.

These four profiling metrics have been selected in the "proof of concept" context reported here, because memories are the most prone to fault occurrence and are therefore a major origin of failures in digital platforms. Our profiling tool also computes other metrics related to the software basic blocks executions that may be considered in further works as discussed in Section VIII. Memory hierarchy and basic block analyses are two new features of the profiling tool compared to the first version presented in [5].

V. DESCRIPTION OF THE EXPERIMENTAL SETUP

In order to validate the methodology introduced in Section III, the four stages of its flow must be applied to a real digital system. Thus, the experimental setup must enable the elaboration of the regressions from reference programs, the computation of the derating formula (Df) applied to another program and the comparison of the resulting estimation with experimental data. Here, the experimental data is obtained after fault injections on a real RISC-V platform. Moreover, to estimate the gain in accuracy offered by this methodology, a derating factor only based on the Life Time metric is also computed, in addition to the one based on the four metrics.

The hardware platform used in this paper is a RISC V based soft-SoC on a flash-based Microsemi Field Programmable Gate Array (FPGA). As presented in Fig. 2, this soft-SoC is based on a MiV core in version v.3.0.1. This core is connected to the main memory and the other peripherals via an AHB/APB bus. This architecture represents a versatile hardware design as it allows the designer to easily change the core or even the peripherals to match his requirements. In our case study, the peripherals are composed of timers, GPIO and UART modules.

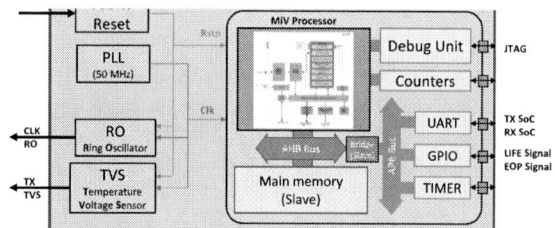

Fig. 2. Simplified block diagram of the DUT used in this study.

Seven programs either custom or extracted from the Mibench [23] test suite have been selected in the context of this study as benchmark set for the case study. Some of them are used as reference to compute the regression formulas and the others are used to quantify the precision of our estimations. The bitcount algorithm has been declined in seven different versions (B1-B7). Moreover, B6, CRC, Dijkstra and Sha algorithms have been compiled with three different optimization flags (o0, o1, o3). Other programs have been compiled only with o0. For each of these programs, the main memory can be separated in 5 different zones. The Text section contains the instructions. The Data section contains data that are initialized at the beginning of the program. The Bss section is reserved for variables that are not initialized at the beginning of the program execution. Finally, the Heap and the Stack sections are reserved for local variables during execution of functions. As these zones have completely different behaviors, regression formulas of the methodology have been computed on each section separately.

The execution of a benchmark is divided into three different parts (initialization, execution, checksum). First the initialization phase puts the DUT in the same state for each program. Then, during the execution phase, the core of the benchmark is executed (e.g., the matrix multiplication is performed for the MxM program). During this phase, a life signal is periodically toggled. Finally, during the checksum phase, an End Of Program (EOP) output is raised and the life signal is turned off. The Adler checksum of the result is computed and sent by UART in order to be compared to a golden reference (execution without fault injection). This global execution is repeated within a loop.

From this behavior, four types of fault effects can be identified by the fault classification system: (1) No impact observed on the result; (2) T-SDC a Transient (one loop) Silent Data Corruption of the result checksum occurs; (3) P-

SDC a Permanent Silent Data Corruption of the checksum impacts multiple computation loops; (4) SEFI a Single Event Functional Interrupt, also known as crash or hang. In our case study, only the percentage of injected faults with no impact on benchmark execution is used for robustness characterization and methodology validation. The dominant class of the other faults depends on the section and the register in GPR.

In order to ensure an accurate assessment of the proposed methodology, fault injections were performed on the real hardware platform. As mentioned in Section II, using fault injections during Spike simulations would have been easier but not representative of actual behaviors. The fault injection tool that has been developed will not be discussed in details in this paper due to length restrictions. A preliminary version of this tool is presented in [5]. As a quick summary, this tool uses the debugging interface offered by Smartdebug to modify a bit at a specific location in MM or GPR after a given instruction has been executed in the soft-SoC. This tool is fully automated and coupled to a script allowing the generation of a list of faults meeting statistical criteria as in [24]. It was developed for Microsemi FPGA boards, and has been upgraded from [5] with the ability to inject faults into both MM and GPR.

VI. APPLICATION OF THE METHODOLOGY IN MM

In order to compute regressions for each section in MM, a separate Statistical Fault Injection (SFI) was performed in each section of each program, meeting the SFI requirements. The parameters used are an error margin of 3% and a confidence level of 95%, resulting in an injection of 1068 faults per memory zone per benchmark.

An analysis of the repartition of fault effects in each section for the whole set of benchmarks reveals that each section has a predominant type of effect. As illustrated in Fig. 3, in the Text section, most faults having an impact lead to a SEFI while in the Data section, the predominant effect is P-SDC. This is because a fault in the Text section is very likely to corrupt instruction encoding, resulting in a crash while a fault in the Data section will corrupt the constant input data of the benchmark and thus change its output for multiple execution loops. In the same manner, the predominant effect in Bss and Stack sections is T-SDC because these sections contain temporary data like local variables of functions. The Heap section, on its side, contains only no impact faults as this section is not used by the compiler for all programs considered in the case study.

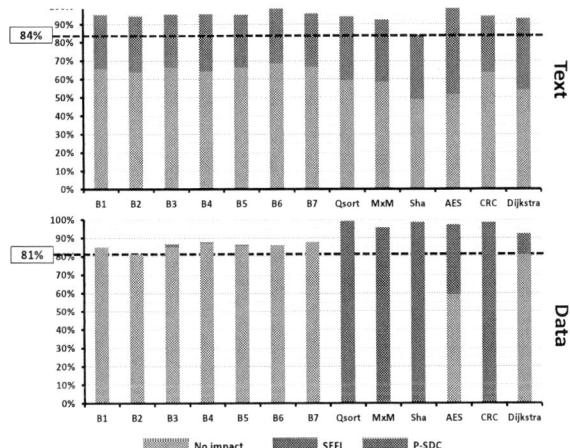

Fig. 3. Faults repartition in Text and Data sections.

The percentage of no impact faults in each section was then plotted against the value obtained for each profiling metric in order to determine the regression equations mentioned in the methodology summary. This is illustrated in Fig. 4 for the Rrat metric in the Text section. In this figure, we show the values for all benchmarks but only some of them are selected to compute the regression equations; the others are used to make the predictions (estimated robustness Er) based on the methodology and evaluate the differences with the fault injection results. In this case study, B1, B3, B7, MxM, Sha_o0, AES, CRC, and Dijkstra were selected to compute the regressions in order to cover a large spectrum of the metric values, as it would be the case in our methodology when using the reference database.

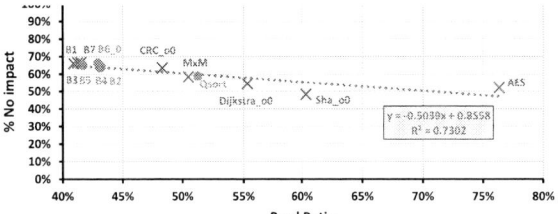

Fig. 4. Percentage of no impact faults in the Text section as a function of the Read Ratio (Rrat) for all o0 benchmarks. Benchmarks used to compute the regression equation are indicated with a blue cross and those used for assessment are indicated with a green circle.

As it can be seen in Fig. 4, the percentage of no impact faults in the Text section as a function of the mean Read Ratio follows a linear regression with a slope factor of -0.5039 and an offset of 0.8558. This behavior is not surprising as an increase in the Read Ratio can be translated into an increase in the number of Read accesses, which tends to increase data lifetimes and therefore reduces the likelihood of a fault having no impact. In the same manner, a negative slope has also been observed in linear regression of the percentage of no impact faults as a function of the Lt profiling metric. On their side, the Nacc and MDbwA profiling metrics follow a linear regression with a positive slope.

The robustness estimated with our methodology (Er) corresponds to the estimated percentage of all possible faults resulting in a no impact outcome in the whole memory, considering a given fault model (in this case study, all possible single bit-flips at any time during the program execution). In this case study, we also considered only linear regressions so the Df of each section can be expressed as presented in Eq. 2. Er is obtained for MM from Eq. 1, taking into account only the five sections in MM.

$$Df_{Text} = \frac{1}{4}\left(\sum_i a_i \times M_i + b_i\right) \quad (2)$$

Eq. 2. Expression of the derating formula (Df) of a section when considering only linear regressions. The four a_i and b_i coefficients are obtained from the regression equations and the four M_i correspond to the four profiling metrics values for the program under robustness assessment.

To validate the methodology, another SFI campaign was performed for each benchmark, this time with random faults uniformly distributed in the whole memory. Moreover, to estimate the gain in accuracy offered by the methodology presented here, an estimation of the percentage of no impact fault was also performed based only on the Life Time metric. Results are shown in Table II, with the estimation differences between our methodology and SFI. Let us recall that SFI was made for an error margin of 3% resulting in 23874 faults injected per benchmark. Last but not least, our estimations based on the derating formulas are obtained 400x to 500x faster than the SFI results on the hardware platform.

Table II. Robustness assessment in MM from SFI experiments in the global MM and relative estimation differences with the proposed methodology using derating formulas and one or four metrics. Benchmarks used to calibrate deratings are indicated by the * marker.

Program	SFI robustness evaluation	Difference (our method, lifetime only)	Difference (our method, 4 metrics)
B1*	67.0 %	4.7%	0.1 %
B2	66.8 %	4.4 %	0.4 %
B3*	66.8 %	4.8 %	- 0.2 %
B4	66.9 %	4.2%	-0.4 %
B5	66.8 %	4.3 %	-0.6 %
B6	66.9 %	4.5 %	-0.4 %
B7*	66.1 %	5.4 %	-1.3 %
Qsort*	37.4 %	8.0 %	6.7 %
MxM	73.8 %	- 14.8 %	-18.0 %
Sha*	46.1 %	11.1 %	4.7 %
AES*	61.9 %	-5.6 %	-3.5 %
CRC*	29.5 %	20.2 %	5.0 %
Dijkstra*	72.2 %	-12.6 %	-0.4 %

For most of the tested benchmarks, the new methodology proposed in this work significantly improves the estimation of robustness compared to the version that only takes into account the Life Time. Except in the case of MxM benchmark (that was part of the reference set), the error in estimating the non-impact fault rate with our methodology is below 5% with respect to SFI. This also holds for the benchmarks outside the reference set, meaning that a good estimation of the metric values for a new program leads to a correct estimation of the robustness. The result with MxM, and to a lesser extent with Qsort, mainly shows that this program should not be in the same group of programs as the others to apply the methodology, so should be in a different section of the reference database as mentioned in section III.

The derating formula previously calibrated with a reference set only composed of programs compiled with the o0 option has been applied to the four benchmarks available in different optimization levels. Results are listed in Table III.

For the four benchmarks used in this case study, the difference in the robustness assessment increases with the optimization but remains only slightly affected. This is mainly due to the fact that these benchmarks only occupy few kilobytes in a main memory of 1MB. We expect a more significant impact on the accuracy of robustness assessment for programs with a larger memory size. This is why we recommend calibrating the derating formula with a reference set using the same optimization than for the target program.

Table III. Impact of compilation options on methodology accuracy (MM)

Program	Difference (o0)	Difference (o1)	Difference (o3)
B6	-0.4 %	0.7 %	0.9 %
Sha	4.7 %	4.9 %	5.1%
CRC	5.0 %	5.4 %	5.6 %
Dijkstra	-0.4 %	-0.6 %	- 0.7 %

VII. APPLICATION OF THE METHODOLOGY ON GPR

For GPR, faults were injected during the execution phase of B6, Sha, CRC and Dijkstra only. The parameters used during SFI are an error margin of 1% and a confidence level of 99.8% resulting in around 22000 faults injected per program. Like in MM, the derating factor computed only with the Life Time criterion is compared to the one computed with the four metrics. Only CRC and Sha have been used to perform the regressions. Results are presented in Table IV.

Table IV. Robustness assessment in GPR with o0 compiltation option from SFI experiments and relative estimation differences with the proposed methodology using derating formulas and one or four metrics. Benchmarks used to calibrate deratings are indicated by the * marker

Program	Measured robustness	Difference (our method, lifetime only)	Difference (our method, four metrics)
B6	90.8 %	9.2%	-2.5 %
Sha*	91.4 %	0.0 %	0.0 %
CRC*	89.7 %	0.0 %	0.0 %
Dijkstra	88.5 %	3.8%	-0.3 %

As for MM, the estimated robustness obtained with the four metrics for GPR is more precise than the one obtained only with Life Time metric. The speed up for estimations with our method is around 400x compared to SFI.

In order to analyze the impact of optimization options on the accuracy of the proposed methodology applied to GPR, three derating formulas (Df) have been calibrated. These three Df, referred as Dfo0, Dfo1 and Dfo3 have been respectively calibrated with Sha and CRC benchmarks compiled with the o0, o1 and o3 optimization options. Estimated robustness obtained with these three Df have been compared to the robustness evaluated by SFI on the o1 version of the test programs. Results are illustrated in Fig. 5.

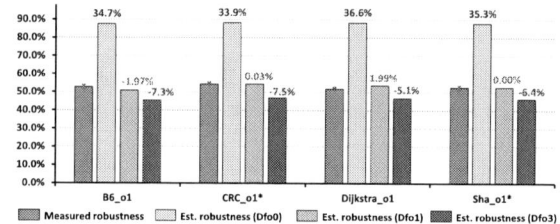

Fig. 5. Impact of compilation options on methodology accuracy (GPR)

For GPR, the impact of compilation options on the robustness assessment accuracy is much more pronounced than for MM. Using the Df calibrated with o0 programs to assess the robustness of o1 programs results in an overestimation around 35% where using the Df calibrated with o3 programs results in an underestimation around 6%. Results obtained with Dfo1 are clearly much accurate. This results stems from the fact that the regression formula (Rf) for these three optimization levels are different.

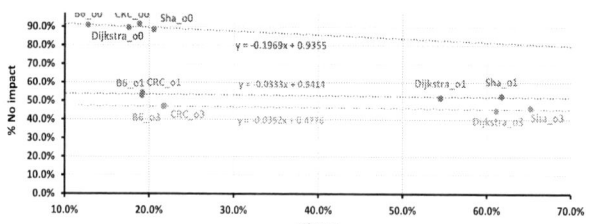

Fig. 6. Illustration of Rf associated to different compilation options

As an example, Fig. 6 presents in dotted lines the Rf used for the Life Time metric in the three Df. These three Rf follow the same trend i.e, a decrease in robustness with an increase in Life Time. Rf1 and Rf3 are close to each other compared to Rf0, indicating a very different trend for o0.

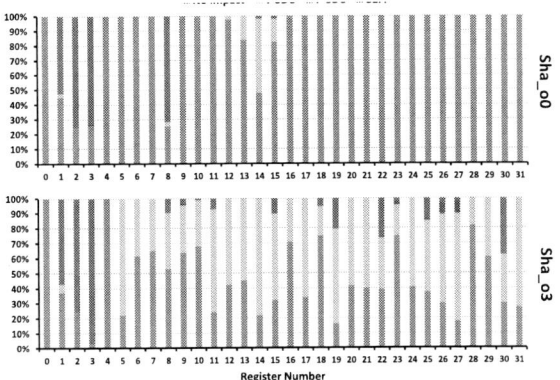

Fig. 7. Repartition of the 22k faults in the GPR of Sha_o0 and Sha_o3

In fact, as illustrated in Fig. 7, when a benchmark is compiled with the o0 optimization flag, only the registers x1, x2, x3, x8 and x12-15 are used. As the registers x12 to x15 are mainly used for containing temporary data during computations, the faults injected generally result in T-SDC. The registers x1, x2 and x3 are more prone to SEFI because they contain the return address, the stack pointer and the global pointer. When compiled with the o3 flag, almost all registers are used and it is almost the same with o1. In consequence, the percentage of no impact faults noticeably decreases. This explains the discrepancy observed in Fig. 5, when the robustness of a program compiled with o1 is estimated using a regression calibrated with programs compiled with o0. In other words, the choice of the compilation options must be done before any estimation for the GPR, and regressions must be based using data corresponding to the same compilation flag.

VIII. CONCLUSIONS AND PERSPECTIVES

A generic methodology has been proposed to early estimate the robustness level of digital systems running application software. It does not require to know the exact hardware implementation or CPU, but just the targeted ISA and if possible the compilation options to achieve more accurate estimations, especially for GPR. Accurate and very quick assessments are achieved from a few high-level profiling metric estimations and a database with profiling and robustness data of a reference program set. Results obtained combining several profiling metrics proved to be more precise than those obtained considering only the Life Time metric.

Only the four most preponderant profiling metrics have been considered here. However, the highly versatile aspect of the methodology allows adding other metrics covering generic hardware mechanisms (e.g., memory caches) or ISA specific mechanisms (e.g., register windows in the SPARC ISA) as long as they are observable at high level. In future works, we plan to explore such extensions and achieve deeper analysis of the robustness with respect to the software patterns.

ACKNOWLEDGEMENT

This work was funded by the French Defence Innovation Agency (AID) as part of a PhD thesis.

REFERENCES

[1] G. Lentaris et al., "High-Performance Embedded Computing in Space: Evaluation of Platforms for Vision-Based Navigation", Journal of Aerospace Information Systems, vol. 15, no.4, pp. 178–192, Feb. 2018.

[2] G. Furano et al., "Roadmap for On-Board Processing and Data Handling Systems in Space", in Dependable Multicore Architectures at Nanoscale, edited by M. Ottavi, D. Gizopoulos, and S. Pontarelli, Springer, Cham, Switzerland, pp. 253–281, 2018.

[3] I. Chatterjee, "From MOSFETs to FinFETs - The Soft Error Scaling Trends", RADNEXT, November 2020 [Online]. Available: https://radnext.web.cern.ch/blog/from-mosfets-to-finfets.

[4] B. James et al., "Investigating How Software Characteristics Impact the Effectiveness of Automated Software Fault Tolerance" in IEEE Transactions on Nuclear Science, vol. 68, no. 5, pp. 1014-1022, 2021.

[5] L. Noizette et al., "Understanding the Link Between Complex Digital Devices Soft Error Rate and the Running Software" in IEEE Transactions on Nuclear Science, vol. 70, no. 8, pp. 1747-1754, 2023.

[6] A. Höller et al., "FIES: A Fault Injection Framework for the Evaluation of Self-Tests for COTS-Based Safety-Critical Systems" in 15th International Microprocessor Test and Verification Workshop, pp. 105-110, Dec. 2014.

[7] M. Kaliorakis et al., "Differential Fault Injection on Microarchitectural Simulators" in IEEE International Symposium on Workload Characterization, pp. 172-182, Oct. 2015.

[8] J. -C. Baraza et al., "Enhancement of Fault Injection Techniques Based on the Modification of VHDL Code," in IEEE Transactions on Very Large Scale Integration Systems, vol. 16, no. 6, pp. 693-706, June 2008

[9] L. A. Aranda et al., "ACME: A Tool to Improve Configuration Memory Fault Injection in SRAM-Based FPGAs," in IEEE Access, vol. 7, pp. 128153-128161, 2019

[10] Q. Lu et al., "LLFI: An Intermediate Code-Level Fault Injection Tool for Hardware Faults," in IEEE International Conference on Software Quality, Reliability and Security, pp. 11-16, August 2015.

[11] R. Velazco et al., "Predicting error rate for microprocessor-based digital architectures through C.E.U. (code emulating upsets) injection," IEEE Trans. Nucl. Sci., vol. 47, no. 6, pp. 2405–2411, Dec. 2000.

[12] S. Houssany et al., "Microprocessor Soft Error Rate Prediction Based on Cache Memory Analysis," in IEEE Transactions on Nuclear Science, vol. 59, no. 4, pp. 980-987, Aug. 2012.

[13] T. S. Hsu et al., "Processor SER Estimation with ACE Bit Analysis," in 21th European Conference on Radiation and Its Effects on Components and Systems (RADECS), pp. 1-5, Sept. 2021.

[14] P. R. Bodmann et al., "Soft Error Effects on Arm Microprocessors: Early Estimations versus Chip Measurements," in IEEE Transactions on Computers, vol. 71, no. 10, pp. 2358-2369, Oct. 2022.

[15] K. Hoste et al., "Microarchitecture-Independent Workload Characterization," IEEE Micro, vol. 27, no. 3, pp. 63-72, June 2007.

[16] R. Panda et al., "Data analytics workloads: Characterization and similarity analysis," IEEE 33rd International Performance Computing and Communications Conference (IPCCC), pp. 1-9, Dec. 2014.

[17] Z. Hongping et al., "BenchSubset: A framework for selecting benchmark subsets based on consensus clustering", Intl. Journal of Intelligent Systems, vol. 37, no. 8, pp. 5248-5271, Aug. 2022.

[18] E. Deniz et al. "Using software architectural patterns for synthetic embedded multicore benchmark development" IEEE Intl. Symposium on Workload Characterization (IISWC), pp. 89-99, Nov. 2012.

[19] Y. Wang et al., "Clone morphing: Creating new workload behavior from existing applications," in IEEE International Symposium on Performance Analysis of Systems and Software, pp. 97-108, 2017

[20] H. Quinn et al., "Using Benchmarks for Radiation Testing of Microprocessors and FPGAs" in IEEE Transactions on Nuclear Science (TNS), vol. 62, no. 6, pp. 2547-2554, Dec. 2015.

[21] H. Cho, "Impact of Microarchitectural Differences of RISC-V Processor Cores on Soft Error Effects" in IEEE Access, vol 6, pp. 41302-41313, 2018.

[22] L. Noizette et al., "Using Application Profiling based on a Virtual Platform for SoC Fault Tolerance Assessment," in 17th Conference on Ph.D Research in Microelectronics and Electronics, pp. 225-228, 2022.

[23] M. R. Guthaus et al., "MiBench: A free, commercially representative embedded benchmark suite" Proceedings of the Fourth Annual IEEE International Workshop on Workload Characterization pp. 3-14, 2001.

[24] R. Leveugle et al., "Statistical fault injection: Quantified error and confidence" in Design, Automation & Test in Europe (DATE) Conference & Exhibition, pp. 502-506, April 2009.

Reliability Analysis of a Low-Cost CCSDS 123 Hyperspectral Image Compressor

Wesley Grignani*, Douglas A. Santos†, Luigi Dilillo†, and Douglas R. Melo*

*LEDS, University of Vale do Itajaí, Itajaí, Brazil
†IES, University of Montpellier, CNRS, Montpellier, France
wesley.grignani@edu.univali.br, {douglas.santos, luigi.dilillo}@umontpellier.fr, drm@univali.br

Abstract—Remote sensing techniques in space applications use images that provide a large volume of information to collect data about the Earth. Hyperspectral images (HSIs) present a high volume of data and affect restrictions on the storage capacity and processing in space systems, highlighting the demand for compression. Systems that operate in space are susceptible to faults due to adverse conditions and require the implementation of protection techniques to mitigate these faults and ensure correct operation. This work implements a fault-tolerant CCSDS 123 HSI compressor using Triple Modular Redundancy (TMR) and Hamming Error Correcting Code (ECC) techniques to protect from Single Event Upsets (SEUs). We present the compressor in a standard version, followed by partially-protected and fully-protected versions, by applying hardening in different circuit components. We performed a reliability analysis through a fault injection campaign considering all implementations. The standard version showed 97.9% of executions with errors, which was reduced in the versions partially hardened, reaching no error propagation in the fully hardened version. In addition, all implementations accelerated the application, achieving a performance increase of up to 24× compared to the software solution.

Index Terms—Systems-on-Chip, Hardware Accelerator, Hyperspectral Images, CCSDS 123.0-B-2, Fault Tolerance.

I. INTRODUCTION

Remote sensing in space applications is a crucial technology that allows data about the Earth to be collected from satellites and space probes. Among these techniques is hyperspectral imaging. Hyperspectral images (HSIs) capture information in hundreds of spectral bands, allowing the detection of very specific characteristics of the materials observed. This proves valuable for applications such as Earth image acquisition, climate analysis, and forest environment surveillance. Due to their capability of capturing several spectral bands, HSIs present a large volume of data [1].

The high number of bands in HSIs makes it a data-intensive processing structure for space systems, impacting ground station storage, processing, and transmission capabilities. To handle the high data volume of HSIs and enhance transmission

This work was supported in part by the Foundation for Support of Research and Innovation, Santa Catarina (FAPESC-2021TR001907), the Brazilian National Council for Scientific and Technological Development (CNPq - process 50794/2023-5), the Brazilian National Coordination of Superior Level Staff Improvement (CAPES/PROSUC), the EU project RADNEXT - Horizon 2020 (grant 101008126), and Project HARV in the framework of the action "Accelerateur d'innovation" from the University of Montpellier.

and processing efficiency, the Consultative Committee for Space Data Systems (CCSDS) developed the CCSDS 123 compression algorithm [2]. This algorithm is designed for low-complexity hardware implementation, offering lossless and near-lossless compression in its latest release (B-2).

Some works implemented hardware accelerators for HSI compression in FPGA (Field Programmable Gate Array). The work [3] presents a low-cost solution for the prediction step of the CCSDS standard. A high-performance solution is presented in [4], while the works [5] and [6] focused on fault-tolerant solutions using SG (Space Grade) FPGAs. The work [7] focused on a reliability analysis duplicating and triplicating the accelerator developed in [6], using EDAC (Error Detection and Correction) in some internal memories.

In this work, we extend our previous project [8], focusing on implementing a fault-tolerant CCSDS 123 hardware accelerator version. We designed the accelerator in different configurations by combining spatial and information redundancies to improve its reliability. We present the accelerator in a standard version, followed by partially-protected and fully-protected versions. We apply techniques such as TMR (Triple Modular Redundancy) to protect the control units and Hamming ECC (Error Correcting Code) to protect memory elements.

II. CCSDS 123 COMPRESSOR

In our previous project [8], we implemented the compressor in HDL (Hardware Description Language) and HLS (High-Level Synthesis) using the AXI4-Lite communication bus in a SoC (System-on-Chip) integration. In this work, we designed a fault-tolerant version of the accelerator, presented in Fig. 1. The architecture was improved by implementing a sample buffer inside the accelerator and the SoC communication architecture to use an AXI4-Stream bus with DMA (Direct Memory Access). These changes were made to eliminate the need for samples to be buffered outside the processor and to reduce the communication bottleneck in the previous system.

The fault-tolerant version was implemented to protect from SEUs (Single Event Upsets) affecting a single bit. This technique was applied considering the results in [9], which observed that most of the events in the experiments consisted of single-bit upsets in the memory elements. We designed the accelerator in different hardened configurations, where TMR and Hamming ECC were applied to different components.

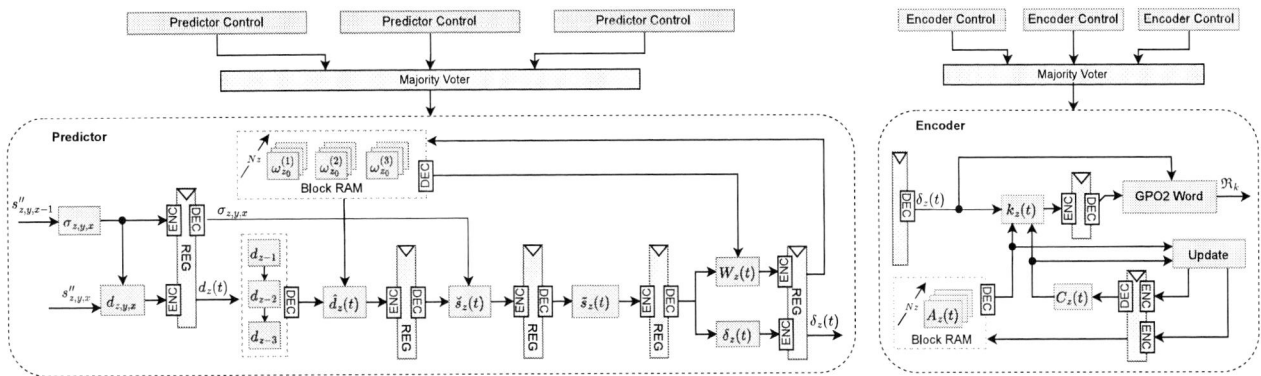

Fig. 1. Hardened compressor design overview.

These configurations were compared in terms of cost, performance, and reliability using different HSI configurations.

The accelerator was designed to minimize resource usage. This includes the local sum $\sigma_{z,y,x}$ in column-oriented mode with BIP (Band Interleaved by Pixel) processing orders, allowing the vector size of local differences $U_z(t)$ to be based on the number of previous bands P_z, reducing resources compared to other processing orders. The size of weights vectors $W_z(t)$ is based on the number of Nz bands in the image.

The sample buffer is implemented considering the local sum mode and the processing order. As the local sum mode is column-oriented, the only value that needs to be stored is above the current sample. In addition, the BIP processing order goes through all the samples in that band and then moves on to the next band. Thus, the size of the sample buffering is based on the dimensions Nx and Nz of the image.

A. Hardware Implementation

The components were designed to work sequentially, and some steps have been parallelized to reduce the cycles needed to process a sample. As the encoder only needs the output of the predictor, it works in a pipeline with the predictor.

The first steps of the predictor are local sum and local difference, which are set to execute simultaneously. Next, the central prediction multiplies the values of the weights and local difference vector. The high-resolution prediction uses the values calculated in the central prediction and local sum step. The double-resolution prediction is next performed, and the mapped quantized index and weights update step are executed simultaneously at the final step.

The encoder starts by registering the information from the predictor. Next, the index $k_z(t)$ is calculated, defining the number of bits to encode the mapped residual. Next, is calculated the GPO2 $\Re_k(\delta_z(t))$ word, with its output formatted in the proper form of the standard. In the end, the accumulator $A_z(t)$ and counter $C_z(t)$ memories are updated.

We created the local difference vector using registers to allow reading all values in a single cycle. The weight vectors were created using block RAMs. As one sample of each band is processed at a time, the z index drives the weight vector selection. In the encoder, simultaneous access to all accumulator values is unnecessary, allowing the block RAMs to implement these vectors and reduce resource usage.

We used a Zedboard Zynq-7000 with an ARM Cortex-A9 to prototype the developed compressor. The DMA transfers data between the memory and the compressor and receives it back to store in memory. The AXI interconnect interfaces between master and slave signals, with one interconnect used for AXI4-Lite and the other for the AXI4-Stream Interface.

B. Fault Tolerance

We created 6 different configurations to harden different compressor components:

- **STD**: the accelerator without fault tolerance techniques.
- **TMR_CTRL**: TMR applied in the predictor and encoder controllers.
- **HAM_BUF**: protects the buffers with Hamming ECC. The buffers in this circuit represent the Weight, Accumulator, and BIP memories.
- **HAM_REG**: protects the registers with Hamming ECC. The registers store intermediate calculations in the predictor and the encoder.
- **HAM_BUF_REG**: both buffers and registers are protected using Hamming ECC.
- **FULL_FT**: represents the compressor protecting the registers and buffers with Hamming ECC and triplicating the predictor and encoder controllers.

The controllers were hardened using TMR with a simple bit-wise majority voter system. The Hamming ECC was used solely for error correction.

III. RELIABILITY EVALUATION

We created a simulation setup using an AVIRIS [10] image cropped in 4 different sizes. The size of the weight and accumulator vector are based on the Nz bands of the image, while the size of the BIP memory is based on the Nx and Nz size. Thus, for a reliability evaluation, we can analyze the impact of these memories changing the spectral resolution (Nz), the spatial resolution (Nx, Ny), and both spectral and spatial resolution.

This work used a simulation-based injection method to perform the fault injection campaign. The solution from [11] performs SEU injections into the registers and was customized to operate on the designed compressor.

The fault injection involves initially simulating without fault injections to obtain a golden run. The next stage involves listing all the registers in the circuit and randomly choosing one to perform the fault injection at a random time. After that, the simulation continues up to the total simulation time.

If the output of any external port differed from the golden run, it was assumed that the fault resulted in an error in the compressed image. For each of the image configurations, 1000 simulations were performed.

IV. RESULTS

We used the Xilinx 2020.1 tool to collect the synthesis and performance results, targeting the Zynq-7020 FPGA. The fault injection campaign was performed by Modelsim simulations executed on a computer running a Linux operating system with an Intel Core i7-12700 processor and 16 GB of RAM.

A. Compressor Synthesis

Table I presents the results in terms of LUTs (Look-up-Tables), FFs (Flip-Flops), BRAMs (Block RAMs), and DSPs (Digital Signal Processors), along with the maximum operating frequency and estimated power dissipated for all compressor versions. The synthesis results considered the image ($128\times128\times40$) configuration.

TABLE I
COMPRESSOR SYNTHESIS RESULTS FOR THE 128X128X40 IMAGE.

Configuration	LUTs	FFs	DSPs	BRAMs (32Kb)	Fmax (MHz)	Power (mW)
STD	1,172	443	4	6.0	102.78	119
TMR_CTRL	1,201	469	4	6.0	102.32	123
HAM_BUF	1,294	443	4	8.0	80.14	138
HAM_REG	2,196	588	4	6.5	72.17	143
HAM_BUF_REG	2,370	588	4	8.0	67.98	151
FULL_FT	2,563	643	4	8.0	67.40	159

The STD version has the lowest resource utilization and highest frequency because it does not implement any redundancy. In the other configurations, we observed that resource usage increased as different protection configurations were applied, as well as a reduction in the maximum frequency. In the full hardened configuration (FULL_FT), the resources increased 118% in LUTs and 45% in FFs, and the maximum frequency decreased by 34% compared to the non-hardened configuration (STD).

B. Compressor Performance

Table II presents the performance results of each compressor version. The throughput consider the number of cycles to process a sample at the maximum operating frequency of the circuit. The difference in throughput is because the maximum frequency of the circuit decreases as the fault tolerance techniques are applied. Energy consumption also changes, reaching an increase of up to $2\times$ compared to the STD version.

TABLE II
PERFORMANCE RESULTS OF COMPRESSOR CONFIGURATIONS.

Configuration	Throughput (MSa/s)	Energy[1] (mJ)
STD	20.55	5.80
TMR_CTRL	20.47	6.00
HAM_BUF	16.02	8.60
HAM_REG	14.44	9.90
HAM_BUF_REG	13.59	11.10
FULL_FT	13.48	11.79

[1] Estimated energy consumption to process 1MSa.

C. Compressor SoC

Table III presents the SoC synthesis results, where we observed that LUTs and FFs increase by an average of $2.6\times$ and $7.6\times$, and the BRAMs increases by $1.33\times$ compared to the isolated compressor. Also, the maximum frequency of the SoC system slightly decreases in all compressor configurations.

TABLE III
SoC SYNTHESIS RESULTS FOR THE 128X128X40 IMAGE.

Configuration	LUTs	FFs	DSPs	BRAMs (32Kb)	Fmax (MHz)	Power (W)
STD	3,763	3,890	4	8.0	100.58	1.701
TMR_CTRL	3,825	3,940	4	8.0	100.07	1.702
HAM_BUF	3,879	3,879	4	10.0	78.93	1.704
HAM_REG	4,767	4,035	4	8.5	71.11	1.711
HAM_BUF_REG	4,943	4,035	4	10.0	66.74	1.713
FULL_FT	5,072	4,130	4	10.0	66.21	1.717

The resource increase is explained by introducing the DMA in the system, which also adds interconnection components for the AXI4-Lite and AXI4-Stream interfaces, which communicate with the ARM processor.

To evaluate the acceleration of the system, we obtained the execution time of all versions of the compressor and compared it with the software solution running on the ARM processor at a frequency of 667 MHz. The results obtained showed an acceleration of $24\times$ for the STD configuration. We observed an increase in the processing time of the other configurations, which leads to a $16\times$ acceleration in the FULL_FT version.

D. Reliability Analysis

We performed 1000 simulations for each configuration of the compressor and image sizes, totaling 300 hours. Table IV presents the simulation results.

The STD configuration showed 705 errors in the first image configuration, increasing to 974 in the last image configuration. We observed that most errors occurred in the BIP, weight, and accumulator memories. As the size of the image increases, the size of these memories also increases, concentrating most of the fault injections in these elements.

979-8-3503-6689-1/24 $31.00 © 2024 IEEE

TABLE IV
RELIABILITY RESULTS FOR DIFFERENT IMAGE SIZES.

Nx,Ny,Nz	20,20,5	20,20,20	128,128,20	128,128,40
STD	705	884	970	979
TMR_CTRL	700	859	967	974
HAM_BUF	34	13	3	1
HAM_REG	499	788	928	965
HAM_BUF_REG	7	2	0	0
FULL_FT	0	0	0	0

A fault in the weight memory impacts compression directly since it is multiplied by the local difference vector. In BIP memory, a fault affects compression for a sample, as the local sum step utilizes the incorrect value. However, errors don't propagate to subsequent compressions since this memory position is overwritten with a new sample. This explains the elevated error count in the STD configuration.

In the TMR_CTRL version, the number of errors starts at 700 and increases to 974 in the final image configuration. Here, we see slightly fewer errors compared to the STD version due to the TMR applied to the controllers. As the controllers use few memory elements compared to the rest of the circuit, applying TMR only to the controllers is not sufficiently effective in reducing errors.

We observed a significant reduction in the number of errors in the HAM_BUF version. In the first image configuration, the number of errors is 34, representing a reduction of $20\times$ compared to the STD version. The number of errors decreases in the other configurations, reaching only 1 error in the last image configuration. The HAM_REG version presented an error reduction of about 28% in the first image configuration compared to both STD and TMR_CTRL versions, but the errors increase as the image size increases. This reinforces the argument that these memories are subject to a higher incidence of faults as the image size increases.

The combined HAM_BUF_REG configuration shows some errors in the first image configuration, reducing to no errors in the last one. As this configuration protects both buffers and registers, the errors observed occurred in the controllers. They were reduced to 0 due to increased memories, reducing the incidence of faults in the controllers. The FULL_FT configuration was the only one that presented no errors in every image configuration since all the memorization components of the compressor are protected in this version.

E. Discussion

The techniques applied showed different results in the compressor configurations. It was also possible to notice that most faults occurred in the weight, accumulator, and BIP memories due to the high use of memorization elements compared to the rest of the circuit.

The different image configurations allowed for verifying the effectiveness of each protection configuration. In addition to the FULL_FT version, the partially hardened configurations, such as HAM_BUF and HAM_BUF_REG, proved to be effective as the image size increased and can be considered alternatives to a fully protected version.

Compared to the FULL_FT configuration, the HAM_BUF_REG shows a reduction of 7% in LUTs and 8% in FFs with almost the same throughput. On the other hand, the HAM_BUF configuration uses 49% less LUTs and 32% less FFs with higher throughput than the FULL_FT.

Among these results, the HAM_BUF version reduced errors as the image size increased despite not fully protecting the circuit. It could be an intermediate solution as a trade-off to maintain low resource utilization and high throughput.

V. CONCLUSION

In this work, we compared the performance and resource utilization of the different configurations. We also used different image sizes to perform a fault-injection campaign for a reliability analysis. The results show good resilience against SEUs, and all configurations accelerated the compression application compared to the software solution.

For future work, we plan to extend the fault injection campaign with different images and perform an experimental analysis in particle accelerators. We also plan to improve the performance of the compressor by implementing a pipelined architecture to increase the throughput and integrate it into a future multi-core system.

REFERENCES

[1] F. Viel, R. C. Maciel, L. O. Seman, C. A. Zeferino, E. A. Bezerra, and V. R. Q. Leithardt, "Hyperspectral image classification: An analysis employing CNN, LSTM, Transformer, and Attention Mechanism," *IEEE Access*, vol. 11, pp. 24835–24850, 2023.

[2] CCSDS, "Low-complexity lossless and near-lossless multispectral and hyperspectral image compression," Available at: https://public.ccsds.org/Pubs/123x0b2c3.pdf. Accessed: February 06, 2024, p. 102, 2021, cor.3.

[3] L. M. Pereira, D. A. Santos, C. A. Zeferino, and D. R. Melo, "A low-cost hardware accelerator for CCSDS 123 predictor in FPGA," in *2019 IEEE International Symposium on Circuits and Systems (ISCAS)*. IEEE, 2019, pp. 1–5.

[4] J. Fjeldtvedt, M. Orlandić, and T. A. Johansen, "An efficient real-time FPGA implementation of the CCSDS-123 compression standard for hyperspectral images," *IEEE Journal of Selected Topics in Applied Earth Observations and Remote Sensing*, vol. 11, no. 10, pp. 3841–3852, 2018.

[5] A. Tsigkanos, N. Kranitis, G. Theodorou, and A. Paschalis, "A 3.3 Gbps CCSDS 123.0-B-1 multispectral & hyperspectral image compression hardware accelerator on a space-grade SRAM FPGA," *IEEE Transactions on Emerging Topics in Computing*, vol. 9, no. 1, pp. 90–103, 2018.

[6] Y. Barrios, A. J. Sánchez, L. Santos, and R. Sarmiento, "SHyLoC 2.0: A versatile hardware solution for on-board data and hyperspectral image compression on future space missions," *Ieee Access*, vol. 8, pp. 54269–54287, 2020.

[7] L. A. Aranda, A. Sánchez, F. Garcia-Herrero, Y. Barrios, R. Sarmiento, and J. A. Maestro, "Reliability analysis of the SHyLoC CCSDS123 IP core for lossless hyperspectral image compression using COTS FPGAs," *Electronics*, vol. 9, no. 10, p. 1681, 2020.

[8] W. Grignani, D. A. dos Santos, L. Dilillo, F. Viel, and D. R. de Melo, "A low-cost hardware accelerator for CCSDS 123 lossless hyperspectral image compression," in *DFT 2023-36th IEEE International Symposium on Defect and Fault Tolerance in VLSI and Nanotechnology Systems*, 2023.

[9] D. A. Santos, A. M. P. Mattos, D. R. Melo, and L. Dilillo, "Characterization of a fault-tolerant RISC-V system-on-chip for space environments," in *2023 IEEE International Symposium on Defect and Fault Tolerance in VLSI and Nanotechnology Systems (DFT)*, 2023, pp. 1–6.

[10] NASA, "Airborne Visible InfraRed Imaging Spectrometer," Available at: https://aviris.jpl.nasa.gov/. Accessed: February 9, 2024.

[11] D. R. Melo, C. A. Zeferino, L. Dilillo, and E. A. Bezerra, "Maximizing the inner resilience of a network-on-chip through router controllers design," *Sensors*, vol. 19, no. 24, p. 5416, 2019.

Dependable Systems and AI in Critical Infrastructures: A Case Study in European Earth Observation Missions

Valentina Zancan
Department of Management Engineering
Politecnico di Milano
Milan, Italy
valentina.zancan@polimi.it

Filomena Decuzzi
Data Systems Division
European Space Agency
Noordwijk, The Netherlands
filomena.decuzzi@esa.int

Gianluca Furano
Data Systems Division
European Space Agency
Noordwijk, The Netherlands
gianluca.furano@esa.int

Lorenzo Canese
Department of Electronic Engineering
University of Rome Tor Vergata
Rome, Italy
canese@ing.uniroma2.it

Abstract—**This paper explores the design principles for creating dependable systems in critical space applications, focusing on new technologies like autonomous systems and Artificial Intelligence (AI). Although AI holds great promise, its lack of transparency ("black box" nature) challenges the traditional "safety first" approach in space design.**

We emphasize the importance of a thorough requirements identification process. Using a case study from the European Space Agency (ESA), we show how institutional policies and user needs are translated into a Mission Requirements Document (MRD) for the Copernicus Programme's Sentinel missions. This approach ensures that user needs are met throughout the development process.

The case study highlights the need for well-defined dependability scenarios, especially for systems using new technologies like AI. We acknowledge the changing commercial space landscape driven by AI and stress the importance of unique dependability scenarios for functions such as low-latency security Earth Observation (EO). Defining these scenarios is crucial for the safe and reliable operation of new AI-based space systems.

Index Terms—**Dependable AI Safety-Critical Systems Requirements Engineering Commercial Space**

I. INTRODUCTION

The space sector is undergoing remarkable transformations driven by several key factors: increased participation of the private sector, innovative satellite-based services, reductions in launch costs, digitalization, and the pursuit of infrastructural designs conducive to asset reusability. New Space actors, originating from both adjacent and distant sectors to the space industry, introduce innovative expertise, capabilities, and practices [1], [2].

In the domain of Earth Observation (EO), recent advancements fuel from companies specializing in data analytics, alongside entities operating within the IT and information services and solutions sector. This trend indicates a gradual transition of the EO segment into a data industry, prompting the space industry to integrate novel solutions from other sectors to innovate technologies and maximize application opportunities.

In this context, the pioneering role of space agencies remains significant for reference missions, technology development, the establishment of engineering standards, and oversight in high-risk activities of low TRL technologies. However, as the New Space pools a vast array of actors, both public and private, space agencies and manufacturing companies necessitate considering new user bases, transcending conventional institutional and scientific ones. This step is crucial to ensure widespread and competitive adoption and commercialization of technologies, following the respective internally-focused design and development.

Targeting a private and commercial user base demands exploration and understanding of new needs and application domains. In 2021 the market for data generated in space achieved estimated revenues of 1.78B\$, with almost 70% of it coming from the military sector. On the other hand, the market for VAS (Value Added Service) achieved revenues of 2.86B\$. The military sector still holds a dominant role (21%), but the revenues come from a wider range of domains, like the natural resources (14%) and the energy (11.5%) sectors. Projections for the following decades show that the VAS market will evolve with a much faster growth with respect to the "raw data" market, 4.91B \$ and 2.74B \$ respectively in 2032 [3]. Justification of space investments often revolves around commercial applications and value-added services stemming from space assets and analytics and processing of space data. For example, the Copernicus Program provides data for thematic services aimed at benefiting society, while the Galileo

Program offers commercial opportunities leveraging satellite navigation capabilities.

However, institutional user requirements do not guarantee seamless adaptation and adoption by commercial and private users. Evidence of this friction is observed in the New Space trend where companies leverage Copernicus data as a starting point for new product ideation, yet address its limitations (e.g., low revisit, limited accuracy) by deploying proprietary constellations tailored for commercial needs. Similarly, initiatives like virtual constellations [4] integrate diverse data sources to address specific market verticals, complementing Sentinel data (for me is wise to add a reference about virtual constellations).

Educating private users about the value of space-based solutions and recognizing their distinct needs compared to institutional ones is essential. Private users often necessitate real-time data, actionable insights, and cost-effective solutions to address respective business problems. Hence, innovative technologies such as on-board Artificial Intelligence for data processing stand as valuable solutions, enabling users to directly task satellites for real-time actionable data, mitigating latency issues traditionally linked to legacy infrastructures.

The increasing complexity and volume of data generated by space-based systems requires innovative approaches to data management. On-board data processing emerges as a pivotal solution, offering several key advantages. By reducing the amount of data transmitted to Earth, on-board processing significantly mitigates latency and transmission costs, thereby enhancing the accessibility and cost-effectiveness of space-based solutions.

Moreover, on-board data processing opens up new market opportunities by enabling the creation of low-latency data products. For instance, in the realm of precision agriculture, real-time analysis of satellite imagery can provide farmers with critical insights for optimizing crop management and resource allocation. Similarly, in maritime monitoring, on-board processing can facilitate the detection of illegal activities, such as fishing and smuggling, by analyzing vessel traffic patterns and radar data.

Ensuring the trustworthiness of data collected and processed through space-based systems is paramount, especially if as in the examples above, the context of civil security is touched. Robust data security measures must be implemented throughout the entire data chain, from on-board processing to final user delivery. Blockchain technology, with its principles of immutability, transparency, and decentralization, offers a promising approach to securing space-based data. By creating an auditable and tamper-proof record of data transactions, blockchain can bolster trust and confidence in the information used for critical decision-making in areas such as disaster response, border security, and environmental monitoring.

II. Dependability in Space: A combination of Reliability and Availability

One of the perceived limitations of using Artificial Intelligence (AI) in onboard data processing and space control systems is the possible unforeseen impact on dependability.

AI systems, while powerful, can introduce new failure modes that are difficult to predict and test for. Within the European Cooperation for Space Standardization (ECSS) framework, dependability is a composite concept encompassing two key pillars: reliability and availability. Let's delve into these concepts through practical examples from the space domain:

Reliability (ECSS-E-ST-10-02): The ability of a system or component to perform its intended function without failure for a specified period under given conditions.

Example: Imagine a critical spacecraft thruster responsible for orbital manoeuvres. Reliability translates to the thruster consistently delivering the required thrust for the duration of the mission, ensuring precise spacecraft positioning for crucial manoeuvres like docking or avoiding debris.

Availability (ECSS-E-ST-10-01): The readiness of a system or component to perform its intended function at a given instant when needed.

Example: Consider a satellite communication system relaying vital telemetry data back to Earth. Availability ensures the system is operational and can transmit data whenever needed, allowing ground control to monitor spacecraft health and mission progress in real-time.

However, it's important to acknowledge that current Space Engineering standards primarily focus on hardware reliability and system uptime. In the specific domain of data integrity, these standards might be inadequate. The increasing complexity of space systems and the introduction of AI introduce new challenges in ensuring the trustworthiness and accuracy of data throughout its lifecycle, from onboard processing to ground control reception.

Now, let's see how these two aspects intertwine to create a dependable space system:

Combined Impact A highly reliable thruster (low failure rate) might still be unavailable if it undergoes scheduled maintenance. Conversely, a communication system with high availability (always operational) might be unreliable if it transmits corrupted data due to internal errors.

A. How Dependability is deployed in institutional mission designs

Redundancy for Dependability: ECSS standards often emphasize redundancy as a key strategy for achieving dependability. For example, a spacecraft might have multiple redundant thrusters or communication modules. If one fails, the others can take over, ensuring overall mission success. This approach enhances both reliability (by mitigating the impact of a single failure) and availability (by maintaining operability despite potential failures).

Dependability and Mission Phases: Dependability requirements might differ throughout a mission. During the critical launch phase, absolute reliability is paramount. However, as the mission progresses, planned maintenance downtime might be acceptable, impacting availability but not necessarily compromising overall success.

979-8-3503-6689-1/24 $31.00 © 2024 IEEE

B. Safety and Security

Dependability in space systems is not only related to reliability and availability. A dependable space system has to assure a certain level of safety and security. Safety and security, while seemingly distinct notions, are becoming increasingly intertwined in the space domain. Safety refers to protecting life, property, and the environment from the hazards associated with space missions. This includes preventing harm during launch, on-orbit operations, and re-entry. Security, on the other hand, focuses on safeguarding space assets from deliberate actions that could compromise their functionality or lead to data breaches.

The rise of complex cyber-physical systems and the growing number of actors in space are blurring the lines between safety and security. For instance, a cyberattack on a satellite's control system could not only disrupt operations (security issue) but also lead to a loss of control and a potential collision (safety issue). This necessitates a holistic approach that considers both safety and security throughout the entire space system lifecycle, from design and development to operation and decommissioning.

III. IMPACT OF USE OF AI IN SPACE SYSTEMS

The emphasis on dependability in space systems significantly hinders the use of AI for onboard data processing and control. While powerful, AI introduces peculiar and highly context-dependent failure modes compared to traditional software. This is especially true in space environments with high soft error rates, where AI's reliance on large near-memory computations becomes a vulnerability. These novel failure modes are often difficult to predict and test for using traditional methods.

Verification and validation of AI systems, particularly complex ones, pose a significant challenge. Traditional testing methods might not be sufficient to uncover all potential edge cases or unexpected inputs that could lead to catastrophic AI malfunctions. In critical space scenarios, understanding why an AI system makes a particular decision is crucial. However, some AI models, particularly deep learning architectures, can be opaque in their decision-making processes. This lack of explainability creates significant trust concerns for mission controllers who rely on the AI's outputs.

The infancy of AI dependability in space presents a double-edged sword. Recent focus on Explainable AI (XAI) Techniques can address trust and understanding concerns by utilising inherently explainable models or implementing techniques to make decision-making processes more transparent. However, this can come at a significant performance cost [5].

Artificial Neural Networks are composed by a multilayer architecture of linear and nonlinear operations parametrized by weights and biases. Due to the huge number of parameters and operations involved is not possible to trace the integrity of signals inside of the network nor have an insight of what the outputs will be given certain inputs. Some others ML paradigms as tabular reinforcement learning algorithms, such as Q-learning [6], or geometrical clustering techniques as Self Organizing Maps (SOM) or K-means store expected values in a table or reduce the dimensionality of the input data respectively, making it easier for humans to directly interpret it and to predict what the output would be in a given context.

Developing robust testing frameworks specifically tailored to AI systems is crucial. This might involve using diverse datasets, simulating various space-borne failure scenarios, and employing formal verification techniques. These frameworks are still in their early stages of development.

Integrating AI with traditional rule-based control systems offers a promising approach. This leverages the strengths of both paradigms: AI handles complex data analysis, while pre-programmed rules provide a safety net and ensure predictable behavior in critical situations [7].

The path forward involves continued research in dependable AI. As AI models become more reliable, transparent, and verifiable, their adoption in space applications will likely increase. However, ensuring dependability will remain paramount, and rigorous safety measures will be essential for mission-critical tasks.

IV. INSTITUTIONAL EARTH OBSERVATION VS. PRIVATE

The widespread adoption of AI within institutional space missions has been so far hampered by the inherent structure of space agencies and their procurement methods. These organizations, often preoccupied with stakeholder satisfaction and risk mitigation, tend to prioritize missions that generate well-defined data products, such as time series in Earth observation. This conservative approach overlooks the transformative potential of AI, which excels when applied to rich, diverse datasets. Moreover, by focusing on "minimum risk" missions, space agencies may fail to establish rigorous operational requirements, particularly concerning actionable products with low data latencies. This lack of clarity regarding real-time or near-real-time insights hinders the development of AI systems capable of unlocking the full potential of space data for mission decision-making and scientific discovery.

The Earth observation sector is undergoing a significant paradigm shift, with private companies emerging as prominent players alongside traditional space agencies. While NASA and ESA concentrate on long-term scientific research and global datasets, private actors demonstrate greater agility in targeting specific market needs. This enables them to develop innovative solutions and collect high-resolution data for applications such as precision agriculture and disaster monitoring. However, space agencies remain indispensable for setting standards, conducting fundamental research, and ensuring global access to Earth observation data.

A. Agency's role - Mitigating Risks of AI in Spacecraft

The European Space Agency should prioritise mitigating risks associated with using Artificial Intelligence (AI) on spacecraft. To achieve this, a well-defined roadmap with high-profile project activities is essential to build confidence and showcase successes. An activity to quickly identify high value use cases in currently running/about to be adopted projects

979-8-3503-6689-1/24 $31.00 © 2024 IEEE

should be initiated. Priority should be given to the cases that showcase strengths and capabilities of ESA.

B. Phased Approach and Core Expertise

This roadmap should outline a phased approach, similar to the one used with ESA's PhiSat-1 mission [8], [9], focusing on manageable risks and demonstrating achievements. Notably, the roadmap should leverage ESA's core and irreplaceable expertise in space hardware development, which was instrumental in PhiSat-1's success rather than generic development of base Artificial Intelligence technology. Additionally, the roadmap should incorporate robust data federation strategies. Data federation allows for seamless integration and analysis of data from various sources (e.g., spacecraft sensors, ground stations, external missions) without physically moving the data. This is crucial for AI applications that require vast amounts of diverse data for training and operation.

C. Focus on Core Activities

Space agencies struggle with AI research. Their structure makes them split AI development into many smaller tasks, mapping internal communication and management structures. Some of these tasks could be better handled by outside companies (e.g. cloud services) or are too complex for them (e.g. large language models). Instead, space agencies should focus on what they do best: one relevant example is building specialized computer hardware for space.

D. Modernizing Engineering Standards for AI

Artificial intelligence presents a disruptive force, particularly for software design, development, and verification within satellites. To cultivate innovation in this domain, a substantial revision of the European Space Agency's (ESA) engineering standards is necessary. For instance, if on-board processing is a requirement, a shift from packet-based communication to file-based communication becomes crucial. This ensures data integrity for straightforward retrieval and processing, eliminating the need for additional data packet recombination. The inherent opacity of high-performing AI models necessitates the definition of a protocol for functional verification, complementing existing formal methods. As shown in Figure 1, the readiness of a specific algorithm or AI system for high-criticality (I-II) satellite missions could be assessed through an incremental risk evaluation process using a heuristic approach. This approach could focus on past performance data gathered during lower-criticality (V-IV) missions. Furthermore, it's recommended to introduce the concept of risk modularity in high-criticality missions, implemented down to the functional level. For example, if the AI system solely generates insights on collected data without influencing critical operations, a less stringent level of dependability could be deemed acceptable.

Furthermore, current procurement practices employed by space agencies, which reimburse cost overruns, limit the focus on development speed. This, combined with the high level of technological readiness (TRL) required for components at the beginning of development, hinders the adoption of innovation, particularly in the realm of AI technologies.

E. Actionable Earth Observation Data

One potential showcase target could be enabling low-latency, actionable data services for Earth observation in the context of civil security [10], [11]. This could be technically integrated with existing New Generation (NG) Copernicus missions with minimal effort, requiring primarily political support for implementation.

One of the biggest hurdles for AI adoption in institutional space missions might be a consequence of their own organizational structures. Conway's Law, which states that "organizations which design systems ... are constrained to produce systems which are copies of the communication structures of these organizations" is particularly relevant here.

Space agencies, like the ESA, often focus on stakeholder service and minimizing risk, and tend to favour missions that generate well-understood data products, like time series in Earth observation. This approach aligns with their siloed, risk-averse structure, which incentivizes the creation of missions that fit internal priorities. However, it overlooks the potential of AI, which thrives on rich, diverse datasets. Furthermore, prioritizing "minimum risk" missions can lead to inadequate definition of operational requirements, particularly around actionable data latencies. This lack of clear requirements for real-time or near-real-time insights hinders the development of AI systems that could unlock the full potential of space data for mission decision-making and scientific discovery. The path forward for AI adoption in ESA requires not just technological advancements but also an organizational change. By embracing a more data-centric approach with well-defined operational needs, including latency requirements, ESA can break free from the limitations imposed by its current structures. This shift would pave the way for a future where AI plays a transformative role in space exploration and utilization. In essence, successful AI adoption necessitates a cultural shift within ESA, fostering collaboration across departments and a willingness to take calculated risks to unlock the full potential of AI for space missions.

F. Agency's role for the ecosystem

Clear data exchange rules are essential for EO data lakes and edge AI. These rules ensure different satellites, instruments and agencies can share information easily and applications based on "data fusion" can be built [12]. This eliminates complex data conversion, both on board, where many instruments are still based on legacy "packet store" data formats, instead of files and allows data lakes to collect information from many sources. Consistent data structure makes it easier to combine and analyze information within the data lake, creating a unified platform for various EO datasets.

Data standards also improve data quality and access. They require specific details like sensor type, capture time and calibration, platform ancilalry data, making the data more reliable and enabling better decisions from edge AI models. Standardized formats make finding and accessing data in the data lake effortless. Researchers and developers can easily

Class type	I	II	III	IV	V
Mission Criteria and Marking					
Criticality to Agency strategy (Flagship mission, Internationnal cooperation, Impact on ESA strategic goals, and image)	Extremely high Criticality	High Criticality	Medium Criticality	Low Criticality	Educational purposes
Marking					
Mission Objectives (Directorate priority and purpose, e.g in orbit demonstration, educational)	Extremely high Priority	High Priority	Medium Priority	Low Priority	Educational purposes
Marking					
Cost (Cost at Completion, Including Phase E1)	>700 M€	200 - 700M€	50 - 200M€	1- 50M€	< 1M€
Marking					
Mission Lifetime (Nominal mission life duration)	> 10 years	5-10 years	2-5 years	1-2 years	1 year
Marking					
Mission Complexity (Design interfaces unique payloads, New technology development)	High	High to Medium	Medium	Medium to Low	Low

Fig. 1. Classification of space mission by criticality level

locate relevant datasets for their edge AI applications without struggling with incompatible formats.

These standards are crucial for deploying edge AI at scale. Standardized data allows reliable pre-training of AI models in the cloud using data from the data lake. These models can then be deployed on edge devices for real-time analysis, even in remote areas with limited bandwidth. Consistent data formats ensure edge AI models work seamlessly with different EO data sources, enabling wider deployment and scalability across the Earth observation field.

In EO, data comes from many sensors and platforms. As an example, standards like CF (Climate and Forecast Metadata) and COG (Cloud-Optimized GeoTIFF) ensure consistent representation of geospatial data. This allows applications like land cover classification or disaster monitoring to leverage data from various sources within the data lake and train effective edge AI models for near real-time analysis on edge devices.

A reference for these standards can be found for example on NASA Earth Science Data and Information System (ESDIS) website. This website [13] discusses the importance of using consistent formats to archive and share Earth science data. It details formats like HDF5, NetCDF, and ASCII that allow researchers to easily access and use data from different sources. ESDIS also uses specific metadata standards to describe its data, making it easier for users to find data.

V. CONCLUSIONS

This paper has explored the critical challenges and opportunities presented by the increasing use of AI in space systems. We have highlighted the importance of ensuring dependability in such systems, particularly given their critical role in various

applications. By addressing the challenges associated with AI's black box nature and data trustworthiness, we can unlock its full potential to revolutionize space exploration and Earth observation.

Key highlights from our analysis include the advantages of on-board data processing, the potential of blockchain technology for ensuring data integrity, and the need for robust testing frameworks for AI systems. We have also identified the importance of integrating AI with traditional control systems, developing functional verification protocols for AI models, and standardizing data formats for geospatial data.

Future research should focus on addressing these challenges and exploring new avenues for leveraging AI in space applications. AI algorithms must be developed and refined to address the unique demands of space environments. This includes algorithms that can operate with limited computational resources, handle noisy or incomplete data, and adapt to changing conditions. Additionally, research should focus on developing algorithms that are transparent and explainable, enabling us to understand and trust their decision-making processes.

Dependable AI is essential for mission-critical space applications. This requires research into techniques for verifying and validating AI models, ensuring their robustness against adversarial attacks, and developing fault-tolerant AI systems. Furthermore, exploring methods for quantifying and managing uncertainty in AI-driven decisions is crucial.

Software engineering for AI in space systems must be advanced. This involves developing standards, tools and methodologies for designing, implementing, and testing AI-powered

979-8-3503-6689-1/24 $31.00 © 2024 IEEE

software, as well as addressing challenges such as integration with legacy systems and ensuring software quality and security.

Hardware co-design with AI is vital. This entails designing hardware architectures that are optimized for AI workloads, such as specialized accelerators for neural networks. Additionally, research should explore the integration of AI into hardware components like sensors and actuators, enabling more autonomous and intelligent space systems.

ACKNOWLEDGEMENTS

The PhD of the corresponding author, Valentina Zancan, has been fully funded by ThalesAleniaSpace Italia SpA.

REFERENCES

[1] McKinsey, "How will the space economy change the world?" 2023. [Online]. Available: https://www.mckinsey.com/industries/aerospace-and-defense/our-insights/how-will-the-space-economy-change-the-world

[2] Deloitte, "Future of space economy," 2023. [Online]. Available: https://www.deloitte.com/global/en/our-thinking/insights/industry/defense-security-justice/future-of-the-space-economy.html

[3] C. Alexis, D. Sylvan, C. Guillaume, P. Romane, B. Badia, and C. Croison, *Earth Observation Data and Services Market*. Eurocounsult, 2023.

[4] CEOS, "Virtual constellation definition." [Online]. Available: https://ceos.org/ourwork/virtual-constellations/

[5] Y. Rong, T. Leemann, T.-T. Nguyen, L. Fiedler, P. Qian, V. Unhelkar, T. Seidel, G. Kasneci, and E. Kasneci, "Towards human-centered explainable ai: A survey of user studies for model explanations," *IEEE Transactions on Pattern Analysis and Machine Intelligence*, vol. 46, no. 4, pp. 2104–2122, 2024.

[6] C. J. C. H. Watkins and P. Dayan, "Q-learning," *Machine Learning*, vol. 8, no. 3, pp. 279–292, 1992.

[7] F. Tambon, G. Laberge, L. An, A. Nikanjam, P. S. N. Mindom, Y. Pequignot, F. Khomh, G. Antoniol, E. Merlo, and F. Laviolette, "How to certify machine learning based safety-critical systems? a systematic literature review," *Automated Software Engineering*, vol. 29, no. 2, 2022.

[8] G. Furano, G. Meoni, A. Dunne, D. Moloney, V. Ferlet-Cavrois, A. Tavoularis, J. Byrne, L. Buckley, M. Psarakis, K.-O. Voss *et al.*, "Towards the use of artificial intelligence on the edge in space systems: Challenges and opportunities," *IEEE Aerospace and Electronic Systems Magazine*, vol. 35, no. 12, pp. 44–56, 2020.

[9] G. Giuffrida, L. Fanucci, G. Meoni, M. Batič, L. Buckley, A. Dunne, C. Van Dijk, M. Esposito, J. Hefele, N. Vercruyssen *et al.*, "The ϕ-sat-1 mission: The first on-board deep neural network demonstrator for satellite earth observation," *IEEE Transactions on Geoscience and Remote Sensing*, vol. 60, pp. 1–14, 2021.

[10] L. Parra Garcia, G. Furano, M. Ghiglione, V. Zancan, C. Clemente, C. Ilioudis, E. Imbembo, and P. Trucco, "Advancements in on-board processing of synthetic aperture radar (sar) data: enhancing efficiency and real-time capabilities," *IEEE Journal of Selected Topics in Applied Earth Observations and Remote Sensing*, 2024.

[11] G. Furano, C. U. Ortega, M. Tali, B. Guesmi, D. Moloney, M. Dean, N. Longepede, P.-P. Mathieu *et al.*, "The use of ai in operational space weather missions," in *104th AMS Annual Meeting*. AMS, 2024.

[12] S. Salcedo-Sanz, P. Ghamisi, M. Piles, M. Werner, L. Cuadra, A. Moreno, E. Izquierdo-Verdiguier, J. Muñoz, A. Mosavi, and G. Camps-Valls, "Machine learning information fusion in earth observation: A comprehensive review of methods, applications and data sources," *Information Fusion*, vol. 63, pp. 256–272, 07 2020.

[13] N. Earthdata, "Standards and practices — earthdata," 2024. [Online]. Available: https://earthdata.nasa.gov/standards-practices

Safe Satellite Electronics Design utilizing COTS Components, FDIR Techniques, LCL Protections, and Thorough Qualifications

Bojan Kotnik
SkyLabs d.o.o.
Maribor, Slovenia
bojan.kotnik@skylabs.si

David Selčan
SkyLabs d.o.o.
Maribor, Slovenia
david.selcan@skylabs.si

Matic Erker
SkyLabs d.o.o.
Maribor, Slovenia
matic.erker@skylabs.si

Tomaž Rotovnik
SkyLabs d.o.o.
Maribor, Slovenia
tomaz.rotovnik@skylabs.si

Dejan Gačnik
SkyLabs d.o.o.
Maribor, Slovenia
dejan.gacnik@skylabs.si

Gianluca Furano
ESA - ESTEC
Noordwijk, The Netherlands
gianluca.furano@esa.int

Iztok Kramberger
Faculty of Electrical Engineering and
Computer Science, University of Maribor
Maribor, Slovenia
iztok.kramberger@um.si

Abstract—This paper presents SkyLabs' concept of advanced and safe satellite electronics design utilizing COTS components. The recent ESA mission class classification paradigm allows the use of COTS components in Class 5, 4, and 3 missions under specific conditions. SkyLabs proposes the utilization of hierarchical FDIR topology using autonomous Latch-up Current Limiter (LCL) on the subsystem level and controlled LCLs on the component level, fault tolerant digital design concepts, and thorough execution of MAIT (manufacture, assembly, integration, and test) process, including TID and SEE characterizations. The proposed concepts are demonstrated on SkyLabs high-performance RISC-V based on-board computer named NANOhpm-obc.

Index Terms—Latch-up current limiter, fault tolerant processor IP, MAIT process, SEE robustness.

I. INTRODUCTION

The European Space Agency (ESA) classifies its missions according to a system that ranges from Class 1 to Class 5 [1]. Class 1 missions are the most ambitious and complex missions, often flagship, decades-long projects representing significant investments and requiring cutting-edge technology and paramount reliability. Class 1 missions usually involve exploration of outer planets, major scientific investigations, or large-scale observatories. Examples include the Rosetta mission, and JUICE - Jupiter Icy Moons Explorer probe.

Class 2 missions are still high-profile and scientifically significant but are somewhat less ambitious or complex compared to Class 1 missions. They may involve medium-scale scientific investigations, technology demonstrations, or missions with a more focused scope. Class 2 missions include missions like the Gaia space observatory, which is mapping the Milky Way in unprecedented detail.

The traditional model of space exploration, represented by missions of Class 1 and 2, often referred to as "Old Space", is characterized by space agency led programs, extensive bureaucracies, and strict following of the standards governed by European Cooperation for Space Standardization (ECSS) in order to assure extremely high reliability and long operational life-times.

In recent years, ESA has been at the forefront of embracing the principles of "New Space" for less-critical missions of Classes 3 to 5, which advocate for increased commercialization, innovation, and collaboration within the space sector. One significant aspect of this transition involves the utilization of Commercial Off-The-Shelf (COTS) components in ESA missions [1]. This shift represents a departure from the traditional reliance on bespoke, highly customized hardware towards a more agile, cost-effective approach that leverages off-the-shelf technology to accelerate mission development and reduce costs.

COTS components that are readily available for purchase from commercial vendors, play a central role in enabling the New Space paradigm. These components encompass a wide range of technologies, including processors, sensors, actuators, communication systems, power systems, propulsion systems, and more. By incorporating COTS components into less-critical Class 3-5 mission designs, space agencies can capitalize on the extensive R&D investments made by commercial vendors, tapping into a vast ecosystem of innovation and expertise. Main benefits of utilizing COTS components can be summarized as follows:

Cost Savings: One of the primary advantages of COTS components is their affordability compared to custom-built solutions. By leveraging off-the-shelf hardware and software, ESA can significantly reduce development and procurement costs, enabling more missions within constrained budgets.

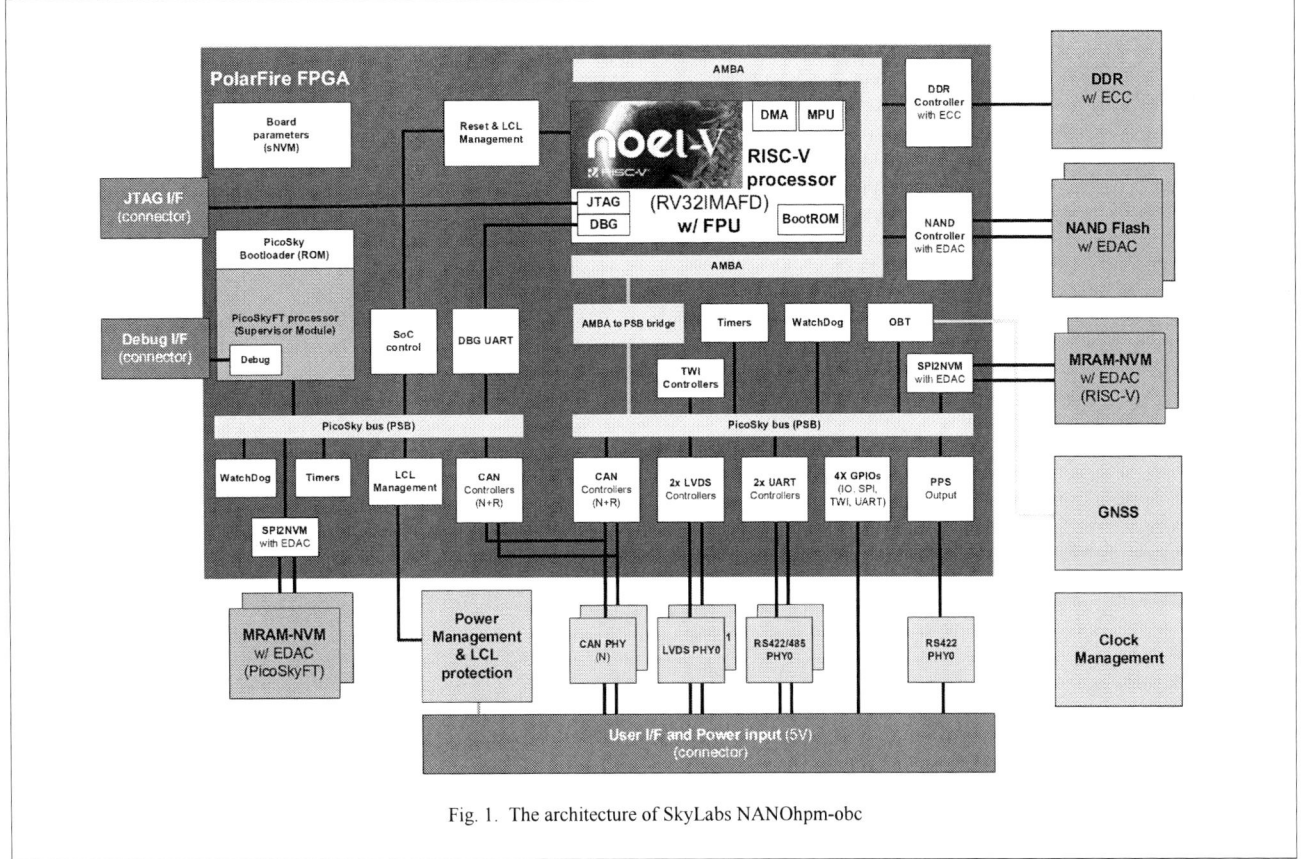

Fig. 1. The architecture of SkyLabs NANOhpm-obc

Faster Development Cycles: COTS components are readily available and well-tested, allowing for faster integration and deployment in mission architectures. The lead times of COTS components is significantly shorter than that of space qualified components, with typical lead times of 24-50 weeks. This accelerated development cycle is crucial for meeting tight deadlines and rapidly evolving mission requirements.

Access to Advanced Technology: Commercial vendors invest heavily in R&D to develop state-of-the-art components for various industries. By incorporating COTS components into missions, ESA gains access to cutting-edge state-of-the-art technology.

However, when utilizing COTS components in space missions, several preconditions must be carefully considered to ensure mission success, reliability, and safety:

Reliability and Qualification Testing: COTS components may not be inherently designed for the harsh conditions of space. Therefore, rigorous reliability and qualification testing are essential to verify the performance and durability of these components in the space environment. Testing should include thermal, vacuum, radiation (total ionising dose (TID), and single event effects (SEE)), vibration, and other environmental stressors to simulate the conditions they will encounter in space. The testing can be performed on component level, or at the whole subsystem level. In addition to traditional heavy-ions based SEE testing, also the new ways of testing based on neutrons and high-energy hadrons can be a good Weibull distribution approximated indicator of radiation robustness of high-complexity systems, where well-defined and implemented FDIR techniques can prevent catastrophic failures and also restore the full functionality of the whole complex system.

Redundancy and Fault Tolerance: Given the critical nature of space missions, redundancy and fault tolerance mechanisms must be implemented to mitigate the risk of component failures. Built-in redundancy features can help ensure mission continuity even in the event of component failures. A thorough risk assessment should be conducted to identify potential vulnerabilities or failure modes associated with the use of COTS components. Mitigation strategies, such as redundancy, contingency plans, and system-level fault management, should be implemented to minimize the impact of potential failures on mission success.

Long-Term Reliability: Space missions often have extended lifetimes, spanning several years. Space electronics utilizing COTS components should demonstrate long-term reliability and stability to maintain performance over the duration of the mission without degradation or failure.

Traceability and Documentation: Comprehensive documentation and traceability of COTS components are essential for quality assurance and configuration management. This includes detailed specifications, test reports, certification documentation, and records of any modifications or customizations made to the components.

Fig. 2. PCB of SkyLabs NANOhpm-obc

available NVM space is 512MB. In addition, TM data and logs are kept in a separate MRAM NVM storage (two banks) with total user-available capacity of 256kB. This storage is additionally EDAC protected.

NANOhpm-obc is implemented in a standard PC104 form factor (see Fig. 2.). It integrates a GNSS receiver with 1PPS signal on-board and provides the following interfaces:

- 2x CAN bus (N+R)
- 2x LVDS channels (configurable to 2x RS422/RS485)
- 1x 1PPS RS422-compliant I/O interface
- 1x GNSS antenna connector
- 2x RS422/RS485 full-duplex channels
- GPIOs

III. FDIR TECHNIQUES IMPLEMENTED IN NANOHPM-OBC

High reliability of NANOhpm-obc is ensured via carefully selected electronic components and combined with an advanced FDIR approach that supervises the whole system. The core of FDIR policy presents the Supervisor Module (SM) which is implemented by SkyLabs' proprietary PicoSkyFT fault tolerant IP core which resides in the same FPGA as the main RISC-V [3] processor (see Fig. 1.). PicoSkyFT core itself has been radiation qualified through several test campaigns, including proton and neutron irradiations.

The SM provides a supervision of booting procedure of RISC-V (SM manages the transfer of RISC-V booting FW image to working memory). SM also has a dedicated channel for low-level TM/TC over CAN. It takes care for a proper RISC-V processor reset, and implements the low-level housekeeping telemetry operations. SM provides a low-level access to all memories, meaning that the NANOhpm-obc bootloader or application software can be reprogrammed also during the mission (e.g. from other platform OBC if present, or via ground segment control). This function is available even if NANOhpm-obc cannot boot up nor to nominal nor to safe mode operation. SM also provides a complete FDIR management in case of detected SEE: control of multiple component-level Latchup Current Limiters (LCL) which protect SEL-susceptible COTS components, and processing of EDAC DED events on RISC-V processor core.

In this paper we present the concept of safe satellite electronics design utilizing COTS components, FDIR techniques, LCL protections, and thorough qualifications, demonstrated in a SkyLabs' off-the-shelf onboard computer **NANOhpm-obc** product (Fig. 1. and Fig. 2.).

II. ABOUT NANOHPM-OBC

NANOhpm-obc is a high-performance microcontroller in a single-board computer, designed for LEO applications. NANOhpm-obc provides a versatile design in terms of variety of resources, extension possibilities and available interfaces. Fig. 1. depicts the detailed architecture diagram of NANOhpm-obc. The core of the system is SoC in a commercial grade PolarFire FPGA (SEU immune FPGA fabric, TID on fabric > 300 kRad [2]), which runs 32-bit Frontgrade Gaisler NOEL-V FT RISC-V (RV32IMAFD) processor IP, running at 80 MHz. The system is supported by 256MB ECC-protected DDR3 RAM, and mass NVM EDAC-protected storage of 2GB. User

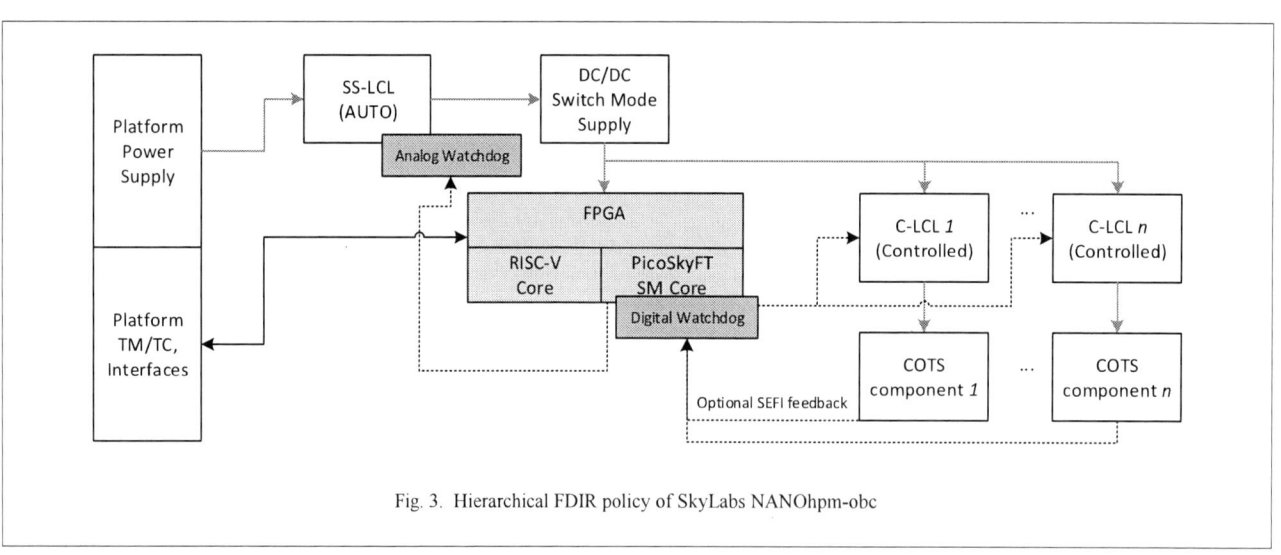

Fig. 3. Hierarchical FDIR policy of SkyLabs NANOhpm-obc

IV. SEE Protection Mechanism

NANOhpm-obc makes extensive use of a SEE protection approach, which consists of a hierarchical FDIR policy, utilizing multiple types of Latching Current Limiters (LCLs) [4], each fulfilling a different role (see Fig.3.). The LCLs are further augmented by the use of watchdog timers, protecting against other SEFI and SEU events. The approach is centered around the use of a FPGA, protected by an autonomous SS-LCL, which in turn controls other COTS components with the help of controlled C-LCL circuits and watchdog timers [5]. The LCL circuits used in NANOhpm-obc (and in other SkyLabs products) are designed with discrete component and were radiation tested using a proton beam at the the Paul Scherrer Institute (PSI) and a neutron beam at the ChipIR ISIS Neutron and Muon Source in England. No SEE events were observed during the tests. Nevertheless, the discrete implementation of 14 LCLs in PC104 form-factor NANOhpm-obc consumes substantial PCB area. Therefore, SkyLabs is developing a compact, integrated LCL ASIC solution (SKY-IC-RHLCLA1) to address this issue for future highly-integrated and complex satellite subsystems [6].

V. NANOhpm-obc Qualification

SkyLabs utilizes the standard model philosophy in satellite electronics development and qualification before reaching the flight worthiness of individual units. Fig. 4. depicts three stages and model strategy of satellite electronics development, and corresponding tests. The engineering model (EM) is typically the first version of the subsystem built. It serves primarily for design validation and early testing in order accomplish a positive Critical Design Review (CDR) status. EM may not have the exact specifications or components that will be used in later qualification- and the final flight model but is used to validate the design concepts and engineering principles. The qualification model (QM) is built after the EM and is used to validate that the final design meets all necessary requirements and can withstand the harsh conditions of space. QM undergoes rigorous testing to simulate the conditions the spacecraft will encounter during launch, and later in space. Qualification testing includes environmental testing (thermal vacuum, vibration, shock, TID, SEE, etc.) and functional testing to ensure that the spacecraft can perform its intended functions reliably (see Fig. 4.). While the TID characterization has been performed at Co60 facility at ESA-ESTEC, the SEE testing has been performed at CERN High Energy Accelerator Mixed-field (CHARM) facility utilizing high-energy hadrons [5]. This testing has been proven as a useful, convenient testing of complex space electronics systems, during which the potential weak points in the design and/or FDIR implementation can be identified. The flight model (FM) is the version of the satellite electronics that is actually utilized in space. FM undergoes final testing before being integrated in the target satellite. The hardware and firmware of FM is equal to QM; however, it undergoes less stressful testing than QM.

VI. Conclusion

Satellite electronics for ESA missions of Class 3, 4, and 5 can be based on COTS components if reliability and fault tolerance are assured and proved according to mission specifications. SkyLabs designs safe satellite electronics utilizing COTS Components, protected by FDIR techniques, SEU/SEE mitigation techniques (fault tolerance), and by thorough qualification processes, including TVAC and radiation testing campaigns. These concepts were demonstrated on SkyLabs NANOhpm-obc product.

References

[1] V. Gupta, "ESA Mission Classification: focus on RHA tailoring & recommendations for COTS projects" in *2023 European Data Handling & Data Processing Conference (EDHPC)*, Juan Les Pins, France: IEEE, Oct. 2023, pp. 1–5.

[2] J. J. Wang *et al.*, "Radiation Characteristics of Field Programmable Gate Array Using Complementary-Sonos Configuration Cell", Microchip, San Jose, CA, 2020.

[3] M. J. Cannizzaro *et al.*, "Evaluation of RISC-V Silicon Under Neutron Radiation," 2023 IEEE Aerospace Conference, Big Sky, MT, USA, 2023, pp. 1-9, doi: 10.1109/AERO55745.2023.10115689.

[4] D. Selčan *et al.*, "Nanosatellites in LEO and beyond: Advanced Radiation protection techniques for COTS-based spacecraft", in Acta Astronautica, Volume 131, February 2017, pp. 131–144, DOI: https://doi.org/10.1016/j.actaastro.2016.11.032.

[5] A. Bernhard *et al.*, "Radiaton tolerant ATTM-WRTU wireless infrastructure for radiation harsh terrestrial applications", in RADECS 2021, pp. 1-4,
DOI: 10. 1109/ RADECS53308. 2021. 9954581

[6] B. Kotnik *et al.*, "Advanced and Safe Satellite Electronics Design by LCL Utilisation on Component Level," in *2023 European Data Handling & Data Processing Conference (EDHPC)*, Juan Les Pins, France: IEEE, Oct. 2023, pp. 1–5, DOI: 10.23919/EDHPC59100.2023.10396176.

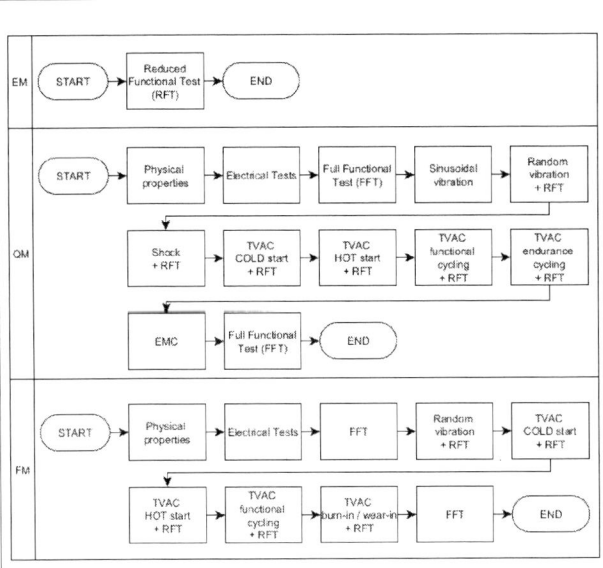

Fig. 4. Typical SkyLabs model philosophy (Top: EM/Engineering Model, middle: QM/Qualification Model, bottom: FM/Flight Model)

Special Session: Exploring the Potential of Versal ACAP: Advancing Onboard Edge AI for Spacecraft

Carlo Ciancarelli*, Davide di Ienno*¶, Renato Trois†, Luca Scandelli†, Catriel de Biase‡,
Paolo Serri‡, Antonio Leboffe*, Dario Pascucci*, David Steenari§, Gianluca Furano§||

*Thales Alenia Space Italia S.p.A., Rome, Italy
†Thales Alenia Space Italia S.p.A., Gorgonzola, Italy
‡Thales Alenia Space Italia S.p.A., L'Aquila, Italy
§European Space Agency, Noordwijk, The Netherlands
¶davide.diienno-somministrato@thalesaleniaspace.com
||Gianluca.Furano@esa.int

Abstract—In the space domain, the necessity for Artificial Intelligence (AI) technologies and computationally demanding applications has become increasingly evident. This need arises from the growing complexity of space missions and the corresponding objective to enhance spacecraft decision-making capabilities. Traditional approaches have faced limitations in terms of flexibility as well as computational constraints, demanding a new perspective. With a focus on addressing these challenges, this investigation delves into the potential of the Versal Adaptive Compute Acceleration Platform (ACAP) developed by Xilinx. This hardware solution promises to enable efficient data processing and real-time inference for Edge AI applications on board spacecraft. The final objective of this ongoing study is to explore Versal ACAP capabilities on a range of use cases, including AI-driven telemetry anomaly detection and Synthetic Aperture Radar (SAR) data processing. AI-based telemetry anomaly detection occupies a critical space in spacecraft systems, as it promises to improve Fault Detection, Isolation, and Recovery (FDIR) functionalities. Traditionally, SAR data processing has been executed on the ground, whereas the Versal ACAP presents the opportunity to perform preliminary SAR data analysis onboard, enabling rapid extraction of information and facilitating more agile mission planning and decision-making. While the Versal ACAP promises to advance Edge AI applications in space, it also presents relevant challenges: power consumption and thermal dissipation pose a critical concern as spacecraft systems operate within strict power budgets and limited dissipation capabilities. Striking a delicate balance between the computational requirements of AI algorithms and spacecraft design constraints becomes a crucial consideration for successful deployment. This work presents an initial investigation into the Versal ACAP through the development of a prototype board, auxiliary to traditional On Board Computers, designed to enhance on board processing capabilities. The available software (SW) tools and the proposed SW architecture required to run the application SW on the board are also discussed. Finally, the plan for the successive validation of the prototype through the use cases mentioned above is outlined.

Index Terms—Artificial intelligence, Machine Learning, Edge Computing, Versal ACAP, Prototype board

I. INTRODUCTION

The combination of advances in Artificial Intelligence (AI) and the increasing complexity of space missions have highlighted the need for novel and tailored hardware (HW)

solutions, to complement spacecraft on board processing units and enhance their capabilities [1]–[3]. Traditional space-grade On Board Computers (OBC) and Instrument Control Units ICU are not equipped with specific HW capable of performing real-time inference of high volume payload data with AI methods such as Neural Networks (NN). More efficient solutions to accelerate on board inference are based on Field Programmable Gate Arrays (FPGA) [4], but they still lag behind when compared with Commercial Off-The-Shelf HW available on ground. A new opportunity to tackle these challenges is represented by the AMD Xilinx Versal Adaptive Compute Acceleration Platform (ACAP) System-On-Chips (SoCs) family [5]. Versal SoCs are highly configurable and provide adaptable processing, acceleration engines (AI and DSP engines) and Programmable Logic (PL); this kind of SoCs provide superior performance/watt when compared with traditional solutions [6]. The configuration possibilities are not limited to connectivity. Some engines can be deactivated if they are not needed to meet computational requirements of the application. These adjustments can help meet the power requirements of the Platform (P/F) and further increase computational efficiency. In fact, Versal ACAP immediately attracted the attention of engineers in the space domain for its excellent performance [7]: benchmarks with NNs often used for this purpose in the AI domain [8], such as MobileNetV1 [9], ResNet-50 [10], and GoogLeNet [11], showed promising results. A few works have already proposed applications featuring space-powered used cases exploiting Versal SoCs. In [4] the objective was to develop a redundant Data Processing Unit capable of storing, handling and analyzing data on board. A different approach was followed in [12] where Xilinx Versal capabilities were exploited to design an high-performance mass memory unit. As will be discussed later, the Versal allows the accommodation of several high-speed interfaces, such as multiple SpaceFibre [13] connections, a new standard that is expected to get attention from space system designers in the coming years. Another perspective is the one described in [14], where the authors evaluated Versal AI Core performances in different configurations, benchmarking the SoC with a U-Net, i.e. a U-shaped Convolutional Neural Network (CNN)

979-8-3503-6689-1/24 $31.00 © 2024 IEEE

[15]. They also discussed possible use cases, ranging from Failure Detection, Isolation and Recovery (FDIR) to Machine Learning (ML) methods applied to telecommunications.

In this paper, the architecture of a new prototype board, targeting a Technology Readiness Level (TRL) [16] of 4, is presented. In the space domain, the TRL is a scale used to assess the maturity of a technology. It ranges from 1 to 9, where 1 represent the lowest level (basic principle) and 9 the highest (flight proven). TRLs and their criteria of assessment are defined by ECSS, including requirements, verification, test environments and necessary documentation. In order to reach TRL 4, it is necessary to demonstrate element functional performance by breadboard testing in laboratory environment, producing a report with breadboard definition, test plan and test results demonstrating functional performance verification [16]. The board is currently under active development, featuring a selected AMD Xilinx Versal ACAP SoC. It is designed for on board processing for computationally demanding tasks, including those based on ML methods. Comprehensive work on both HW and SW components is ongoing, with the long-term objective to exploit Versal capabilities in near-future Low Earth Orbit constellations. The paper is organized with a first section presenting what are the applications which have been used to guide the design of the board. The same applications will be also used to verify that the manufactured board satisfy the performance requirements. A dedicated section is included to outline the proposed HW architecture for this development. The following section is dedicated to the software (SW) solutions that allow to access and exploit all the resources available within the board. Such SW components are designed taking into account the proposed applications and the related constraints. Finally, a summary of the content discussed and the outlook of the work.

II. BENCHMARKING APPLICATIONS

The prototype board capabilities will be evaluated using three diverse applications. Each application has specific requirements that guide both HW and SW design, setting the target performances expected at project completion. The applications considered are:

- ML models for satellite health monitoring.
- Radio Frequency Machine Learning (RFML).
- Synthetic Aperture Radar (SAR) Processing.

The first application is part of the broader effort that Thales Alenia Space Italia (TASI) is making to bring innovative AI-based techniques on board the satellite to improve traditional FDIR [17], [18]. These techniques are tailored to detect anomalies through intelligent telemetry time series analysis. ML methods for anomaly detection range from classical ML approaches to more recent NN-based Autoencoders, particularly those using convolutional kernels. The Versal's efficiency in performing convolutional operations with AI engines is crucial for monitoring multiple subsystems simultaneously. Additionally, the impact of model quantization on critical applications like FDIR needs to be verified. RFML is an emerging technology that merges principles from ML and

wireless communication [19]. It is of interest as it can enable intelligent decision-making on board, based on RF data analysis. Even with input downsampling, RFML applications are more computationally demanding than telemetry anomaly detection, therefore they would benefit from powerful SoCs. In the context of this work, this application concerns the use of a CNN, designed to process RF signals in the form of In-phase and Quadrature (IQ) samples [20], with the task of determining the identity of the transmitter. SAR processing allows to generate high-resolution images of the Earth surface starting from radar signals captured from Satellite using highly specialized SAR payloads. SAR processing is a complex and computationally intensive task, normally performed on ground, that can leverage the parallel processing power and adaptability of the Xilinx Versal board to execute algorithms efficiently. To summarize, the presented applications serve as a broad basis for testing the capabilities of the Xilinx Versal, especially AI and DSP engines.

III. HW ARCHITECTURE

In the frame of this work, the XCVC1902-1MSIVSVA2197 has been selected from the Xilinx catalog [6], this component provides, among other features:

- 400 AI Engine Tiles
- 1968 DSP Engines
- Programmable logic (1968 System Logic Cells)
- Dual-core Arm® Cortex®-A72 application processing unit
- Dual-core Arm® Cortex®-R5F real-time processing unit
- 44 High Speed Serial Links (HSSL) up to 26.5625 Gb/s.

A significant advantage of this product lies in the extensive flexibility afforded to hardware engineers for customization, allowing for tailored adaptations to meet specific design requirements. The designers are basically encouraged to find a strategic equilibrium between computational power and thermal limitations. These kinds of constraints are usually imposed by the P/F and must be respected throughout the design process. Besides the Versal ACAP, key components and subsystems include the System Controller (SC), Power Management System (PMS), Platform Management System (PLMS), and Clock Management System (CMS). The block diagram in Figure 1 shows these subsystems and their interfaces. External interfaces are mainly provided through a backplane connector, with an additional front panel connector. The following sections discuss the architecture of the prototype board, describing each subsystem.

A. System Controller

The first concern of a space design is the radiation environment. In order to cope with the reliability and system availability requirements, a rad-hard Cortex®-M4 processor has been selected to take charge of system monitoring, boot loading and ACAP patch and dump task. The SC directly increases the reliability of the solution. It is able to continuously monitor the status of the ACAP and perform recovery actions whenever needed. It also allow to have tightly controlled power

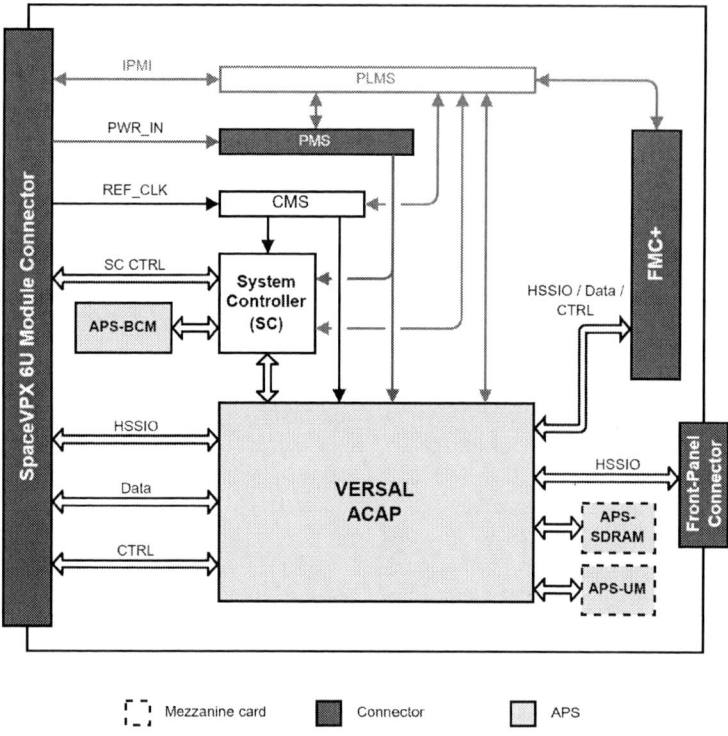

Fig. 1. Versal board block scheme.

modes, especially at start up, where only the controller is lit up, waiting for a comprehensive built in test for the full system start up. To start the Versal boot sequence the SC can access a non-volatile Boot and Configuration Memory (BCM).

B. Power Management

To guarantee the maximum computational power of the ACAP SoC, the core supply current can exceed 100 A while maintaining a core voltage below 1 V. This high current demand, coupled with low voltage operation, presents significant challenges in power delivery and integrity. Therefore, to comply with the SpaceVPX [21] standard, which specifies a 12 V main supply rail, a sophisticated multiple-stage DC/DC converter must be designed. This converter is responsible for supplying the Versal core—by far the most demanding power rail—and other essential power rails required by the board.

Future iterations of this system might consider avoiding the use of a backplane altogether. This approach would simplify both the connection and thermal dissipation design, leading to a less complex PCB layout. Without the constraints imposed by a backplane, the design can more easily accommodate the power and ground planes required by these FPGAs. These FPGAs necessitate multiple very thick power and ground planes, which may be incompatible with thin vias. By eliminating the backplane, the design can incorporate thicker planes more effectively, with less placement constraints, thereby enhancing power integrity and reducing electrical resistance [22].

The PMS is tasked with the crucial role of ensuring stable power delivery to the SoC. This includes regulating voltage levels and managing power distribution across the various components of the system. Given the stringent requirements of the ACAP core, the voltage must be maintained within a ±3% tolerance. Achieving this level of precision necessitates thorough power integrity analysis. Such analysis is critical and must be factored in during the selection of the optimal PCB stack-up. Additionally, it is essential during the design of the decoupling network, which helps in filtering out noise and maintaining a stable voltage supply.

C. Interfaces

The XCVC1902 provides 44 HSSL up to 26.5625 Gb/s. Careful selection of external connectors and signal integrity analysis are crucial, especially during the early stages of the project when designing the PCB stack-up. The main connector is represented by the SpaceVPX 6U [21] Module Connector, which includes power delivery and other relevant connections with the P/F. The board also features a Front-Panel Connector for high data rate exchanges.

D. Thermal Management

When the maximum performance is achieved more than 100 W are dissipated on the board thus the thermo-mechanical design shall consider an high-performance solution (e.g. heat pipes) to be compliant with missions requirements.

979-8-3503-6689-1/24 $31.00 © 2024 IEEE

E. Memory

The board features an Adaptable Processing System (APS). The APS-SDRAM, which will be DDR4 type, and APS-UM (User Memory) will be placed on a separate mezzanine board. This solution implies a more complex signal integrity but allows the user to explore different solutions in terms of number and typology of memory chips. This solution also provides flexibility during the board qualification and acceptance process. The User Memory can be exploited by the designer to implement a wide range of different applications (e.g. using a NAND Flash mezzanine card to implement a local mass-memory to store data processed by the ACAP). Having more memory available near the SoC can, for instance, reduce data exchanges with other subsystems, but the actual sizing of volatile and non-volatile memory will be dictated by the specific needs of the target space mission.

F. Monitoring and Clock

The PLMS is based on the Intelligent Platform Management Interface (IPMI) specifications [23] and provide an hardware-level interface to monitor key board parameters. Most of the monitoring by the SC is done through PLMS. The last subsystem present in the block diagram is the CMS, which distribute board clocks from external reference clock signals.

IV. SW ARCHITECTURE

AI applications are on the rise in various sectors, and the space domain is no exception. There is a growing interest in deploying AI efficiently on spacecraft, considering limitations on resources and time. Versal SoCs, with their powerful accelerators and complex architecture, offer advantages for space applications. However, efficiently deploying AI algorithms on this platform presents unique challenges. This work introduces a meticulously designed SW framework to address these challenges. The framework enables the seamless execution of AI algorithms while adhering to strict timing constraints in space missions. Furthermore, this framework allows for experimentation with the Versal SoC's computational capabilities in real-world space scenarios. The powerful processing units within Versal ACAP SoCs make it possible to integrate various functionalities on a single platform. This includes not only high-performance AI algorithm execution and monitoring but also onboard command and control management for the spacecraft. The SW Framework's Key Features and Advantages are:

- **AI Payload Algorithms:** The framework is able to integrate and execute in an efficient way different types of algorithms. Within the study, three cutting-edge AI payload algorithms tailored for specific applications have been integrated, encompassing tasks such as Anomaly Telemetry Detection, RF Analytics and SAR processing. These algorithms leverage the power of Versal accelerators to deliver high-performance AI capabilities. However, adjustments to the models are needed to ensure full compatibility with the framework, including verifying that NN kernels are supported by AI engines. Another crucial aspect is quantization, which can negatively impact model performance.

- **Accelerator Allocation:** Through a SW abstraction layer, the framework is able to execute the AI payload algorithms into the most suitable accelerators within the Versal platform. This ensures optimal utilization of hardware resources while minimizing execution latency and meeting timing constraints. The framework design takes into account the goal to decouple the framework management from the HW interaction with the ambitious target to be easily ported on a different HW accelerator platform.

The framework uses the Xilinx framework made up by: Xilinx Runtime (XIR) and Vitis AI [24]. Both bring specific advantages: thanks to the XIR it is possible to exploit efficiently the board ecosystem. Vitis AI lets to leverage the power of the IP integrated Deep Processing Unit for an optimal AI algorithm execution and deployment. The Versal capabilities are managed and abstracted by Petalinux [25], a toolset provided by AMD Xilinx to automate the process of building, configuring, and deploying Linux OS on selected AMD Xilinx products. Petalinux can be configured to be tailored to specific accelerators, drivers and SoC peripheral interfaces. The SW Framework for AI Payload Algorithms on Versal represents a significant advancement in the deployment of AI solutions on complex Versal heterogeneous platforms. By seamlessly orchestrating the execution of AI algorithms across accelerators while ensuring adherence to timing constraints, the framework empowers developers to harness the full potential of Versal for AI applications. The possibility to monitor and easily deploy AI algorithms (or in general high performance algorithm) can permits in the future to dynamically adjusts the execution of AI payloads based on changing workload demands and hardware availability. This enables adaptive resource allocation and efficient utilization of Versal computational resources and heralds a new era of efficiency and performance in AI deployment on heterogeneous architectures, which could be used in the space domain.

V. CONCLUSION

In this paper, a new architecture for a high-performance prototype board based on the Xilinx Versal AI Core Series was presented. The HW architecture was reported in detail, highlighting its more relevant building blocks, discussing the rationale behind the chosen solution. Design choices are indeed a direct consequence of in-depth trade-offs. It was also possible to present the SW framework that is in charge of managing and dynamically allocating diverse applications on the most suitable resource, in order to maximize the performance per watt of the board. The SW is indeed designed to be flexible and manage diverse applications, as the ones presented as benchmarks. Such applications, that initially provided guidelines for the architecture design, will be, in the near future, used to verify the prototype performances through dedicated test sessions. Completion of the work is expected by the end of 2025.

ACKNOWLEDGMENT

The work presented in this paper has been carried out within the framework of Edge Processing Image Inference Processing Unit (DIEPU), Contract No. 4000142524/23/NL/AS, supported by ESA General Support Technology Programme (GSTP).

REFERENCES

[1] G. Furano, G. Meoni, A. Dunne, D. Moloney, V. Ferlet-Cavrois, A. Tavoularis, J. Byrne, L. Buckley, M. Psarakis, K.-O. Voss *et al.*, "Towards the use of artificial intelligence on the edge in space systems: Challenges and opportunities," *IEEE Aerospace and Electronic Systems Magazine*, vol. 35, no. 12, pp. 44–56, 2020.

[2] G. Furano, C. U. Ortega, M. Tali, B. Guesmi, D. Moloney, M. Dean, N. Longepede, P.-P. Mathieu *et al.*, "The use of ai in operational space weather missions," in *104th AMS Annual Meeting*. AMS, 2024.

[3] G. Giuffrida, L. Fanucci, G. Meoni, M. Batič, L. Buckley, A. Dunne, C. Van Dijk, M. Esposito, J. Hefele, N. Vercruyssen *et al.*, "The φ-sat-1 mission: The first on-board deep neural network demonstrator for satellite earth observation," *IEEE Transactions on Geoscience and Remote Sensing*, vol. 60, pp. 1–14, 2021.

[4] P. Kuligowski, P. Wozny, R. Czerwinski, and W. Sladek, "Serval: A new chapter of on-board data processing with versal acap-based units," in *2023 European Data Handling & Data Processing Conference (EDHPC)*, 2023, pp. 1–7.

[5] B. Gaide, D. D. Gaitonde, C. Ravishankar, and T. Bauer, "Xilinx adaptive compute acceleration platform: Versaltm architecture," *Proceedings of the 2019 ACM/SIGDA International Symposium on Field-Programmable Gate Arrays*, 2019. [Online]. Available: https://api.semanticscholar.org/CorpusID:67870688

[6] "Versal architecture and product data sheet: Overview," AMD, 2024, accessed: 2024-05-14, Release Date: 2024-04-24. [Online]. Available: https://docs.amd.com/v/u/en-US/ds950-versal-overview

[7] N. Perryman, C. Wilson, and A. George, "Evaluation of xilinx versal architecture for next-gen edge computing in space," in *2023 IEEE Aerospace Conference*, 2023, pp. 1–11.

[8] S. Bianco, R. Cadene, L. Celona, and P. Napoletano, "Benchmark analysis of representative deep neural network architectures," *IEEE Access*, vol. 6, pp. 64 270–64 277, 2018.

[9] A. G. Howard, M. Zhu, B. Chen, D. Kalenichenko, W. Wang, T. Weyand, M. Andreetto, and H. Adam, "Mobilenets: Efficient convolutional neural networks for mobile vision applications," 2017.

[10] K. He, X. Zhang, S. Ren, and J. Sun, "Deep residual learning for image recognition," 2015.

[11] C. Szegedy, W. Liu, Y. Jia, P. Sermanet, S. Reed, D. Anguelov, D. Erhan, V. Vanhoucke, and A. Rabinovich, "Going deeper with convolutions," 2014.

[12] F. Prautzsch, C. Spindeldreier, D. Smith, T. Dirkes, K. Grürmann, and J. Rust, "Architecture design of a high-performance mass memory unit based on xilinx versal fpga for future space applications," in *European Data Handling & Data Processing Conference for Space 2023*, ser. Proceedings of the 2023 European Data Handling & Data Processing Conference for Space (EDHPC 2023), 2024. [Online]. Available: http://hdl.handle.net/20.500.12738/15242

[13] European Cooperation for Space Standardization (ECSS), *SpaceFibre – Very high-speed serial link*, Std., May 2019, eCSS-E-ST-50-11C. [Online]. Available: https://ecss.nl/standard/ecss-e-st-50-11c-spacefibre-very-high-speed-serial-link/

[14] M. Petry, A. Koch, M. Werner, U. Hoch, T. Helfers, and R. Wiest, "Machine learning on telecommunication satellite," *DATA SYSTEMS IN AEROSPACE - 2023 DASIA*, 2023.

[15] O. Ronneberger, P. Fischer, and T. Brox, "U-net: Convolutional networks for biomedical image segmentation," 2015.

[16] European Cooperation for Space Standardization (ECSS), "Technology readiness level (trl) guidelines," ECSS-E-HB-11A, March 2017, https://ecss.nl/home/ecss-e-hb-11a-technology-readiness-level-trl-guidelines-1-march-2017/.

[17] C. Ciancarelli, E. Mariotti, F. Corallo, S. Cognetta, L. Manovi, A. Marchioni, M. Mangia, R. Rovatti, and G. Furano, "Innovative ml-based methods for automated on-board spacecraft anomaly detection," in *The Use of Artificial Intelligence for Space Applications*, C. Ieracitano, N. Mammone, M. Di Clemente, M. Mahmud, R. Furfaro, and F. C. Morabito, Eds. Cham: Springer Nature Switzerland, 2023, pp. 213–228.

[18] S. Cognetta, C. Ciancarelli, F. Corallo, E. Mariotti, and A. Leboffe, "Spacecraft on-board anomaly detection: computational constrained machine learning approaches," in *17th International Conference on Space Operations*. Dubai, United Arab Emirates: Mohammed Bin Rashid Space Centre (MBRSC) on behalf of SpaceOps, 2023, additional Information.

[19] L. J. Wong, W. H. Clark, B. Flowers, R. M. Buehrer, W. C. Headley, and A. J. Michaels, "An rfml ecosystem: Considerations for the application of deep learning to spectrum situational awareness," *IEEE Open Journal of the Communications Society*, vol. 2, pp. 2243–2264, 2021.

[20] A. Jagannath, J. Jagannath, and P. S. P. V. Kumar, "A comprehensive survey on radio frequency (rf) fingerprinting: Traditional approaches, deep learning, and open challenges," *Computer Networks*, vol. 219, p. 109455, 2022. [Online]. Available: https://www.sciencedirect.com/science/article/pii/S1389128622004893

[21] VMEbus International Trade Association, "Spacevpx (vita 78)," VITA, ANSI/VITA Standard 78.0-2022, 2022, available from VITA: https://www.vita.com/Standards.

[22] I. Corporation, "Pcb stackup design considerations for intel® fpgas," Intel, Tech. Rep. [Online]. Available: https://cdrdv2-public.intel.com/666750/an613-683883-666750.pdf

[23] I. Corporation, H.-P. Company, N. Corporation, and D. Inc., "Intelligent platform management interface specification, second generation v2.0, rev. 1.1," 2015.

[24] *Vitis AI Development Environment*, Xilinx, 2024, accessed: 2024-05-16. [Online]. Available: https://github.com/Xilinx/Vitis-AI

[25] *PetaLinux Software Development Kit*, Xilinx, 2024, accessed: 2024-05-16. [Online]. Available: https://www.xilinx.com/products/design-tools/embedded-software/petalinux-sdk.html

Special Session: SE-UVM, an Integrated Simulation Environment for Single Event Induced Failures Characterization and its Application to the CV32E40P Processor

Marcello Barbirotta, Marco Angioli, Antonio Mastrandrea, Francesco Menichelli, Abdallah Cheikh, Mauro Olivieri
Dept. of Information Engineering, Electronics and Telecommunications, Sapienza University of Rome, Italy,
Email: {name.surname}@uniroma1.it

Abstract—Fault injection tests are created and designed to reproduce real-world faults and errors in controlled environments to evaluate the robustness and reliability of digital systems across various domains. There are three primary types of fault injection mechanisms: simulation-based, hardware emulation-based and physical-based. Each approach presents distinct advantages, challenges, and specific testing requirements and scenarios. In this work, we propose a novel, easy-to-use open-source fault injection simulation environment, written following the Universal Verification Methodology (UVM) Standard, and able to verify any digital design under different fault scenarios like Single Event Upsets (SEU) and Single Event Transients (SET), collecting, and producing as output, failure probabilities on target signals according to five types of output errors, offering detailed insights into the nature and severity of potential failures caused by injected faults.

As a case study, we present an integration of this environment on the CV32E40P processor, obtaining an accurate fault injection analysis on the pipeline units for the *Helloworld* and *Coremark* benchmarks, with up to more than 23680 faults for each of the 9231 tested architectural bits.

Index Terms—fault injection simulation, fault tolerance, RISC-V, microprocessors

I. INTRODUCTION

Computer systems can be vulnerable to different events, such as environmental conditions, faults, errors or failures that may affect them and change the expected system behaviour [1] [2]. A fault is a physical defect in a hardware or software component, caused by environmental conditions which may cause an error leading to system failure [3].

Errors in electronic devices can be caused by Single-Event Effects (SEE), which can be divided into Hard Errors (non-recoverable) and Soft Errors (recoverable). Soft Errors have in common the possibility of being detected and possibly solved by hardware or software protection measures. For this reason, they need to be tested and validated in fault injection test campaigns, ensuring the detection of weaknesses, adherence to standards, and mitigation of risks linked to failures. In this work, we will refer only to faults that appear at the hardware level internally to a digital electronic system. Among the main methods used to test circuits against faults, we can recognize:

- Physical hardware-based fault injection approaches;
- Hardware emulation-based fault injection approaches;
- Simulation-based fault injection approaches.

A physical fault injection process aims to mimic real-world fault conditions, but it comes with its own set disadvantages related to the specialized equipment and facilities it requires as well as safety measures to ensure accurate and reliable fault injection. Unlike physical fault injection, which involves inducing real faults in physical hardware components, fault injection Hardware Emulation replicates fault conditions using specialized debug hardware and software tools to flip bits without the need for expensive or complex setups. Simulation-based methods are less expensive than Hardware-based approaches, but their main advantage is to accurately reproduce the behaviour of the target system under normal and fault conditions. On the other side they require no additional hardware and offer greater availability and flexibility in experiments at the expense of computational overhead and simulation time.

Approaches at different level of abstraction provide different levels of accuracy (e.g. bit accurate or cycle accurate) and can be categorized into *gate level*, *register-transfer level* (RTL), and *system level* simulations [4]. This work focuses on gate level and RTL approaches, proposing an easy-to-use simulation-based fault injection environment called SE-UVM, written in the UVM standard. The current work extends the previous one described in [5], where the Time Frame Spanning approach was described using the UVM standard as an alternative fault injection technique to Monte Carlo analysis. The current work distinguishes from it in the detailed description of the entire UVM simulation environment and the application of the same to the RISCV CVA32E40P processor as a case study, illustrating the integration process and analyzing fault injection outcomes in the pipeline units using two distinct benchmark examples. The environment allows the injection of SEU and SET anywhere in the circuit during simulation, creating Single Bit Upset (SBU) or Multi Bit Upset (MBU). Users can specify target signals in the design, obtaining failure probabilities and error distributions. This means that users can analyze

979-8-3503-6689-1/24 $31.00 © 2024 IEEE

the impact of injected faults on specific signals or paths and determine the likelihood of failures caused by errors occurring in those areas. Additionally, SE-UVM can distinguish between five different types of failures, providing detailed insights into the nature and severity of potential failures resulting from injected faults. This level of granularity enables the analysis and optimization of the design fault tolerance and reliability.

The rest of the article is organized as follows: Section II discusses the related works about fault injection approaches; Section III and IV discuss the details of the proposed design; Section V reports the integration case example and performance evaluation methodology on the CVA32E40P microarchitecture and finally Section VI summarizes the conclusions.

II. RELATED WORKS

Generally speaking, a simulation platform for fault injection must provide flexible and controllable mechanisms for injecting faults into the target system, including the type, timing, duration, and severity along with the affected system components. Moreover it should be validated and verified against the expected behaviour under fault-free conditions and known fault scenarios.

According to [6], simulation-based fault injection approaches can be divided into two main categories:

- dynamic approaches, where single or multiple faults are injected into the circuit and then the circuit is simulated with different input vectors to compute the number of errors;
- static approaches, where, by using matrices or graphs that allows to model the effects of triple masking, probability rules are used to evaluate their impacts.

Among the dynamic approaches, it is possible to find the work in [7], where authors utilize the HSPICE tool to simulate the circuit, aiming to derive the probability distribution of transient pulses under various simulated conditions of particle strikes on gates. In contrast, the method presented in [8] aims to streamline the simulation process by classifying and selecting the most efficient abstraction level for different types of injected faults, enhancing simulation speed.

In some static approaches like [9], the circuit under study and the transient pulses are modeled as a matrix, and the transient pulse effect is investigated using operators on these matrices ,while in other approaches like [10], a fault is modeled by a Binary Decision Diagram (BDD) in order to study the logical masking effect of the transient pulse. Another distinction can be done considering generic and specific tools [11]. Specific tools are normally tightly coupled to a single target or simulator, example of them are extensions of QEMU [12], which is a virtual machine capable of running complete operating systems, or GeFIN, a fault injection framework built on top of the Gem5 micro-architectural simulator for ARM processors [13]. Generic tools, on the other hand, allow for broader applicability, often giving up specific analysis functions. Authors in [14] propose GOOFI as a generic architecture that facilitates the implementation of new system targets and new fault injection techniques. Fidalgo et al. [15]

propose a tool for real-time fault injection through built-in debug circuitry included in real processors, while David and Campbell [16] adopt a similar approach, using the GDB debugger interface. MiFIT [17] is a modular, open-source fault injection tool designed for microprocessor architectures. This tool enables the simulation of faults in both specific and general-purpose registers. However, it lacks the support for simulating faults in the microprocessor main memory.

III. FAULT INJECTION

This work presents an open-source bit-level cycle-accurate fault injection environment implemented in a pure UVM structure, falling into the category of generic [11] and dynamic [6] approaches, targeting the fault tolerance characterization of RTL-based digital architectures. For the sake of conciseness, we will limit in this paper its use to the specific task of testing microprocessors. However, the environment can be configured to perform fault injection experiments in any digital architecture. In [18], the authors demonstrate that UVM encompasses all the desirable requirements of a fault injection environment, emphasizing scalability and reusability. Their analysis relies on UVM and uses statistically estimated error paths with complex ad-hoc tools not integrated into standard simulation environments. Since UVM facilitates code reusability across different DUTs or different stimuli by simply modifying sequences, we chose to use UVM to implement our fault injection environment, fully leveraging its modular paradigm.

Universal Verification Methodology is an industrial verification standard for complex integrated circuits and systems-on-chip (SoCs) that contains all the SystemVerilog guidelines, methodologies and libraries used to facilitate the creation of modular, reusable verification components, such as test benches, test cases, and verification IPs, which are key components in modern digital design verification. The foundational structure of an UVM system enhances reusability, scalability, and efficiency in verification environments. It is based around a test environment known as the *Test* class, which comprises all the elements that manage communication interfaces with the DUT and stimulus production. Our environment does not contain all the characteristics of UVM environments but only those necessary and useful to create a lightweight and complete fault injection tool. This means that, given the great modularity of the uvm standard, it can be easily integrated into any existing verification environment.

The SE-UVM structure is depicted in Figure 1, it is completely based on a single UVM package named *tb_se_uvm_pkg.sv*, which contains the main *Test* class that launches, configures and activates the whole environment. The package can be configured through initial parameters that enable different error-injecting procedures, like the classical statistical Monte Carlo approach and the Time-Frame-Spanning approach [5]. Moreover, these parameters allow to choose the fault type, ranging from SBU faults up to MBU faults, as well as SET faults with customizable widths. Initial parameters can configure the optional compare to a golden memory image or exception signal monitoring, useful in case the DUT is

979-8-3503-6689-1/24 $31.00 © 2024 IEEE

a processor. Finally, the initial parameters allow indicating the DUT internal signal names and sizes, that represent the deterministic target registers or nets of the architecture under test, saved by the user in separate files (i.e. *signal_paths.txt* and *signal_sizes.txt*). Additionally, output files where results are saved (i.e. *result_logs.txt* and *error_logs.txt*).

Following the structure of UVM environments, each *Test* class contains an *Agent*. Within each *Agent*, a *Sequencer* can be instantiated to generate *Sequences*, which are then passed to a *Driver* responsible for organizing and transmitting the data (Transaction Level Modeling stimuli, i.e. TLM stimuli) to the DUT through a customized Virtual Interface. The same interface should be created on both sides (test class and DUT test bench) to properly set up the fault injection package and it represents the only part that should be added to the DUT test bench to use the entire SE-UVM tool.

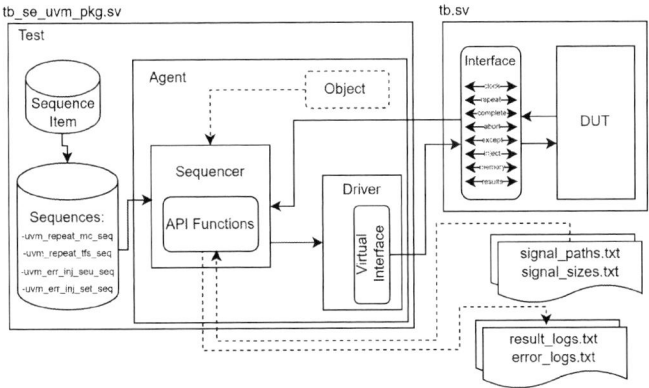

Fig. 1. SE-UVM Architecture

The *Sequencer* normally manages the entire environment. It contains a series of APIs specifically created to perform functions such as: the creation of log files, the reading of external input files or the creation of time intervals for fault injection. The last function is implemented using object classes capable of randomizing time values between UVM components through transaction classes, which exchange time information on specific ranges, pre-configured at parametric level and dependent on the benchmark under consideration.

Among the most important API functions are those relating to injecting faults within the architectural signals inside single FFs, nets, or entire registers, exploiting *uvm_hdl_force*, *uvm_hdl_release* and *uvm_hdl_deposit* functions. The sequences that manage the type of test under examination will call these functions. The primary sequence, launched by the *Test* class, is the repeat sequence chosen during the test setup phase and regards the statistical fault injection approach that can be applied, choosing between Monte Carlo (i.e. *uvm_repeat_mc_seq*) or Time Frame Spanning approach (i.e. *uvm_repeat_tfs_seq*) [5] [19]. The repeat sequence manages the entire fault injection test by synchronizing it with the DUT test bench through control signals exchanged by the interfaces and the *Sequencer*, such as repeat, complete or abort signals of the test, or error injection signal, launched iteratively through secondary sequences. The repeat sequences then allow the

control of the DUT with specific watchdog tasks, in case the DUT hangs due to injected errors or when certain exceptions occur, monitored through specific architecture signals. The secondary sequence launches the fault injection method after a certain time interval regulated by the object classes described above. They read the path and target bits from the *Sequencer* that called and managed them, implementing SEU (SBU or MBU) faults (e.g., *uvm_err_inj_seu_seq*) or SET faults (e.g., *uvm_err_inj_set_seq*). Regarding SET faults, as anticipated, their introduction occurs by varying their duration but not their amplitude, as this is impossible in RTL simulation. It is necessary to specify that for a correct simulation of SET, the DUT must be simulated with post-implementation timing simulation.

IV. OUTPUT CLASSIFICATION

There are many different ways in which a transient error can be propagated or masked inside the circuit. The tool may not be able to keep track of any electrical or timing masking effects that may occur when a transient fault is inserted, however, by knowing in advance the origin of the fault and its logical path, any possible masking phenomena can easily be reconstructed. When a fault is not masked, there are different effects that can be generated and they should be considered when developing a fault injection simulation environment. In our case, five distinct output categories have been identified [20]:

- Incorrect termination: the program execution ends before completion.
- Incorrect data: despite normal termination, the data memory content deviates from the error-free execution (referred to as silent data corruption).
- Crash with exception: the execution triggers an unrecoverable processor exception.
- Crash with hang: the program enters an infinite loop, failing to conclude.
- Correct outcome: the program concludes successfully despite the injection of an error, with the data memory matching that of the error-free execution.

V. THE CV32E40P CASE STUDY

This work contains, in addition to the description of the developed UVM environment, a detailed code integration and fault characterization example performed on the CV32E40P processor based on the RI5CY architecture [21] supported by the PULP platform team until February 2020 and now maintained by the OpenHW Group. The CV32E40P is a small and efficient, 32-bit in-order RISC-V 4-stage pipeline core that implements the RV32IM[F]C instruction set architecture, fully compatible with the Pulpissimo SOC platform [22] in which is tested and verified.

A. Code Integration

Once configured, the fault injection environment can be called within the DUT test bench (i.e. the Pulpissimo test bench *tb_pulp.sv*) by including the files and the various UVM packages into the test bench module as listed below:

```
// include the uvm test class
`include "uvm_macros.svh"
`include "tb_se_uvm_pkg.sv"
```

As anticipated, the test bench has to instantiate the interface with the communication signals between the modules, along with parameters used to perform the memory comparisons, such as a memory path that indicates where the memory is located inside the design, memory size, and the number of instantiated banks:

```
parameter int MEM_SIZE  = 32768;
parameter int NUM_BANKS = 4;
'define MEMORY_PATH_0
    tb_pulp.i_dut. ... .l2_ram_i.CUTS[0].bank_i.sram;
...
interface tb_sv2uvm_if(input w_clk_ref );
  logic [3:0][7:0] mem_if [NUM_BANKS][MEM_SIZE];
  bit set_except_condition_if;
  logic is_sv_execution_completed = 1'b0;
  logic sv_execution_repeat = 1'b1;
  logic sv_execution_abort = 1'b0;
  logic fault = 1'b0;
  int results;

  //declaring a clocking clock to synch signals
  clocking cb @(posedge w_clk_ref );
      input is_sv_execution_completed;
      input sv_execution_repeat;
      input sv_execution_abort;
      input fault;
      output results;
  endclocking

  modport drv(clocking cb);
  modport mnt( input fault, input is_sv_execution_completed,
               input sv_execution_repeat,
               input sv_execution_abort, output results);
  modport acc( input mem_if, input set_except_condition_if );
endinterface
```

Once declared, the interface should be instantiated and saved inside the uvm database *uvm_config_db*:

```
module tb_pulp;            //top test bench module
import uvm_pkg::*;         //import the uvm package
import tb_se_uvm_pkg::*; //import the new defined package
...
//instantiate the interface
tb_sv2uvm_if tb_sv2uvm_if(w_clk_ref);
  assign tb_sv2uvm_if.mem_if[0] = 'MEMORY_PATH_0;
...
initial
 begin
    uvm_config_db#(virtual tb_sv2uvm_if)::set(
                          uvm_root::get(),
                          "*uvm_test_err_inj",
                          "tb_sv2uvm_if_vi",
                          tb_sv2uvm_if
                          );
    run_test("uvm_test_err_inj");
end
```

Furthermore, the test bench must be able to start the *Test* class routine and must be able to be appropriately modified to assert the proper communication signals, such as complete or repeat signals:

```
initial //start pulpissimo tb
begin
    do begin : pulpissimo_loop
        ...
    //tb starting
    wait(tb_sv2uvm_if.mnt.sv_execution_repeat === 1'b0);
    #1
    tb_sv2uvm_if.mnt.is_sv_execution_completed=1'b0;
        ...
    //tb completed
    tb_sv2uvm_if.results = exit_status;
    tb_sv2uvm_if.mnt.is_sv_execution_completed=1'b1;
        ...
    end //pulpissimo loop
    while(1);
end //end pulpissimo tb
```

Finally, the test bench should contain a specific control loop able to monitor the abort signal coming from the interface and perform the reset routine to restart the correct funtionality of the DUT:

```
initial // start control tb
begin
do begin : control_loop
    wait(tb_sv2uvm_if.sv_execution_abort === 1'b1);
    tb_sv2uvm_if.mnt.sv_execution_abort = 1'b0;
    disable pulpissimo_loop;
    //reset routine
    tb_sv2uvm_if.results = 1;
    tb_sv2uvm_if.mnt.is_sv_execution_completed=1'b1;
end //control loop
while(1);
end //end control tb
```

B. Benchmark Setup and Results Comparisons

In a fault tolerant system, complete protection against faults in every bit of the core might be excessive. Research has demonstrated that the overall impact of local faults in microprocessors depends on the program being executed. Therefore, examining the fault injection resilience of a processor hardware design using an application-oriented methodology is crucial. For this reason, we decided to launch and characterize the core using two different application scenarios: the classical Helloworld and the well-known Coremark benchmark, using SEU fault sequences, creating SBU faults. Both executions were tested using the Time Frame Spanning approach described in [5] and chosen for its ability to produce an upper bound of the error resilience for each target signal in less time than the classical Monte Carlo approach, thanks to the division of the benchmark duration into time frames and the insertion of deterministic faults within them.

This methodology, outlined in [5], evaluates the average number of Architecturally Correct Execution (ACE) bits in a cycle, dividing the total execution time into m intervals, referred to as time frames, and conducting the entire RTL simulation during the benchmark execution for each frame, with a fault injected on a target bit j of the processor for the duration of a time frame. By definition, a single bit of the hardware microarchitecture is ACE in a certain clock cycle if a value change occurring on that bit in that clock cycle causes a program failure [20]. The fault injection on bit j indicates a system failure when faults are injected only in m_F time frames out of the total m time frames. Assuming that SEU physical events in the real system occur with a uniform time distribution during the whole program execution, we can estimate the probability that bit j is in an ACE state when the fault occurs, as:

$$Pe(j) \triangleq \Pr \{ \text{ a SEU hitting } j^{th} \text{ bit causes a system fault } \}$$
$$\leq m_F/m \tag{1}$$

This analysis accurately identifies the time frames during program execution in which the system remains resilient to faults on bit j. Each interval is calculated as one-tenth of the benchmark execution time, facilitating the estimation of the likelihood that the target bit is in an ACE state, represented as a percentage in increments of 10%. Table I reports the benchmark setup. The Helloworld was tested with a total of 4720 faults/bit for each simulation run, while Coremark, having a longer duration, was tested with 23680 faults/bit for each simulation run.

TABLE I
TEST SETUP WITH REQUIRED CLOCK CYCLES, BIT QUANTITY, AVFs, NUMBER OF FRAMES [5] AND TOTAL FAULTS PER FRAME.

Pipeline Units	Helloworld			Coremark		
	IF	ID	IE	IF	ID	IE
Bits Quantity	1302	4046	3883	1302	4046	3883
AVFs	0.23	0.41	0.04	0.78	0.92	0.25
Total clock cycles	1771602			8886647		
#frames	10			10		
Faults/frame	472			2368		
Deterministic fault rate	1 every 40 cycles			1 every 40 cycles		

The analyzed components include the Fetch, Decode and Execute units, with sizes of 1302, 4046, and 3883 bits, respectively. These units also include the signals of the Register File located within the Decode unit, totalling 9231 bits examined across the whole architecture. We opted to visually present only a subset of signals from each architectural unit for conciseness, as showed in Figure 2, leaving the entire results log available under request. The analysis was conducted at bit level, and the resulting failure probability values were arithmetically averaged to determine the failure probabilities of the registers, allowing a comprehensive assessment of the microprocessor fault tolerance. As depicted in the graphs in Figure 2, for each signal, the failure values may differ between benchmarks. This observation underscores the variability in microprocessor architecture behaviour depending on the running application. The failure probabilities provide an upper limit of the signal actual fault tolerance described for N_{ACE} as outlined in (1) [23], providing a valuable starting point for implementing fault tolerance methodologies. Additionally, the number of ACE bits can be used to derive the Architectural Vulnerability Factors (AVF), which represent a crucial metric for fault-tolerant systems.

$$N_{ACE} = \sum_{j=1}^{N} P_F(j), \qquad \text{AVF} = \frac{N_{ACE}}{N} \qquad (1,2)$$

By following (2) it is possible to calculate the AVFs for each unit and for each benchmark. Table I also contains these results. Values close to 1 indicate strong architectural vulnerability while values close to zero indicate low vulnerabilitys. Comparing the results, as well as observing Figure 2, the *Helloworld* benchmark has lower values than *Coremark* due to its simplicity and shorter time duration. Another notable aspect of the environment is its capability to furnish details regarding the distribution of faults within the architectural units for each benchmark, helping the design of error correction methods. Table II provides a comprehensive summary of the output distribution with different error categories. From this table, it can be seen that the most frequent error type concerns the processor *Hang*. Specifically, for both benchmarks, the Decode unit emerges as the most critical component since this is responsible for the type of instruction executed, and any fault inside the unit can lead to erroneous jumps to instructions or incorrect routine execution, resulting in software crashes. Notably, the Exception error type was not activated due to a design choice since, in a non-redundant processor setup,

identifying Exception errors is frequent, and considering them would mask the count of other error types.

TABLE II
OUTPUT DISTRIBUTION ON THE DIFFERENT PIPELINE UNITS

Output Types [%]	Helloworld			Coremark		
	IF	ID	IE	IF	ID	IE
Correct	76.865	59.428	95.813	21.019	7.80	75.17
Inc. Data	0.039	0.019	0.069	0.007	0	0.0154
Inc. Termination	0.0242	0.0171	0.161	0.367	0.085	0.551
Exception	-	-	-	-	-	-
Hang	22.617	50.527	3.956	78.37	92.113	24.285

VI. CONCLUSIONS

This work describes a simulation-based fault injection environment named SE-UVM. The tool facilitates the injection of different types of emulated faults in the circuit during simulation, quickly ranging from SEU (SBU or MBU) and SET. Moreover, differently from Hardware Emulation and Physical methods, it allows a deeper inspection of specific internal target signals within the DUT, also allowing the distinction between five different types of output errors. The presented case study shows the integration of the entire environment within the CV32E40P processor, demonstrating how the analysis can target all the internal signals of the architecture by getting the failure probabilities and evaluating the design fault tolerance. Overall, the analysis included the test of 9231 bits, each with more than 23680 SBU faults, one every 40 clock cycles, representing one of the most complete fault injection simulation approaches in literature. Regarding performances in time required to perform the analysis, the environment itself does not influence the duration of the simulation tests, except for the chosen approach: Time Frame Spanning or Monte Carlo. This topic has already been discussed in the work [5], where the first method presents less than 50% of simulation time concerning a normal Monte Carlo simulation on a x86 64-bit Intel(R) Xeon(R) 4116 CPU @2.10GHz 32GB RAM.

REFERENCES

[1] M. Barbirotta, F. Menichelli, A. Cheikh, A. Mastrandrea, M. Angioli, and M. Olivieri, "Dynamic triple modular redundancy in interleaved hardware threads: An alternative solution to lockstep multi-cores for fault-tolerant systems," *IEEE Access*, vol. 12, pp. 95 720–95 735, 2024.

[2] R. Della Sala and G. Scotti, "A novel fpga implementation of the nand-puf with minimal resource usage and high reliability," *Cryptography*, vol. 7, no. 2, 2023. [Online]. Available: https://www.mdpi.com/2410-387X/7/2/18

[3] M. Barbirotta, A. Cheikh, A. Mastrandrea, F. Menichelli, M. Ottavi, and M. Olivieri, "Evaluation of dynamic triple modular redundancy in an interleaved-multi-threading risc-v core," *Journal of Low Power Electronics and Applications*, vol. 13, p. 2, 12 2022.

[4] W. Sheng, L. Xiao, and Z. Mao, "An automated fault injection technique based on vhdl syntax analysis and stratified sampling," in *4th IEEE International Symposium on Electronic Design, Test and Applications (delta 2008)*. IEEE, 2008, pp. 587–591.

[5] M. Barbirotta, A. Mastrandrea, F. Menichelli, F. Vigli, L. Blasi, A. Cheikh, S. Sordillo, F. D. Gennaro, and M. Olivieri, "Fault resilience analysis of a risc-v microprocessor design through a dedicated uvm environment," in *33rd IEEE International Symposium on Defect and Fault Tolerance in VLSI and Nanotechnology Systems, DFT 2020*. Institute of Electrical and Electronics Engineers Inc., 10 2020.

[6] M. Eslami, B. Ghavami, M. Raji, and A. Mahani, "A survey on fault injection methods of digital integrated circuits," *Integration*, vol. 71, pp. 154–163, 2020.

[7] Y.-H. Kuo, H.-K. Peng, and C. H.-P. Wen, "Accurate statistical soft error rate (sser) analysis using a quasi-monte carlo framework with quality cell models," in *2010 11th International Symposium on Quality Electronic Design (ISQED)*. IEEE, 2010, pp. 831–838.

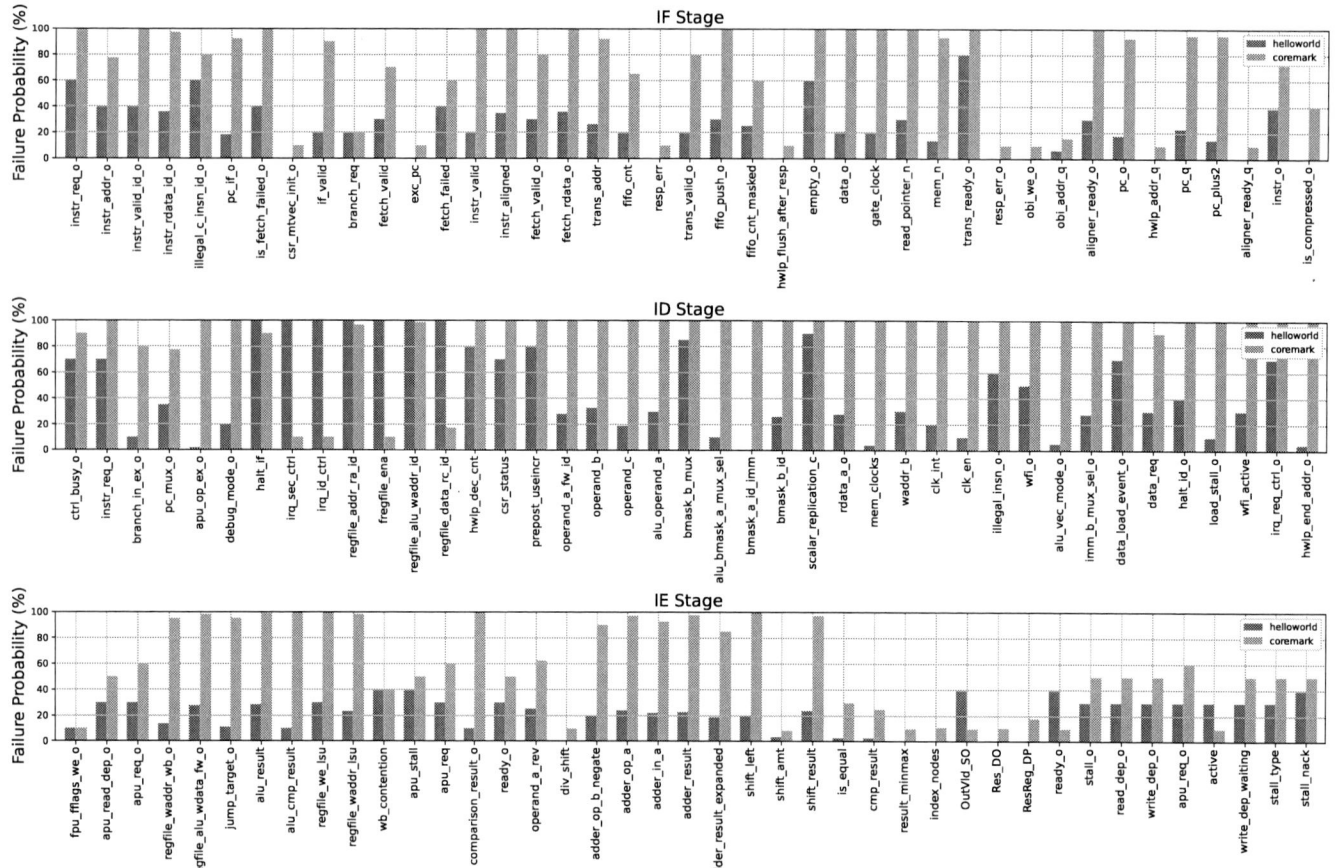

Fig. 2. Helloworld and Coremark benchmark results for SBU fault injection. The x-axis represents some signals inside the various pipeline units, while the y-axis represents the Failure Percentage (%) results obtained after the described fault injection campaign.

[8] P. R. Maier, U. Sharif, D. Mueller-Gritschneder, and U. Schlichtmann, "Efficient fault injection for embedded systems: as fast as possible but as accurate as necessary," in *2018 IEEE 24th International Symposium on On-Line Testing And Robust System Design (IOLTS)*. IEEE, 2018, pp. 119–122.

[9] L. Entrena, M. Garcia-Valderas, R. Fernandez-Cardenal, A. Lindoso, M. Portela, and C. Lopez-Ongil, "Soft error sensitivity evaluation of microprocessors by multilevel emulation-based fault injection," *IEEE Transactions on Computers*, vol. 61, no. 3, pp. 313–322, 2010.

[10] B. Zhang and M. Orshansky, "Symbolic simulation of the propagation and filtering of transient faulty pulses," in *Proc. SELSE Workshop, Urbana-Champaign*. Citeseer, 2005.

[11] A. Aponte-Moreno, J. Isaza-González, A. Serrano-Cases, A. Martínez-Álvarez, S. Cuenca-Asensi, and F. Restrepo-Calle, "Evaluation of fault injection tools for reliability estimation of microprocessor-based embedded systems," *Microprocessors and Microsystems*, vol. 96, p. 104723, 2023.

[12] F. Bellard, "Qemu, a fast and portable dynamic translator." in *USENIX annual technical conference, FREENIX Track*, vol. 41, no. 46. California, USA, 2005, pp. 10–5555.

[13] M. Kaliorakis, S. Tselonis, A. Chatzidimitriou, N. Foutris, and D. Gizopoulos, "Differential fault injection on microarchitectural simulators," in *2015 IEEE International Symposium on Workload Characterization*, 2015, pp. 172–182.

[14] J. Aidemark, J. Vinter, P. Folkesson, and J. Karlsson, "Goofi: generic object-oriented fault injection tool," in *2001 International Conference on Dependable Systems and Networks*, 2001, pp. 83–88.

[15] A. Fidalgo, M. Gericota, G. Alves, and J. Ferreira, "Using nexus compliant debuggers for real time fault injection on microprocessors," in *Proceedings of the 19th annual symposium on Integrated circuits and systems design*, 2006, pp. 214–219.

[16] F. M. David and R. H. Campbell, "Building a self-healing operating system," in *Third IEEE International Symposium on Dependable, Autonomic and Secure Computing (DASC 2007)*, 2007, pp. 3–10.

[17] A. Aponte-Moreno, F. Restrepo-Calle, and C. Pedraza, "Mifit: A fault injection tool to validate the reliability of microprocessors," in *2019 IEEE Latin American Test Symposium (LATS)*. IEEE, 2019, pp. 1–5.

[18] D. Lohmann, F. Maziero, E. J. dos Santos, and D. Lettnin, "Extending universal verification methodology with fault injection capabilities," in *2018 IEEE 9th Latin American Symposium on Circuits & Systems (LASCAS)*, 2018, pp. 1–4.

[19] M. Barbirotta, A. Cheikh, A. Mastrandrea, F. Menichelli, M. Angioli, S. Jamili, and M. Olivieri, "Fault-tolerant hardware acceleration for high-performance edge-computing nodes," *Electronics*, vol. 12, no. 17, 2023. [Online]. Available: https://www.mdpi.com/2079-9292/12/17/3574

[20] M. Barbirotta, A. Cheikh, A. Mastrandrea, F. Menichelli, and M. Olivieri, "Design and evaluation of buffered triple modular redundancy in interleaved-multi-threading processors," *IEEE Access*, vol. 10, pp. 126 074–126 088, 2022.

[21] P. Davide Schiavone, F. Conti, D. Rossi, M. Gautschi, A. Pullini, E. Flamand, and L. Benini, "Slow and steady wins the race? a comparison of ultra-low-power risc-v cores for internet-of-things applications," in *2017 27th International Symposium on Power and Timing Modeling, Optimization and Simulation (PATMOS)*, 2017, pp. 1–8.

[22] P. D. Schiavone, D. Rossi, A. Pullini, A. Di Mauro, F. Conti, and L. Benini, "Quentin: an ultra-low-power pulpissimo soc in 22nm fdx," in *2018 IEEE SOI-3D-Subthreshold Microelectronics Technology Unified Conference (S3S)*, 2018, pp. 1–3.

[23] I. Oz and S. Arslan, "A survey on multithreading alternatives for soft error fault tolerance," *ACM Computing Surveys (CSUR)*, vol. 52, no. 2, pp. 1–38, 2019.

Special Session: Testing of Digital Computing-In Memories with MAC Function

Jin-Fu Li

Advanced Reliable System (ARES) Lab.
Department of Electrical Engineering
National Central University
Taoyuan, Taiwan 320

Abstract—**Digital computing-in-memory (DCIM) with MAC function has been considered as a good alternative for the computation deep neural networks (DNNs). A basic DCIM unit with MAC function consists of a bit-multiplication memory (BMM) and an adder tree. The architecture and bit-cell structure of DCIM is very different from those of conventional SRAMs and digital CIMs with logic operations. This means that DCIM testing is more difficult than SRAM testing and different from DCIMs with logic operations. Fault models and march-like tests for SRAMs and DCIMs with logic functions are not enough for the DCIMs with MAC operation. In this embedded tutorial, therefore, we introduce the fault analysis and testing methods for DCIMs with MAC operation. Some perspectives on testing challenges of DCIMs with MAC operation are also raised.**

Index Terms—**Computing-in memories, static random access memory, fault analysis, testing, March test.**

I. INTRODUCTION

Modern von-Neumann computing architecture separating the memory storage and processing cores is not energy efficient for data-intensive computation. More and more applications are data-intensive, such as artificial intelligence, deep neural networks (DNNs), biological systems, and so on. Efficient computing architectures thus are eager to be developed. Computing-in memory (CIM) architecture which moves the processing logic into the memory array is considered as a good computing architecture for data-intensive computing [1]–[11]. The supported computing function of CIMs can be divided into logic operations and arithmetic operations. Since the CIM is more complicated than the conventional memories, the testing of CIMs is more difficult than that of conventional memories [12].

Recently, many works on the testing of CIMs with logic function were reported, e.g., [12]–[22]. Testing techniques for 8T-SRAM CIMs are introduced in [12]–[14]. The testing of CIMs must be performed in memory and computing modes, since a defect or a range of defect sizes might cause the CIM to malfunction in memory mode only or computing mode only. In computing mode, especially, two wordlines are activated for the two-operand logic operations, which shrinks the sensing margin of sensing circuit. That is, the computing operations are more sensitive to parametric faults [12], [14]. A test algorithm for CIMs with logic operations usually consists of Read/Write operations and logic operations. For example, a test algorithm

for 8T-SRAM CIMs with logic operations comprises Write, Read, and NAND operations [12], [14]. In [21], [22], the design-for-testability and reliability issues of CIMs are also introduced.

Testing techniques for resistance-based CIMs are reported in [15]–[19]. In [16], testing strategies including the testing of cell array and peripheral circuits for resistance-based CIMs are introduced. The equivalent resistance in computing mode for sensing is smaller than that in memory mode since multiple wordlines are activated in computing mode. Therefore, the fault analysis of CIMs should be executed in memory mode and computing mode [15], [17], [18]. Fault analysis results show that some computation specific faults exist. Also, the memory operations and computing operations might activate different defect ranges. Furthermore, test algorithms are also developed to cover the specific faults and conventional faults [15], [17], [18]. For the resistance-based CIMs, the I-V characteristics of the storage device is nonlinear. To accurately model the defective behavior of memristive device, a device-aware fault modeling method is needed for the fault analysis [19].

CIMs with multiply-and-accumulate (MAC) operation have been considered as a good alternative for the acceleration of DNN computation [7]–[11], [23]. However, existing works for the testing of CIMs target the CIMs with logic operations. The memory cell and peripheral circuit of CIMs with MAC operation are much different from those of CIMs with logic operation. The testing of CIMs with MAC operation thus are needed to be investigated. In this embedded tutorial, the fault analysis and testing of digital CIMs (DCIMs) with MAC operation are discussed. Also, testing challenges of DCIMs with MAC operation are raised.

The rest of this paper is organized as follows. Section II introduce the concept of DCIMs with MAC function. Section III describes the fault analysis of DCIMs and defines computing faults. Section IV introduces the testing method of DCIMs and a March-DC algorithm is introduced. Section V summarizes possible test challenges. Finally, Sec. VI concludes this paper.

II. DIGITAL CIM WITH MAC FUNCTION

CIMs can be divided into two categories: digital and analog CIMs. Analog CIMs typically can achieve very high energy efficiency, but low computing accuracy due to the influence

979-8-3503-6689-1/24 $31.00 © 2024 IEEE

of variation of PVT. On the other hand, many works demonstrate digital CIMs can achieve high throughput and energy efficiency without sacrificing computing accuracy, e.g., [7]–[11]. In this paper, therefore, we focus on the testing of DCIMs with MAC function. A DCIM with MAC function is composed of multiple DCIM units. A DCIM unit consists of a bit-multiplication memory (BMM) and an adder tree [7]–[11]. Fig. 1 shows the block diagram of a DCIM unit which consists of $m \times n$-bit BMM and $log_2 m$-stage adder tree. In the BMM, each cell executes one-bit multiplication of the stored data and the computing input data. Each input data is applied bit by bit. For each clock cycle, each word of the BMM executes $I_i[j] \times W_i[n-1:0]$ and the adder tree executes $\Sigma_{i=0}^{m-1} I_i[j]W_i[n-1:0]$. The accumulator can execute the function of shift and add. After n clock cycles, the DCIM unit completes the computation of Acc = $\Sigma_{j=0}^{n-1} \Sigma_{i=0}^{m-1} I_i[j]W_i[n-1:0]$.

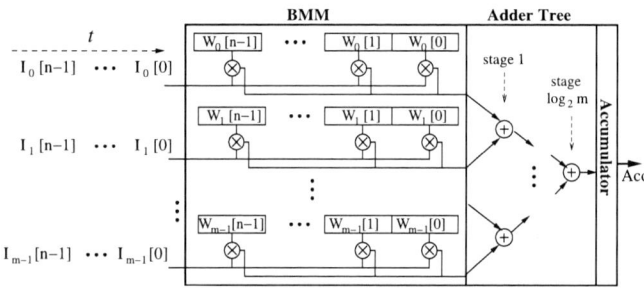

Fig. 1: Block diagram of a digital CIMs with MAC function.

A straightforward method to realize the BMM cell is to use an 6T SRAM cell and an 2-input AND gate. Various BMM cells have been proposed to minimize the cell area [7], [8], [23]. For example, an 6T SRAM cell and NOR gate is proposed in [7], utilizing NOR gates instead of AND gates is achieved through the inversion of inputs of the NOR gate. Fig. 2 shows the DCIM unit reported in [7]. The BBM cell is composed of the 6T SRAM-based CIM and an NOR gate. The inputs of the NOR fate include the complement computation input (IB) and the complement of the stored data. Similar to conventional memories, the data in BMM can be written word by word in advance. So, a row decoder is required for the wordline selection (WLi). Also, the input drivers are required for the computation inputs. The Read/Write IO circuit is responsible for the Read/Write operations of BMM.

III. FAULT ANALYSIS OF DIGITAL CIMS

A digital CIM mainly consists of two parts: the CIM array and the adder tree. Since the adder tree can be considered a random logic circuit, existing fault models can be used for its testing. Therefore, we execute the fault analysis of the CIM array. As aforementioned, each cell of the CIM array consists of a SRAM cell and a bit-multiplication logic (BML) circuit as shown in Fig. 3. The BML can be an AND gate, NOR gate, or other circuits. The short, resistive bridge, resistive open, and transistor stuck-on and stuck-open defects are considered for the fault analysis. Since the fault analysis of SRAM cell has

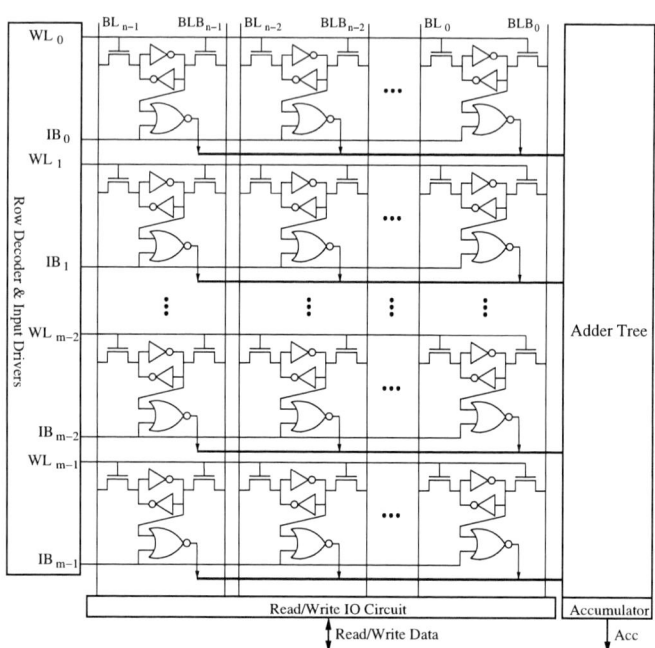

Fig. 2: A digital SRAM-based CIM with MAC function [7].

been well investigated, only the defects within the BML and between the BML and the SRAM cell are analyzed. For the defects between the BML and the SRAM cell, only resistive bridge defects are analyzed since the other defects cannot affect two nodes/lines.

Similar to conventional digital circuits, defects within the BML can be modeled as stuck-at faults at inputs and output. Small defects within the BML can also be modeled as delay faults. If the BML is realized with CMOS logic, the transistor stuck-open causes a sequential fault and two patterns should be used to detect the sequential fault. Comprehensive fault analysis can be executed in a systematic approach [15]. In the sequel, we only focus on the fault analysis for the defects between the BML and SRAM cell. As aforementioned, various logic gates can be used to realize the BML. Without loss of generality, we assume that the BML realizes an logic AND operation for the fault analysis. That is, $P=IW$.

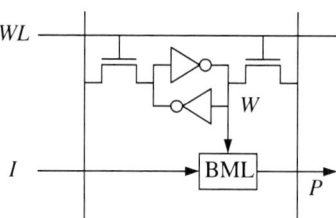

Fig. 3: A SRAM cell with bit-multiplication logic (BML) circuit.

A. Computing Input Disturbance Fault

Consider a resistive bridge defect $d1$ exists between the computing input I and the false node $\overline{W}$ of the SRAM cell as shown in Fig. 4. Since the computing input is connected to many BMLs, a large driving capability buffer is needed to drive the computing input signal. Thus, the driving capability of the input buffer is much larger than the inverters of the SRAM cell. Therefore, if the stored data W of SRAM cell and the computing input data I are the same, the stored data W is flipped. On the other hand, when the computing input I is the complement of W, the stored data of SRAM cell is not influenced since the two nodes between the defect have the same voltage. According to the discussion above, the $d1$ defect leads to a *Computing Input Disturbance Fault (CIDF)*. Similarly, if a resistive defect exists between I and W, the defect causes a CIDF when $I \neq W$. That is, when $(I, W)=(0, 1)$ or $(1, 0)$, the W is forced to 0 or 1, respectively.

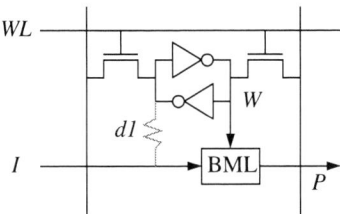

Fig. 4: A cell with a bridge defect $d1$.

Subsequently, the test requirement for the CIDF is discussed. According to the discussion above, a CIDF can be activated when 0/1 are written into the cell (i.e., W=0 and W=1) for a given input I. Then, a read operation is performed to observe the fault effect.

B. Computing Output Disturbance Fault

Consider a resistive bridge defect d_2 between the output and the false node of the SRAM cell as shown in Fig. 5. If the output P is different from the state of the false node of the SRAM cell, i.e., $P!=\overline{W}$, the cell state might be flipped. There are two possible fault behaviors. First, if I=1, the resistive bridge defect might cause the output P oscillate. The reason is that if W=0, the P=0 which forces $\overline{W}$ (W) to 0 (1). Then P becomes 1, which forces $\overline{W}$ (W) to 1 (0). Then P becomes 0. Thus, the P generates 0/1 oscillation signals. Second, if I=0, P=0 regardless of W is 0 or 1. When W=0, the resistive bridge might cause the stored data is flipped since P=0 which forces $\overline{W}$ to 0 and W is changed to 1. For the first case, the oscillation behavior causes the data of either the storage or output to alternatively switch between 0 and 1. According to the fault duration, the fault belongs to an intermittent fault. For the second case, the stored data is flipped. Therefore, we model the resistive bridge defect d_2 as a *Computing Output Disturbance Fault (CODF)*. Similarly, if a resistive bridge defect exists between the output and the true node of the SRAM cell, it might cause a CODF when I=0. If I=1, the resistive bridge defect will not cause a fault since P=W.

According to the discussion above, the activation requirement of CODF is as follows: the computing input I=0 and write-1 and write-0 must be performed. Then a read operation is executed after the write operation to observe the fault effect.

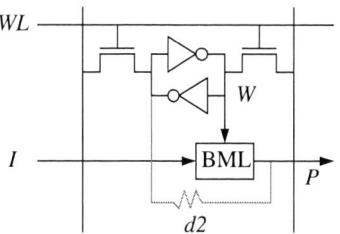

Fig. 5: A cell with a bridge defect $d2$.

IV. TESTING OF DCIMs

A DCIM unit consists of a BMM and an adder tree. The testing of DCIM can be divided into the testing of BMM and the adder tree. Although the testing of BMM is similar to that of conventional memories, the BMM has the computation input and output in addition to read/write data IOs. The computing outputs are connected to the adder tree which output is the output of the accumulator.

A. Testing of BMM

Similar to a conventional SRAM, a BMM consist of a cell array and peripheral circuits. The peripheral circuits include an address decoder, computing input drivers, and read/write I/O circuits. The testing of the address decoders and read/write circuit is similar to that in conventional memories. Most of faults in read/write circuit can be mapped to cell array faults. Therefore, a test detecting the cell array faults implicitly detects the faults in read/write circuit as well. Conventional address decoder faults can be used for the testing of address decoder in the BMM by using read/write test operations.

A test algorithm should be developed for the testing of cell array of BMM according to the modeled faults. Since a BMM cell consists of a SRAM cell and a BML, conventional memory faults should also be targeted. For the conventional memory faults, only read and write test operations are needed for the fault detection. As Section III describes, new computing faults, e.g., CIDF and CODF, might occur in the BMM. For the computing faults, computing input, read, and write test operations are needed for the fault detection.

March tests are widely used for memory fault detection due to their linear time complexity. Therefore, a test algorithm for the BMM can be developed based on a March test. For example, March C− is widely used to test SRAMs [24]. March C− test algorithm is as follows: $(\{\updownarrow (w0); \Uparrow (r0, w1); \Uparrow (r1, w0); \Downarrow (r0, w1); \Downarrow (r1, w0); \updownarrow (r0)\})$, where wd and rd denote the Write operation with data d and the Read operation with expected data d, respectively; $\Uparrow$, $\Downarrow$, and $\updownarrow$ denote the addressing sequence of the memory under test is ascending, descending, and either ascending or descending sequence.

We propose a March-DC to cover the conventional memory faults and computing faults based on March C−. March-DC test algorithm is as follows:

$$\{\updownarrow (w0); \Uparrow (r0, w1, I0, r1); \Uparrow (r1, w0, I0, r0); \Downarrow (r0, w1);$$
$$\Downarrow (r1, w0); \updownarrow (r0)\},$$

979-8-3503-6689-1/24 $31.00 © 2024 IEEE

where $I0$ denotes the computing input $I=0$ of the corresponding addressed word. Please note that computing input and write/read operations can be executed simultaneously. That is, it cannot cause additional test complexity. Therefore, the test complexity of March-DC is $12N$ for a BMM with N words. As Section III describes, the test requirements for CIDF and CODF are as follows:

- CIDF—for a given computing input I, $(w0, r0)$ and $(w1, r1)$ are performed.
- CODF—for the computing input $I=0$, $(w0, r0)$ and $(w1, r1)$ are performed.

In summary, the test requirement for covering the two faults is $(w0, r0)$ and $(w1, r1)$ are performed when the computing input $I=0$. Since the second march element ($\Uparrow (r0, w1, I0, r1)$) and the third march element ($\Uparrow (r1, w0, I0, r0)$) of March-DC can meet the test requirement of CIDF and CODF, we conclude that March-DC can cover conventional memory faults and computing faults defined in Section III.

Subsequently, the testing of BML is discussed. As mentioned in Section III, defects within the BML can be modeled as stuck-at faults at inputs and output. Of course, small defects within the BML can also be modeled as delay faults. If the BML is realized with CMOS logic, the transistor stuck-open causes a sequential fault and two patterns should be used to detect the sequential fault. For the testing of BML, the test pattern should be applied through the computing input and the SRAM cells as shown in Fig. 2 and the test responses should be observed through the output of the accumulator. For example, assume that the BML is realized an AND gate and the stuck-at faults of the inputs and output of the BML are considered. The test patterns for detecting the stuck-at faults of a two-input AND gate are $\{11, 01, 10\}$. If we want to apply $\{11\}$ test pattern, the BMM should be written as all-1 data and then all-1 data are applied through the computing inputs as well.

B. Testing of Adder Tree

The adder tree is a regular random logic circuit, i.e., an iterative logic array. Clearly, we can use automatic test pattern generation (ATPG) methods to generate the test patterns for the adder tree. Given the targeted fault models, the commercial ATPG tools can generate efficient test patterns for the combinational circuits. As Fig. 6(a) shows, however, the inputs of the adder tree for a $m \times n$-bit BMM are $I_0W_0[n - 1 : 0]$, $I_1W_1[n - 1 : 0]$, ..., $I_{m-1}W_{m-1}[n - 1 : 0]$. For applying a test pattern for the adder tree, m write operations should be executed to load the test pattern to the BMM cells and all-1 pattern should be applied to $(I_0I_1 \ldots I_{m-1})$. Therefore, if the ATPG generates k test patterns for the adder tree, $m \times k$ write operations should be executed.

We also can use DFT techniques for the testing of the adder tree. As Fig. 6(b) shows, we can insert a scan chain between the BMM and the adder tree. Then, the ATPG generated test patterns for the adder tree circuit can be applied through the scan chain. For a $m \times n$ BMM, the number of required scan flip flops is $m \times n$. Therefore, the test application time for k ATPG

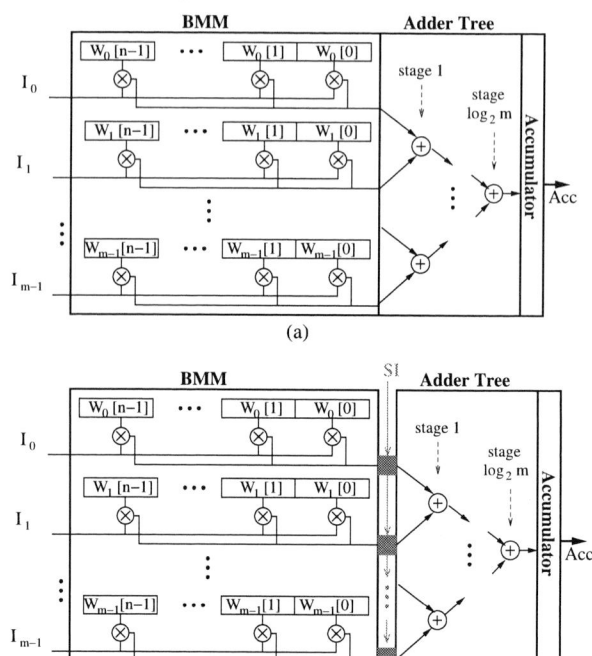

Fig. 6: (a) A conceptual block diagram of a DCIM. (b) A conceptual block diagram of a DCIM with scan insertion.

test patterns will be $m \times n \times k$ clock cycles. Clearly, the test time is larger than that of test application through the BMM. Although we can use multiple scan chains to reduce the test time, the required number of scan input/output is increased. Also, additional $m \times n$ scan flip flops should be needed. But, the scan chain will be benefit to the diagnosis.

Another DFT approach is to modify the scan flip flops as a linear feedback shift register (LFSR) in test mode. Then, the LFSR can generate pseudo-random test patterns for the adder tree. Due to the regularity, the adder tree can be partitioned into multiple blocks. Then, each block is tested by a corresponding LFSR. All the blocks of the adder tree are tested by multiple LFSRs simultaneously such that the test time can be reduced. We also can modify the LFSR to generate pseudo-exhaustive test patterns for the adder tree.

V. CHALLENGES OF DCIM TESTING

Built-in self-test (BIST) techniques are widely-used to test the SRAMs. DCIM typically is embedded in a chip for the acceleration of computation. Clearly, BIST techniques are needed for the testing of DCIMs. Since the DCIM is used for data-intensive computations, the number of DCIM blocks in a DCIM is large and all the DCIM blocks usually are activated simultaneously. As a result, the number of bits of computing inputs, computing outputs, and read/write IOs is huge. This causes that the area cost of BIST circuit will be high. One approach is to design a BIST circuit which tests the DCIM block by block. However, this cannot test the DCIM under normal operation condition, which impacts the test quality.

979-8-3503-6689-1/24 $31.00 © 2024 IEEE

Cost-effective BIST schemes for ensuring the test quality of DCIMs are needed.

As aforementioned, the test response of BMLs of the BMM should be observed through the output of the accumulator if no DFT insertion is considered. This means that the fault location of BMM cells with defective BML is difficult to locate. If a scan chain is inserted between the BMM and the adder tree, the fault location of the cells with faulty BML is easy to locate. But, this incurs large area cost and fault location time. Cost-effective fault-location techniques thus are needed for the DCIM.

In a DCIM, the adder tree typically represents a significant portion of the area of DCIM. To improve the yield of DCIM, redundancies should be added in the BMM and the adder tree. Although the adder tree is a regular circuit, designing cost-effective redundancies for the adder tree is difficult, since the reconfiguration circuit incurs large area cost. To replace a faulty adder in the adder tree, efficient fault-location techniques are needed to identify the faulty adder.

VI. CONCLUSIONS

Fault models and test algorithms for SRAMs and DCIMs with logic function are not enough for the DCIMs with MAC operation. In this embedded tutorial, we have executed the fault analysis for BMMs of DCIMs with MAC operation. Two computing faults are defined and a March-DC test algorithm has been proposed. March-DC test algorithm required 12N test complexity to cover conventional memory faults and computing input disturbance and computing output disturbance faults for a DCIM with N words. Some perspectives on the testing challenges of DCIMs with MAC operation also have been raised.

ACKNOWLEDGMENT

This work was supported in part by the National Science and Technology Council (NSTC), Taiwan, R.O.C., under Contracts MOST 112-2221-E-008-096-MY3, MOST 112-2218-E-002-025-MBK, and NSTC 113-2640-E-008-001.

REFERENCES

[1] S. Jeloka, N. Bharathwaj, and D. Sylvester, "A 28nm configurable memory (TCAM/BCAM/SRAM) using push-rule 6T bit cell enabling logic-in-memory," *IEEE Jour. of Solid-State Circuits*, vol. 51, no. 4, pp. 1009–1021, Apr. 2016.

[2] J. Zhang, Z. Wang, and N. Verma, "In-memory computation of a machine-learning classifier in a standard 6T SRAM array," *IEEE Jour. of Solid-State Circuits*, vol. 52, no. 4, pp. 915–924, Apr. 2017.

[3] Q. Dong, S. Jeloka, M. Saligane, Y. Kim, M. Kawaminami, A. Harada, S. Miyoshi, M. Yasuda, D. Blaauw, and D. Sylvester, "An 4T+2T SRAM for searching and in-memory computing with 0.3-V V_{DDmin}," *IEEE Jour. of Solid-State Circuits*, vol. 53, no. 4, pp. 1006–1014, Apr. 2018.

[4] Y. Zhang, L. Xu, Q. Dong, J. Wang, D. Blaauw, and D. Sylvester, "Recryptor: a reconfigurable cryptographic Cortex-M0 processor with in-memory and near-memory computing for IoT security," *IEEE Jour. of Solid-State Circuits*, vol. 53, no. 4, pp. 995–1005, Apr. 2018.

[5] A. Agrawal, A. Jaiswal, C. Lee, and K. Roy, "X-SRAM: enabling in-memory boolean computations in CMOS static random access memories," *IEEE Trans. on Circuits and Systems-I: Regular Papers*, vol. 65, no. 12, pp. 4219–4232, Dec. 2018.

[6] A. Agrawal, A. Jaiswal, D. Roy, B. Han, G. Srinivasan, A. Ankit, and K. Roy, "Xcel-RAM: accelerating binary neural networks in high-throughput SRAM compute arrays," *[Online]. Available: https://arxiv.org/abs/1802.08601*, 2018.

[7] Y.-D. Chih and et al., "A 89TOPS/W and 16.3TOPS/mm^2 all digital SRAM-based full-precision compute-in memory macro in 22nm for machine-learning edge applications," in *Proc. IEEE Int'l Solid-State Cir. Conf. (ISSCC)*, 2021, pp. 252–253.

[8] H. Fujiwara and et al., "A 5nm 254-TOPS/W 221-TOPS/mm^2 fully-digital computing-in-memory macro supporting wide-range dynamic-voltage-frequency scaling and simultaneous MAC and write operations," in *Proc. IEEE Int'l Solid-State Cir. Conf. (ISSCC)*, San Francisco, Feb. 2022.

[9] H. Mori and et al., "A 4nm 6163-TOPS/W/b 4790-TOPS/mm^2/b SRAM based digital-computing-in-memory macro supporting bit-width flexibility and simultaneous MAC and weight update," in *Proc. IEEE Int'l Solid-State Cir. Conf. (ISSCC)*, San Francisco, Feb. 2023.

[10] C.-T. Lin and et al., "DIMCA: an area-efficient digital in-memory computing macro featuring approximate arithmetic hardware in 28nm," *IEEE Jour. of Solid-State Circuits*, vol. 59, no. 3, pp. 960–971, Mar. 2023.

[11] A. Sridharan, J. Saikia, Anupreetham, F. Zhang, J.-S. Seo, and D. Fan, "PS-IMC: a 2385.7-TOPS/W/b precision scalable in-memory computing macro with bit-parallel inputs and decomposable weights for DNNs," *IEEE Solid-State Circuits Letters*, vol. 7, pp. 102–105, Feb. 2024.

[12] T.-L. Tsai, J.-F. Li, C.-L. Hsu, and C.-T. Sun, "Testing of in-memory-computing 8T SRAMs," in *IEEE Int. Symp. on Defect and Fault Tolerance in VLSI Systems (DFT)*, Netherlands, Oct. 2019, pp. 1–4.

[13] J.-F. Li, T.-L. Tsai, C.-L. Hsu, and C.-T. Sun, "Testing configurable 8T SRAMs for in-memory computing," in *Proc. IEEE Asian Test Symp. (ATS)*, Malaysia, Nov. 2020.

[14] T.-L. Tsai, J.-F. Li, C.-L. Hsu, and C.-T. Sun, "Testing of in-memory-computing memories with 8T SRAMs," *Microelectronics Reliability*, vol. 123, August 2021.

[15] Y.-C. Yang and J.-F. Li, "Fault modeling and testing for RRAM-based computing-in memories," in *IEEE Int. Test Conf. in Asia (ITC-Asia)*, Taipei, Aug. 2022.

[16] S. Hamdioui, M. Fieback, S. Nagarajan, and M. Taouil, "Testing computation-in-memory architectures based on emerging memories," in *Proc. Int'l Test Conf. (ITC)*, Washington, Nov. 2019.

[17] S. M. Nair, C. Munch, and M. Tahoori, "Defect characterization and test generation for spintronic-based compute-in-memory," in *Proc. IEEE European Test Symp. (ETS)*, Tallinn, May 2020.

[18] M. Fieback, S. Nagarajan, R. Bishnoi, M. Tahoori, M. Taouil, and S. Hamdioui, "Testing scouting logic-based computation-in-memory architectures," in *Proc. IEEE European Test Symp. (ETS)*, Tallinn, May 2020.

[19] R. Bishnoi, L. Wu, M. Fieback, C. Munch, S. M. Nair, M. Tahoori, Y. Wang, H. Li, and S. Hamdioui, "Emerging memristor based memory and CIM architecture: test, repair and yield analysis," in *Proc. IEEE VLSI Test Symp. (VTS)*, San Diego, Apr. 2020.

[20] L. Xia, M. Liu, X. Ning, K. Chakrabarty, and Y. Wang, "Fault-tolerant training enabled by on-line fault detection for RRAM-based neural computing systems," *IEEE Trans. on Computer-Aided Design of Integrated Circuits and Systems*, vol. 38, no. 9, pp. 1611–1642, Sept. 2019.

[21] J.-F. Li, "Testing and reliability of computing-in memories: solutions and challenges," in *IEEE Int. Test Conf. in Asia (ITC-Asia)*, Taipei, Aug. 2022.

[22] ——, "Testing of computing-in memories: faults, test algorithms, and design-for-testability," in *IEEE Int. Symp. on Defect and Fault Tolerance in VLSI Systems (DFT)*, Juan-Les-Pins, Oct. 2023, pp. 1–6.

[23] D. Wang and et al., "DIMC: 2219TOPS/W 2569F^2/b digital in-memory computing macro in 28nm based on approximate arithmetic hardware," in *Proc. IEEE Int'l Solid-State Cir. Conf. (ISSCC)*, San Francisco, Feb. 2022.

[24] A. J. van de Goor, *Testing Semiconductor Memories: Theory and Practice*. Chichester, England: John Wiley & Sons, 1991.

Special Session: Overcoming Transient Faults and Aging Effects in Digital Computing-in-Memory Architectures: Detection, Tolerance, and Mitigation Strategies

Yu-Guang Chen
Department of Electrical Engineering
National Central University
Taoyuan, Taiwan
andyygchen@ee.ncu.edu.tw

Ting-Yi Wu
Department of Electrical Engineering
National Central University
Taoyuan, Taiwan
hyes96087@gmail.com

Abstract—Digital computing-in-memory (DCIM) is a promising solution to the von Neumann bottleneck. By utilizing bit-multiplication memory (BMM) cells and adder trees, DCIM efficiently performs MAC operations. However, both BMM cells and adder trees are susceptible to reliability issues, such as transient faults and aging effects. Transient faults can cause single-event upsets (SEUs), leading to data loss, while aging can degrade circuit performance over time, potentially resulting in functional errors. These reliability threats can impact the accuracy of MAC computation results. In this tutorial, we aim to comprehensively investigate the influence of transient faults and aging effects on DCIM. Additionally, we will explore detection, tolerance, and mitigation strategies to enhance the reliability of DCIM.

Index Terms—Computing-in-memory, circuit aging, transient fault.

I. INTRODUCTION

Artificial intelligence (AI) and machine learning (ML) have transformed various industries with their remarkable ability to handle vast amounts of data. Applications that depend on AI and ML algorithms will require intensive data processing, like autonomous vehicles [1] and chatbots [2]. However, as illustrated in Fig. 1, the traditional von Neumann architecture often leads to data flow bottlenecks because of the constant movement of large volumes of data between memory and the CPU. This bottleneck negatively impacts system performance and increases power consumption. To tackle this issue, researchers have proposed the concept of computing-in-memory (CIM). CIM allows computations to be performed directly within memory units, thereby eliminating the need for extensive data transfer between memory and the CPU. By integrating processing capabilities into memory, CIM boosts computational efficiency, lowers power consumption, and reduces the performance degradation caused by communication overhead.

This work was supported in part by the National Science and Technology Council, Taiwan (R.O.C) under grants NSTC 112-2221-E-008-097-MY3, 110-2221-E-008-099-MY3, 113-2218-E-007-020, 113-2640-E-008-001, and 113-2640-E-006-001.

AI applications require billions of Multiply-and-Accumulate (MAC) operations to process vast amounts of data efficiently. Due to the feasibility of process technology, SRAM CIMs are extensively researched for implementing energy efficient multibit MAC operations [3][4]. The SRAM CIM can be categorized into Analog CIM (ACIM) [3] and Digital CIM (DCIM) [4] based on their computing mechanisms for MAC operations. While ACIM achieves high energy efficiency and high speed, it suffers from accuracy issues due to a low signal-to-noise ratio (SNR). Therefore, applications that demand high accuracy may find ACIM less suitable. On the other hand, as digital operations are less susceptible to noise and variations, DCIM offers higher accuracy and precision compared to ACIM. However, DCIM has lower energy efficiency and low speed [5].

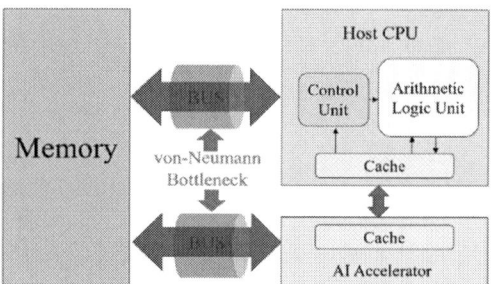

Fig. 1.Traditional von Neumann architecture and the bottleneck

DCIM has better accuracy than ACIM, ensuring more reliable and repeatable results. The DCIM architecture that we discuss in this paper is 6T SRAM-based all-digital CIM[4], composed of bit-multiplication memory (BMM) cells and adder trees as Fig. 2 (a) shows. This CIM macro can operate in SRAM and CIM modes. SRAM mode is used to preload the weights into the 6T SRAM. CIM mode is used to perform MAC operation. The BMM cells consist of 6T SRAM and NOR gate, as illustrated in Fig. 2 (b). Four 6T SRAM cells are grouped to store a 4-bit weight. Each SRAM cell is accompanied by a 4T NOR gate used for multiplication. Input activations are fed in a bit-serial manner from CI[n-1:0], resulting in a 4-bit partial product. These 256 partial products are then fed into a multi-level adder tree for accumulation.

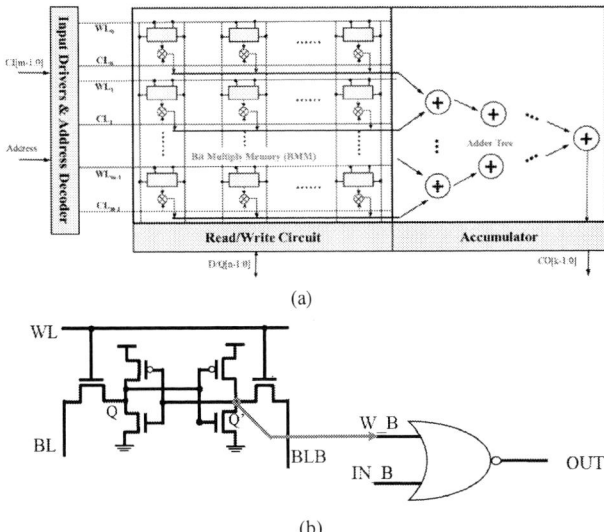

(a)

(b)

Fig. 2. (a)SRAM-based CIM macro block diagram and bit cell layout (b)BMM cells architecture

While DCIM offers advantages such as reduced data movement and improved energy efficiency, it must contend with reliability challenges like aging effects and transient faults. Aging effects occur as semiconductor devices degrade over time due to environmental stressors and usage patterns, leading to changes in transistor characteristics and increased leakage currents. These changes can compromise the integrity of stored data and the accuracy of computations performed within memory. Transient faults, on the other hand, manifest as temporary disruptions caused by external factors like electromagnetic interference, radiation, or voltage fluctuations. Although transient faults do not permanently damage memory cells, they can induce errors in data processing and computation when they occur. Effective mitigation strategies, including robust error detection and correction mechanisms, are essential for maintaining the reliability and performance of DCIM systems amidst these challenges. Addressing aging effects and transient faults ensures the dependable operation of DCIM, supporting its potential applications in advanced computing paradigms.

In this paper, our primary objective is to provide an in-depth tutorial on the various reliability challenges associated with DCIM architectures. We will meticulously identify the multiple sources of unreliability, rigorously examine detection methods, and thoroughly discuss reliability-conscious DCIM design from different perspectives. By addressing these critical aspects, we aim to offer valuable insights for designing robust and reliable DCIM circuits, which are essential for a wide range of applications in machine learning and artificial intelligence. Specifically, our focus is on the transient faults and aging effects that can cause faults in 6T SRAM-based all-digital CIM circuits used for MAC operations. We will delve into the methods for tolerating, detecting, and mitigating these

faults, providing a detailed analysis and practical solutions to enhance the reliability of these circuits.

II. TRANSIENT FAULT ON DCIM

Transient faults are types of soft errors that can cause errors in electronic devices, occurring when radiation induces errors in circuits caused by ionization resulting from cosmic rays and alpha ray particles [6]. These faults may disturb or momentarily disrupt the data state of a memory cell, register, latch, or flip-flop. Such disruptions can lead to temporary errors in the output of a combinational logic circuit, thus affecting the system's functionality. Moreover, transient faults can result in single-event upset (SEU), which occurs when a radiation event causes a flip in the data state of a memory cell or logic component.

When there is an SEU in DCIM, it indicates the presence of faults in either the adder tree, the BMM cell, or potentially both. In BMM, an SEU can lead to incorrect results in the product matrix, as binary operations are sensitive to even small errors. This error can propagate through the entire matrix computation, compromising data integrity, especially in high-density, high-functionality devices. Similarly, in an adder tree, a bit flip can cause errors in the final sum due to the cumulative nature of additions. Errors introduced at any level can propagate through subsequent levels, leading to an incorrect final result. Additionally, SEU in an adder tree can increase computation delays and power consumption due to the need for error detection and correction. Thus, effective error mitigation techniques are crucial to maintaining the reliability of electronic systems in such applications.

Significantly, the occurrence of SEUs in both the adder tree and the BMM cell simultaneously is generally more severe than when SEU is present in only one of these components. This is due to the compounded impact on data integrity and computation accuracy. Therefore, it is crucial to consider the tolerance and mitigation methods for both the adder tree and the BMM cell to ensure the reliability and correctness of the overall system.

III. TOLERANCE METHOD OF TRANSIENT FAULT

In this section, we discuss the tolerance and mitigation method of transient fault on adder tree and BMM cells respectively.

A. Adder Tree

An adder tree structure utilizes multiple adders to perform operations in parallel, typically using half adders (HA) or full adders (FA). Consequently, we focus on addressing transient faults in individual adders and extend this approach to the entire adder tree, as discussed in [7]-[15]. To mitigate transient faults in adders, several tolerance methods and mitigation approaches have been developed, with one of the most prominent being Triple Modular Redundancy (TMR) [7][8]. TMR involves duplicating the critical components of a circuit three times and using a majority-voting system to determine the correct output, thereby ensuring high reliability. However,

979-8-3503-6689-1/24 $31.00 © 2024 IEEE

the use of TMR is generally restricted to circuits where reliability is critical, such as in aerospace or medical applications. This restriction arises because TMR incurs a significant overhead, effectively tripling the resources required, which would render it impractical for widespread use in all circuits due to the substantial increase in cost and complexity. In this section, we discuss a self-repairing adder, which is described in [11], as an alternative approach to addressing transient faults while avoiding the substantial overhead associated with TMR.

Self-repairing adders can detect the faults and repair them with a little overhead [11][12]. As illustrated in Fig. 3(a), The architecture of the self-repairing adder is designed using the self-checking adder by the addition of two XOR gates [11]. First, use the self-checking adder, depicted in Fig. 3(b), to check whether the FA has faults or not. Fc and FS indicate the fault status of SUM and Cout in the full adder. If Fs and Fc both are high, then these indicate a fault is present in SUM and Cout respectively. Then Fs and SUM apply to one XOR gate, and Fc and Cout are applied to another XOR gate. This design repairs both transient and permanent faults using XOR gates instead of secondary full adders, achieving 100% fault repair. In operation, This configuration ensures correct SUM and Cout outputs, achieves substantial area savings, and provides completely fault-free outputs.

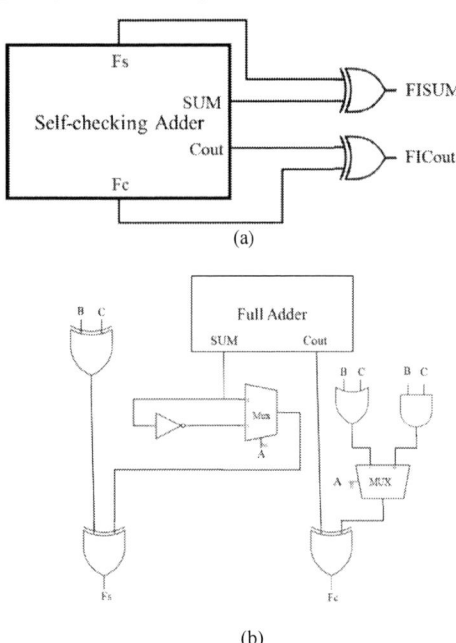

Fig. 3.(a) Self-repairing adde (b)Self-checking adder [11]

B. BMM Cell

The transient faults on a BMM cell will occur on 6T SRAM or NOR gate or both. If the transient faults occur on both the 6T SRAM and NOR gate, the accuracy will drop dramatically. Thus, we need to consider the tolerance method for both the memory cell [16]-[22] and the NOR gate[18]-[20].

1) 6T SRAM

Earlier generations of SRAMs were more robust due to high voltages and bistable circuit design. As technology scaled down, the SRAM junction area decreased, but the operating voltage decreased as well. SEU initially increased with each generation but has saturated in a deep submicron regime. However, overall system SEU continues to increase with increasing memory density.

Error correction codes (ECC) are widely used for mitigating data corruption and ensuring data integrity in memory systems to address soft errors induced by radiation events like SEU [17]-[19]. ECC works by adding extra bits to data vectors, ensuring an "information distance" of at least three between any two possible data vectors, which facilitates error detection and correction. The ECC algorithm detects errors through parity checks or more advanced error-detection codes like Hamming codes [20] or Reed- Solomon codes [21]. Once errors are detected, ECC algorithms can correct them using the redundant information, improving the reliability and accuracy of computations performed in SRAM.

Non-volatile SRAM (NVSRAM) has been proposed to degrade the storage node charge compared with a 6T SRAM core [16][22]. In [16], they add a novel two-level arrangement for achieving SEU tolerance, where the first level involves hardening the cells and the second level involves utilizing non-volatile storage for error correction. Compared to the 6T SRAM, the NVSRAM reduces the delay time, as shown in Table I. The NVSRAM also reduces complexity by leveraging the high critical charge of cell designs, decreasing errors at memory outputs, and reducing error detection hardware overhead. However, the NVSRAM has a larger area than the 6T SRAM. The total area that adds the error detection hardware with both schemes needs to be considered.

Table I. PERFORMANCE OF NVSRAM AND 6T SRAM

Memory scheme (cell)	Operation (time)	Performance (ps)			
		45 nm	32 nm	22 nm	16 nm
NVSRAM	Read (T_R)	45.78	39.52	29.42	19.21
	Error Detection (T_{ED})	557.5	487.2	457.9	399.5
	Restore (Tr)	34.85	29.78	24.67	21.26
6T SRAM	Read (T_R)	43.63	38.62	28.27	17.14
	Error Detection (T_{ED})	977.8	854.5	803.1	700.7
	Error Detection & Correction ($T_{ED} + T_{EC}$)	1086	953.7	901.1	787.4
	Write Back (T_{WI})	31.30	24.59	15.96	7.035

2) NOR gate

Model prediction is used to analyze logic gates susceptibility [23][25]. Models generate probabilistic matrices used in Probabilistic Transfer Matrix (PTM), Signal Probability Reliability (SPR), or Signal Probability Reliability Multi-Pass (SPR-MP) for circuit reliability estimation, excluding SEU masking conditions in single-stage analysis. Crucially, susceptibility values aid in selecting optimal logic functions and circuit designs.

IV. AGING EFFECT ON DCIM

Beyond transient faults, aging mechanisms such as bias temperature instability (BTI) [26] and hot carrier injection HCI [27] also present significant challenges to the reliable operation of DCIM.

BTI notably impacts DCIM reliability. It includes Negative Bias Temperature Instability (NBTI), affecting pMOS transistors, and Positive Bias Temperature Instability (PBTI), affecting nMOS transistors. NBTI induces unstable Si-H bonds in pMOS, causing interface-trapped charges and fixed charges in the oxide layer. Fig. 4(a) illustrates the NBTI process, this increases pMOS threshold voltage and decreases current over time. Similarly, PBTI affects nMOS by raising its threshold voltage and reducing current. PBTI has been observed to cause similar effects on nMOS transistors in processes below 45nm. Long-term conduction in nMOS leads to higher threshold voltage and reduced current passing ability over time. Fig. 4(b) demonstrates the PBTI process.

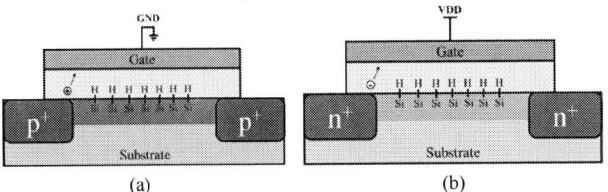

Fig. 4.The schematic diagram of (a) NBTI (b)PBTI

HCI occurs when electrons or holes in the channel of a MOS transistor gain sufficient energy to overcome the potential barrier and induce impact ionization. Some electrons become trapped in the oxide as fixed charges, while bond breakage at the oxide/silicon interface may form interface states. Additionally, carriers may flow into the substrate, generating substrate current. Fig. 5 illustrates the HCI process.

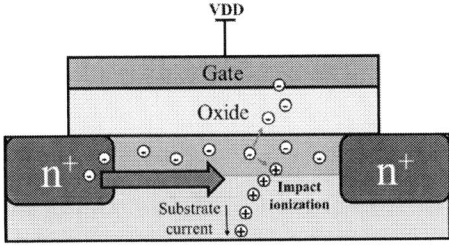

Fig. 5. The schematic diagram of the HCI effect.

These changes alter the threshold voltage of the MOS transistor, as shown in Fig. 6, which consequently reduces the current that can pass through at the same gate voltage and affects the long-term reliability of various circuits, as mentioned in references [28]-[32] for memory and CIM circuits, and [32]-[34] for adders. Due to the detrimental effects of aging on circuit performance and reliability, [28] proposed an innovative technique to tolerate these aging effects specifically in DCIM circuits, thereby ensuring more stable and reliable operation over extended periods.

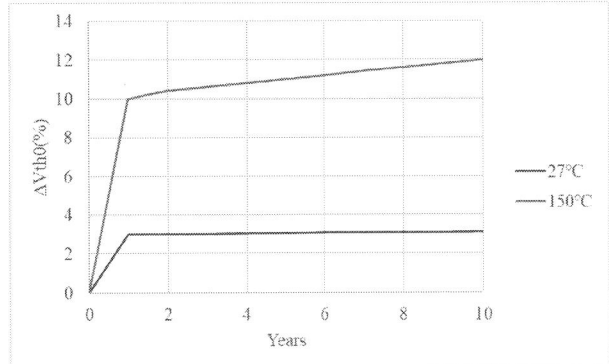

Fig. 6. HCI and NBTI induced physical parameters degradation in (a) NMOS and (b) PMOS transistors [33].

V. MITIGATION AND TOLERANCE METHOD OF AGING EFFECTS

An aging-aware device-to-circuit evaluation method for low-temperature polycrystalline silicon and oxide (LTPO) CMOS thin-film transistors (TFTs) large-scale integrated circuits have been proposed in [28]. Fig. 7 demonstrates that LTPO-CMOS logic circuits exhibit superior stability performance during aging compared to other process circuits. Specifically, LTPO-CMOS logic experiences significantly less impact from threshold voltage drift, maintaining a consistently high noise margin value as the device ages, whereas pseudo-CMOS logic circuits exhibit a pronounced decrease in noise margin over time, highlighting the enhanced durability and reliability of LTPO-CMOS technology. Furthermore, as illustrated in Fig. 8(a) and 8(b), LTPO-CMOS logic circuits achieve remarkable improvements in energy consumption and computing speed on the DCIM circuit, with reductions in energy consumption by a factor of 122.7 and increases in computing speed by a factor of 3.19, showcasing the significant advancements in both efficiency and performance. Consequently, the implementation of LTPO-CMOS technology in realizing the DCIM circuit not only effectively mitigates the adverse effects of aging but also substantially enhances overall

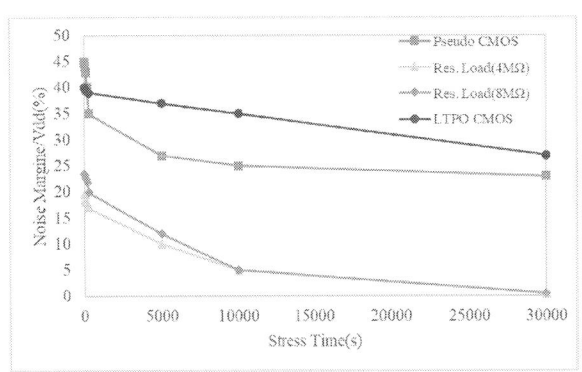

Fig. 7.Simulation results of SNM with PBTI [28].

performance, making it a highly promising solution for future large-scale integrated circuit applications where longevity and efficiency are critical.

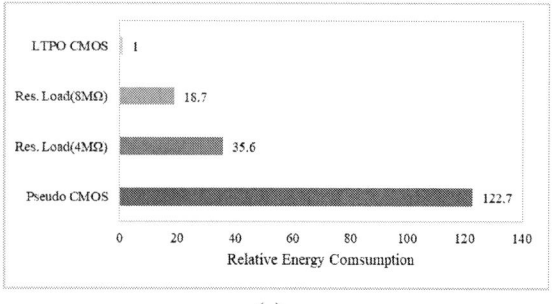

(a)

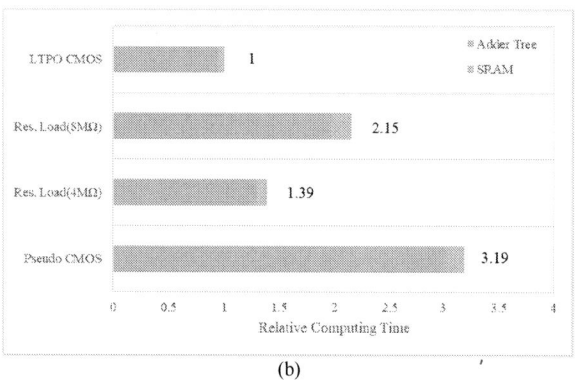

(b)

Fig. 8. Simulation results of different processes in DCIM (a) energy consumption (b) relative computing time [28]

Besides using CMOS technology to mitigate aging effects, altering the architecture can also mitigate aging. Instead of using full adders to implement the adder tree, using alternative adders such as Carry Lookahead Adders (CLA), Carry Select Adders (CSA), Kogge-Stone Adders (KSA), and SKlansky Adders (SKA) in an adder tree can offer significant advantages in terms of speed and power efficiency. However, these benefits often come with trade-offs in terms of increased area, power consumption, design complexity, and aging awareness. Table II shows a comparison of different adders, can easily be observed that the fast adders are more vulnerable to HCI than NBTI while the opposite is true for low-speed adders [33]. As a result, SKA can better tolerate the aging effects than other adders.

Table II. PERFORMANCE COMPARISON OF 4-BIT ADDERS

Adder Architecture	Area (μm²)	Delay (ps)	Power	Delay Degradation over 10 years (%)	
				NBTI	HCI
CSA	96.7	372.7	7.24	0.33	0.37
CLA	64.5	348.2	4.71	0.52	0.49
KSA	64.0	350.4	3.87	0.23	0.34
SKA	57.7	416.6	3.59	0.2	0.3

Optimizing the duty cycle can also mitigate the aging effect on the DCIM circuit [32]. Different weights that are stored in the 6T SRAM will cause various levels of aging. The cells that are used frequently will accelerate aging [31]. The proposed solution in [32] involves using a sophisticated micro-architecture with memory-write and read transducers for optimal duty cycle management, ensuring that aging of the SRAM cells are balanced over time. By effectively managing the duty cycle, the system can reduce the frequency of high-stress conditions on individual cells, thereby extending the overall lifespan of the memory array. This approach not only addresses the immediate issue of aging but also maintains the performance and energy efficiency of the system, providing a robust solution for long-term reliability in DCIM circuits.

VI. CONCLUSIONS

This paper provides a comprehensive exploration of the transient faults and aging effects encountered by DCIM architectures. It introduces effective methods for tolerance, detection, and mitigation of these issues. By highlighting the reliability challenges of DCIM, the paper aims to encourage widespread engagement and enhance its dependability for critical applications such as autonomous vehicles and medical monitoring. Addressing these challenges requires collaborative efforts from researchers, industry experts, and policymakers to ensure the successful and trustworthy deployment of DCIM. Improving DCIM's reliability can significantly contribute to advancing safer autonomous systems and enhanced healthcare monitoring, thereby unlocking its full potential across various fields.

REFERENCES

[1] M. Dikmen and C. Burns, "Trust in autonomous vehicles: the case of Tesla autopilot and summon," in 2017 IEEE International Conference on Systems, Man, and Cybernetics (SMC), 2017, pp. 1093-1098.

[2] R. Goel, D. K. Arora, V. Kumar and M. Mittal, "A machine learning based medical chatbot for detecting diseases," in 2022 2nd International Conference on Innovative Practices in Technology and Management (ICIPTM), Gautam Buddha Nagar, India, 2022, pp. 175-181.

[3] M. Ali, A. Jaiswal, S. Kodge, A. Agrawal, I. Chakraborty and K. Roy, "IMAC: In-Memory Multi-Bit Multiplication and ACcumulation in 6T SRAM Array," in IEEE Transactions on Circuits and Systems I: Regular Papers, vol. 67, no. 8, pp. 2521-2531, Aug. 2020, doi: 10.1109/TCSI.2020.2981901.

[4] Y.-D. Chih et al., " 16.4 An 89TOPS/W and 16.3TOPS/mm 2 all-digital SRAM-based full-precision compute-in memory macro in 22nm for machine-learning edge applications," IEEE Int. Solid-State Circuits Conf. (ISSCC) Dig. Tech. Papers, vol. 64, pp. 252-254, Feb. 2021.

[5] Joonhyung Kim and Jongsun Park, "The Quantitative Comparisons of Analog and Digital SRAM Compute-In-Memories for Deep Neural Network Applications," 19th International SoC Design Conference (ISOCC), 2022.

[6] Robert C. Baumann, "Radiation-Induced Soft Errors in Advanced Semiconductor Technologies", IEEE Transactions on Device and Materials Reliability, vol. 5,no. 3, Sep. 2005.

[7] Whitney J. Townsend, Jacob A. Abraham, Earl E. Swartzlander, Jr., "Quadruple Time Redundancy Adders," in 18th IEEE International Symposium on Defect and Fault Tolerance in VLSI Systems (DFT'03), p. 250, Nov. 2003.

[8] Jie Han, Jianbo Gao, Yan Qi, Pieter Jonker, Jose A.B. Fortes, "Toward Hardware-Redundant, Fault-Tolerant Logic for Nanoelectronics," IEEE Design and Test of Computers, vol. 22, no. 4, pp. 328-339, Jul./Aug. 2005.

[9] M. Valinataj and S. Safari "Fault tolerant arithmetic operations with multiple error detection and correction," Proc. 22nd IEEE Int. Symp. Defect Fault Tolerance VLSI Syst., pp. 1-9, 2007.

[10] Akbar, Muhammad Ali, and Jeong-A. Lee. "Self-repairing adder using fault localization." Microelectronics Reliability 54, no. 6 (2014): 1443-1451.

[11] S. Gupta, A. Jasuja and R. Shandilya, "Real-time fault tolerant full adder using fault localization," 2018 IEEE International Students' Conference on Electrical Electronics and Computer Science (SCEECS), pp. 1-6, 2018.

[12] Prachi Palsodkar, Prasanna Palsodkar and Rupali Giri, "Multiple Error Self Checking-Repairing Fault Tolerant Adder-Multiplier," 2018 IEEE Region 10 Humanitarian Technology Conference (R10-HTC).

[13] P. Ndai, Shih-Lien Lu, Dinesh Somesekhar and K. Roy, "Fine-Grained Redundancy in Adders," Quality Electronic Design 2007. ISQED '07. 8th International Symposium, pp. 317-321, March 2007.

[14] Jyothi Velamala, Robert LiVolsi, Myra Torres and Yu Cao, "Design sensitivity of Single Event Transients in scaled logic circuits," 2011 48th ACM/EDAC/IEEE Design Automation Conference (DAC), June 2011.

[15] M. Glorieux et al., "DAMSEL - Dynamic and Applicative Measurement of Single Events in Logic", IEEE Transactions in Nuclear Science, vol. 6, no. 1, 2018.

[16] W. Wei, K. Namba, Y.-B. Kim and F. Lombardi, "A novel scheme for tolerating single event/multiple bit upsets (SEU/MBU) in non-volatile memories," IEEE Trans. Comput., vol. 65, no. 3, pp. 781-790, Mar. 2016.

[17] R.C. Baumann, "Radiation-induced soft errors in advanced semiconductor technologies," IEEE Transactions on Device and Materials Reliability, vol. 5, no. 3, pp. 305-316, Sept. 2005.

[18] Yan, Z., Shi, Y., Liao, W., Hashimoto, M., Zhou, X., Zhuo, C. "When single event upset meets deep neural networks: observations, explorations, and remedies," in 2020 25th Asia and South Pacific Design Automation Conference (ASP-DAC), pp.163–168. IEEE (2020).

[19] Li, J. "Testing and reliability of computing-in memories: solutions and challenges," in 2022 IEEE International Test Conference in Asia (ITC-Asia).

[20] C. Hillier and V Balyan. "Error Detection and Correction On-Board Nanosatellites Using Hamming Code" in Journal of Electrical and Computer Engineering, 2019.

[21] M. Sudan "Decoding of Reed Solomon Codes beyond the Error-Correction Bound," in Journal of Complexity Volume 13, Issue 1, March 1997, Pages 180-193.

[22] W. Wei, K. Namba, and F. Lombardi, "Design and analysis of nonvolatile memory cells for SEU tolerance," in Proc. 17th IEEE Symp. Defect Fault Tolerance VLSI Nanotechnol. Syst., Amsterdam, The Netherlands, Oct. 2014, pp. 69–74.

[23] R. Schvittz, D.T. Franco, L. Soares and P.F. Butzen, "A Simplified Layout-Level method for Single Event Transient Faults Susceptibility on Logic Gates," IEEE/IFIP Int. Conf. VLSI Syst. VLSI-SoC, pp. 185-190, 2019, [online] Available: https://doi.org/10.1109/VLSI-SoC.2019.8920333.

[24] X Tang, A Xu, W Li and Z Yang, "Fault Models of CMOS Gates: An Empirical Study Based on Mutation Analysis," Proc. IEEE International Conference on Dependable Autonomic and Secure Computing, pp. 115-120, 2014.

[25] R. B. Schvittz, P. F. Butzen and L. S. da Rosa, "Methods for susceptibility analysis of logic gates in the presence of single event transients," 2020 IEEE International Test Conference (ITC), pp. 1-9, 2020.

[26] A. Kerber and T. Nigam, "Bias temperature instability in scaled CMOS technologies: a circuit perspective," Microelectron. Rel., vol. 81, pp. 31-40, Feb. 2018

[27] K.-L. Chen, S. A. Saller, I. A. Groves, and D. B. Scott, "Reliability effects on MOS transistors due to hot-carrier injection," in IEEE Transactions on Electron Devices, vol. 32, no. 2, pp. 386-393, Feb. 1985.

[28] Shuaidi Zhang et al., "Aging-aware LTPO DTCO for large-scale integrated circuit-driven flexible intelligent sensing system," in IEEE Transactions on Electron Devices, vol. 71, no. 5, pp. 3322-3328, May 2024.

[29] M. Karimi, N. Rohbani and S. Miremadi, "A Low Area Overhead NBTI/PBTI Sensor for SRAM Memories," in Proc. of IEEE Transactions on Very Large Scale Integration (VLSI) Systems, vol. 25, no. 11, pp. 3138-3151, 2017.

[30] H. Mostafa, M. Anis and M. Elmasry, "Adaptive body bias for reducing the impacts of NBTI and process variations on 6T SRAM cells," IEEE Transactions on Circuits and Systems I: Regular Papers, vol. 58, no. 12, pp. 2859-2871, 2011.

[31] C. Dilopoulou and Y. Tsiatouhas, "BTI aging influence and mitigation in neural networks oriented in-memory computing SRAMs," Proc. 12th Int. Conf. Modern Circuits Syst. Technol. (MOCAST), pp. 1-4, Jun. 2023.

[32] M. A. Hanif and M. Shafique, "Dnn-life: An energy-efficient aging mitigation framework for improving the lifetime of on-chip weight memories in deep neural network hardware architectures," 2021 Design Automation & Test in Europe Conference & Exhibition (DATE), pp. 729-734, 2021.

[33] T. An, H. Cai and L. A. de Barros Naviner, "Simulation study of aging in CMOS binary adders", 2014 37th International Convention on Information and Communication Technology Electronics and Microelectronics (MIPRO), pp. 51-55, 2014.

[34] H. Amrouch, B. Khaleghi, A. Gerstlauer and J. Henkel, "Towards aging-induced approximations", 54th ACM/EDAC/IEEE Design Automation Conference (DAC'17), pp. 1-6, June 2017.

[35] F. J. H. Santiago, H. Jiang, H. Amrouch, A. Gerstlauer, L. Liu and J. Han, "Characterizing approximate adders and multipliers for mitigating aging and temperature degradations", IEEE Transactions on Circuits and Systems I: Regular Papers, vol. 69, no. 11, pp. 4558-4571, 2022

Special Session: Enhancing Reliability in Digital Computing-In-Memory Architectures through Approximation and Fault Tolerance Methods

Shih-Hsu Huang, Chih-Li Hsiao and Wei-Che Cheng

Department of Electronic Engineeering
Chung Yuan Christian University
Taoyuan, Taiwan
{shhuang,g11276017,g11176053}@cycu.edu.tw

Abstract—In recent years, although there have been many works on digital computing-in-memory (DCIM), they have mainly focused on improving computational efficiency, while neglecting the issue of reliability enhancement. However, during circuit manufacturing or operation, defects (caused for reasons) may arise in DCIM circuits, impacting the accuracy of results. In this paper, we explore overcoming these defects through approximation and fault tolerance techniques to enhance the reliability of DCIM circuits. Given that a DCIM circuit comprises two components – a memory array and digital logic (adder tree), we present established approximation techniques and fault tolerance techniques separately for each component. Furthermore, we discuss possible research directions for designing approximation and fault tolerance techniques in DCIM circuits. Through this paper, an in-depth exploration of approximation and fault tolerance issues in DCIM circuits and potential solutions is provided.

Keywords—Adder tree, Low power, In-memory-computing, Memory array, Redundancy design.

I. INTRODUCTION

Today, artificial intelligence and machine learning are widely applied in tasks requiring intelligent analysis and decision-making, such as image recognition and machine vision. However, these technologies are accompanied by substantial computational demands. The traditional von Neumann architecture has significant drawbacks such as the memory wall, low energy efficiency, and high latency. Therefore, in-memory computing is considered an effective solution to address these issues. In-memory computing architectures can reduce the number of memory accesses and achieve higher efficiency compared to traditional von Neumann architectures. Among these, SRAM, a well-established memory process technology, has garnered much attention for SRAM-based computing-in-memories (CIM) [1].

Early SRAM-based CIMs were implemented using analog circuits. Computations are directly performed in the memory, enabling high-speed, low-power data processing. The fundamental principle involves carrying out multiplication or multiply-accumulate operations in SRAM, followed by conversion of the signal via an analog-to-digital converter for further processing or storage. However, analog CIMs are susceptible to noise interference, process variations, and fluctuations in voltage and temperature, thus facing reliability issues.

In recent years, several studies have proposed digital CIM designs (DCIM) that perform multiplication operations within SRAM cells [2]. SRAM-based DCIM offers the advantage of high throughput while also avoiding interference from PVT (process, voltage and temperature) variations. The DCIM architecture [2] primarily consists of two parts: the calculation of product terms within the memory array (i.e., the SRAM cells), and the summing operations performed by an adder tree, as shown in Figure 1.

Current DCIM research primarily focuses on improving computational efficiency while neglecting reliability issues. However, to enhance the yield of chips (the yield of DCIM), reliability cannot be overlooked. Generally, to improve circuit reliability, fault-tolerance mechanisms must be added to address potential defects. For neural network applications, since minor computational errors do not influence inference outcomes, approximate designs also can be employed to overcome potential defects or to reduce area and power consumption. In this paper, we will explore approximate and fault-tolerant designs for DCIM circuits to enhance their reliability and thereby improve chip yield.

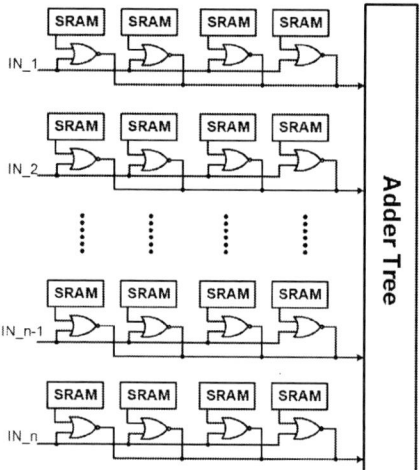

Figure 1: The DCIM architecture [2].

The DCIM architecture primarily comprises two parts: the memory array and the adder tree. In this paper, we explore approximate and fault-tolerant designs for these two components to enhance the reliability of DCIM designs.

● Memory Array. Since the calculation of product terms is completed within the memory array, any malfunction in the array can lead to discrepancies between the computed results and the correct outcomes. Therefore, it is crucial for the memory array to possess fault-tolerance capabilities [3-5]. Additionally, we discuss approximate design techniques for memory cells [6-8] to reduce power consumption.

● Adder Tree. The adder tree is responsible for performing the summation of product terms. If a fault occurs in the adder tree, it can lead to discrepancies between the computed results and the correct outcomes, making fault tolerance essential for the adder tree as well. Additionally, since the adder tree involves substantial computational load, the area and power consumption of the adder tree are critical issues to address. To reduce the area and power consumption of the adder tree, adopting approximate design approaches is a viable direction [9-14].

In this paper, in addition to discussing the memory array and adder tree components separately, we also explore approximate and fault-tolerant designs for DCIM circuits. Recently, our research team (our laboratory) has engaged in the fault-tolerant and approximate design of DCIM circuits and has achieved some preliminary research results. Therefore, in this paper, we will also briefly demonstrate our current preliminary findings. Simultaneously, we will also point out possible research directions for designing approximation and fault tolerance techniques in DCIM circuits.

The rest of this paper is organized as follows. In Section II, we discuss the fault tolerance and approximation techniques for the DCIM memory array. In Section III, we explore the fault tolerance and approximation techniques for the DCIM adder tree. Section IV briefly introduces the recent research progress on fault-tolerant and approximate design in DCIM circuits at our laboratory. Finally, in Section V, we make some concluding remarks.

II. APPROXIMATION AND FAULT TOLERANCE TECHNIQUES FOR MEMORY ARRAY

SRAM is a widely used memory technology. This section introduces the approximation and fault tolerance techniques for SRAM proposed in previous literatures.

A. Approximation Techniques for SRAM

In recent years, as deep learning has been widely applied, many accelerators have been proposed to enhance chip performance in response to the substantial training workloads. The most common approach is to optimize general matrix multiplications in deep neural networks to improve overall training performance.

During matrix operations, because matrix multiplication calculations are typically parallelized, this process involves reading a large amount of data from memory and sending them to the processor for computation, resulting in significant power consumption and causing a bottleneck on the data bus. To address these issues, the CIM method has been proposed. The CIM approach performs calculations directly in or near memory, reducing the impact of memory access bottlenecks on the overall system [15,16].

In [15,16], a bit-serial method is used for matrix-vector multiplication. This bit-serial method processes one bit at a time, accumulates the results after each bit is processed, and ultimately produces the complete output. In the bit-serial

approach, because calculations are performed on one bit at a time, only simple logic circuits are needed for implementation. This method significantly reduces power consumption during computations, especially when dealing with highly sparse data. The effectiveness of the bit-serial method is enhanced because it can skip zero values, reducing unnecessary computations and further improving computational efficiency.

In addition to using CIM to reduce power consumption, approximate arithmetic can also be used to decrease the number of operations and enhance overall computational performance. In neural networks, these minor errors can be corrected through the inherent resilience of neural networks.

In [17], an approximate multiplier within SRAM is proposed that can perform multiplication operations within memory to reduce the time and energy required for data transfer between memory and the processor. The literature [17] achieves approximate multiplication by activating multiple word lines in a bit-parallel manner. This type of multiplier can be directly utilized in conventional SRAM, allowing for minimal design modifications to be implemented in existing systems.

Figure 2 illustrates the multiplier concept proposed in [17], assuming multiplicand a = 1101 and multiplier b = 0110. Initially, the multiplicand 1101 is stored in the memory array of the SRAM. Then, according to each bit of the multiplier 0110, the corresponding word lines (WL) are activated. If a bit of the multiplier is 1, the corresponding row in the SRAM is read. For example, as shown in Figure 2, the least significant bit (LSB) of multiplier b is 0, so the corresponding SRAM is not activated, resulting in 0000. Since the second bit is 1, the corresponding row is activated, read, and then shifted left by one position, resulting in 11010. The third bit is also 1, so the corresponding SRAM row is read again and shifted left two positions, resulting in 110100. The highest bit is 0, so the SRAM is not activated, resulting in 0000000. Subsequently, all the above results are subjected to an OR operation, resulting in 0111110. This outcome closely approximates the actual multiplication result of 1001110.

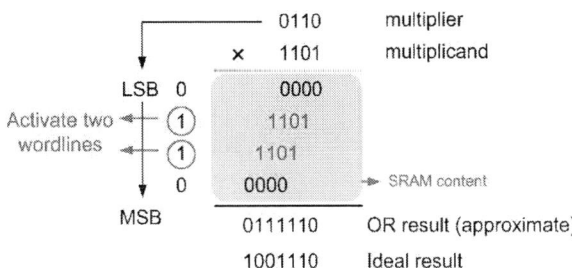

Figure 2: The approximate multiplication concept in [17].

In this computational mode, since it only requires performing an OR operation on all values of each bit to calculate the approximate result, there is no need to consider carry issues. Therefore, calculations can be performed on any selected segment, reducing the number of bits involved and thus enhancing computational efficiency.

Figure 3 illustrates the architecture proposed in [17]. In this architecture, data is first fetched and then stored in a register. Subsequently, it is sequentially fed into the SRAM memory array via an address decoder. The address decoder activates the corresponding SRAM word lines based on the address of the input data. Each piece of input data simultaneously undergoes multiplication operations with all kernel elements in the same row. The results of these multiplications are then sent to an accumulation unit for summation. Each row has its own accumulation unit, which is used to sum the computational results within the current SRAM. The final results are stored in an output buffer for subsequent use.

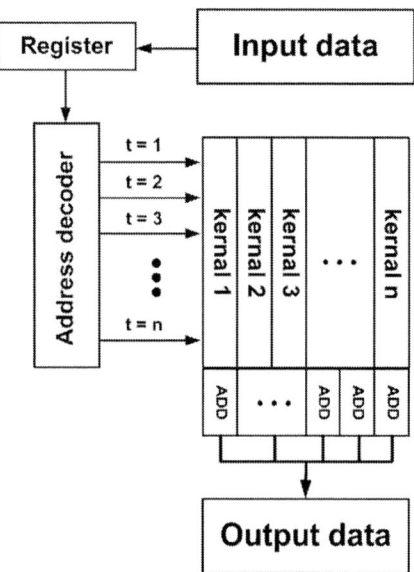

Figure 3: The architecture proposed in [17].

B. Fault Tolerance Techniques for SRAM

As the transistor density in memory circuits increases, the likelihood of faults occurring within the circuits also rises. These faults include static faults, dynamic faults, coupling faults, data retention faults, neighborhood pattern sensitive faults, and random faults. To ensure the correct operation of circuits, it is necessary to detect and repair these faults. Built-in self-test (BIST) circuits are used for fault detection, and built-in self-repair (BISR) circuits are employed for fault repair. For some irreparable faults, additional rows or columns need to be added as redundant elements within the circuit to replace the faulty units.

Redundant elements [5] can be added in multiple ways. If only spare rows or columns are added within the memory, this is known as 1-D redundancy. If both spare rows and columns are added simultaneously, it is referred to as 2-D redundancy. 2-D redundancy is more suitable than 1-D redundancy because it offers a larger number of redundant elements, which provides more options for the routing algorithm, facilitating the bypassing of faulty elements during reconfiguration within the memory. Additionally, redundancy can also be added within the memory array itself. In this case, the entire memory is divided into multiple blocks. If a block contains faults, that block will be replaced by a redundant block.

In addition to the common faults encountered in memory, in space environments, memory is also susceptible to the effects of energetic particle radiation, leading to single-event upsets (SEU). SEU causes data errors in SRAM cells, thereby affecting the reliability of the entire electronic system. With continuous advancements in CMOS technology, the length of transistors is becoming increasingly smaller, which further increases the likelihood of soft errors in SRAM cells. To address this issue, radiation-hardened-by-design (RHBD) technology is widely applied to enhance the radiation tolerance capacity of circuits [18-20].

In [18], the authors propose a type of RHPD-12T SRAM cell. The RHPD-12T circuit includes 2 PMOS transistors and 10 NMOS transistors. This design approach utilizes a higher number of NMOS transistors to mitigate the charge-sharing effect, thereby enhancing the circuit's SEU robustness. By designing appropriate transistor sizes, the RHPD-12T SRAM cell achieves efficient write and read operations and maintains stability under low voltage conditions.

Compared to other RHPD SRAM cells such as We-QUATRO [19] and DICE [20], the RHPD-12T SRAM [18] shows a significant improvement in write speed. Moreover, through Monte Carlo simulations, the RHPD-12T SRAM maintains a lower write failure probability under various process variation conditions, showing its reliability in extreme environments.

III. APPROXIMATION AND FAULT TOLERANCE TECHNIQUES FOR ADDER TREES

The adder tree is a core part of DCIM, making its optimization extremely important. The performance of the adder directly affects the area, delay, and power efficiency of the adder tree. Therefore, in this section, we will introduce approximation and fault tolerance techniques for adders to reduce the impact of the adder tree on the area and power consumption of DCIM circuits, or to further enhance the reliability of DCIM circuits.

A. Fault Tolerance Techniques for Adders

In literature [21], the fault-tolerant design of adders is discussed, with structural improvements made to the traditional triple modular redundancy (TMR) approach. This design achieves reliability comparable to TMR while reducing the significant area overhead caused by conventional TMR. Note that TMR is a common fault-tolerance technique used to enhance a circuit's resistance to errors. In TMR, the same function is computed three times, and the final output is determined through majority voting. This method effectively improves the circuit's reliability in the face of SEU and other faults [22-24]. However, TMR faces the challenge of high hardware costs, as it requires three times the hardware resources to implement the redundant calculations, leading to increased power consumption and efficiency issues. Therefore, literature [21] proposes a technique called partial triple modular redundancy (PTMR), which aims to reduce hardware costs and power consumption.

Here we use a ripple carry adder (RCA) design to explain TMR. For given two binary data inputs, each bit calculation in the RCA is independently executed three times. This means that the inputs of each full adder are duplicated, and three full adders operate simultaneously. The final sum is then fed into a majority logic circuit, as shown in Figure 4. This ensures that if one of the computations has an error, as long as the other two are correct, the final output will not be erroneous.

Next, we explain the PTMR for a RCA. As shown in Figure 5, PTMR adopts the principles of TMR but applies them only to the most significant portion of the data rather than to all bits. This optimization is possible because an error in the MSB has a significant impact on the final result due to its high weight. For example, in a 16-bit unsigned number, the weight of the MSB is 32,768 times greater than that of the LSB. Therefore, errors occurring in different significant bits do not have an equal impact. Overall, this PTMR approach reduces the hardware cost and computational complexity of TMR while still maintaining a certain level of fault tolerance.

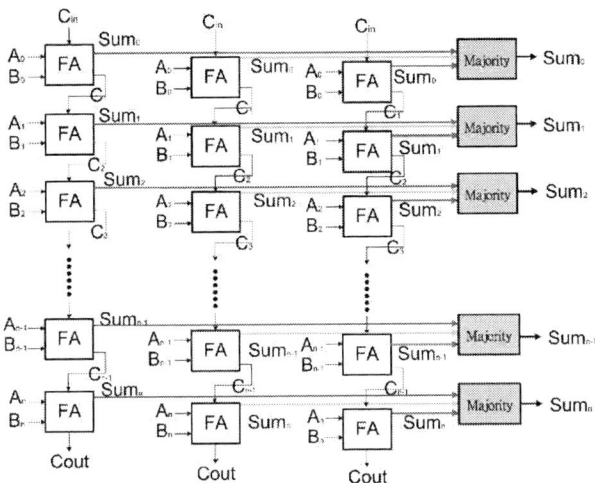

Figure 4: TMR design for a ripple carry adder.

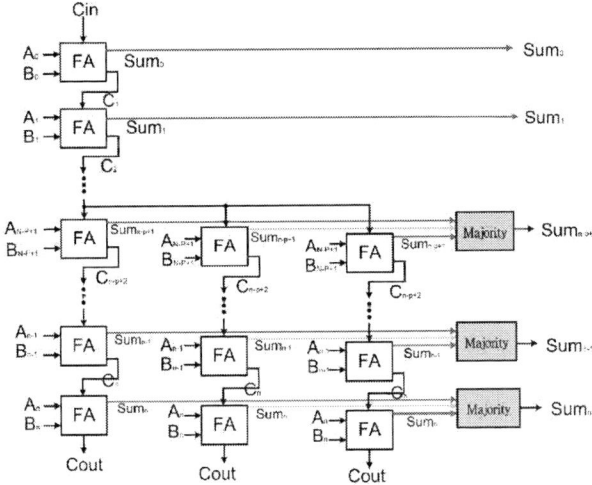

Figure 5: PTMR design for a ripple carry adder [21].

B. Approximation Techniques for Adders

Traditional adders require three inputs to be processed through multiple logic gates simultaneously to produce two outputs (Sum and Cout). In contrast, approximate full adders (AFA) reduce the total number of logic gates used by adjusting or simplifying part of the output logic. The most typical example is the majority logic-based approximate full adder (MLAFA) [25].

In [26], low-power approximate adder designs are explored, including the SXAFA approximate full adder and the SOAFA approximate full adder. Compared to precise full adders, both SXAFA and SOAFA approximate full adders can significantly reduce the number of logic gates. To balance accuracy and power consumption, the literature [26] also proposes a hybrid approximation scheme, using SXAFA adders for higher bits and SOAFA adders for lower bits.

In literature [13], several new approximate adder designs are also proposed, such as the selector based fault tolerant adder-I (SBFTA-I), selector based fault tolerant adder-II (SBFTA-II), and OFTA. The common feature of these three designs is that they all improve circuit area and power consumption by reducing hardware complexity.

We have compiled the truth tables of the aforementioned various approximate adder designs, as displayed in Table I. By analyzing Table I, we can understand the potential errors associated with different approximate adder designs.

In literature [25], a method using truncated inputs is proposed to achieve a fixed-width adder tree design. This design estimates bias through probabilistic methods to compensate for the errors caused by input truncation. Experiments show that this fixed-width adder tree design saves a significant proportion of the area-delay product compared to existing truncation methods when the input vector size is 8 and 16, while the computed output remains almost as accurate as that of the post-truncation fixed-width adder tree.

Truncated input involves using only a portion of the input vector bits (usually the higher significant bits) in the adder tree design while ignoring some of the less significant bits. This approach can significantly reduce the number of adders and overall hardware resources required, as fewer bits are involved in the computation. The fixed-width adder tree (FX-AT) design is mainly divided into two truncation methods: fixed-width post-truncated adder tree (FX-AT-PT) and fixed-width direct-truncated adder tree (FX-AT-DT). FX-AT-PT truncates the least significant bits of the result after the entire addition calculation is completed. This method typically maintains higher computational accuracy. On the other hand, FX-AT-DT truncates the least significant bit directly at each addition stage. This reduces computational complexity and hardware requirements but may increase the overall error accumulation. Both two methods have their advantages and disadvantages, making them suitable for different application requirements and performance goals.

In literature [25], the proposed improved truncated fixed-width adder-tree (ITFX-AT) is a novel approximate adder tree design aimed at enhancing computational efficiency and reducing errors caused by input truncation. This design achieves its objectives through more precise control of the truncation process and improved error compensation methods.

The core design improvement of ITFX-AT lies in dividing the input bit-matrix into three parts: the most significant part (MSP), and the least significant part (LSP) which is further divided into the major-part (MJP) and the minor-part (MNP). The function σ_{major} is used to accurately compute the contribution of the MJP. The function σ_{minor} is used to estimate the contribution of the MNP, which is estimated as 2 for N=8.

TABLE I
Truth tables for different full adder designs.

INPUTS			Precise full adder		SXAFA		SOAFA		MLAFA		OFTA		SBFTA-I		SBFTA-II	
a	b	Cin	Cout	Sum	Cout	Sum	Cout	Sum	Cout	Sum	Cout	Sum	Cout	Sum	Cout	Sum
0	0	0	0	0	0	0	0	0	0	0	0	0	0	1	0	1
0	0	1	0	1	1	0	1	0	1	0	0	1	0	1	0	1
0	1	0	0	1	0	1	0	1	0	1	0	1	0	1	0	1
0	1	1	1	0	1	1	1	1	1	0	0	0	1	0	1	0
1	0	0	0	1	0	1	0	1	0	1	0	0	0	1	0	1
1	0	1	1	0	1	1	1	1	1	0	0	1	1	0	1	0
1	1	0	1	0	1	0	0	1	0	1	1	1	1	0	1	0
1	1	1	1	1	1	0	1	1	1	1	1	0	1	0	1	0

IV. OUR RELATED RESEARCH

Most DCIM literature focus on improving computational efficiency. However, reliability is also an important issue. Therefore, our research team (our laboratory) is conducting research on approximation and fault tolerance techniques for DCIM circuits. In this section, we would like to share some preliminary ideas.

Our current research [27] involves adding redundant memory to DCIM to enhance the fault tolerance of the computational memory. By incorporating spare rows into the DCIM architecture, as shown in Figure 6, the system can maintain computational accuracy even when some cells in the memory fail. To ensure the correctness of the multiply-accumulate (MAC) operation, the outputs of spare rows are also connected to the adder tree.

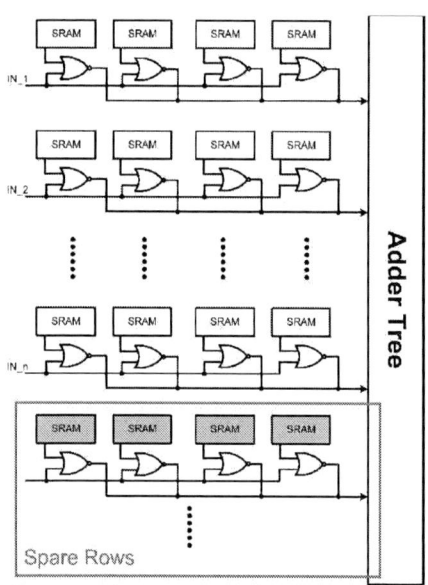

Figure 6: Incorporating spare rows into the DCIM circuit.

Additionally, we are exploring approximate fault-tolerant techniques for SRAM-based MAC operations [28]. In the event of an MSB failure, we utilize the memory units responsible for calculating the LSB to substitute for the MSB calculation, as illustrated in Figure 7, ensuring the correctness of the MSB value. This approximate fault-tolerant approach requires only a small additional area to mitigate errors caused by memory unit failures.

Finally, we also explore truncation error compensation improvements for fixed-width multipliers to reduce truncation errors. This multiplier requires two n-bit inputs and uses compensation logic to reduce truncation errors, resulting in an n-bit output. For unsigned number calculations, the compensation logic, as shown in Figure 8(a), applies input correction (IC) to the most significant part (MSP) to achieve the final result [29]. For signed number calculations, based on the Baugh-Wooley multiplier, we modify the IC part, as shown in Figure 8(b). This approach allows for operations on signed inputs, effectively addressing the limitations and error issues traditional methods may encounter with signed multiplication [30].

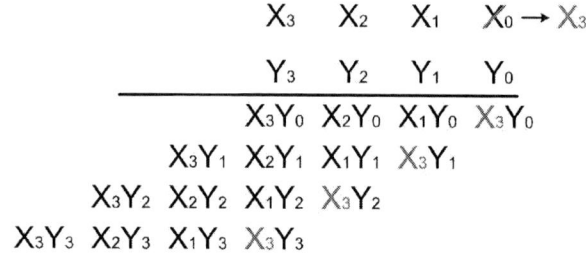

Figure 7: Applying approximate fault tolerance to SRAM-based MAC operations.

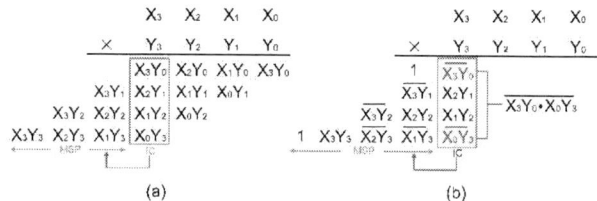

Figure 8: The fixed width multiplier design. (a) Unsigned fixed-width multiplier. (b) Singed fixed-width multiplier.

V. CONCLUSION

This paper explores approximation and fault tolerance techniques to enhance the reliability of DCIM circuits. We analyzed the two main components of DCIM circuits, namely the memory array and the digital logic (adder tree), and introduced the existing approximation and fault tolerance techniques for each component. These techniques can effectively reduce the impact of defects during manufacturing and operation on result accuracy, thereby significantly improving the reliability of DCIM circuits.

Furthermore, the paper discusses potential research directions for designing and implementing approximation and fault tolerance techniques within DCIM circuits. Some preliminary results of our research team are also presented in this paper.

Through this paper, we hope to provide readers with a deeper understanding of approximation and fault tolerance issues in DCIM circuits and their solutions, offering valuable insights for future research and practical applications. Ultimately, we believe that as these technologies continue to evolve and improve, the reliability of DCIM circuits will be further enhanced, providing a more robust foundation for efficient and reliable digital computing.

ACKNOWLEDGEMENTS

This work was supported in part by National Science and Technology Council, Taiwan, under grant numbers 112-2221-E-033-050-MY3 and 113-2640-E-008-001.

REFERENCES

[1] A. Biswas et al., "CONV-SRAM: An Energy-efficient SRAM with in Memory Dot-product Computation for Low-power Convolutional Neural Networks," IEEE J. Solid-State Circuits, vol. 54, no. 1, pp. 217–230, Jan. 2019.

[2] Y.-D. Chih et al., "An 89TOPS/W and 16.3TOPS/mm2 All-Digital SRAM-Based Full-Precision Compute-In Memory Macro in 22nm for Machine-Learning Edge Applications," Proc. of IEEE International Solid-State Circuits Conference ISSCC, 2021.

[3] K. Wang, L. Chen and J. Yang, "An Ultra Low Power Fault Tolerant SRAM Design in 90nm CMOS," Proc. of IEEE Canadian Conference on Electrical and Computer Engineering, 2009

[4] T. Yoon, J. Park, and H. Jeong, "Design of Static Random-Access Memory Cell for Fault Tolerant Digital System," Applied Science, vol. 12, no. 22, 2022,

[5] S. Tak and M. Mali, "Fault Tolerant SRAM by Redundant Array Structure," Proc. of IEEE International Conference on Pervasive Computinga, 2015,

[6] S. Ataei and J. E. Stine, "A 64 kB Approximate SRAM Architecture for Low-Power Video Applications," in IEEE Embedded Systems Letters, vol. 10, no. 1, pp. 10-13, 2018

[7] Y. Chen, X. Yang, F. Qiao, J. Han, Q. Wei and H. Yang, "A Multi-accuracy-Level Approximate Memory Architecture Based on Data Significance Analysis," Proc. of IEEE Computer Society Annual Symposium on VLSI, 2016

[8] L. Yang and B. Murmann, "Approximate SRAM for Energy-Efficient, Privacy-Preserving Convolutional Neural Networks," Proc. of IEEE Computer Society Annual Symposium on VLSI, 2017

[9] S. E. Fatemieh, S. S. Farahani, and M. R. Reshadinezhad, "LAHAF: Low-power, Area-efficient, and High-performance Approximate Full Aadder based on Static CMOS," Sustainable Computing: Informatics and Systems, vol. 30, pp. 1–12, Jun. 2021.

[10] S. Salavati, M. H. Moaiyeri, and K. Jafari, "Ultra-efficient Nonvolatile Approximate Full-adder with Spin-hall-assisted MTJ Cells for In-memory Computing Applications," IEEE Trans. on Magnetics, vol. 57, no. 5, pp. 1–11, May 2021.

[11] Y. He et al., "An RRAM-Based Digital Computing-in-Memory Macro With Dynamic Voltage Sense Amplifier and Sparse-Aware Approximate Adder Tree," IEEE Trans. on Circuits and Systems II: Express Briefs, vol. 70, no. 2, pp. 416-420, Feb.

2023,

[12] S. Boroumand and P. Brisk, "Approximate Adder Tree Synthesis for FPGAs," Proc. of IEEE International Conference on ReConFigurable Computing and FPGAs, 2019,

[13] T. Mendez and S. G. Nayak, "Performance Evaluation of Fault-Tolerant Approximate Adder," Proc. of IEEE International Conference on Devices, Circuits and Systems, 2022

[14] F. Ahmadi., M.R. Semati, H. Daryanavard, A. Minaeifar, "Energy-efficient Approximate Full Adders for Error-tolerant Applications", Computers and Electrical Engineering, vol. 110, Article 108877, 2023.

[15] J.-H. Kim et al., "Z-PIM: A Sparsity-aware Processing-in-memory Aarchitecture with Fully Variable Weight Bit-precision for Energy-efficient Deep Neural Networks," IEEE JSSC, vol. 56, no. 4, pp. 1093–1104, Jan 2021.

[16] J. Heo et al., "T-PIM: An Energy-efficient Processing-in-memory Accelerator for End-to-end On-device Training," IEEE JSSC, vol. 58, no. 3, pp. 600–613, Nov 2023.

[17] L. Sonnino, S. Shresthamali, Y. He and M. Kondo, "DAISM: Digital Approximate In-SRAM Multiplier-Based Accelerator for DNN Training and Inference," Proc. of IEEE Design, Automation & Test in Europe Conference & Exhibition (DATE), 2024.

[18] Q. Zhao, C. Peng, J. Chen, Z. Lin and X. Wu, "Novel Write-Enhanced and Highly Reliable RHPD-12T SRAM Cells for Space Applications," in IEEE Trans. on Very Large Scale Integration (VLSI) Systems, vol. 28, no. 3, pp. 848-852, March 2020.

[19] L. D. T. Dang, J. S. Kim, and I. J. Chang, "We-Quatro: Radiationhardened SRAM cell with parametric process variation tolerance," IEEE Trans. on Nuclear Science., vol. 64, no. 9, pp. 2489–2496, Sep. 2017.

[20] Q. Zheng et al., "The Increased Single-event Upset Sensitivity of 65-nm DICE SRAM Induced by Total Ionizing Dose," IEEE Trans. on Nuclear Science, vol. 65, no. 8, pp. 1920–1927, Aug. 2018.

[21] R. Parhi, C. H. Kim and K. K. Parhi, "Fault-tolerant Ripple-Carry Binary Adder using Partial Triple Modular Rdundancy (PTMR)," Proc. of IEEE International Symposium on Circuits and Systems (ISCAS), 2015

[22] H. Quinn, P. Graham, and B. Pratt, "An Automated Approach to Estimating Hardness Assurance Issues in Triple-modular Redundancy Circuits in Xilinx FPGAs," IEEE Trans. on Nuclear Science, vol. 55, no. 6, pp. 3070–3076, December 2008.

[23] P. K. Samudrala, J. Ramos, and S. Katkoori, "Selective Triple Modular Redundancy (STMR) based Single-event Upset (SEU) Tolerant Synthesis for FPGAs," IEEE Trans. on Nuclear Science, vol. 51, no. 5, pp. 2957–2969, October 2004.

[24] I. Polian and J. P. Hayes. "Selective hardening: Toward Cost-Effective Error Tolerance," IEEE Design and Test of Computers, vol. 28, no. 3, pp. 54–62, 2011.

[25] B. K. Mohanty, "Efficient Fixed-Width Adder-Tree Design," IEEE Trans. on Circuits and Systems II: Express Briefs, vol. 66, no. 2, pp. 292-296, Feb. 2019

[26] C. He et al., "LSAC: A Low-Power Adder Tree for Digital Computing-in-Memory by Sparsity and Approximate Circuits Co-Design," in IEEE Trans. on Circuits and Systems II: Express Briefs, vol. 71, no. 2, pp. 852-856, Feb. 2024.

[27] H.-C. Lu and S.-H. Huang, "Fault-Tolerant Near-Memory MAC Design with Redundant Memories," Proc. of IEEE ICCE-TW, 2023.

[28] Y.-C. Lin and S.-H. Huang, "An Approximate Fault-Tolerance Mechanism for SRAM-Based Near-Memory MAC Units," Workshop on Synthesis And System Integration of Mixed Information Technologies, 2024.

[29] E.-H. Chang and S.-H. Huang, "A Simple Yet Accurate Method for The Unsigned Fixed-Width Multiplier Design," Proc. of IEEE ICCE-TW, 2020.

[30] E.-H. Chang and S.-H. Huang, "Low-Power Low-Error Fixed-Width Multiplier Design for Digital Signal Processing," Proc. of IEEE International Conference on Consumer Electronics (ICCE), 2021.

979-8-3503-6689-1/24 $31.00 © 2024 IEEE

Special Session: Architecture-Level DCIM Technologies for Edge AI Computing Applications

Chun-Lung Hsu, Hsuan-Yu Chen and Yi-Lin Chen
National Central University (NCU), Taoyuan 320317, Taiwan

Abstract — Nowadays, deep neural networks (DNNs) and artificial intelligence (AI) are widely used in image recognition, autonomous vehicles, speech recognition, and natural language processing. However, the Von-Neumann bottleneck slows down data retrieval from memory, consuming significant time and energy. The technique of computing in memory (CIM) (including analog CIM (ACIM) and digital CIM (DCIM)) has emerged as a solution, integrating computing logic into memory to improve power efficiency by reducing data movement. Despite CIM's advantages, it still faces challenges like accuracy, adaptability and dataflow flexibility due to the computing complexity. This paper addresses the architecture-level digital computing in memory (DCIM) framework to discuss the abovementioned issues, ensuring the features of low-power, high-precision, reconfigurability, and repairability across diverse DNN applications. Additionally, for large-scale language model applications like LLMs and Transformers, a scalable DCIM chiplet architecture is introduced, leveraging 2.5D/3D heterogeneous packaging technologies to achieve flexible scalability, meeting various edge AI computing requirements.

Index Terms — DNN, AI, ACIM, DCIM, low-power, high-precision, reconfigurable, scalable

I. INTRODUCTION

Over the past decade, the rapid advancement of deep learning algorithms (deep neural networks, DNN) has been driven by increases in processor speeds and the emergence of big data [1]-[3]. These advancements have led to breakthroughs in various fields, most notably in image classification and localization, object detection, semantic segmentation and instance segmentation. (see Fig. 1). In some cases, these technologies have even surpassed human recognition accuracy. However, this progress comes at the cost of significantly increased computational demands, introducing two new challenges for hardware system implementation on device endpoints: real-time operation and low power consumption. Thus, in edge device systems, there is a need for high-precision operations with low computational power demands. To achieve this, many new design approaches have emerged which are different from traditional ones.

Fig. 1 DNN application fields

In order to execute deep learning algorithms on edge devices, the development of low-power, low-latency deep learning accelerators (DLA) to implement AIoT (artificial intelligence of things) device systems has become a key design consideration (see Fig. 2) [4], [5]. Different scenarios have their own specific customization requirements. For example, face ID applications require privacy considerations, autonomous vehicle systems need real-time detection and hazard prediction capabilities, and smart wearable devices must detect signals such as EEG/ECG signals from users while maintaining medical privacy and ensuring data reliability. Additionally, smart glasses, drones, and smart speakers are all moving towards the design of application-specific integrated circuits (ASICs). However, as data volume and computational demands grow, how to effectively configure hardware will become a problem that needs to be addressed and explored in the customization of these chips.

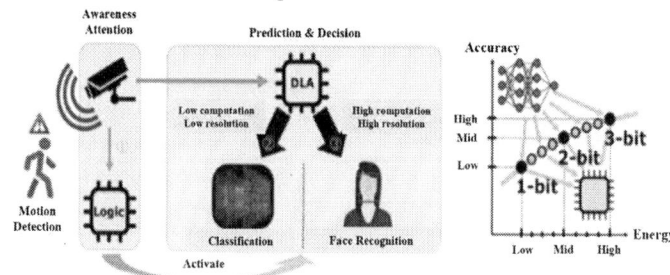

Fig. 2 Concept diagram of smart AIoT device system

Due to the development trends of algorithms, accompanied by increasing computational demands and memory capacity on hardware, the traditional Von Neumann architecture, which processes data by transferring it between the processor and memory, not only limits speed but also consumes additional power. Moreover, the computational performance is constrained by memory bandwidth and hardware computational power allocation. Therefore, to improve data transfer speed and reduce power consumption, researchers have begun developing various new system architectures, ranging from memory hierarchy configurations to computational architecture designs. Executing AI computations on edge devices has become an emerging research topic. Additionally, the co-computing architecture between sensors and memory will not be the only solution. Considering energy efficiency and signal processing flexibility, a multitude of architectures will emerge to address customization needs and achieve optimal system design. Generally, the issues of "low latency" and "power consumption" of multiply-accumulate computing (MAC) operation are two critical challenges for current edge devices. The CIM architectures, including ACIM and DCIN [6]-[10] were presented to overcome the limitation of the traditional Von Neumann architecture and meet the low latency/low-power requirements for edge device computing applications (see Fig. 3).

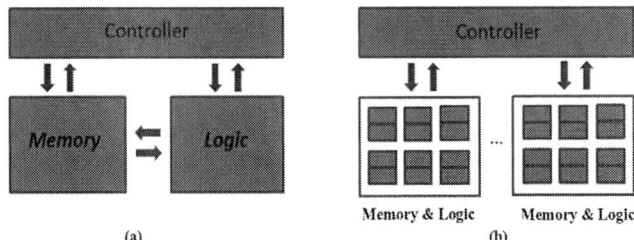

Fig. 3 (a) Von Neumann architecture (b) CIM architectures

By using the CIM architecture to execute MAC operations, this processing method can extend binary to multi-bit MAC operations in both volatile and non-volatile CIM architectures. Based on this concept, combining software and hardware co-design and simplifying algorithm computations can overcome the design bottlenecks of traditional thinking. Although hardware accelerators based on CIM architecture can provide the good performance in terms of latency and power consumption, there are still many challenges in designing CIM systems. First, CIM architectures must consider the co-design of software and hardware. Second, analog-to-digital converters (ADCs) and digital-to-analog converters (DACs) will increase power consumption and area. Third, IR drop and the capacitance on the bit lines will limit the storage capacity of the matrix. Fourth, the variability of memory components. These are the key challenges that the CIM architectures will need to address in the future. This paper aims to address and analyze the architecture-level DCIM technologies for application-oriented edge AI devices. Previous works on ACIM and DCIM are addressed in Section II. An architecture for a low-power, high-precision, reconfigurable, repairable DCIM systems with scalable features is presented in Section III. Section V shows the performance evaluation. Finally, the conclusions are given in Section V.

II. PREVIOUS WORKS

In recent years, CIM has emerged as a promising alternative to process element (PE)-based accelerators by performing MAC operations directly near or in memory cells. This greatly reduces access overheads and enables massive parallelization, potentially improving energy efficiency and throughput by orders of magnitude. Most recent CIM designs focus on ACIM

where computation occurs in the analog domain [7], [8]. This method ensures high energy efficiency and massive parallelization. However, the analog nature of the computation, along with intrinsic circuit noise and mismatches, compromises output accuracy. On the other hand, the DCIM technique is gaining interest as an alternative due to its noise-free computation and flexible spatial mapping [9], [10]. It offers accurate computation and flexibility but at the cost of reduced energy efficiency. The new opportunities from ACIM and DCIM have led to many studies focused on design in hardware architecture, array dimensions, and silicon technology. Generally, research on CIM circuit design follows two main paths: optimizing computing mechanisms and improving memory devices. Efforts to optimize computing mechanisms focus on enhancing energy and area efficiency for DNN operators using both analog and digital technologies. Due to their feasibility in process technology, SRAM CIMs are widely studied for implementing energy-efficient multi-bit MAC operations. The SRAM CIM can be divided into ACIM and DCIM according to the computing mechanism of MAC operations, as shown in Fig. 4.

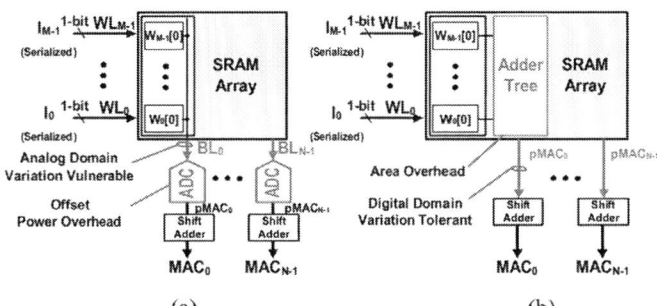

Fig. 4. The structure of CIM (a) ACIM (b) DCIM

In the ACIM structure (Fig. 4(a)), SRAM cells storing the weight information perform multiplication by activating multiple word-lines representing activation information. The accumulation results appear as a bit-line analog voltage, which is then converted into a digital value using an ADC. Despite its area-efficient structure, ACIM faces challenges due to the hardware cost of ADCs (large area and power consumption) and the PVT (process, voltage, temperature) variations on bit-lines, which cause significant errors in the digital MAC value. To achieve in-memory multiplication and accumulation, various ACIM circuits including current-based, charge-based and time-based which are addressed here and illustrated in Fig. 5(a)-(c).

The current-based ACIM shown in Fig. 5(a) used modulated currents through bit cell transistors for summation. This approach can be implemented with 6T SRAM cells eliminating the need for additional multiplication circuits. The accumulated modulated currents on the bit-line naturally perform MAC operations following Kirchhoff's Law. Despite its high density, this method faces challenges such as PVT variations and read disturbance. Fig. 5(b) illustrated the charge-based ACIM, which achieves multiplication using controlled charges on capacitors and accumulation through charge redistribution. Generally, charge-based ACIM introduces latency and additional charge injection overhead due to the pre-charging phase and extra switches. Additionally, the charge-coupling approach directly controls the voltage on the bottom plate of capacitors for accumulation onto the top plate, without needing shorting switches. While the charge-based CIM achieves high MAC linearity, its accuracy is limited by quantization noise in the ADC. Time-based ACIM (see Fig. 5(c)) uses delay units to control latency for various computations and accumulates results through a delay line. Compared to current-based and charge-based approaches, a significant advantage of time-based ACIM is its larger sense margin, although this comes with increased latency due to sequential processing. Additionally, time-based computation potentially offers higher efficiency due to its lower toggle rate. However, like the current-based approach, time-based ACIM is also affected by PVT variations, which can degrade its accuracy.

To address the limitations of ACIM, recent studies have proposed DCIM (Fig. 4(b)), where multiplication is performed inside SRAM cells. In other words, the adder-tree circuits are used in the accumulation operation to enable direct multiplication and dimension reduction through digital logic (see Fig. 5(d)). The robustness and ease of architectural integration make DCIM a promising alternative for modern spatial architectures. However, the significant overhead of full-precision digital circuits limits the efficiency of DCIM. To address the area and energy overhead associated with digital logic, the work in [9] integrated CMOS logic-based adders with transmission gate logic-based adders. Additionally, the work in [10] suggested using look-up-table-based adders to reduce the dynamic power consumption of the initial adder tree stage.

	Current	Charge	Time	Digital
Pros	High density	PVT robustness	Larger sense margin	High throughput & Accuracy
Cons	PVT sensitive	Larger bitcell	Lower throughput	Limited efficiency
Example	(a)	(b)	(c)	(d)

Fig. 5. Different types of CIM circuits [6]

As observed the concepts structures from the abovementioned ACIM and DCIM, although the DCIM has the limitation of efficiency, good performance in throughput and accuracy could be obtained. This paper addresses the DCIM architecture-level design methodologies for edge AI computing applications, aims at the following characteristics,

(1) Low-power sparse-aware multiplying unit and adder-tree design for PE-array MAC operations.
(2) Design a data fusing scheme to integrate the data of integer (INT) type and float-point (FP) type for high-precision operation requirement.
(3) Propose the multi-model DCIM architectures, including pipelines, homogeneous, heterogeneous and repairable for dynamic reconfigurable in various application-oriented edge AI computing needed.
(4) Develop the scalable chiplet DCIM architecture to support the advanced interposer-based 2.5D side-by-side and through-silicon via (TSV)-based 3D stacking technologies.

III. ARCHITECTURE-LEVEL DCIM DESIGN DISCUSSION

Figure 6 illustrates the block diagram of the architecture-level DCIM design including: low-power sparse-aware MAC design, high-precision INT/FP integration design, reconfigurable multi-model design and scalable chiplet design.

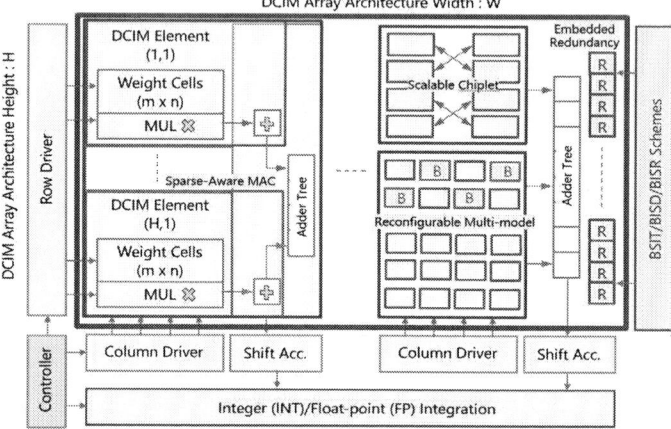

Fig. 6. The block diagram of architecture-level DCIM design

A. Low-Power Sparse-Aware MAC Design

Sparsity refers to the proportion of zeros in the activation or weight data of neural network layers. Skipping zero operations can save both energy and execution time. Implementing the issue of sparsity on DCIM architecture is more challenging. The work in [11] presented a CIM-friendly block-wise sparsity (BWS) architecture to improve the energy efficiency, however, this approach in [11] cannot be effectively applied to a DCIM architecture. Here, a novel DCIM architecture-level low-power sparse-aware strategy is discussed to efficiently utilize both weight and activation sparsity. Figure 7 shows the low-power sparse-aware MAC design, which consists of the activation sparsity controller (ASC), weight sparsity controller (WSC), sparsity-aware driver (SAD), ping-pong SRAM cell and low-power adder tree (LPAT).

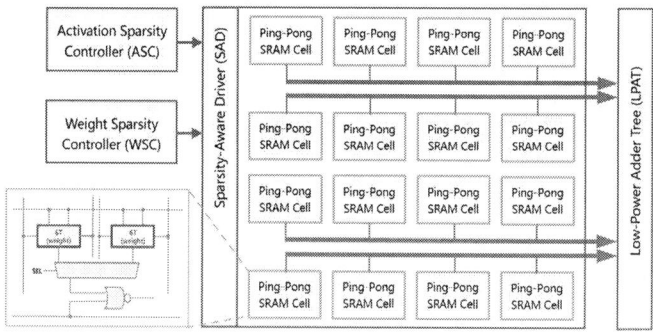

Fig. 7. Low-power sparse-aware MAC architecture

The main design innovations in Fig. 7 are given as follows.
(1) *ASC and WSC:* These two controllers are used to control the SAD for sending the sparsity instructions to the ping-pong SRAM cell for MAC operations. Significantly, the activation sparsity and weight sparsity mainly come from the non-linear ReLU function and the weight pruning technique, respectively.
(2) *SAD:* The SAD is used to store the information both of activation sparsity and weight sparsity.

(3) *Ping-pong SRAM cell:* Two 6T SRAM cells, one multiplexer (MUX) and one NOR gate are used to group the ping-pong SRAM cell to enable weight updates and avoid bit-line pre-charge. In other words, by using the SEL signal, the outputs from 6T SRAM cells will be selected for MAC operation or weight updating.

(4) *LPAT:* The function of LPAT is designed to compute the final accumulate MAC results. Significantly, the bit-precision and computing power can be effectively reduced because of the sparse-aware MAC architecture.

B. High-Precision INT/FP Data Integration Design

Generally, the high-efficiency neural network inference in edge AI tasks is getting more and more important. Thus, to meet the application-oriented edge AI computing, this paper addresses an INT/FP data integration design to provide the high-precision flexible MAC support (see Fig. 8). The main component in Fig. 8 is the exponent pre-alignment scheme, which is interleaving the tightly coupled exponent alignment and INT mantissa MAC to totally avoid alignment in the DCIM. Generally, the standard FP format consists of three parts: a sign bit (S), exponent bits (E), and mantissa bits (M). For example, both BF16 and FP32 have eight exponent bits. They have 7 and 23 mantissa bits, respectively. Figure 8 clearly seen that the FP MACs are simultaneously processed a group including both the pre-align activation (input) and weight exponents. In this way, the exponents of the product and partial sum are always the same in this group. In other words, no need to perform exponent alignment during INT mantissa MAC, which is friendly to DCIM implementation.

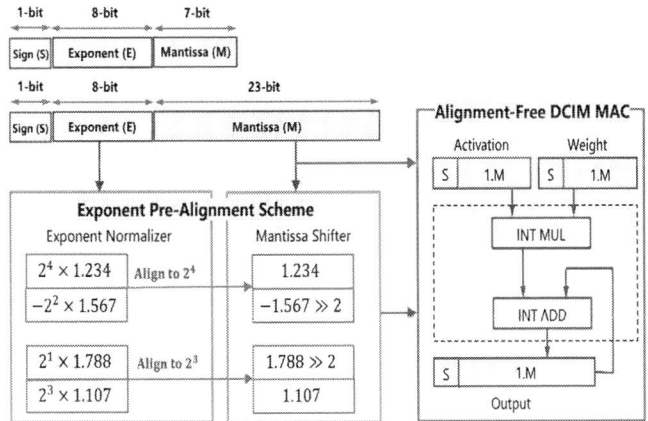

Fig. 8. INT/FP data integration design for alignment-free DCIM MAC

C. Reconfigurable Multi-Model Design

This paper addresses several types of reconfigurable multi-model architecture-level DCIM design including pipeline, homogeneous, heterogeneous and repairable architecture designs, as shown in Fig. 9. The pipeline DCIM architecture (see Fig. 9 (a)) focuses on workload mapping and integrating multiple DCIM arrays, connected by simple surrounding arithmetic circuits and buffers. Most computational workloads are processed in the DCIM arrays, with small buffers transferring temporal data between them. The pipeline architecture is ideal for multiple cascaded matrix-vector

multiplications.

Compared to the pipeline DCIM architecture, the homogeneous DCIM architecture (see Fig. 9(b)) integrates more diverse components, including DCIM blocks, subarray driver, shift accumulator, controller and buffer. Significantly, the homogeneous DCIM architecture features relatively large input/output buffers, mainly composed of SRAM. This is because input data are accessed repeatedly over multiple iterations, and intermediate results need to be stored and accumulated before being sent to the next neural network layer. Also, in the homogeneous DCIM architecture design, complex control circuits are needed for iterative operations involving image pixels, weight kernels, and input/output channels.

Figure 9(c) illustrates the heterogeneous DCIM architecture design, which integrates several functional cores alongside different sizes of DCIM cores with multiple DCIM arrays, enabling end-to-end algorithms and applications. Significantly, the heterogeneous cores may have their own local memory and share a common global on-chip memory. The heterogeneous DCIM architecture can be used to co-optimize the different sizes of cores with a RISC-V CPU for optimizing system-level metrics, such as flexibility, accuracy, performance, and energy efficiency, by assigning different workloads to the specific optimized cores.

The redundant cores for implementing a built-in self-repair (BISR) technique in the DCIM architecture design are also presented in this paper, as shown in Fig. 9(d). Generally, the BISR technique uses additional hardware (the redundant cores or the spare cores) instead of traditional software diagnostics, minimizing the computation time. In other words, by using the redundant cores in the homogeneous/heterogeneous DCIM architecture design, the defect cores can be replaced, if required, by an appropriate connection configuration to ensure the function work correctly. However, the designers should pay attention to the trade-off between hardware overhead and fault coverage of the repairable DCIM architecture-level design.

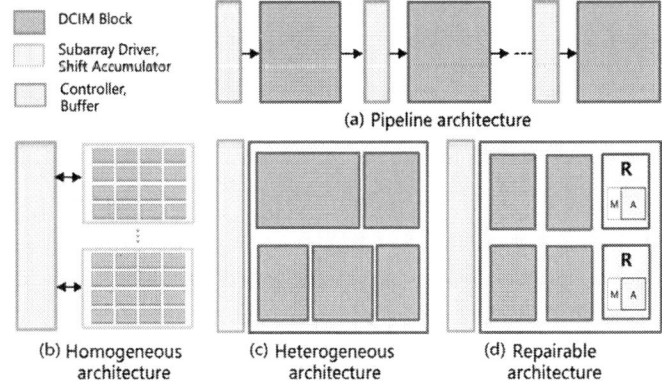

Fig. 9. DCIN architecture reconfigurable multi-model design

D. Scalable Chiplet Design

The rise of 2.5D/3D-integration is driving exploration into chip-level CIM architecture. Multiple CIM chiplets can be integrated to scale performance as required. In other words, the large neural network models can be split across several chiplets, storing most weights and intermediate data internally. Minimal intermediate data transfer between chiplets, the power consumption can be effectively reduced by using the advanced packaging technology. A DCIM architecture scalable chiplet design is illustrated in Fig. 10, which could support both the side-by-side 2.5D with interposer and 3D stacking with TSV chiplet system design. Significantly, current CIM chiplet research is constrained by fabrication technology and high costs. However, advancements in 2.5D/3D integration suggest future DCIM chiplets will feature more advanced packaging and accommodate multiple chiplets per package.

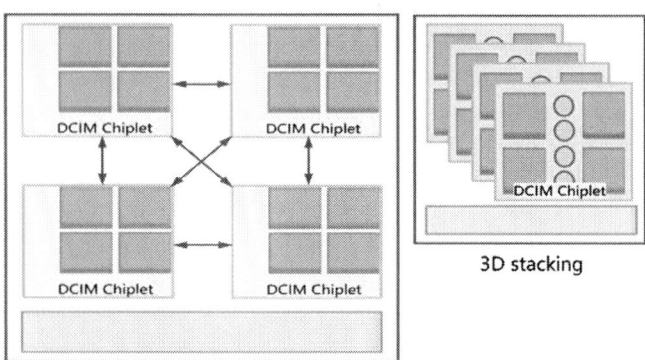

Fig. 10. DCIM architecture scalable chiplet design

IV. PERFORMANCE EVALUATION

This paper addresses the architecture-level DCIM design discussion for different hierarchies including low-power sparse-aware MAC design, high-precision INT/FP integration design, reconfigurable multi-model design and scalable chiplet design. Compare with the previous works [9], [10], [12]-[15], the good features of the proposed approaches are described as follows and summarized in Table 1-4.

(1) *Low-power sparse-aware MAC design:* To enhance the efficiency of sparse PE array utilization and improve the energy efficiency of DCIM MAC operations without sacrificing accuracy (see Table 1).

(2) *High-precision INT/FP integration design:* To integrate INT/FP operations into a unified system, supporting simultaneous the data format of FP-32, FP-16, BF-16, INT-16, and INT-8 (see Table 2). In other words, the approach presented in this paper could effectively overcome the issues of low utilization of resources such as frequent exponent alignment and mantissa addition.

(3) *Reconfigurable multi-model design:* To overcome the issues such as low flexibility in DCIM architecture-level designs and limited applicability to various types of neural network models (see Table 3).

(4) *Scalable chiplet design:* For large-scale NN models applications, supporting advantage technologies in interposer-based 2.5D side-by-side and 3D stacking DCIM operation structures (see Table 4).

Table 1. Comparisons of DCIM MAC operations

	[10]	[12]	This Work
Sparse Aware	NA	Available	Available
Array Utilization	Low	Medium	High
Reconfigurable	NA	NA	Available

Table 2. Comparisons of DCIM operation precision

	[9]	[10]	This Work
INT Precision	NA	Available	Available
FP Precision	Available	NA	Available
BF Precision	NA	NA	Available

Table 3. Comparisons of reconfigurable DCIM design

	[9]	[13]	This Work
Pipeline Arch.	NA	Available	Available
Heterogeneous Arch.	Available	NA	Available
Repairable Arch.	NA	NA	Available

Table 4. Comparisons of scalable DCIM design

	[14]	[15]	This Work
2.5D side-by-side	NA	Available	Available
3D stacking	Available	NA	Available

On the other hand, since CIM architectures are widely used in big data and AI nowadays, prompting numerous studies on neural network deployment. However, the absence of a standard cell library leads to extended simulation times with traditional SPICE tools. To expedite the chip development and enhance design efficiency, the architecture-level DCIM design approaches presented in this paper could be integrated to develop a synthesis design toolchain, as shown in Fig. 11. For a specific application-oriented edge AI computing, the design processes shown in Fig. 11 could fully support the DCIM design from unit-level, block-level to architecture-level design simulation and performance (area, latency, energy-efficiency) evaluation.

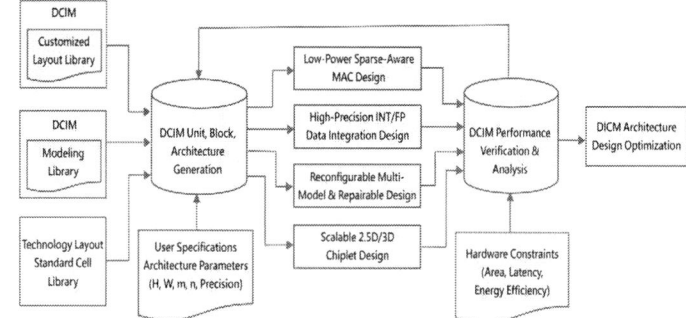

Fig. 11. DCIM synthesis design toolchain

V. CONCLUSION

This paper addresses architecture-level DCIM designs for the application-oriented edge AI computing. In other words, the features of low-power sparse-aware MAC operations, high-precision INT/FP data integration, reconfigurable multi-model scheduling of pipeline/homogeneous/heterogeneous/repairable approaches and scalable designs of interposer-based 2.5D side-by-side and TSV-based 3D stacking are presented in this paper for the different architecture-level DCIM designs requirement. The synthesis toolchain for the proposed DCIM approaches mapping and scheduling is also addressed aiming to identify the design challenges and guide for future DCIM designs.

Acknowledge: The authors would like to thank the NSTC 113-2640-E-008-001- for funding this study.

REFERENCES

[1] Mingcan Fang, "Intelligent Processing Technology of Cross Media Intelligence Based on Deep Cognitive Neural Network and Big Data", *Int. Conf. Machine Learning, Big Data and Business Intelligence (MLBDBI)*, pp., 505-508, 2020.

[2] Philip Derbeko, Shlomi Dolev and Ehud Gudes, "Deep Neural Networks as Similitude Models for Sharing Big Data", *IEEE Int. Conf. Big Data*, pp., 5728-5736, 2019.

[3] F. A. Pontes, M. Schukat and E. Curry, "Uncertainty-Aware Optimisation for Sustainable Multimedia Event Processing in Big Data Streams", *IEEE Int. Conf. Big Data*, pp., 5338-5345, 2023.

[4] Weiheng Liu and Xucan Chen, "An Efficient Computational Performance Analysis Method for CGRA-based Deep Learning Accelerator", *Asia Conf. Advanced Robotics, Automation, and Control Engineering (ARACE)*, pp., 11-16, 2023.

[5] T. Zhao, Y. Xie, Y. Wang, J. Cheng, X. Guo, B. Hu and Y. Chen, "A Survey of Deep Learning on Mobile Devices: Applications, Optimizations, Challenges, and Research Opportunities", *IEEE Proceedings*, Vol. 110, No. 3, pp. 334-354 , Mar. 2022.

[6] Wenyu Sun et. al., ei, "A Survey of Computing-in-Memory Processor: From Circuit to Application", *IEEE Open Journal of the Solid-State Circuits Society*, 2022.

[7] Runxi Wang and Xinfei Guo, "A Hierarchically Reconfigurable SRAM-Based Compute-in-Memory Macro for Edge Computing", *IEEE Int. Conf. Artificial Intelligence Circuits and Systems (AICAS)*, 2023.

[8] Joonhyung Kim and Jongsun Park, "The Quantitative Comparisons of Analog and Digital SRAM Compute-In-Memories for Deep Neural Network Applications", *Int. SoC Design Conference, ISOCC*, pp. 129-130 , Mar. 2022.

[9] Yu-Der Chih et al., "16.4 An 89TOPS/W and 16.3TOPS/mm2 All-Digital SRAM-Based Full-Precision Compute-In Memory Macro in 22nm forMachine-Learning Edge Applications", *IEEE Int. Conf. Solid-State Circuits (ISSCC)*, pp. 252-254, 2021.

[10] Chia-Fu Lee et al., "A 12nm 121-TOPS/W 41.6-TOPS/mm2 All Digital Full Precision SRAM-based Compute-in-Memory with Configurable Bit-width For AI Edge Applications", *IEEE Symp. VLSI Technology and Circuits*, pp. 24-25, 2022.

[11] Jinshan Yue et al., "STICKER-IM: A 65 nm Computing-in-Memory NN Processor Using Block-Wise Sparsity Optimization and Inter/Intra-Macro Data Reuse", *IEEE Journal Solid-State Circuits*, Vol. 57, No. 8, pp. 2560-2572, Aug. 2022.

[12] Chaojie He et al., "LSAC: A Low-Power Adder Tree for Digital Computing-in-Memory by Sparsity and Approximate Circuits Co-Design", *IEEE Trans. Circuits and Systems-II: Express Brief*, Vol. 71, No. 1, pp. 852-856, Feb. 2024.

[13] Fengbin Tu et al., "TranCIM: Full-Digital Bitline-Transpose CIM-based Sparse Transformer Accelerator With Pipeline/Parallel Reconfigurable Modes", *IEEE Journal Solid-State Circuits*, Vol. 58, No. 6, pp. 1798-1809, June. 2023.

[14] Haozhe Zhu et al., "COMB-MCM: Computing-on-Memory-Boundary NN Processor with Bipolar Bitwise Sparsity Optimization for Scalable Multi-Chiplet-Module Edge Machine Learning", *IEEE Int. Conf. Solid-State Circuits (ISSCC)*, 2022.

[15] Fengbin Tu et al., "TensorCIM: A 28nm 3.7nJ/Gather and 8.3TFLOPS/W FP32 Digital-CIM Tensor Processor for MCM-CIM-Based Beyond-NN Acceleration", *IEEE Int. Conf. Solid-State Circuits (ISSCC)*, 2023.

Dual-Modular-Redundancy Voting Circuits for Single-Event-Transient Mitigation

Marcello Barbirotta, Marco Angioli, Antonio Mastrandrea,
Francesco Menichelli, Abdallah Cheikh, Mauro Olivieri
Dept. of Information Engineering, Electronics and Telecommunications, Sapienza University of Rome, Italy,
Email: {name.surname}@uniroma1.it

Abstract—Single Event Transient faults pose an increasing challenge in reliability design especially regarding internal nodes of combinational voting circuits, as device dimensions shrink and working frequencies boost in modern technologies. This work proposes a novel voting structure for Dual Modular Redundancy lock-step architectures, made of a comparator with parity and recovery signal, able to reduce the failure rate down to 6.6% in case of internal SET faults, achieving the lowest value in the literature when compared to the 33% achieved by conventional DMR lock-step comparators and the 68.75% of DMR self-voter approaches without filtering methods. The fault resilience performance comes at the cost of only a slight increase in hardware utilization, power consumption and frequency degradation.

Index Terms—Fault Resilience, Single Event Transient, Fault Tolerance, Double Modular Redundancy, Fault Injection

I. INTRODUCTION

THe presence of faults within the internal nodes of a system, such as a digital circuit or a computer processor, depends on various factors in practice, such as designs and operating environments [1]. High power consumption and voltage fluctuations can cause thermal problems, while elevated operating temperatures can speed up component aging, leading to more thermal-induced faults [2]. Complex circuits, with their increasing number of components and interconnections, are more likely to experience faults, even if they are designed to be faster and more efficient in power consumption [3]. Smaller technology nodes tend to be more susceptible to issues like leakage currents and variability [4], and the transistor size in a design affects the vulnerability to Single Event Effects (SEEs). SEEs occur when a single energetic particle hits an electronic device, generating electron-hole pairs. In the deep submicron era [5], [6], the problem of SEEs aggravates as a side effect of CMOS fabrication technology advancement [7], [8]. In a combinational logic circuit, a SEE manifests as a glitch in the logic value of a signal [9], which can last from a few tens of ps up to about 1-2 ns [10], and as such is called Single Event Transient (SET). The most serious consequence of SETs occurs when the transient pulse reaches a register, and the clock edge arrives before the transient pulse extinguishes (*SET latching*). While transient pulses which do not overlap with the window of vulnerability of a storage element will not cause an error [11], [12], the likelihood of *SET latching* is proportional to the clock frequency [13]. Moreover, recent studies have reported that spatial redundancy techniques might increase the susceptibility to SET due to the increased area [14], [15].

Fault Tolerant (FT) techniques have been proposed to prevent errors caused by SETs at the microarchitecture levels [16], [17], yet not reaching 100% coverage of possible faults. They either try to eliminate the SET pulse from the circuit by modifying fabrication processes or, as this option is more expensive, they try to eliminate the effects of the SET pulse from the circuit through circuit design [18].

At the gate level, Triple Modular Redundancy (TMR) offers the benefit of possibly protecting all the sequential cells and immediately producing error correction but requires triple hardware resources or execution time [19]. Consequently, traditional implementations of TMR incur more than double the area overhead of a non-redundant system [20], which may be too costly in typical scenarios with infrequent faults. In such cases, a Double Modular Redundancy (DMR) solution can be a more convenient choice. Since DMR solely provides error detection, researchers explored different voting mechanisms, like *self-voters*, able to create the majority voting thanks to internal feedback paying in terms of SET fault coverage, or *comparators with recovery*, implementing relatively expensive checkpoint/recovery procedures to recover the correct architectural state.

This study analyzes the behaviour of voting structures based on *comparator with recovery* and presents a new voting structure specifically designed to significantly enhance the SETs faults within both the internal combinational and final leaf nodes of the circuit, i.e. the endpoint of a combinatorial chain without any further branches or child nodes, proposing a comparator with parity and a recovery signal, enhancing fault resilience at the expense of an acceptable worsening in terms of area occupation and performances compared to prevailing DMR standard-cell solutions documented in the literature. The solution applies to all DMR processors, such as those in the dual-core lock-step category [21]–[23], which are a type of fault-tolerant design used in digital processors to enhance reliability and ensure system correctness, especially in high-reliability applications like aerospace, automotive, and critical distributed environments [24], using redundancy to detect and correct faults that may occur in the system, executing instructions in a lock-step mode producing identical results.

The work is organized as follows: Section II discusses the background of DMR voting systems, focusing on the

979-8-3503-6689-1/24 $31.00 © 2024 IEEE

widely used self-voter approach, while Section III explain the motivation of the work with a theoretical analysis on the reliability of the DMR voting systems. Section IV presents the proposed voting architecture, while Section V and Section VI summarize the results obtained by comparing the proposed design with the state of the art regarding FT performances, frequency, area and power consumption. Finally, Section VII concludes the work.

II. BACKGROUND AND RELATED WORKS

The first work to promote methodologies against SETs inside DMR architectures was [25]. The technique, called temporal sampling, addresses both traditional SEUs and SET-induced errors, inserting two extra-temporal sampling times through delays of around 200 ps to prevent the latching of transient errors below this threshold width. For a circuit operating at low frequencies, adding the temporal latch results in a small reduction in speed. However, for a circuit operating at 500 MHz, the speed penalty escalates to nearly 20% [25], making the method unsuitable for high-performance applications.

The temporal sampling technique can also be added to circuits known in the literature as Guard-Gates, also known as Muller-C elements or simply self-voters. During the SET pulse at one of its inputs, the Guard-Gate output enters a high impedance state, maintaining its voltage value. The Guard-Gate-based method combines temporal and spatial schemes to eliminate SET pulses from combinational logic circuits by using multiple paths to latch inputs and a delayed signal to eliminate the SET pulses.

A. Guard-Gates in Custom Designs

In recent years, various methods have been developed to enhance the performance of Guard-Gates in speed, power consumption and FT, often employed in conjunction with other techniques, however, many of them are custom design applications, specifically when trying to solve the leaf node issues.

Authors in [18] introduce Guard-Gates inside latch as custom designs with minimal increases in area and power compared to the prevalent SET mitigation techniques. As anticipated, the reduction is insignificant for circuits operating at low frequencies (below 100 MHz). However, it becomes as large as 33% at 1 GHz to eliminate a SET pulse width of less than 500 ps duration. The work in [26] introduces a selective SET hardening technique connecting additional transistors in parallel to critical nodes, boosting resilience against SETs without resolving the problem of charged particle strikes on leaf nodes. Differently, authors in [7] propose two techniques to perform Positive SET Mitigation (PSM) from logic values '0' to '1' and Bidirectional SET Mitigation (BSM) to filter SET at leaf nodes, still adding custom modifications.

B. Self-Voters in Standard-Cell Designs

The self-voter can also be used in standard-cell FT applications. A self-voter circuit (depicted in Figure 1) is a 3-input

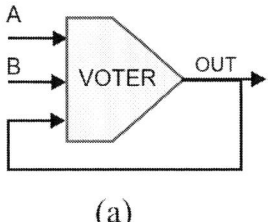

(a)

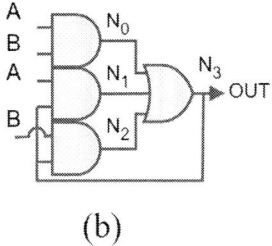

(b)

Fig. 1. Self-voting circuit (a) symbol, (b) standard-cell circuit [27]

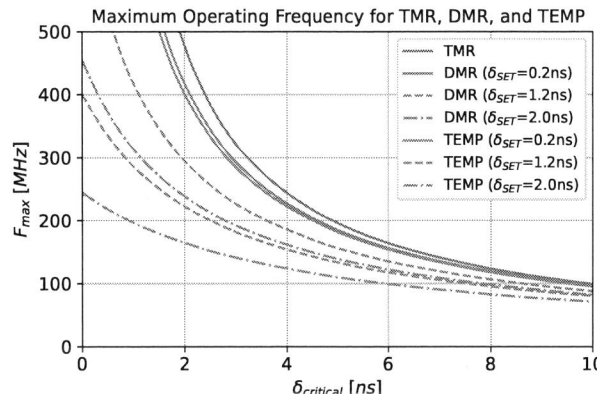

Fig. 2. Maximum Operating Frequency between TMR, DMR, and temporal-redundant (TEMP) from [27]. $\delta_{critical}$ is the maximum circuit delay through the datapath logic, and δ_{SET} is the pulse width of the striking particle.

majority voter configured to vote on two external inputs and its current output [27], and works as an asynchronous state-holding circuit. Like the majority voter, in the case of single events, a SET glitch on any one input will not propagate to the output, and an SET striking inside of the self-voter can cause a glitch on the output without causing a permanent upset of the stored state [27].

Since Flip-flop (FF) circuits are edge-triggered devices, inserting self-voters inside DMR circuits would not resolve the problem of SETs latched on clock edges. The self-voters would maintain their values from the previous clock cycle and refrain from updating to the accurate value. Furthermore, using self-voters in DMR circuits requires majority voters, clock delays, and additional FFs to store the self-voter results.

Figure 2 depicts the frequency curves obtained from [27],

979-8-3503-6689-1/24 $31.00 © 2024 IEEE   217

where TMR, DMR with self-voters, and temporal sampling techniques [25] are compared depending on SETs width.

DMR is a performance compromise between TMR and temporal redundancy techniques. However, it compromises the overall speed of the system, which could be a major problem for time-critical applications [28]. Moreover, again, using self-voter combined with delays reduces performances, especially at high frequencies, without resolving the problem of leaf nodes.

Authors in [12] propose two different approaches. The first is the Temporal Self-Voting (TSVL) approach with a simplex datapath, in which only the FFs are replicated and latched on different clock lines, realizing a power consumption improvement of 20.1% and 35.55% over the Self-Voting DMR and conventional TMR approach, imposing less area overhead than those of 22.02% and 36.84%. The second standard-cell approach, the Hybrid Spatial and Temporal Redundancy Double-Error Correction (HSTR-DRC), utilizes self-voters enclosing better resilience features to Multi-Bit-Upset (MBU) reducing area and power overheads by approximately 18.2% and 16.83%, respectively, while enhancing performance by 20.17% compared to conventional TMR. Despite both approaches resolve several fault cases internal to voter logic and FFs, a SET that hits the final leaf node will cause a momentary glitch in the final output.

To the best of our knowledge, the published design techniques targeting standard-cell implementation focus on mitigating SETs either at the input node or at intermediate nodes, without considering that the final leaf node output of the circuit remains susceptible to charged particle strikes. This possibly leads to incorrect data values stored to the final output flip-flops of a circuit [7].

III. THEORETICAL ANALYSIS

Self-voters are normally used in conjunction with a delay element, slowing down the performances of the circuit, for that they represent an optimal mitigation technique for Commercial Of The Shelf (COTS) components used in small applications in space [28]. On the other hand, voting systems inside DMR architectures with checkpointing and restore procedures, usually recognized as lock-stepped cores, are made of comparators with recovery signal, obtaining reduced overhead at the cost of a relatively expensive procedure to restore the accurate architectural state.

Figure 3 depicts a classical DMR voting structure with comparators. The correct result is directly taken by one of the inputs, while the recovery signal is obtained through a parallel *xor* and *or* chain. Unlike the self-voter, this structure does not contain filtering mechanisms for SETs, regardless of their entity. However, by inspecting the circuit structure in Figure 3, we see that the possible presence of faults within the comparison logic that generates the recovery signal would lead to unintended activation of the recovery procedure, generating an overhead on the total benchmark execution time without providing any incorrect results. Nevertheless, this masking does not occur when the fault in the internal logic

is combined with an error in the input A due to previous faults. Despite having a low probability, this scenario can arise when processors execute long latency instructions, such as vector instructions, or receive data from external hardware (memories, sensors, or other processors). In this situation, it can be easily deduced that if the fault affects any of the *or* gates connected to the initial *xor* gate that detects the erroneous bit, there is an erroneous activation of the recovery signal that would no longer reveal the presence of a mismatch on the inputs.

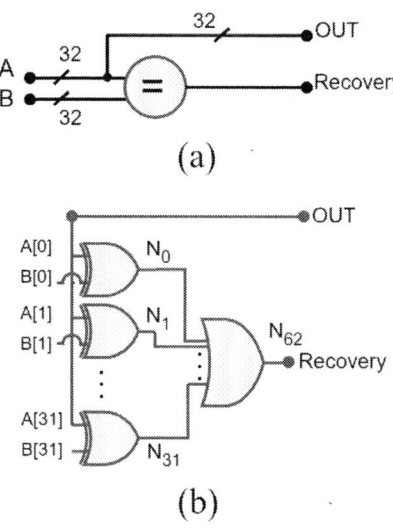

Fig. 3. DMR comparator with recovery circuit (a) symbol, (b) standard-cell circuit

Any fault occurring on any internal node may be combined with errors in the input bits associated with the logical chain linked to that node. Table I summarizes the fault cases for both internal logic solely and internal logic plus errors in the input bits. To illustrate, in the case of a 32-bit input, there are 63 internal nodes. The final node could exhibit 32 distinct error combinations, while the second-to-last node could manifest 16 combinations. The number of error combinations related to this situation can be easily calculated by multiplying the number of input bits of signal A (connected to out) by the depth of the logic chain. In the case of a 32-bit signal, there would be a total of $32 \times 6 = 192$ possible error combinations, where 6 is the number of nodes connected to each pair of bits. It is worth noting that if the erroneous input were B, the presence or absence of errors would not affect the correctness of the output connected to input A.

In a self-voter, any fault occurring on the internal nodes that persists longer than the pre-determined SET width delay could propagate an error at the output depending on the input state of the voter. Since each pair of input bits operates independently, we can analyze the single one-bit self-voter depicted in Figure 1. If a fault affects any of the internal nodes *N0*, *N1*, and *N2*, an output error only occurs when both inputs are '0', as the

TABLE I

FAULT TABLE FOR THE DMR COMPARATOR WITH RECOVERY. '0' MEANS
NO FAULT, '1' MEANS ACTIVE RECOVERY SIGNAL, '-' MEANS DO NOT
CARE ABOUT THE OUTPUT BECAUSE THE RECOVERY PROCEDURE IS
ACTIVE, 'F' MEANS FAULT AND '✗' MEANS ERROR.

A[32-bit]	B[32-bit]	N_0	N_1	...	...	N_{62}	Out	Recovery
0	0	0	0	0	0	0	A	0
0	0	0	0	0	0	F	-	1
0	0	0	0	0	:	0	:	1
0	0	0	0	:	0	0	:	1
0	0	0	F	0	0	0	-	1
0	0	F	0	0	0	0	-	1
0	F	0	0	0	0	0	-	1
0	F	0	0	0	0	F	A	0
0	:	0	0	0	:	0	:	1
0	:	0	0	:	0	0	:	1
0	F	0	F	0	0	0	-	1
0	F	F	0	0	0	0	-	1
F	0	0	0	0	0	0	-	1
F	0	0	0	0	0	F	✗	0
:		0	0	0	:	0	:	1
:		0	0	:	0	0	:	1
F	0	0	F	0	0	0	✗	1
F	0	F	0	0	0	0	✗	1

final *or* gate masks the voting mechanism when inputs are '1'. However, a fault on the last node, *N3*, would result in failures regardless of the input values. Each bit has five potential error scenarios, leading to 160 error combinations for 32-bit input.

In the above scenario, with input errors and internal SET fault, the self-voter presents a 50% chance of obtaining a correct result based on his ability to maintain the previous state when the inputs are correct. This scenario is depicted in Table II for the simple 1-bit case. For each bit, it is possible to count two error cases and two correct cases. Hence, for a 32-bit input, there would be 32 x 2 = 64 error combinations, which, combined with the previous case, would lead to a total of 160 + 64 = 224 potential error combinations. Consequently, assuming the duration of the internal SETs exceeds the established logic delay width, the self-voter exhibits a worse level of fault tolerance to that of a comparator with recovery.

TABLE II

1-BIT TRUTH TABLE SELF-VOTER WITH INTERNAL FAULTS WHEN INPUTS
ARE DIFFERENT.

Correct value	Previous value	out	error
0	0	0	-
0	1	1	✗
1	0	0	✗
1	1	1	-

IV. PROPOSED DESIGN

The proposed circuit is illustrated in Figure 4 and is referred to as the *comparator with parity recovery*. By incorporating two parallel branches of *xor* chains responsible for computing parity as well as the final *or* operation between the comparator output and the *xor* of the computed parities, we effectively reduce all instances of faults internal to the comparator to only cases where its output node holds an incorrect value. The two parity branches on the inputs allows to detect any transient

mismatch, additionally, this structure solves —similar to that of the standard comparator with recovery— all instances of faults within the recovery logic, paying with a slight temporal execution overhead attributed to occasional activations of the recovery signal.

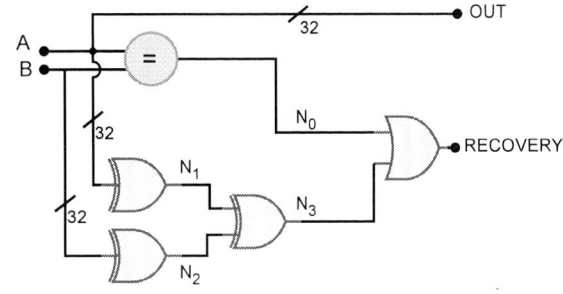

Fig. 4. Proposed DMR comparator with parity recovery standard-cell circuit

Unlike the conventional comparator, the total number of errors resulting from internal logic faults combined with input errors deriving from previous processing is solely contingent upon the final recovery gate and the bit count of the input signal. Consequently, the total number of error cases would be limited to $32 \times 1 = 32$.

TABLE III

FAULT TABLE FOR THE PROPOSED DMR COMPARATOR WITH PARITY. '0'
MEANS NO FAULT, '1' MEANS ACTIVE RECOVERY SIGNAL, '-' MEANS DO
NOT CARE ABOUT THE OUTPUT BECAUSE THE RECOVERY PROCEDURE IS
ACTIVE, 'F' MEANS FAULT AND '✗' MEANS ERROR.

A[32bit]	B[32bit]	N_0	N_1	N_2	N_3	N_4	Out	Recovery
0	0	0	0	0	0	0	A	0
0	0	0	0	0	0	F	-	1
0	0	0	0	0	F	0	-	1
0	0	0	0	F	0	0	-	1
0	0	0	F	0	0	0	-	1
0	0	F	0	0	0	0	-	1
0	F	0	0	0	0	0	-	1
0	F	0	0	0	0	F	A	0
0	F	0	0	0	F	0	-	1
0	F	0	0	F	0	0	-	1
0	F	0	F	0	0	0	-	1
0	F	F	0	0	0	0	-	1
F	0	0	0	0	0	0	-	1
F	0	0	0	0	0	F	✗	0
F	0	0	0	0	F	0	-	1
F	0	0	0	F	0	0	-	1
F	0	0	F	0	0	0	-	1
F	0	F	0	0	0	0	-	1

The validity of the proposed system is demonstrated by Table III, which comprehensively outlines all potential internal faults in the architecture. Notably, the presence of faults within the internal nodes of the architecture does not result in any incorrect output. However, when a fault occurs in one of the inputs, the unique scenario where an incorrect value emerges at the output without being detected by the recovery logic is when the fault affects the final node N_4, corresponding to the ultimate *or* operation within the system. In fact, all the other cases could potentially present an incorrect output, whose usage is avoided by the recovery signal, which detects

TABLE IV
OPERATING FREQUENCY, FAILURE RATE, AREA AND POWER RESULTS OBTAINED IN GLOBALFOUNDRIES 22FDX (GF22FDX) TECHNOLOGY ON SYNOPSYS FUSION COMPILER

	Estimated Frequency [GHz]	Area [μm^2]	Dynamic Power [μW]	Failure Rate [%]
DMR comparator with recovery	2.25	23.30	583	33
Self Voter*	2.5*	53.25*	554*	68.75*
Proposed Design	2.00	49.12	633	6.6

*without considering filtering delays.

the mismatch and activates the recovery procedure. The unique wrong scenario with fault on the final node, supposing 32 bits input, yields only 32 possible error cases, contrasting the 192 cases DMR comparator with recovery and 224 cases in the self-voter.

V. SYNTHESIS RESULTS

The discussed architectures — comparator with recovery, self-voter and comparator with parity — were synthesized using GlobalFoundries 22 nm (GF22FDX) technology on Synopsys Fusion Compiler. Results are presented in Table IV, showing the failure rate compared with the area occupation, power consumption and frequency estimation.

A comparison with the self-voter architecture regarding performances is not feasible in the normal configuration without filtering delays since the decrease in frequency that would occur with delays is not considered. There is no significant difference in frequency when comparing the classical DMR comparator with recovery against the proposed design. However, regarding area occupation, the self-voter architecture emerges as the largest, even without considering the overhead of delay elements. Moreover, the proposed design results in double area compared to the classical DMR comparator. Regarding power consumption, all tests were carried out by measuring the switching activities of the voters with random inputs in the absence of faults, assuming normal operation. The resulting power consumption is similar although the area consumption is higher for the proposed design compared to the DMR comparator with recovery. Despite the slight decrease in frequency compared to the DMR comparator with recovery and the area and power overhead, these costs are justified by the gains in fault tolerance.

VI. FAULT INJECTION RESULTS

The analyzed circuits were subjected to HDL Fault Injection (FI) simulation campaigns, obtaining the failure rate parameters that discriminate their effectiveness. The FI environment was built using an RTL Universal Verification Methodology (UVM) approach directly linked to the Design Under Test (DUT) netlist, mapping every combinational node to reproduce the analytical analysis presented in Section III. Testbenches

were implemented with SET faults of various magnitudes ranging from 200 ps to 1 ns [10], and all the voters were tested with random input values covering the whole 32-bit data width for a total of 10,000 different tests for each type of voter. Across all tested scenarios, in agreement with previous results in the literature, it became evident that the circuits exhibited failures solely when faults were captured from the output register during the rising edge of the clock. For convenience, we only report the results relating to this event. The self-voter, without specific SET width filtering delays, exhibited a failure rate of 68.75% in cases involving internal logic alone and internal logic failures combined with incorrect inputs, proving his ability to mask the faults depending on the input values and previous output state. Conversely, the other circuits displayed a trend comparable with the analysis in Table I and III, since the comparator with recovery figured out failure rates of around 33% ($\frac{1}{3}$ of all possible cases). At the same time, the proposed design exhibited a failure rate of circa 6.6%.

VII. CONCLUSIONS

This study introduced an innovative circuit scheme tailored for DMR architectures incorporating checkpoint/restore methodologies. We provided an overview of competitor voting schemes in existing literature, evaluating their performance and fault tolerance, particularly when integrated as leaf nodes within DMR architectures. Based on this analysis, we proposed the comparator with parity and recovery as a solution to enhance the overall circuit reliability with a slight decrease in performances. The approach significantly reduces the number of error combinations to 32 from 224 and 192, respectively, compared to self-voter and standard comparators with recovery. Despite a slight increase in area occupancy and a limited speed degradation, the failure rate calculated under a specific SET fault injection simulation campaign is reduced to 6.6% from 68.75% and 33% achieved by the self-voter and the classic comparator with recovery, respectively, thus demonstrating that the adopted solution represents the best trade-off between fault tolerance gains and performance degradation.

979-8-3503-6689-1/24 $31.00 © 2024 IEEE

REFERENCES

[1] R. Della Sala and G. Scotti, "A novel fpga implementation of the nand-puf with minimal resource usage and high reliability," *Cryptography*, vol. 7, no. 2, 2023.

[2] R. D. Sala and G. Scotti, "Exploiting the dd-cell as an ultra-compact entropy source for an fpga-based re-configurable puf-trng architecture," *IEEE Access*, vol. 11, pp. 86178–86195, 2023.

[3] M. Angioli, M. Barbirotta, A. Cheikh, A. Mastrandrea, F. Menichelli, S. Jamili, and M. Olivieri, "Design, implementation and evaluation of a new variable latency integer division scheme," *IEEE Transactions on Computers*, vol. 73, no. 7, pp. 1767–1779, 2024.

[4] R. Della Sala, D. Bellizia, and G. Scotti, "A lightweight fpga compatible weak-puf primitive based on xor gates," *IEEE Transactions on Circuits and Systems II: Express Briefs*, vol. 69, no. 6, pp. 2972–2976, 2022.

[5] M. Barbirotta, A. Cheikh, A. Mastrandrea, F. Menichelli, M. Angioli, S. Jamili, and M. Olivieri, "Fault-tolerant hardware acceleration for high-performance edge-computing nodes," *Electronics*, vol. 12, no. 17, 2023.

[6] A. Buzzin, A. Rossi, E. Giovine, G. de Cesare, and N. P. Belfiore, "Downsizing effects on micro and nano comb drives," *Actuators*, vol. 11, no. 3, 2022.

[7] F. M. Sajjade, N. K. Goyal, and B. Varaprasad, "Single event transient (set) mitigation circuits with immune leaf nodes," *IEEE Transactions on Device and Materials Reliability*, vol. 21, no. 1, pp. 70–78, 2021.

[8] R. Della Sala, V. Spinogatti, C. Bocciarelli, F. Centurelli, and A. Trifiletti, "A 0.15-to-0.5 v body-driven dynamic comparator with rail-to-rail icmr," *Journal of Low Power Electronics and Applications*, vol. 13, no. 2, 2023.

[9] E. Vacca, S. Azimi, and L. Sterpone, "A comprehensive analysis of transient errors on systolic arrays," in *2023 26th International Symposium on Design and Diagnostics of Electronic Circuits and Systems (DDECS)*, pp. 175–180, 2023.

[10] V. S. Veeravalli, T. Polzer, U. Schmid, A. Steininger, M. Hofbauer, K. Schweiger, H. Dietrich, K. Schneider-Hornstein, H. Zimmermann, K.-O. Voss, *et al.*, "An infrastructure for accurate characterization of single-event transients in digital circuits," *Microprocessors and microsystems*, vol. 37, no. 8, pp. 772–791, 2013.

[11] V. Ferlet-Cavrois, L. W. Massengill, and P. Gouker, "Single event transients in digital cmos—a review," *IEEE Transactions on Nuclear Science*, vol. 60, no. 3, pp. 1767–1790, 2013.

[12] F. S. Alghareb, M. Lin, and R. F. DeMara, "Soft error effect tolerant temporal self-voting checkers: Energy vs. resilience tradeoffs," in *2016 IEEE Computer Society Annual Symposium on VLSI (ISVLSI)*, pp. 571–576, IEEE, 2016.

[13] B. Narasimham, B. L. Bhuva, W. T. Holman, R. D. Schrimpf, L. W. Massengill, A. F. Witulski, and W. H. Robinson, "The effect of negative feedback on single event transient propagation in digital circuits," *IEEE transactions on nuclear science*, vol. 53, no. 6, pp. 3285–3290, 2006.

[14] O. Amusan, A. Stemberg, A. Witulski, B. Bhuva, J. Black, M. Baze, and L. Massengill, "Single event upsets in a 130 nm hardened latch design due to charge sharing," in *2007 IEEE International Reliability Physics Symposium Proceedings. 45th Annual*, pp. 306–311, IEEE, 2007.

[15] Y. Lin, M. Zwolinski, and B. Halak, "A low-cost, radiation hardened method for pipeline protection in microprocessors," *IEEE Transactions on Very Large Scale Integration (VLSI) Systems*, vol. 24, no. 5, pp. 1688–1701, 2015.

[16] M. Barbirotta, A. Cheikh, A. Mastrandrea, F. Menichelli, and M. Olivieri, "Design and evaluation of buffered triple modular redundancy in interleaved-multi-threading processors," *IEEE Access*, vol. 10, pp. 126074–126088, 2022.

[17] M. Barbirotta, A. Cheikh, A. Mastrandrea, F. Menichelli, M. Ottavi, and M. Olivieri, "Evaluation of dynamic triple modular redundancy in an interleaved-multi-threading risc-v core," *Journal of Low Power Electronics and Applications*, vol. 13, p. 2, 12 2022.

[18] A. Balasubramanian, B. Bhuva, J. Black, and L. Massengill, "Rhbd techniques for mitigating effects of single-event hits using guard-gates," *IEEE Transactions on Nuclear Science*, vol. 52, no. 6, pp. 2531–2535, 2005.

[19] M. Barbirotta, A. Cheikh, A. Mastrandrea, F. Menichelli, F. Vigli, and M. Olivieri, "A fault tolerant soft-core obtained from an interleaved-multi- threading risc- v microprocessor design," in *2021 IEEE International Symposium on Defect and Fault Tolerance in VLSI and Nanotechnology Systems (DFT)*, pp. 1–4, IEEE, 10 2021.

[20] R. E. Lyons and W. Vanderkulk, "The use of triple-modular redundancy to improve computer reliability," *IBM journal of research and development*, vol. 6, no. 2, pp. 200–209, 1962.

[21] M. Barbirotta, F. Menichelli, A. Cheikh, A. Mastrandrea, M. Angioli, and M. Olivieri, "Dynamic triple modular redundancy in interleaved hardware threads: An alternative solution to lockstep multi-cores for fault-tolerant systems," *IEEE Access*, vol. 12, pp. 95720–95735, 2024.

[22] P. R. Nikiema, A. Kritikakou, M. Traiola, and O. Sentieys, "Design with low complexity fine-grained dual core lock-step (dcls) risc-v processors," in *2023 53rd Annual IEEE/IFIP International Conference on Dependable Systems and Networks - Supplemental Volume (DSN-S)*, pp. 224–229, 2023.

[23] X. Iturbe, B. Venu, E. Ozer, J.-L. Poupat, G. Gimenez, and H.-U. Zurek, "The arm triple core lock-step (tcls) processor," *ACM Transactions on Computer Systems (TOCS)*, vol. 36, no. 3, pp. 1–30, 2019.

[24] L. Canese, G. C. Cardarilli, L. Di Nunzio, R. Fazzolari, M. Re, and S. Spanò, "Resilient multi-agent rl: introducing dq-rts for distributed environments with data loss," *Scientific Reports*, vol. 14, no. 1, p. 1994, 2024.

[25] D. G. Mavis and P. H. Eaton, "Soft error rate mitigation techniques for modern microcircuits," in *2002 IEEE International Reliability Physics Symposium. Proceedings. 40th Annual (Cat. No. 02CH37320)*, pp. 216–225, IEEE, 2002.

[26] A. T. Sheikh, A. H. El-Maleh, M. E. Elrabaa, and S. M. Sait, "A fault tolerance technique for combinational circuits based on selective-transistor redundancy," *IEEE Transactions on Very Large Scale Integration (VLSI) Systems*, vol. 25, no. 1, pp. 224–237, 2016.

[27] J. Teifel, "Self-voting dual-modular-redundancy circuits for single-event-transient mitigation," *IEEE Transactions on Nuclear Science*, vol. 55, no. 6, pp. 3435–3439, 2008.

[28] F. Smith and J. Omolo, "Experimental verification of the effectiveness of a new circuit to mitigate single event upsets in a xilinx artix-7 field programmable gate array," *Microprocessors and Microsystems*, vol. 79, p. 103327, 2020.

A Novel Self-Repair Mechanism for Tiled Matrix Multiplication Unit

Chandra Sekhar Mummidi, Sandeep Bal, Sandip Kundu
Department of Electrical and Computer Engineering
University of Massachusetts Amherst
{cmummidi, sbal, kundu}@umass.edu

Abstract—**General Matrix Multiplications (GEMMs) are widely used in scientific and machine learning applications. Convolutional Neural Networks (CNNs) rely primarily on convolution operation, which is often expressed as a GEMM problem. To speed up GEMM operations, modern CPUs include hardware accelerators specifically designed for this purpose. One such hardware accelerator is the Tile Multiplication unit (TMUL), which is supported by the Advanced Matrix Multiplication (AMX®) instruction set in the Intel Sapphire Rapids microarchitecture. However, like other frequently used hardware components, the TMUL is susceptible to permanent or persistent faults due to aging, which can lead to Silent Data Corruption (SDC). Google and Meta have recently reported recurring SDCs. Algorithm-based error detection (ABED) is a highly effective method for detecting persistent faults in GEMM operations.**

Our previous work involved a novel hardware implementation of ABED in TMUL, which could locate faults but only identify the offending matrix column, not the exact cell. In this paper, we present a self-repair technique for persistent faults in GEMM based on the aforementioned ABED approach. Our solution hinges on software-based column avoidance. When the hardware-based ABED detects an error caused by a faulty column in the TMUL, our software redirects computations to circumvent the faulty column. This eliminates the need for additional hardware modifications to ABED. We demonstrate that our proposed approach can perform self-repair for single faults for all matrix sizes, from those smaller than the TMUL dimension to larger matrices. Furthermore, we show that our solution can even correct multiple faults with low-performance overhead, and we report the performance overhead for multiple fault repair using a clustered fault model. Unlike prior solutions, our approach performs online detection and repair, enabling real-time correction of both persistent and intermittent faults.

Index Terms—**GEMM, Machine Learning, Hardware Accelerator, TMUL, Built-In Self Repair**

I. INTRODUCTION

General Matrix multiplication (GEMM) operations are widely used in high-performance computing and machine learning applications. With pervasive use of machine learning (ML), it has become the dominant workload in data centers [1]. GEMM is fundamental to all ML applications, including CNNs, RNNs, and MLPs. As a result, GEMM computations account for the vast majority of data center computing.

As a result, contemporary CPU and GPU manufacturers prioritize hardware acceleration to enhance GEMM perfor-

mance. Major companies such as Google, Microsoft, Intel, and Qualcomm have introduced machine learning accelerators [2], [3] to accelerate GEMM computation. At their core, these accelerators feature an array of processing elements (PE) to perform simultaneous multiply-and-add operations. Depending on dataflow structure, these arrays take various forms. For example, Google's TPU is a tightly-coupled systolic array [4].

In 2021, Intel introduced Intel Advanced Matrix Extensions (AMX), a novel instruction set that aims to accelerate GEMM operations [3]. AMX incorporates a Tile Matrix Multiply Unit (TMUL) to expedite matrix multiplication computations. To avoid the overhead of transferring large matrices into and out of memory, TMUL operates on smaller sub-matrices, known as *tiles*, of the input matrices. This approach not only enhances cache hit rates but also accelerates computation times by leveraging TMUL's parallel computation. In AMX, a tile is 1KB in size that fits into an AMX register. In the AMX instruction set architecture, there are 8 such registers. A TMUL unit receives data from two tiled matrices (AMX registers) and outputs a tiled matrix output that is aligned for storage in an AMX register. AMX instructions are accessible directly by software through intrinsics. Basic Linear Algebra Subprograms (BLAS), Intel's Math Kernel Library (MKL), includes support for GEMM operations.

Machine learning applications are increasingly adopting low-precision arithmetic alongside hardware acceleration for GEMM operations. Low-precision arithmetic offers the advantage of faster and more power-efficient computations without significantly impacting machine learning accuracy. Today's machine learning models commonly employ 8-bit integer and 16-bit floating-point representations (bf16 and fp16). To cater to these specific workloads, TMUL performs GEMM operations on a variety of data types including 8-bit integers and bf16 floating-point. For 8-bit integer data, which is commonly used in machine-learning classification, the TMUL unit performs 1024 simultaneous multiply and add operations with much lower latency compared to a systolic array. While the TMUL unit delivers exceptional performance, its large-scale parallel computations and high power consumption lead to elevated operating temperatures and device aging.

Hardware errors can manifest as *permanent*, *intermittent*, or *transient* faults. Permanent hardware errors arise from device degradation due to aging. Intermittent errors, on the other hand, occur sporadically at a hardware location and persist

---

This work has been supported in part by a grant from the National Science Foundation (NSF) and the Intel Corporation.

979-8-3503-6689-1/24 $31.00 © 2024 IEEE

for some time. Transient errors, lasting for a cycle or two, are typically caused by cosmic or alpha-particle radiation. Recent studies have indicated that these hardware errors are more frequent than previously anticipated. Recently Google and Meta have reported finding persistent execution errors in their data centers [5], [6].

In our previous work, we explored the problem of concurrent error detection in TMUL hardware using an algorithm-based error-detection (ABED) approach [7]. We demonstrated the ability to detect single execution errors with 100% accuracy with minimal area and power overhead. However, that work focused primarily on error detection and did not address error correction due to permanent or persistent faults. In this paper, we present an efficient solution for performing matrix multiplication in the presence of persistent error sources. This solution employs a fault avoidance strategy [8]. We demonstrate the effectiveness of this solution even in scenarios with multiple persistent error sources (faults) in the hardware, provided the cardinality of these faults is low.

The main contributions of this paper are:

- We investigate low-cost repair mechanism for GEMM operation on TMUL with faulty PE(s).
- Unlike previous solutions, where faulty PEs are diagnosed during manufacturing testing, our approach detects faulty PEs at runtime, thus able to diagnose and repair not just permanent faults, but also intermittent faults, when a PE exhibits faulty behavior at elevated temperatures but not at normal temperature of operation.
- The proposed method requires diagnosis of faulty columns which is our replacement unit. It simplifies the overall approach because it does not require diagnosis of the faulty PE.
- No modifications to the hardware is necessary beyond what is required for error-detection because the repair mechanism is entirely software-based. Algorithm-Based Error Detection implemented in hardware has very little overhead.
- The fault effect is contained by a software pre and post-processor, ensuring correct computation of results.

II. BACKGROUND

A. TMUL

Intel's Advanced Matrix Extensions (AMX) include eight 1KB registers for basic data operations. The TMUL instructions utilize these for data tile processing. AMX's integration within the user process space, unlike external accelerators like TPUs, eliminates mode transitions and external memory setup, offering efficiency benefits. However, TMUL's processing is limited to 1KB data slice. Its core comprises Processing Elements (PEs) capable of fused multiply-and-add (FMA) operations, as illustrated in Fig. 1.

During TMUL operation, up to 1024 processing elements (PEs) can execute concurrently, resulting in significant power consumption. To accommodate this high power demand, each processor core is equipped with integrated power controller.

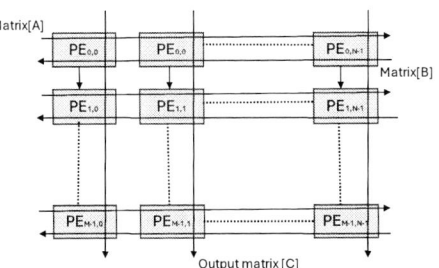

Fig. 1: Fused Multiply and Add array structure of TMUL

Additionally, the processor frequency is dynamically adjusted downward during TMUL operation to effectively manage the power consumption. The frequency is set low enough to meet the power bar, but high enough to provide the best possible throughput. If the frequency is set incorrectly, it may lead to persistent error.

B. Algorithm Based Error Detection for GEMM

Let $C_{M \times N} = A_{M \times K} * B_{K \times N}$ be the multiplication of matrices A and B resulting in product C where M,N,K are matrix dimensions. Let $a_{i,j}$ be an element of matrix $A_{M \times K}$ where $i \in [0, M-1]$ and $j \in [0, K-1]$. $b_{i,j}$ and $c_{i,j}$ may be defined similarly. The traditional ABED approach augments matrix A with an additional row and matrix B with an additional column [9], where:

$$a_{M,j} = \sum_{i=0}^{M-1} a_{i,j} \quad and \quad b_{i,N} = \sum_{j=0}^{N-1} b_{i,j} \qquad (1)$$

$a_{M,j}$ and $b_{i,N}$ are referred to as column checksum of A and row checksum of B. Now C becomes $C_{(M+1) \times (N+1)} = A_{(M+1) \times K} * B_{K \times (N+1)}$. The additional column and row of C are guaranteed to be C's checksum as well [9]. Thus, we have two different ways of calculating the C's checksum.

$$c_{i,N} = \sum_{j=0}^{K-1} a_{i,j} * b_{j,M} \quad \forall i \in [0, M-1] \qquad (2)$$

$$c'_{i,N} = \sum_{j=0}^{N-1} c_{i,j} = \sum_{k=0}^{N-1} \sum_{j=0}^{K-1} a_{i,j} * b_{j,k} \quad \forall i \in [0, M-1] \quad (3)$$

ABED hinges on the equality of $c_{i,N} = c'_{i,N}$ which comes from commutative and associative properties of computations to confirm the accuracy of the multiplication. Any divergence in value signals an error. Further, divergence in column checksum indicates the column location of the error.

C. Algorithm-Based Error Detection in TMUL

To implement error correction in TMUL operations, it is essential to first detect errors. In our earlier paper [7] we presented ABED based error detection in hardware. Even though Intel's TMUL unit design is not disclosed, we make certain conjectures about Intel's TMUL design using the knowledge that is currently accessible embodied in second baseline design, as illustrated in Figure 2 [3]. The size of the

TMUL may be estimated using the allowed data types and the AMX register size. The core TMUL operation and the augmented operation for error detection between the matrices A and B, resulting in C based on deduced dimensions, are $C_{64 \times 64} = A_{64 \times 16} * B_{16 \times 64}$.

In our previous work [7], we introduced ABED for error detection in TMUL operations. Considering Intel's undisclosed TMUL unit design, we developed informed conjectures [3], as shown in Figure 2. This approach enabled us to estimate the TMUL size based on permissible data types and AMX register size, culminating in the core operation formula: $C_{64 \times 64} = A_{64 \times 16} * B_{16 \times 64}$. While the initial designs form the basis of our discussion, the primary focus of this paper is the novel integration of self-repair mechanisms into the enhanced TMUL design, which will be elaborated below.

D. Enhanced TMUL

a) Design: Figure 2 illustrates the Enhanced TMUL design, with PEs arrayed in 64 columns and 16 rows, as described in [7]. To facilitate error detection with ABED, the design includes an Output Checksum (OC) unit and an Input Checksum (IC) unit, which compute checksums for the outputs and inputs of the PEs, respectively. The IC in ETMUL employs 16 basic adders corresponding to the 16 rows of the multiplier unit for input checksum computation. The OC unit features a dedicated adder for each of the 64 PE unit columns.

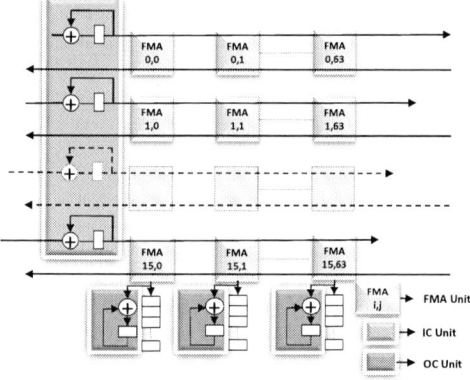

Fig. 2: Enhanced TMUL

b) Operation: The matrix $B_{16 \times 64}$ elements are broadcast to PE array in a single clock cycle, while the transposed $A_{64 \times 16}$ (denoted as $A^T_{64 \times 16}$) is simultaneously queued at the input. Element-wise multiplication begins with shifting the first column of $A^T_{64 \times 16}$ and $B_{16 \times 64}$, generating the first output element in the 17th clock cycle. The process continues, shifting one column of $A^T_{64 \times 16}$ per cycle, completing the shift in 64 cycles. Draining the output starts in the 17th cycle, with the full $C_{64 \times 64}$ matrix computed by the 144th cycle.

Elements of the transposed matrix, $A^T_{64 \times 16}$, are sequentially accumulated row-wise into the Input Checksum (IC) unit's registers as the matrix is loaded. This occurs during the shift process. Simultaneously, the Output Checksum (OC) unit accumulates the elements of each column in the shifted matrix $C_{64 \times 64}$. This concurrent operation optimizes the computation

of the final result. A successful operation is confirmed when the calculated OC matches the precomputed input checksum (OC_{input}).

E. Previous Works

While recent research by other authors has explored error detection in GEMM across various platforms like CPUs, GPUs, and systolic arrays, none have addressed the specific needs of TMUL. We have already mentioned our previous work on TMUL which provides a foundation for this work [7], [10].

Sadi *et al.*, addressed the challenge of yield loss in AI chips as a metric and introduced a purely software-based fault tolerance scheme known as Yield And Accuracy Aware Optimum Test For AI Accelerators (YAOTA). In their approach, they deactivate specific faulty Processing Elements (PEs) in systolic array architectures within an acceptable yield and accuracy threshold [8]. They employ an offline methodology to detect stuck-at faults originating from manufacturing defects but do not account for defects arising in AI chips due to aging.

Santos *et al.*, explored the feasibility of applying Algorithm-Based Fault Tolerance (ABFT) to GPUs without hardware modification or repair, without considering other GEMM accelerators such as the Tensor Processing Unit or TMUL. Zhao et al. and Hari et al. have investigated the utilization of both the Algorithm-Based Error Detection (ABED) technique and ABFT on GPUs, as well as the application of ABFT to Convolutional Neural Networks (CNNs) [11] [12]. In contrast, our work focuses specifically on any General Matrix Multiplication (GEMM) operation executing on the TMUL architecture.

Kosaian *et al.*, explored the idea of applying ABFT in two distinct phases—the offline profiling phase and online fault tolerance phase [13]. In our study, we concentrate on implementing ABFT for GEMM operations in TMUL architectures by directly integrating it into the hardware and incorporating a repair mechanism. Our primary focus is on repairing these TMUL-like architectures using a Built-In Self Repair (BISR) algorithm, achieved through a three-step process: Pre-processing, TMUL Operation, and Post-processing, described in the next section.

Kosaian et al. explored the application of ABFT in two distinct phases: the offline profiling phase and the online fault tolerance phase [13]. Yet, there remains a notable absence of research focused on fault-tolerance mechanisms within the TMUL architecture. In order to address this gap, our work concentrates on implementing a Built-In Self Repair (BISR) algorithm for TMUL-like architectures, which we detail through a three-step process: Pre-processing, TMUL Operation, and Post-processing, as will be described in the next section. This approach is a significant advancement from our previous work in TMUL error detection [7], [10], [14], marking a novel foray into self-repair mechanisms specifically designed for TMUL.

Prior research has investigated error detection techniques for AI accelerators, including methods that deactivate faulty processing elements (PEs) to improve yield despite permanent faults [8]. This involves testing the chip to identify permanent faults and then selectively deactivating the faulty PEs to ensure the remaining operational components function correctly [8]. By doing so, these techniques manage to maintain correct operations and increase the yield of usable chips, as chips with some faulty PEs can still be sold and used rather than discarded.

This work builds upon our previous research that introduced a fault detection mechanism specifically for TMUL accelerators. Building on this foundation, we propose a novel self-repair technique that *dynamically* identifies all faults in a TMUL accelerator and repairs *both permanent and intermittent faults*. A key limitation of prior self-repair research is its focus on permanent faults. Intermittent faults, which occur sporadically, and lasts for several cycles are not typically addressed by these techniques. Our innovative method leverages a combination of hardware-assisted fault localization and software-based repair strategies. By implementing a column avoidance strategy at the software level, we ensure continuous and reliable operation of the TMUL units. This advancement significantly enhances the robustness and fault tolerance of TMUL accelerators.

Next, we introduce our proposed self-repair mechanism tailored for TMUL architectures, followed by its application to the two TMUL designs from our prior work, with hardware augmentations. We then analyze the hardware overhead of this self-repair system. Section IV concludes with the presentation of our results, including the effectiveness and hardware costs of the self-repair system.

III. PROPOSED BUILT-IN SELF REPAIR ALGORITHM

In this section, we refer to the ETMUL architectures discussed in the earlier Section II and how we implement a Built-In Self Repair algorithm on top of it. The extensions will allow us to perform error-free GEMM operations with a fault. In the following subsections, we discuss the three steps required for repairing. For illustrative purposes, we take two matrices, $A_{4 \times 4}$ and $B_{4 \times 4}$ as shown in Figure 3 below:

$$A_{4 \times 4} = \begin{bmatrix} 4 & 2 & 2 & 3 \\ 1 & 3 & 2 & 8 \\ 3 & 5 & 3 & 4 \\ 2 & 1 & 6 & 7 \end{bmatrix} B_{4 \times 4} = \begin{bmatrix} 2 & 3 & 4 & 1 \\ 1 & 6 & 1 & 1 \\ 7 & 3 & 5 & 1 \\ 9 & 2 & 6 & 1 \end{bmatrix} \quad (4)$$

Our focus is on addressing persistent faults in the TMUL hardware, necessitating a clear differentiation between persistent and transient faults. Transient faults, characterized by their non-recurring nature, can typically be rectified through repeated computation. In contrast, a fault that recurs n times, where n is a user-defined threshold, is classified as persistent. Upon reaching this threshold, the fault activates the software correction mechanism outlined below.

A. Pre-processing

Figure 3 shows the location of a single fault in the $[1, 2]$ processing element of the TMUL architecture. Figure 4 demonstrates the pre-processing step. In this step, the matrix $B_{4 \times 4}$ is loaded into the TMUL architecture while avoiding

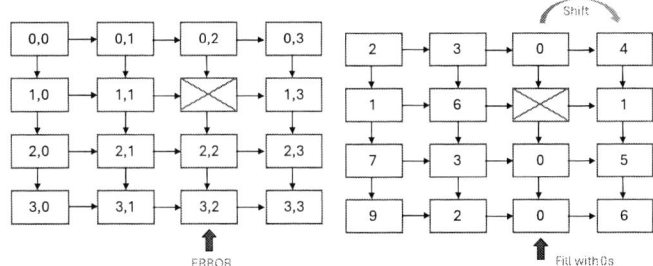

Fig. 3: Single fault in TMUL   Fig. 4: Pre-processing

the column containing the erroneous processing element. The third column of $B_{4 \times 4}$ is strategically shifted and loaded in the fourth column of the TMUL while padding the erroneous column with zeroes.

B. TMUL Operation

Figure 5 demonstrates the TMUL operation that takes place between Tile A and pre-processed Tile B. This operation has already been discussed in detail in the earlier Section II.

$$\begin{bmatrix} 4 & 2 & 2 & 3 \\ 1 & 3 & 2 & 8 \\ 3 & 5 & 3 & 4 \\ 2 & 1 & 6 & 7 \end{bmatrix} * \begin{bmatrix} 2 & 3 & 0 & 4 \\ 1 & 6 & 0 & 1 \\ 7 & 3 & 0 & 5 \\ 9 & 2 & 0 & 6 \end{bmatrix} = \begin{bmatrix} 51 & 36 & 0 & 46 \\ 91 & 43 & 0 & 65 \\ 68 & 56 & 0 & 56 \\ 110 & 44 & 0 & 81 \end{bmatrix}$$

Fig. 5: TMUL Operation

C. Post-processing

Figure 6 demonstrates the post-processing procedure after the intermediate result is generated from the TMUL operation. This step involves shifting the fourth column of intermediate output to the third column and generating the subsequent primary result. The subsequent results for the rest of the GEMM operation will be stored from the fourth column onwards.

The above three steps are repeated until all the columns of the matrix $B_{4 \times 4}$ are accounted for, generating the final output.

$$\begin{bmatrix} 51 & 36 & 0 & 46 \\ 91 & 43 & 0 & 65 \\ 68 & 56 & 0 & 56 \\ 110 & 44 & 0 & 81 \end{bmatrix} \xrightarrow{\text{Shift column to the left}} \begin{bmatrix} 51 & 36 & 46 & 46 \\ 91 & 43 & 65 & 65 \\ 68 & 56 & 56 & 56 \\ 110 & 44 & 81 & 81 \end{bmatrix}$$

Fig. 6: Post-processing

D. Fault Detection and Isolation in Interconnects

In addition to PE faults, interconnect faults can impact matrix multiplication architectures. These faults can occur in both column and row interconnects, each requiring distinct strategies for identification and isolation.

Column Interconnect Faults: Column interconnect faults behave similarly to PE faults, disrupting the data flow along specific columns. These faults can be detected through the ABED process and isolated in the same manner as PE faults as shown in Figure 7, by identifying the faulty column and isolating it to prevent the propagation of errors.

Row Interconnect Faults: Row interconnect faults pose a more complex challenge. Unlike column faults, a row interconnect fault affects multiple columns as data propagates

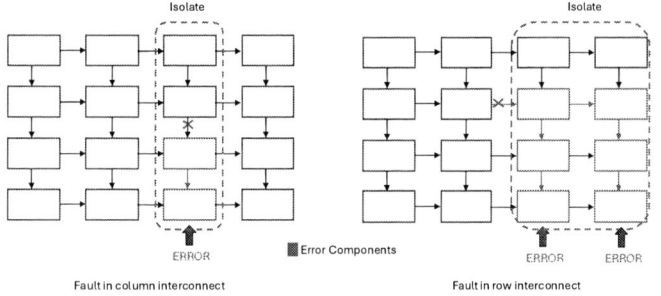

Fig. 7: Column-row interconnects fault isolation

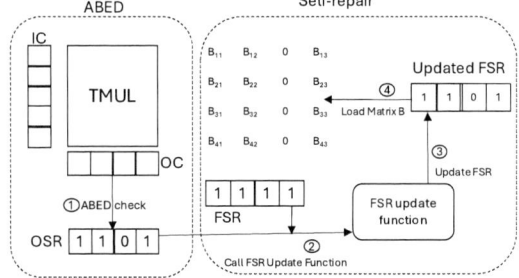

Fig. 8: System overview of self-repair

horizontally. As shown in Figure 7, this fault manifests as sequential errors across columns during the ABED check.

Fault Isolation: Upon detecting a row interconnect fault, all subsequent columns in matrix B are zeroed out, effectively isolating the fault and preventing further propagation of erroneous data. However, this approach has a trade-off: if the fault occurs early in the row, a large number of columns will be isolated, potentially diminishing the overall computational effectiveness. Nonetheless, this method preserves the integrity of the remaining computations.

E. System overview

The system leverages ABED mechanism to monitor the columns processed by the TMUL for errors. When ABED identifies one or more faulty columns, this information is recorded in a hardware register called *Operational Status Register* or OSR. OSR is distinct from the *Fault Status Register* (FSR), which is accessible to the software responsible for pre and post-processing of TMUL data. The FSR keeps a record of all faulty columns. If a mismatch is detected between OSR and FSR, FSR is updated with the new information.

Algorithm 1: FSR Update Function

Input: Operational Status Register (OSR), Fault Status Register (FSR)
if $OSR \neq FSR$ **then**
    Raise Interrupt;
    for $count \leftarrow 0$ **to** n **do**
        Relaunch Computation;
        if $OSR = FSR$ **then**
            return

    Update FSR with OSR;
    Zero and shift subsequent columns in Matrix B;
    Load updated Matrix B into TMUL Register;

FSR Update Function: This function re-launches the computation for the faulty column a specified number of times, n. In Figure 8, the OSR indicates that the 3rd column is faulty, which does not match the current state of the FSR. If the fault persists across all retries, the function updates the FSR as FSR = FSR — OSR. Subsequently, the corresponding columns in matrix B are zeroed out and shifted subsequent columns, effectively isolating the fault. In Figure 8, 3rd column of matrix B is isolated. The updated matrix B is then loaded into the TMUL register, ensuring that the faulty columns are excluded from further computations.

IV. RESULTS

This section details the overhead of a software-based method for real-time repair in ETMUL architecture. We assess this overhead through Processing Element (PE) utilization, reflecting the extra PEs needed for the BISR algorithm relative to fault-free scenarios. The section begins by outlining the fault model used for the study, followed by an analysis of the observed performance overhead in our experiments.

A. Experimental Setup

Our experiments, conducted in a C++ Version 11 simulated environment, replicated the TMUL architecture based on Intel's documentation to accurately mimic its data flow. Each loop iteration was treated as one clock cycle, simulating data movement through the TMUL hardware per clock. We tested the TMUL's functionality using tile-based inputs for matrices of various sizes, fitting them into 1kB registers, with zero-padding for smaller matrices. The simulated TMUL's output was then validated against GEMM operation outputs, achieving 100% correctness.

Additionally, we integrated pre-processing and post-processing algorithms into this simulation, as detailed in Section III. The algorithms' correctness was similarly confirmed by comparing their outputs with GEMM results for various matrices, also achieving 100% accuracy.

B. Fault Injection Model

For simulating the faults in the TMUL architecture, we followed a clustered fault model in view of real-world fault scenarios. This type of fault model hinges on the fact that when faults occur in hardware, they usually tend to be clustered around an area [15], [16]. In this case, clustered faults would manifest by occurring in PEs (Processing Elements) of the TMUL architecture in close proximity.

The clustered fault model that we implement distributes the faults within a PE fault distribution distance from the fault origin. For instance, if the first fault occurs in the PE $[5, 17]$, the next fault will occur in the range $[5\pm2, 17\pm2]$, if we choose multiple faults within a PE fault distribution distance of 2. The number of faults and the PE fault distribution distance can be varied in accordance with the experiment. These experiments are run on the simulation environment explained earlier.

979-8-3503-6689-1/24 $31.00 © 2024 IEEE

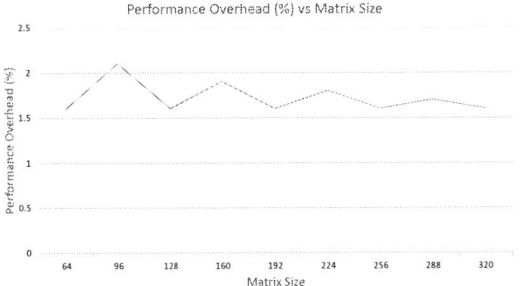

Fig. 9: Performance Overhead vs Matrix Sizes for Single Fault

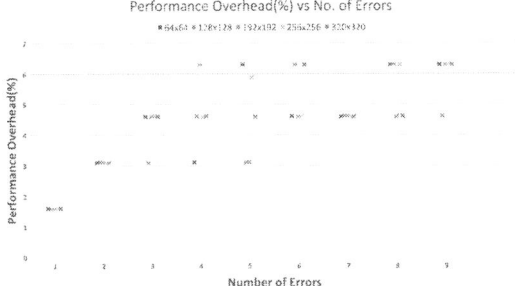

Fig. 10: Performance Overhead (%) vs No. of Faults

C. Overhead Analysis

In this section, we discuss the performance overhead for implementing our proposed approach in the simulated TMUL architecture that we observe from our experiments.

In our initial experiments, we evaluated the performance overhead relative to the initial PE utilization for two input matrices A and B. We varied the column size of matrix B from 32 to 320 and conducted 100 experiments per size. The results, plotted in Figure 9, reveal that the performance overhead is low at 1.6% for a 64-column size in B, decreasing with larger column sizes. The graph's saw-tooth pattern is attributed to lower PE utilization overhead when the matrix size nearly matches the TMUL size, allowing for efficient tile reuse and previously unused column utilization.

We investigated the performance impact of clustered faults in a simulated TMUL architecture with matrix sizes from 64×64 to 320×320, conducting tests with 1 to 9 faults across 100 trials for each case. The performance overhead reaches saturation beyond 9 faults due to our fault model's tendency to cluster faults in processing elements, leading to minimal incremental overhead. This behavior, depicted in Figure 10, initially shows a rising trend but then levels off, demonstrating that even with multiple faults, real-world scenarios can be managed efficiently with reasonable overhead.

V. CONCLUSION

General Matrix Multiplication (GEMM) dominates machine learning and HPC applications, providing motivation for hardware acceleration. However, fixed-size hardware matrix multipliers found in accelerators are typically smaller than target matrices, requiring tiling to resize multiplicands. Unfortunately, faulty behavior in PEs during multiplication has

been observed by Meta and Google. Our previous work to address this problem implements ABED hardware for error detection, but does not address the error correction problem. This paper presents a column-avoidance technique, leveraging ABED based fault localization and repair using software via resized tiling. Simulation experiments show a performance overhead of 2.1% in the worst case, with lower overheads observed under clustered fault conditions. Our proposed method is purely software-based without requiring any additional hardware beyond error detection.

REFERENCES

[1] M. Naumov, J. Kim, D. Mudigere, S. Sridharan, X. Wang, W. Zhao, S. Yilmaz, C. Kim, H. Yuen, M. Ozdal *et al.*, "Deep learning training in facebook data centers: Design of scale-up and scale-out systems," *arXiv preprint arXiv:2003.09518*, 2020.

[2] N. P. Jouppi, C. Young, N. Patil, D. Patterson, G. Agrawal, R. Bajwa, S. Bates, S. Bhatia, N. Boden, A. Borchers *et al.*, "In-datacenter performance analysis of a tensor processing unit," in *Proceedings of the 44th annual international symposium on computer architecture*, 2017.

[3] "Intel® architecture instruction set extensions and future features," 2021. [Online]. Available: https://www.intel.com/content/dam/develop/external/us/en/documents/architecture-instruction-set-extensions-programming-reference.pdf

[4] C. Room, "Tensor processing unit," *machine learning*, vol. 15, no. 54, p. 13, 2021.

[5] H. D. Dixit, S. Pendharkar, M. Beadon, C. Mason, T. Chakravarthy, B. Muthiah, and S. Sankar, "Silent data corruptions at scale," *arXiv preprint arXiv:2102.11245*, 2021.

[6] P. H. Hochschild, P. Turner, J. C. Mogul, R. Govindaraju, P. Ranganathan, D. E. Culler, and A. Vahdat, "Cores that don't count," in *Proceedings of the Workshop on Hot Topics in Operating Systems*, 2021, pp. 9–16.

[7] S. Bal, C. S. Mummidi, V. D. C. Ferreira, S. Srinivasan, and S. Kundu, "A novel fault-tolerant architecture for tiled matrix multiplication," *2023 Design, Automation and Test in Europe Conference (DATE)*, pp. 1–6, 2023.

[8] M. Sadi and U. Guin, "Test and yield loss reduction of ai and deep learning accelerators," *IEEE Transactions on Computer-Aided Design of Integrated Circuits and Systems*, vol. 41, no. 1, pp. 104–115, 2021.

[9] K.-H. Huang and J. A. Abraham, "Algorithm-based fault tolerance for matrix operations," *IEEE transactions on computers*, vol. 100, no. 6, pp. 518–528, 1984.

[10] C. S. Mummidi, V. C. Ferreira, S. Srinivasan, and S. Kundu, "Highly efficient self-checking matrix multiplication on tiled amx accelerators," *ACM Transactions on Architecture and Code Optimization*, 2023.

[11] K. Zhao, S. Di, S. Li, X. Liang, Y. Zhai, J. Chen, K. Ouyang, F. Cappello, and Z. Chen, "Ft-cnn: Algorithm-based fault tolerance for convolutional neural networks," *IEEE Transactions on Parallel and Distributed Systems*, vol. 32, no. 7, pp. 1677–1689, 2020.

[12] S. K. S. Hari, M. B. Sullivan, T. Tsai, and S. W. Keckler, "Making convolutions resilient via algorithm-based error detection techniques," *IEEE Transactions on Dependable and Secure Computing*, vol. 19, no. 4, pp. 2546–2558, 2021.

[13] J. Kosaian and K. Rashmi, "Arithmetic-intensity-guided fault tolerance for neural network inference on gpus," in *Proceedings of the International Conference for High Performance Computing, Networking, Storage and Analysis*, 2021, pp. 1–15.

[14] C. S. Mummidi, S. Bal, B. F. Goldstein, S. Srinivasan, and S. Kundu, "A highly-efficient error detection technique for general matrix multiplication using tiled processing on simd architecture," in *2022 IEEE 40th International Conference on Computer Design (ICCD)*. IEEE, 2022, pp. 529–536.

[15] C. H. Stapper, "On yield, fault distributions, and clustering of particles," *IBM Journal of Research and Development*, vol. 30, no. 3, pp. 326–338, 1986.

[16] T. Ni, Q. Xu, Z. Huang, H. Liang, A. Yan, and X. Wen, "A cost-effective tsv repair architecture for clustered faults in 3-d ic," *IEEE transactions on computer-aided design of integrated circuits and systems*, vol. 40, no. 9, pp. 1952–1956, 2020.

979-8-3503-6689-1/24 $31.00 © 2024 IEEE

An Effective TMR Approach for Low-Latency Configurable-Accuracy Adders

Ioannis Tsounis, Dimitris Agiakatsikas, and Mihalis Psarakis
Department of Informatics, University of Piraeus
{itsounis, dagiakatsikas, mpsarak}@unipi.gr

Abstract—**Low-Latency Approximate Adder (LLAA) is a high-performance type of approximate adder, which can still produce an exact output by integrating proper Approximation Error Detection and Correction (AxEDC) extra circuitry for detecting and correcting the expected approximation error. Furthermore, the hardware accelerators may be used for applications susceptible to hardware (HW) faults, which can produce additional errors to the adders during their lifetime. We propose an effective TMR methodology which exploits the resemblance of the errors caused by HW faults with the approximation errors both either case of AxEDC integration or not. We modify the AxEDC unit to detect and correct both types of errors and triplicate selectively parts of the adder. Our methodology achieves to minimise significantly the area overhead while also increasing the performance gain.**

Index Terms—**Approximate Adders, Low-Latency, Hardware Faults, TMR**

I. INTRODUCTION

Triple modular redundancy (TMR) is used to mitigate Single Points of Failure (SPF) in high-reliability applications at the expense of HW redundancy overheads that make the circuits larger and slower. On the other hand, Approximate Computing Techniques (ACTs) allow an approximation error to occur in trade-off to improve its design metrics, like area, performance and power [1]. Then approximate adders can be used in error-resilient applications, where each individual circuit can tolerate an arithmetic error up to an application-aware limit. ACTs can also be employed in an application requiring accurate arithmetic results. This can be achieved by incorporating an Approximation Error Detection and Correction (AxEDC) unit in the circuit to negate any occurring approximate arithmetic errors. In fact, approximate errors occur only for a small portion of its input combinations [2] and, thus, error correction does not need to be applied to all arithmetic operations, which leads to better performance. For instance, previous works have introduced AxEDC in Low-Latency Approximate Adders (LLAAs) to obtain a result with configurable accuracy, depending on the application's requirements [2], [3]. Nonetheless, HW faults in approximate adders or their AxEDC units may produce an additional HW error that leads to a total error different from the anticipated approximation error; if the overall error of the adder becomes unbounded, the application may fail occasionally or permanently, depending on the nature of the fault. Thus, classic TMR still holds in approximate adders (with or without AxEDC) to mitigate SPFs caused by HW faults.

To the best of our knowledge [4], methods to detect HW faults in approximate arithmetic circuits have been proposed only in [5], [6] and our previous work [7]. In more detail, the authors in [5], [6] proposed methodologies to reveal approximation-redundant faults during the fabrication phase of approximate circuits, and exploited these faults to improve yield during device testing. In our previous work [7], we proposed a technique to selectively apply Double Modular Redundancy (DMR) only to the approximate adder cells that can cause a total error higher than the error limit. We assumed that 1) each approximate adder consists of precise or approximate 1-bit Full Adder (FA) cells, and 2) faults follow the Cell Fault Model (CFM) [8].We define FA cells that, when they become faulty, can cause non-tolerable error as *sensitive* and the rest as *non-sensitive* FA cells, respectively. Thus, since HW faults in the *non-sensitive* FA cells are tolerable, we proposed a generic methodology that selectively applies classical error detection techniques, such as DMR, to any approximate adder architecture. We demonstrated our methodology to four different state-of-the-art approximate adder models, including LLAAs (without AxEDC). TMR can also be used for the same *sensitive* parts if we want error correction. Works [9]–[11] use ACTs to introduce novel methods that reduce the overhead cost of state-of-the-art fault-tolerant schemes(e.g. TMR). However, these works have proposed to insert the correct amount of redundancy at specific locations of precise arithmetic circuits to reduce hardware overheads. Some of these solutions use approximated logic to build the redundancy logic and thus reduce the associated costs [10], [11]. Since, the LLAA with AxEDC produces precise outputs, a TMR approach on it could be similar to them. In this work, we propose an efficient TMR methodology for the LLAAs with AxEDC, that modifies properly the original AxEDC in order to detect any possible HW error as well as the approximation errors. More specifically, the proposed selective triplication of the LLAA combined with the triplicated modified EDC circuit takes advantage of the inherent spatial redundancy of the LLAA design to achieve 100% fault-tolerance against SPF, without negating the performance gains and with less area overhead than the simple full TMR of the original LLAA with AxEDC. Thus, the proposed low-latency TMR for accuracy configurable adders can be used as an alternative to TMR precise adders at safety-critical applications where high performance is needed despite area overhead.

II. BACKGROUND

We focus on the broader category of LLAAs and specifically on the Generic Accuracy Configurable Adder (GeAr) [3]. GeAr covers many of the LLAAs, while Quality-area optimal Low-Latency approximate Adder (QuAd) provides an even more extended design space, that includes all the available LLAAs [12]. The optimisation comes at the cost of an arithmetic approximation error E_{appr}, defined as the difference

---

979-8-3503-6689-1/24 $31.00 © 2024 IEEE

between the output of the approximate adder (O_{appr}) and its respective accurate counterpart (O_{prec}), considering the same input operands: $E_{appr} = O_{prec} - O_{appr}$. E_{appr} takes specific values for the case of the LLAAs [2]. R is the number of the resultant bits contributing to the final output, while P is the number of bits used for carry prediction in each sub-adder unit (i.e., prediction bits). Any GeAr configuration can also be formulated as a QuAd configuration. Although our methodology can be generalised for all the LLAAs, we use the GeAr module as a demonstration vehicle due to its simplicity. The authors in [2] propose the implementation of (approximation) error detection and correction unit (AxEDC), which detects the approximation error and corrects the output of the adder. The E_{appr} of the GeAr(N, R, P) is calculated as follows [2]. For the ith sub-adder: $\wedge_{j=0}^{P-1}(A_j \oplus B_j) = cp_i$. The final carry-out of the previous sub-adder Co_{i-1} is also needed to define the signal that indicates the E_{appr} of the ith sub-adder, defined as: $ED_i = cp_i * Co_{i-1}$. When the signal $ED_i = 1$ for at least one sub-adder, then $E_{appr} \neq 0$. More specifically, for GeAr with two sub-adders: $E_{appr} = 0$ if $ED_2 = 0$, and $E_{appr} = 2^{R+P}$ if $ED_2 = 1$. The same error detection procedure can also be applied to QuAd. In this work, we will use the AxEDC scheme proposed in [2].

III. TMR METHODOLOGY FOR LLAAs

We propose a methodology for designing effective TMR schemes to build fault-tolerant adders relying on the LLAA architectures and targeting different objectives (i.e., latency or area). Our strategy for choosing the most appropriate TMR scheme is demonstrated in the decision diagram of Fig. 1. First, it must be clarified whether the target application is resilient to approximate errors. In this case, an LLAA model can be used instead of a precise adder to achieve lower latency. Next, to protect from faults that produce errors with magnitude higher than the acceptable LLAA's WCE_{appr}, we can choose either the naive full triplication of the original LLAA (TMR-1) or our first proposed solution (TMR-2), which selectively hardens the sensitive cells of the LLAA as they are classified in [7]. If the target application is not resilient to approximate errors, then the straightforward option is to apply full TMR on a precise adder (e.g., a ripple carry adder, RCA) to obtain full protection against SPF (TMR-0). Alternatively, if low latency is the main objective, we can build a fault-tolerant adder based on the LLAA architecture with AxEDC support. Here, it is the main contribution of the current paper: *how to reduce the area overhead of a TMR scheme of the AxEDC-capable LLAA model by exploiting the inherent redundancy of its architecture and the AxEDC logic.* To demonstrate the efficiency of the proposed TMR-4, we compare it with the naive full triplication of the original LLAA with AxEDC support (TMR-3). The proposed TMR schemes for the LLAA models without AxEDC (TMR-2) or with AxEDC support (TMR-4) are described in detail in Section III-A and Section III-B respectively. Table I summarises the characteristics of each TMR scheme.

A. TMR Approaches for the LLAA models without AxEDC

Naive TMR LLAA (TMR-1): We apply full TMR on the original LLAA (without AxEDC) to detect and correct all HW faults. **Proposed TMR LLAA (TMR-2):** We propose an approximation-aware TMR scheme for the LLAAs, which

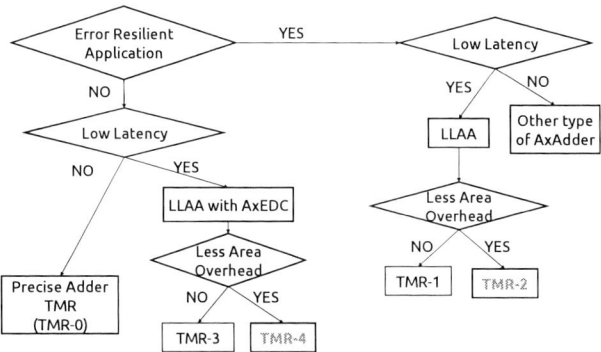

Fig. 1: Decision diagram for TMR scheme for LLAAs.
TABLE I: Characteristics of the Proposed TMR Schemes

TMR Scheme	Approximate adder	AxEDC	Triplication
TMR-0	NO	N/A	Full
TMR-1	LLAA	NO	Full
TMR-2	LLAA	NO	Selective
TMR-3	LLAA	YES	Full
TMR-4	LLAA	YES	Selective

corrects any error of magnitude higher than the WCE_{appr}. Our approach observes that only a subset of FA cells can produce an error higher than the WCE_{appr} limit when a HW fault hits them; these are the non-overlapped FAs of the sub-adders of LLAAs. On the contrary, the overlapped FAs of the sub-adders do not produce errors higher than the limit, either alone or in combination with approximation errors. This is proven analytically and also validated through fault simulations in our previous work [7]; we suggested selective DMR of the sensitive FA cells to achieve an effective error detection. Similarly, here we propose selective triplication of these cells also to support error correction. The proposed TMR LLAA scheme (TMR-2) is shown in Fig. 2 for a LLAA with two sub-adders. It must be noted that the paper's main contribution is the proposed TMR scheme for the LLAA model with AxEDC presented in the next subsection, but we also present the TMR-2 scheme for completeness purposes.

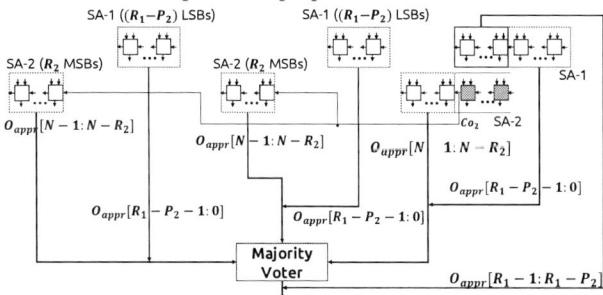

Fig. 2: Proposed TMR-2 for 2-SA LLAA without AxEDC.

B. TMR Approaches for the LLAA models with AxEDC

Naive TMR LLAA with AxEDC (TMR-3): We apply full TMR in every part of the AxEDC-capable LLAA, including its sub-adders and the respective AxEDC, as shown in Fig. 3 for the case of a LLAA model with two sub-adders. We use the GeAr model with two sub-adders and the AxEDC combinational circuit proposed in [2] for the TMR-3 scheme. **Proposed TMR LLAA with AxEDC (TMR-4):** We propose an alternative TMR for the LLAA with AxEDC, which exploits its redundant logic and the AxEDC unit to

979-8-3503-6689-1/24 $31.00 © 2024 IEEE

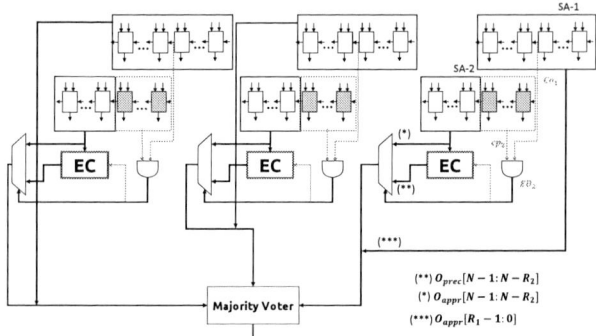

Fig. 3: TMR-3 for 2-SA LLAA with AxEDC [2].

protect the circuit against any type of error imposing less area overhead than the naive TMR-3. Thus, we do the following: a) selectively triplicate the sub-adders of the LLAA (i.e. only the sensitive FA cells are triplicated similar to the TMR-2 scheme), b) enhance its AxEDC unit to detect and correct not only the approximation errors but also the hardware errors affecting the non-sensitive FA cells and c) triplicate the AxEDC logic. Fig. 4 shows the proposed TMR approach for the case of LLAA with two sub-adders. The first sub-adder (i.e., the one adding the LSBs), as well as the part of the R bits of the second sub-adder, are fully triplicated. On the contrary, the P bits of the second sub-adder (i.e., the grey FA cells of Fig. 4) are not. *But how do we protect this part of the adder against HW faults?* First, notice that any single HW fault in these FAs can only affect the Co_2, and thus it can propagate to the output only through the second sub-adder. We modify the AxEDC as follows: it operates as a sub-adder performing the same addition (on the same input operands) with the second sub-adder but differs on the initial carry-in. In the second sub-adder the carry-in is equal to Co_2, while in the AxEDC it is equal to the inverted Co_2. Comparing the Co_2 with the Co_1 (the carry-out of the 1st sub-adder), we can detect any error affecting the Co_2. We can take as granted that Co_1 is error-free since its generating logic, i.e., first sub-adder is protected through TMR. Thus, the comparator (XOR gate) will detect if the Co_2 value is not correct, either due to the approximation logic or due to a HW fault at the prediction (grey) FA cells or even due to a fault in the Co_2 itself. The Co_2 produced by the non-triplicated part is connected to all three replicas. If the modified EDC is activated, then the resultant bits of the second sub-adder will be driven by the EDC. Given that the EDC is also triplicated, these outputs will be error-free. In the corner case, where the Co_2 differs from the Co_1 due to an approximation error, but an HW fault occurs concurrently,

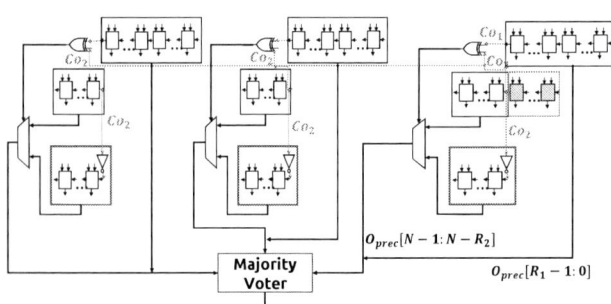

Fig. 4: Proposed TMR-4 for 2-SA LLAA with AxEDC.

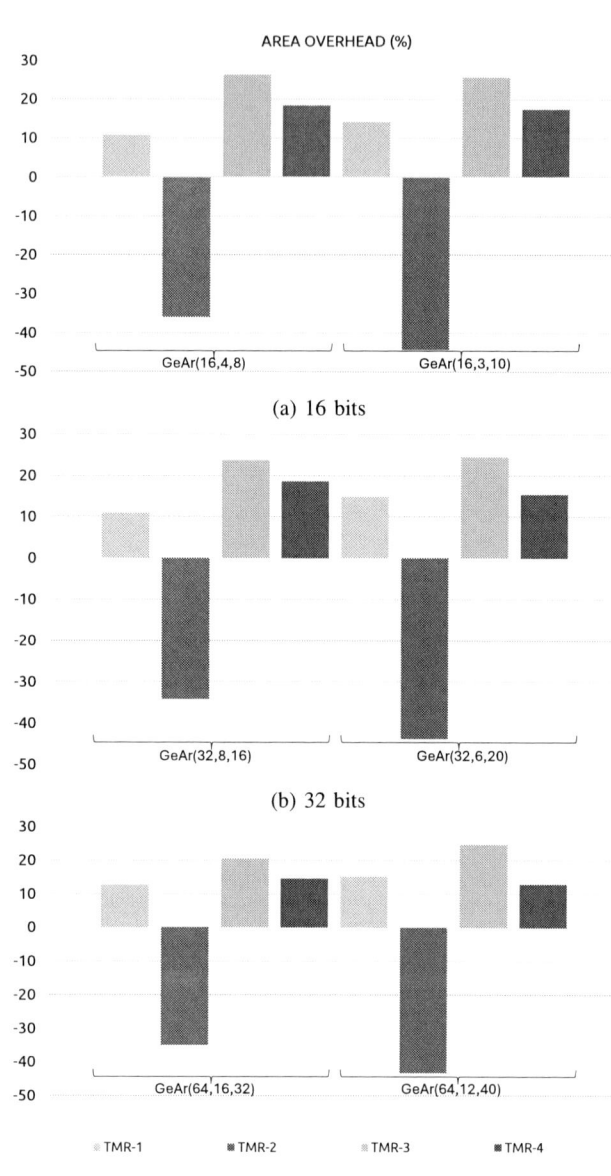

(a) 16 bits

(b) 32 bits

(c) 64 bits

Fig. 5: Area overhead for the TMR implementations

IV. EXPERIMENTAL RESULTS AND CONCLUSION

We developed in Verilog : GeAr(16, 4, 8), GeAr(16, 3, 10), GeAr(32, 8, 16), GeAr(32, 6, 20), GeAr(64, 16, 32) and GeAr(64, 12, 40). Next, we implemented all five TMR schemes: TMR-0 (TMR Precise RCA as baseline scheme), TMR-1, TMR-2, TMR-3 and TMR-4. Finally, we implemented all the TMR GeAr models using Openlane [13]. The proposed TMR-2 and TMR-4 outperform TMR-1 and TMR-3 respectively, in terms of both area overhead and latency. Fig. 5a compares the TMR-1 to TMR-4 in terms of area overhead against the baseline TMR-0 for: (a) 16-, (b) 32-, and (c) 64-bit models. For example, the TMR-1 GeAr(16, 4, 8) has 10% more area than the baseline (i.e., TMR-0 Precise RCA-16), while the TMR-2 GeAr(16, 4, 8) has almost 35% less area than the baseline as shown in Fig. 5a. We can observe the following: (a) the TMR-2

979-8-3503-6689-1/24 $31.00 © 2024 IEEE

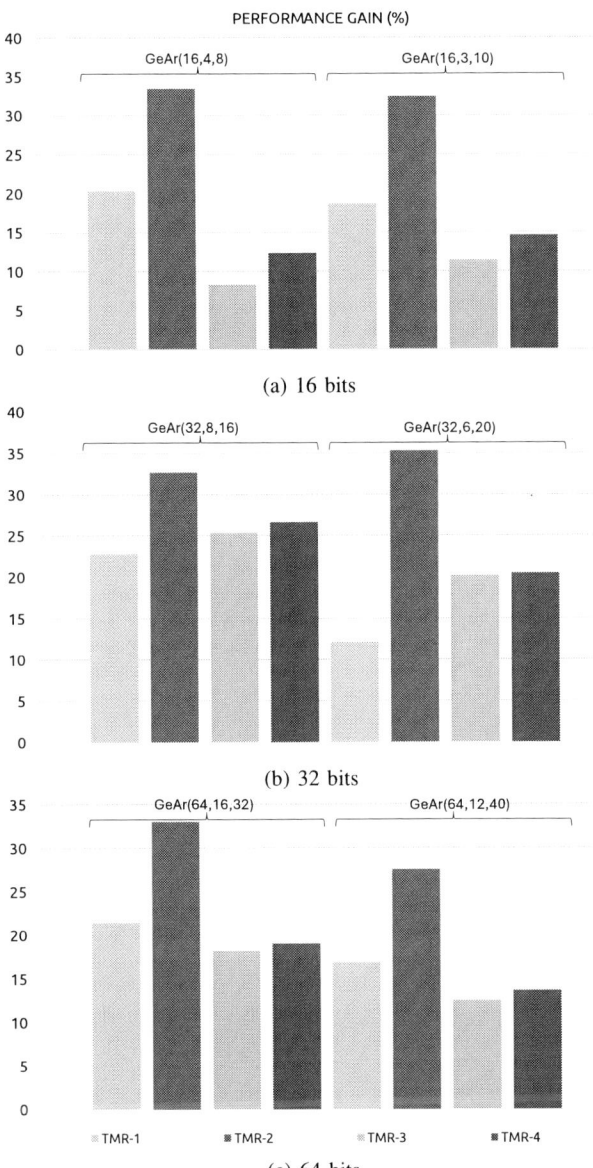

(a) 16 bits

(b) 32 bits

(c) 64 bits

Fig. 6: Performance gain for the TMR implementations

objective of the proposed TMR-2 and TMR-4 was to reduce the area of the triplicated logic, the experimental results showed that they also achieved performance improvements against the naive TMR-1 and TMR-3 for all GeAr models. More specifically, the latency reduction of the TMR-2 scheme ranges between 25% and 35% compared to the TMR-0, while the latency reduction of the TMR-1 ranges between 10% and 23%, since the TMR-2 majority voter is simplified compared to the TMR-1 voter. The TMR-4 achieves better performance than the TMR-3, but only 1-5% higher latency reduction. In brief, we present an approach for implementing efficient TMR schemes on LLAAs. The experimental results proved that the proposed schemes achieve significant gains in terms of area overhead and latency compared to the naive schemes. We aim to explore the deployment of the proposed TMR schemes in applications as in [14], [15].

ACKNOWLEDGMENT

The research work was supported by the Hellenic Foundation for Research and Innovation (HFRI) under the HFRI PhD Fellowship grant (Fellowship Number: 917). This work has been partly supported by the University of Piraeus Research Center.

REFERENCES

[1] A. Bosio, S. Di Carlo, P. Girard, E. Sanchez, A. Savino, L. Sekanina, M. Traiola, Z. Vasicek, and A. Virazel, "Design, Verification, Test and In-Field Implications of Approximate Computing Systems," in *IEEE ETS*, 2020, pp. 1–10.
[2] S. Mazahir, O. Hasan, R. Hafiz, M. Shafique, and J. Henkel, "An Area-Efficient Consolidated Configurable Error Correction for Approximate Hardware Accelerators," in *ACM/IEEE DAC*, 2016.
[3] M. Shafique, W. Ahmad, R. Hafiz, and J. Henkel, "A Low Latency Generic Accuracy Configurable Adder," in *DAC*, 2015.
[4] S. Burel, A. Evans, and L. Anghel, "Techniques for detecting and masking faults in semantic segmentation applications," *Microelectronics Reliability*, vol. 157, p. 115397, 2024.
[5] M. Traiola, A. Virazel, P. Girard, M. Barbareschi, and A. Bosio, "A Test Pattern Generation Technique for Approximate Circuits Based on an ILP-Formulated Pattern Selection Procedure," *IEEE Transactions on Nanotechnology*, vol. 18, pp. 849–857, 2019.
[6] ——, "Maximizing Yield for Approximate Integrated Circuits," in *DATE*, 2020, pp. 810–815.
[7] I. Tsounis, D. Agiakatsikas, and M. Psarakis, "Detecting Hardware Faults in Approximate Adders via Minimum Redundancy," *2023 IEEE 29th International Symposium on On-Line Testing and Robust System Design (IOLTS)*, pp. 1–7, 2023.
[8] M. Psarakis, D. Gizopoulos, and A. Paschalis, "Test Generation and Fault Simulation for Cell Fault Model using Stuck-at Fault Model based Test Tools," *Journal of Electronic Testing*, vol. 13, pp. 315–319, 1998.
[9] A. J. Sanchez-Clemente, L. Entrena, R. Hrbacek, and L. Sekanina, "Error Mitigation Using Approximate Logic Circuits: A Comparison of Probabilistic and Evolutionary Approaches," *IEEE Transactions on Reliability*, vol. 65, no. 4, pp. 1871–1883, 2016.
[10] G. Rodrigues, J. Fonseca, F. Kastensmidt, V. Pouget, A. Bosio, and S. Hamdioui, "Approximate TMR based on successive approximation and loop perforation in microprocessors," *Microelectronics Reliability*, vol. 100, p. 113385, 2019.
[11] B. Deveautour, M. Traiola, A. Virazel, and P. Girard, "QAMR: an Approximation-Based Fully Reliable TMR Alternative for Area Overhead Reduction," in *IEEE ETS*, 2020, pp. 1–6.
[12] M. A. Hanif, R. Hafiz, O. Hasan, and M. Shafique, "QuAd: Design and analysis of quality-area optimal low-latency approximate adders," *54th Annual DAC*, pp. 1–6, 2017.
[13] A. Ghazy and M. Shalan, *Openlane: The open-source digital asic implementation flow*, 2020.
[14] I. Tsounis, A. Papadimitriou, and M. Psarakis, "Analyzing the Impact of Approximate Adders on the Reliability of FPGA Accelerators," in *ETS*, 2021, pp. 1–2.
[15] I. Tsounis, D. Agiakatsikas, and M. Psarakis, "A Methodology for Fault-Tolerant Pareto-Optimal Approximate Designs of FPGA-Based Accelerators," *ACM Trans. Embed. Comput. Syst.*, oct 2022.

scheme achieves significant area savings compared to the TMR-1 scheme in all GeAr models; it reverses the TMR-1 area overhead (10-15%) to TMR-2 area gain (which ranges between -33 and -45%), (b) similarly, the TMR-4 scheme imposes less area overhead (10%-20%) than the TMR-3 scheme (20%-30%), (c) the area savings of the proposed schemes are higher when the ratio P/R increases and especially for larger adders (e.g. the area savings of TMR-4 compared to TMR-3 are higher for the GeAr(64, 12, 40) than the GeAr(64, 16, 32)), since the TMR-2 and TMR-4 exploit the non-sensitive cells of the prediction bits to reduce the area overheads. Fig. 6 compares the TMR-1 to TMR-4 schemes in terms of performance gains against the baseline TMR-0 for: (a) 16-, (b) 32-, and (c) 64-bit models. For example, the TMR-1 GeAr(16, 4, 8) has 20% lower latency than the TMR-0 Precise RCA-16, while the TMR-2 GeAr(16, 4, 8) has 33% lower latency, as shown in Fig. 6a. Although the main

979-8-3503-6689-1/24 $31.00 © 2024 IEEE

AUTHOR INDEX

Agiakatsikas, Dimitris	228
Ahmadilivani, Mohammad H.	32
Al Halabi, Jad	76, 138
Al-Kaf, Ahmed	36
Alonso, Martí	90, 108
Alshaer, Ihab	36
Andreu, David	90
Angioli, Marco	187, 216
Angione, Francesco	64
Arai, Masayuki	80
Azaïs, F.	58
Azarpeyvand, Ali	32
Bal, Sandeep	222
Barbirotta, Marcello	187, 216
Becker, Bernd	132
Becker, Juergen	26
Becker, Steffen	70
Bernardi, Paolo	64
Beroulle, Vincent	36
Bertani, Claudia	64
Bolchini, Cristiana	1
Burelle, R.	58
Canal, Ramon	90
Canese, Lorenzo	172
Cantoro, Riccardo	132
Cassano, Luca	1, 120
Chakma, Amit	42
Chapman, Glenn H.	42
Chatzopoulos, Odysseas	90
Cheikh, Abdallah	187, 216
Chen, Dejiu	7
Chen, Hsuan-Yu	210
Chen, Kuan-Hsun	156
Chen, Yi-Lin	210
Chen, Yizhi	7
Chen, Yu-Guang	198
Chenet, Cristiano	90
Cheng, Wei-Che	204
Ciancarelli, Carlo	182
Costa, J.	90, 108
Costantino, Alessandra	150
Daneshtalab, Masoud	32
Das, Anup	13
De Biase, Catriel	182
Decuzzi, Filomena	172
Deligiannis, Nikolaos I.	132
Devi, Meenakshi	48
Di Carlo, Stefano	90

Di Ienno, Davide	182
Dilillo, Luigi	54, 126, 168
Domanski, Peter	70
Ecker, Wolfgang	76, 138
Egloff, Valentin	36
Ellervee, Peeter	144
Erker, Matic	178
Faehn, E.	86
Faller, Tobias	132
Farahmandpour, Alireza	42
Forlin, Bruno E.	120, 156
Foscale, Tommaso	64
Furano, Gianluca	120, 172, 178, 182
Gacnik, Dejan	178
Gerlin, Nicolas	76, 138
Ghasempouri, Tara	32
Giardino, Nicola D. G.	64
Girard, P.	86
Girones, Andreu	90
Gizopoulos, Dimitris	90
Grignani, Wesley	54, 168
Harbaum, Tanja	26
Helen, Youri	162
Hellebrand, Sybille	96
Hoefer, Julian	26
Hosokawa, Toshinori	80
Hotfilter, Tim	26
Hsiao, Chih-Li	204
Hsu, Chun-Lung	210
Huang, Shih-Hsu	204
Imianosky, Carolina	126
Jabir, Abusaleh	20, 48
Jenihhin, Maksim	144
Johari, Sarah	13
Kaja, Endri	76, 138
Khandelwal, Saurabh	48
Khound, Parthib	7
Koren, Israel	42
Koren, Zahava	42
Kosmidis, Leonidas	150
Kotnik, Bojan	178
Kramberger, Iztok	178
Kreß, Fabian	26
Kritikakou, Angeliki	114
Kundu, Sandip	222
Kunz, Wolfgang	76, 138
Latorre, L.	58
Lazzeri, Elia	120

Leboffe, Antonio 182
Lefevre, F. 58
Leveugle, Régis 162
Li, Jin-Fu 193
Lu, Zhonghai 7
Mastrandrea, Antonio 187, 216
Matinizadeh, Shadi 13
Melo, Douglas R. 54, 126, 168
Menichelli, Francesco 187, 216
Miele, Antonio 1
Miller, Florent 162
Mizota, Momona 80
Mohammadhassani, Arghavan 13
Mohammed, Omar 7
Morancho, Enric 90, 108
Mummidi, Chandra S. 222
Nazari, Samira 32
Nikiema, Pegdwende R. 114
Noizette, Luc 162
Olivieri, Mauro 187, 216
Otero, Beatriz 90
Ottavi, Marco 120, 156
Papadimitriou, George 90
Parchekani, Bahram 32
Pascucci, Dario 182
Passarello, Dario 1
Pflüger, Dirk 70
Plecenik, Tomas 48
Polian, Ilia 70, 102
Prasad, Rajendra 20
Prebeck, Sebastian 76, 138
Psarakis, Mihalis 156, 228
Raik, Jaan 32, 144
Rast, Alexander 20
Ravikumar, Abishaan 20
Reorda, Matteo S. 132
Rodriguez-Ferrandez, Ivan 150
Rotovnik, Tomaž 178
Sadeghi-Kohan, Somayeh 96
Santana, Khalil G. Q. 54
Santos, Douglas A. 54, 126, 168
Sauer, Matthias 70
Savino, Alessandro 90
Scandelli, Luca 182
Schwachhofer, Denis 70
Selcan, David 178
Selg, Hardi 144
Serri, Paolo 182
Shibin, Konstantin 144
Sivaraj, Rajanataraj 20
Smit, Tijmen T. 156
Souvatzoglou, Ioanna 156

Stammler, Matthias 26
Steenari, David 150, 182
Stoffel, Dominik 76, 138
Tahiraga, Ares 76, 138
Tahraoui, K. 58
Tali, Maris 150
Tancorre, Vincenzo 64
Traiola, Marcello 114
Trois, Renato 182
Tsertov, Anton 144
Tsounis, Ioannis 228
Upadhyaya, Devanshi 102
Vayssade, T. 58
Vidiš, Marek 48
Viel, Felipe 126
Virazel, A. 86
Wagner, Stefan 70
Wu, Ting-Yi 198
Wunderlich, Hans-Joachim 96
Xhafa, X. 86
Yang, Xiaohan 20
Yoshimura, Masayoshi 80
Zancan, Valentina 172
Zhu, Wenyao 7